# ADHESION

# MOLECULES

*in*

# ALLERGIC

# DISEASE

# ADHESION

# MOLECULES

# *in*

# ALLERGIC

# DISEASE

*edited by*

## BRUCE S. BOCHNER

*Johns Hopkins Asthma and Allergy Center*
*Johns Hopkins University School of Medicine*
*Baltimore, Maryland*

CRC Press
Taylor & Francis Group
Boca Raton London New York

CRC Press is an imprint of the
Taylor & Francis Group, an **informa** business

CRC Press
Taylor & Francis Group
6000 Broken Sound Parkway NW, Suite 300
Boca Raton, FL 33487-2742

First issued in paperback 2019

© 1998 by Taylor & Francis Group, LLC
CRC Press is an imprint of Taylor & Francis Group, an Informa business

No claim to original U.S. Government works

ISBN-13: 978-0-8247-9836-9 (hbk)
ISBN-13: 978-0-367-40102-3 (pbk)

**Visit the Taylor & Francis Web site at**
**http://www.taylorandfrancis.com**

**and the CRC Press Web site at**
**http://www.crcpress.com**

# Preface

Over the past few decades, our understanding of the pathophysiology of allergic diseases, including bronchial asthma, allergic rhinitis, and atopic dermatitis, has evolved from a primary focus on IgE, mast cells, and their roles in initiating allergic reactions, to the study of more downstream events such as late-phase responses and the inflammation that accompanies clinically significant chronic allergic diseases. We now realize that preferential migration of human eosinophils, basophils and T lymphocytes, especially those of the Th2 subtype, occurs during allergic inflammatory responses in the skin and airways. Further, these cells and their products play a critical role in producing allergic inflammation, and allergic diseases are the net result of IgE-dependent inflammatory cascades involving certain cells and mediators. These discoveries have fueled efforts to understand the mechanisms involved in selective recruitment processes that differ in other forms of inflammation.

Inflammation has classically been viewed as an interplay between cellular and fluid elements in blood with like constituents in tissues. Under this paradigm, local endothelium and epithelium were thought to play relatively inactive roles, functioning exclusively as barriers. However, it has been known for more than a century that structural changes in these cells can occur at sites of inflammatory reactions, in association with the acquisition of an ability to actively adsorb leukocytes to their surfaces. This process is now known to be mediated by adhesion molecules. Over a decade ago, technology was developed to isolate and culture vascular endothelial cells and airway epithelial cells, leading to tremendous advances in our understanding of adhesion molecules and their importance in a variety of normal and pathological responses. The past decade has also witnessed a flood of information on cytokines and chemokines and their effects on adhesion molecules in inflammation. Indeed, most of these molecules were not even known to exist until a few years ago. For adhesion molecules, this new knowledge includes characterization of their molecular structures and counter-ligands, classifications into superfamilies, investigations into signal transduction events, and analyses of the regulation of their expression and function both in vitro and in vivo.

The overall aim of this book is to update the reader on the cells, proteins, and mechanisms involved in allergic inflammation, with a major emphasis on mechanisms of local cell recruitment. With this in mind, the first five chapters present overviews of adhesion molecule biology, familiarizing the reader with the latest lists of adhesion molecules, their nomenclature, and general biological functions. Included are chapters on specific types of adhesion molecules (e.g., integrins), as well as chapters on the adhesive capabilities of endothelial cells, respiratory epithelial cells and homing to mucosal surfaces. The remaining chapters have a more narrow focus. Seven chapters cover adhesion-related biology of cell types felt to be particularly important in allergic inflammatory reactions, including mast cells, basophils and eosinophils. These chapters primarily discuss findings from in vitro studies and summarize cell adhesion phenotype and functional endothelial, epithelial, and extracellular matrix protein ligands for each cell type. Also covered are the effects of adhesion on cell function, and regulation of adhesion molecule expression and function. In the final eight chapters, adhesion molecule expression and function in vivo in various allergic and other immune responses are covered, and, where available, data on adhesion molecule antagonism are presented.

Although other reviews of adhesion molecules have appeared, this is the first book to be devoted exclusively to allergic inflammation. Its publication seems especially timely in that efforts are now underway, in both animals and humans, to antagonize the function and/or expression of these adhesion molecules as therapeutic targets, in an attempt to generate novel anti-inflammatory treatments of allergic disease. Hopefully, future editions of this book will be able to incorporate new information in this area, as well as data in other areas where information is lacking, such as the trafficking of monocytes and macrophages during allergic inflammation.

As editor of this book I am indebted to the contributors, all of whom are experts in their respective fields, and without whose contributions this text would not have been possible. I would also like to thank my family for their never-ending love and support and my mentor and long-time collaborator, Dr. Robert Schleimer, who helped to foster my intense interest in the field, and with whom I have been fortunate to share the excitement and evolution of our work. My thanks to the past and present members of the "Schlochneck" laboratory group for their hard work, ambition, inquisitiveness and enthusiasm, which makes research fun, the staff at Marcel Dekker, Inc., for continual assistance throughout this project, and Bonnie Hebden for excellent administrative and secretarial support.

*Bruce S. Bochner*

# Contents

# Contributors

**Lisa A. Beck** Departments of Dermatology and Medicine, Johns Hopkins Asthma and Allergy Center, Johns Hopkins University School of Medicine, Baltimore, Maryland

**Bruce S. Bochner** Department of Medicine, Johns Hopkins Asthma and Allergy Center, Johns Hopkins University School of Medicine, Baltimore, Maryland

**Michael J. Briskin** Department of Molecular Biology, LeukoSite Inc., Cambridge, Massachusetts

**William W. Busse** Department of Medicine—Allergy Section, University of Wisconsin Medical School, Madison, Wisconsin

**Giorgio Ciprandi** Allergy and Clinical Immunology Service, Department of Internal Medicine, University of Genoa, Genoa, Italy

**Giorgio Walter Canonica** Allergy and Clinical Immunology Service, Department of Internal Medicine, University of Genoa, Genoa, Italy

**Motohiro Ebisawa** Department of Pediatrics, National Sagamihara Hospital, Kanagawa, Japan

**Nicolò Fiorino** Allergy and Clinical Immunology Service, Department of Internal Medicine, University of Genoa, Genoa, Italy

**Steve N. Georas**   Department of Medicine, Johns Hopkins Asthma and Allergy Center, Johns Hopkins University School of Medicine, Baltimore, Maryland

**Kirstin Goldring**   Department of Physiology and Pharmacology, University of Southampton, Southampton, England

**Stephen T. Holgate**   Department of Immunopharmacology, University of Southampton, Southampton General Hospital, Southampton, England

**Stephen W. Hunt III**   Department of Molecular Biology, Parke-Davis Pharmaceutical Research, Division of Warner-Lambert Company, Ann Arbor, Michigan

**Michael D. Ioffreda**   Department of Dermatology, University of Pennsylvania School of Medicine, Philadelphia, Pennsylvania

**Arthur Kavanaugh**   Department of Internal Medicine, The University of Texas Southwestern Medical Center, Dallas, Texas

**Sirid-Aimée Kellermann**   Department of Laboratory Medicine and Pathology, Center for Immunology, University of Minnesota Medical School, Minneapolis, Minnesota

**Hirohito Kita**   Department of Immunology, Mayo Clinic and Mayo Foundation, Rochester, Minnesota

**Paul Kubes**   Department of Physiology and Physics, Immunology Research Group, University of Calgary Medical Center, Calgary, Alberta, Canada

**Steven L. Kunkel**   Department of Pathology, University of Michigan Medical School, Ann Arbor, Michigan

**Donald Y. M. Leung**   Department of Pediatrics, The National Jewish Medical and Research Center, and University of Colorado Health Sciences Center, Denver, Colorado

**Roy R. Lobb**   Biogen, Inc., Cambridge, Massachusetts

**Nicholas W. Lukacs**   Department of Pathology, University of Michigan Medical School, Ann Arbor, Michigan

**Francis W. Luscinskas**   Vascular Research Division, Department of Pathology, Brigham and Women's Hospital, and Harvard Medical School, Boston, Massachusetts

**Kenji Matsumoto**   Department of Medicine, Johns Hopkins Asthma and Allergy Center, Johns Hopkins University School of Medicine, Baltimore, Maryland

**Dean D. Metcalfe**   Laboratory of Allergic Diseases, National Institute of Allergy and Infectious Diseases, National Institutes of Health, Bethesda, Maryland

**Stephen Montefort**   Department of Medicine, St. Luke's Hospital, G'Mangia, Malta

**George F. Murphy**   Department of Dermatology, University of Pennsylvania School of Medicine, Philadelphia, Pennsylvania

**Francesca Paolieri**   Allergy and Clinical Immunology Service, Department of Internal Medicine, University of Genoa, Genoa, Italy

**Giovanni Passalacqua**   Allergy and Clinical Immunology Service, Department of Internal Medicine, University of Genoa, Genoa, Italy

**Louis J. Picker**   Department of Pathology, The University of Texas Southwestern Medical Center, Dallas, Texas

**Albert J. Polito**   Department of Medicine, Johns Hopkins Asthma and Allergy Center, Johns Hopkins University School of Medicine, Baltimore, Maryland

**David Proud**   Department of Medicine, Johns Hopkins Asthma and Allergy Center, Johns Hopkins University School of Medicine, Baltimore, Maryland

**Kirsty Rich**   Department of Physiology and Pharmacology, University of Southampton, Southampton, England

**Sarbjit S. Saini**   Department of Medicine, Johns Hopkins Asthma and Allergy Center, Johns Hopkins University School of Medicine, Baltimore, Maryland

**Robert P. Schleimer**   Department of Medicine, Johns Hopkins Asthma and Allergy Center, Johns Hopkins University School of Medicine, Baltimore, Maryland

**Mary K. Schroth**   Department of Pediatrics, University of Wisconsin Children's Hospital, Madison, Wisconsin

**Antonio Scordamaglia**   Allergy and Clinical Immunology Service, Department of Internal Medicine, University of Genoa, Genoa, Italy

**Julie B. Sedgwick**   Department of Medicine—Allergy Unit, University of Wisconsin Medical School, Madison, Wisconsin

**Yoji Shimizu**   Department of Laboratory Medicine and Pathology, Center for Immunology, University of Minnesota Medical School, Minneapolis, Minnesota

**James M. Stark**   Department of Pulmonary Medicine, Children's Hospital Medical Center, Cincinnati, Ohio

**Robert M. Strieter**   Department of Internal Medicine, University of Michigan Medical School, Ann Arbor, Michigan

**Harissios Vliagoftis**   Laboratory of Allergic Diseases, National Institute of Allergy and Infectious Diseases, National Institutes of Health, Bethesda, Maryland

**Garry M. Walsh**   Department of Respiratory Medicine, University of Leicester, School of Medicine, Glenfield Hospital, Leicester, England

**Andrew J. Wardlaw**   Department of Respiratory Medicine, University of Leicester, School of Medicine, Glenfield Hospital, Leicester, England

**Jane A. Warner**   Department of Physiology and Pharmacology, University of Southampton, Southampton, England

# ADHESION

# MOLECULES

*in*

# ALLERGIC

# DISEASE

# 1

# Overview of Cell Adhesion Molecules and Their Antagonism

**Arthur Kavanaugh**   *The University of Texas Southwestern Medical Center, Dallas, Texas*

## I.  INTRODUCTION

In the course of reviewing developments within their own disciplines, investigators are wont to expound on "exciting progress" and "dramatic breakthroughs." Such phrases often seem trite, but they are hardly sufficient to describe the extraordinary advancements this past decade has witnessed in the study of cell surface adhesion molecules. These diverse molecules, which bind to specific ligands expressed on cells and extracellular matrix (ECM) molecules, have recently been the subject of intense investigation by numerous investigators. Not only have many adhesion molecules been discovered in recent years, but novel and intriguing characteristics of these molecules continue to be defined on an ongoing basis.

Adhesion molecules have been demonstrated to be capable of effecting various prominent interactions. This includes the adhesion of circulating leukocytes to the vascular endothelium, the transendothelial migration of these cells, retention of cells at extravascular sites, and activation of immunocompetent cells (1–10). By mediating these events, adhesion molecules play a critical role in both inflammatory responses as well as immunosurveillance. Although these processes are essential to salubrious activities such as the elimination of infectious agents, they also underlie various pathologic states. Thus, adhesion molecule–mediated interactions are integral to the initiation and propagation of allergic diseases. Accompanying the substantial developments in our understanding of adhesion molecules has been the expectation that they may serve as valuable therapeutic targets. In addition to having considerable theoretical appeal, the concept of targeting adhesion receptors has been successfully tested in both animal models and human diseases.

## II. HISTORY

Progress in our understanding of adhesion molecules has followed developments in the appreciation of the endothelium as a dynamic organ. In both fields, many years of slow progress have given way to the current exponential growth in both comprehension and interest (Fig. 1). It was over a millennium before William Harvey's description of the circulatory system supplanted Galen's archaic concept of the tidal ebb and flow of blood elements (11). Later, in the 1600s, Malphigi identified the capillary bed. By the 18th century, it became accepted that lymph was derived from blood. Although all components of the circulation were thus apparent by this time, the nature of its endothelial lining layer could not be properly appreciated until cell theory was promulgated, nearly a century later. During the 19th century, several seminal observations helped provide the foundation for studies that would ultimately lead to the discovery of adhesion molecules. Using intravital microscopy, Dutrochet, Metchnikoff, Cohnheim, and other investigators described the interaction of circulating cells with the vessel wall (12). It was observed that leukocytes passing through postcapillary venules at high velocity initially slow down and roll along the vessel wall. Subsequently, these cells become flatter and more tightly adherent to endothelial cells. Finally, some migrate through the vascular endothelium into the inflammatory site. Although the mechanisms underlying these observations would not be defined until well into the next century, these investigators formed prescient hypotheses on the dynamic nature of the interaction between leukocytes and endothelial cells.

Approximately 40 years ago, it was noted that lymphocytes possess a pattern of recirculation distinct from other leukocytes; they enter lymphoid tissue from the bloodstream via postcapillary venules, traverse lymph vessels back to the circulation, and again recirculate to lymphoid tissue (13). Moreover, this lymphocyte recirculation was organ specific. Lymphocytes isolated from a particular lymphoid tissue would return or "home" back to that same tissue after intravenous injection. This initiated the concept of specific "homing receptors" on lymphocytes that interact with distinct "vascular addressins" on the vascular endothelium within particular organs, thus directing this specific recirculation.

The initial descriptions of cell surface "adhesion molecules" in the early 1980s ushered in an era of exponential discovery (Fig. 1) (14). Since then, not only have numerous adhesion molecules been defined, but myriad functions and diverse interactions mediated by these versatile molecules continue to be described.

## III. ADHESION RECEPTORS

On the basis of structural homology, many adhesion molecules can be classified into one of three important families: the integrins, the selectins, and the immunoglobulin superfamily members.

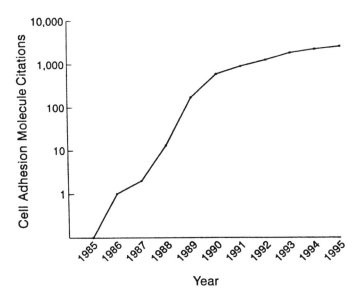

**Figure 1** Citations retrieved by the MEDLINE reference base, searching for the term cell adhesion molecule during the years indicated.

## A. Integrins

The integrins (Table 1) are large, heavily glycosylated, heterodimeric proteins composed of one of at least 15 distinct $\alpha$-subunits in noncovalent linkage with one of at least eight $\beta$-subunits. These adhesion receptors, which can bind ligands expressed on cell surfaces, ECM molecules, and soluble molecules, are phylogenetically ancient. Integrins can be subcategorized based on their $\beta$-subunit usage (Table 1).

Although their intracytoplasmic segments are relatively short, integrins interact with extracellular cytoskeletal components, such as actin and talin. The name integrins derives from the concept that these molecules "integrate" information from the extracellular milieu to intracellular compartments. Recently, it has been established that integrins are capable of bidirectional signaling, as they also transduce signals "inside out" (15). Thus, many integrins constitutively have minimal ability to bind their ligands. Upon activation of the cell, integrins undergo conformational changes that permit binding of divalent cations by the $\alpha$-subunit and markedly increase their avidity for ligand. The presence of conformationally dissimilar forms has been confirmed by the description of monoclonal antibodies (mAb) specific for individual integrins at distinct states of activation.

As is true of other adhesion receptors, integrins exhibit pleiotropy and redun-

**Table 1** Integrin Family

| Receptor | Counter-receptor | Distribution | MW (kD) [α/β] chain | Regulation |
|---|---|---|---|---|
| *β1 (CD29)* | | *Very Late Activation (VLA) Antigens* | | |
| α1β1/VLA-1 (CD49a/CD29) | Laminin (Lm), collagen (Co) | Activated T cells, fibroblasts, mesangial cells, hepatic sinusoids | 210/130 | Expression increased by antigen, mitogen |
| α2β1/VLA-2 (CD49b/CD29) | Co, Lm[a] | Activated T cells, endothelial cells (EC), platelets, basophils | 165/130 | Expression increased by antigen, mitogen |
| α3β1/VLA-3 (CD49c/CD29) | Lm, Co Fibronectin (Fn)[a] | Glomerulus, thyroid, basement membrane, many cell lines | 135/130 | |
| α4β1/VLA-4 (CD49d/CD29) | VCAM-1 (domains 1 and 4), Fn (CS-1 domain; LDV motif), (also, with activation, weaker binding to Fn via RGD motif on Hep II site) | Lymphocytes[a], monocytes, eosinophils, basophils, mast cells, NK cells (not PMN) | 150/130 | Expression/activity increased by many stimuli (antigen mitogen, etc.) |
| α5β1/VLA-5 (CD49e/CD29) | Fn[a] | Lymphocytes, monocytes, EC, basophils, mast cells, fibroblasts | 160/130 | Activity increased by antigen |
| α6β1/VLA-6 | Lm | Platelets, T cells, eosinophils, monocytes, EC | 130/130 | Activity increased by antigen |
| α9β1 | Tenascin (non-RGD site) | Basal keratinocytes, hepatocytes, airway epithelial cells, smooth and skeletal muscle cells | 130/130 | |
| αvβ1 (CD51/CD29) | Fn, vitronectin (Vn)[b] | Platelets, B cells | 135/130 | |

| | | | *Leukocyte Integrins* | |
|---|---|---|---|---|
| **β2 (CD18)** | | | | |
| LFA-1 (CD11a/CD18) | ICAM-1, 2, 3 | 180/95 | All leukocytes | Activity increased by many stimuli (antigen, mitogen, etc) |
| Mac-1, CR3 (CD11b/CD18) | ICAM-1, fibrinogen (Fb), iC3b, LPS, factor X, ICAM-2 (A domain) | 170/95 | Granulocytes, monocytes, LGL | Expression/activity increased by cytokines, other stimuli |
| p150,95, CR4 (CD11c/CD18) | iC3b, Fb | 150/95 | Monocytes, granulocytes, LGL, B-cell subset, platelets | Expression increased by TNF-α |
| αD/CD18 | ICAM-3 | | Tissue macrophages, monocytes, CD8 + T cells, granulocytes | Constitutive |
| | | | *Cytoadhesins* | |
| **β3 (CD61)** | | | | |
| gp IIb/IIIa CD41/CD61 | Fb, Vn, Fn, vWF[c] | 130/105 | Platelets, EC | Expression increased by many stimuli |
| αv/IIIa CD51/CD61 | Vn, Fb, vWF, Lm, Fn, thrombospondin, osteopontin[c] | 135/105 | Platelets, many non-hematopoietic cells | Expression increased by various stimuli |
| **β7** | | | | |
| α4β7[d], LPAM-1[e] CD49d/β7 | MAdCAM-1[d], VCAM-1, Fn (CS-1 domain) | 160/130 | Subset of memory T cells, eosinophils, basophils, EC | |
| αEβ7 CD103/β7 | E-cadherin, other ligands | | >95% intestinal intraepithelial lymphs [<7% circulating T cells] | Expression increased by TGF-β (decreased by TNF-α, IL-1, IFN-g) |

[a] Provides costimulation to T cells (in part by a focal adhesion kinase $pp125^{FAK}$); VLA-4 can also support rolling of lymphocytes on EC.

[b] Provides costimulation to T cells.

[c] Binds the amino acid sequence RGD (arginine–glycine–aspartic acid). The RGD motif is part of the recognition site for several extracellular matrix (ECM) molecules, including Fn, Fb, vWF, Vn, collagen, and osteopontin.

[d] α4β7, which mediates homing to Peyer's patch and mesenteric lymph nodes, recognizes a protein-based epitope of MAdCAM-1 (the MECA-367 mAb determinant).

[e] LPAM-1 = lymphocyte Peyer's patch adhesion molecule-1.

Other integrins include α7β1, α8β1, αvβ5, αvβ6, αvβ8, α6β4 (β4 = CD104).

dancy. Thus, most integrins are expressed on more than one cell type, and most cells express more than one integrin on their surface. Moreover, several molecules function as ligand for more than a single integrin, although their avidity for that ligand may be variable. Although many integrins have been shown to play some role in the pathophysiology of allergic diseases, two may be of particular importance: LFA-1 (CD11a/CD18) and VLA-4 (CD49d/CD29). These molecules have been shown to be central to the adhesion and transendothelial migration of T cells, eosinophils, and other leukocytes (1,2,6,8). Because antigen-driven T cells serve a key role in the orchestration of allergic diseases, and eosinophils play an important pathophysiologic role in these diseases, these adhesion receptors would be expected to be of particular importance. Moreover, LFA-1 and VLA-4 serve as accessory or costimulatory molecules for T cells (1,2,16,17). In conjunction with stimulation via the antigen-specific T-cell receptor, these molecules are capable of providing a "second signal" and thereby driving a productive immunologic response. As will be discussed subsequently, therapy directed at these adhesion molecules might therefore be expected not only to inhibit the accrual of T cells and other leukocytes at inflammatory sites but also to attenuate the activation of cells that is integral to the immune response.

## B.   Selectins

The selectins (Table 2), so named because they are *select*ively expressed on cells related to the vasculature and contain a *lectin*-binding domain, are composed of three extracellular domains: one of complement regulatory protein–like short consensus repeat units, an epidermal growth factor (EGF)-like domain, and the amino-terminal lectin domain that confers binding specificity. The EGF and complement-like domains may function primarily as a scaffold—optimally positioning the lectin domain above the cellular glycocalyx, thereby facilitating its interaction with ligand. Selectin counter-receptors are typically sialylated, fucosylated carbohydrate moieties, such as the sialyl-Lewis blood group oligosaccharides (e.g., sialyl-Lewis X [sLe$^X$]) (18). Glycosylation may be a key regulatory point for the selectins. For example, L-selectin is differentially glycosylated on lymphocytes and neutrophils, with resultant distinct binding profiles.

Some confusion concerning the selectins and their ligands originates historically from the various means by which this knowledge was derived. For example, ligands of the mouse homologue of L-selectin were named according to functional characteristics (e.g., the "peripheral lymph node addressin") or by mAb binding specificity (Table 2). As ligand characteristics yield to molecular analyses, the confusion surrounding these complex adhesion receptors may abate.

**Table 2** Selectin Family

| Receptor | Counter-receptor | Distribution | MW (kd) | Regulation |
|---|---|---|---|---|
| E-Selectin[a] CD62-E (ELAM-1) | Sialyl-Lewis[x] (sLe[x], CD15s) (recognizes sLe[x] presented on various molecules, e.g. L-Selectin, CLA[b], LFA-1, CD66); sialyl-Lewis A, ESL-1[c], PSGL-1[d] | Activated EC | 107–115 (various glycosylated forms) | Synthesized and expressed (hours) by IL-1, TNF-α, LPS |
| L-Selectin[e] CD62-L (LECAM-1, Leu-8, Mel-14 Ag) | PNAd[f]: GlyCAM-1[g] (sulfated glycoprotein [sgp]50), CD34 (sgp90), sgp200, MAdCAM-1[h], E-selectin, P-selectin, other ligands at inflammatory sites | Resting leukocytes | 74 (lymph) 90–100 (PMN) | Rapidly cleaved upon activation |
| P-Selectin[i] CD62-P (GMP-140, PADGEM) | Sialyl-Lewis[x] L-Selectin, Lewis[x] (CD15), PSGL-1[†] | Activated EC, activated platelets | 140 | Rapidly redistributed to cell surface (minutes) by PAF thrombin, histamine, LTC$_4$, LTB$_4$, H$_2$O$_2$ |

[a]E-selectin serves as a vascular addressin for skin-homing T cells, and also mediates the adhesion of various leukocytes to endothelium during rolling; E-selectin appears to recognize extended chain sLe[x], such as sialyl-dimeric Lewis X.

[b]CLA = the cutaneous lymphocyte antigen, a sialylated oligosaccharide that defines a subset of memory T cells that exhibit tropism for skin (present on ∼ 7–20% of circulating T cells).

[c]ESL-1 = E-selectin ligand-1 (homologous to fibroblast growth factor receptor) (see refs. 8–10).

[d]PSGL-1 = P-selectin glycoprotein ligand-1 (see refs. 8–10).

[e]L-Selectin, a protein with homology to C-type lectins, mediates adhesion of various leukocytes to endothelium during rolling, and homing of lymphocytes to peripheral and mesenteric lymph nodes via Ca$^{2+}$-dependent recognition of specific carbohydrate residues (e.g., sialic acid, fucose, sulfate) expressed on certain endothelial glycoprotein ligands.

[f]PNAd = the peripheral node addressin. Referred initially to the binding specificity of the mAb MECA-79, which recognizes various L-selectin ligands on peripheral lymph node HEV (see refs. 8–10).

[g]GlyCAM-1 = glycosylation-dependent cell adhesion molecule-1; it is expressed on lymph node HEV and mammary gland epithelium. GlyCAM-1 has no transmembrane region and therefore may function as a soluble or circulating receptor. CD34 is widely expressed on vascular endothelium and hematopoietic stem cells. GlyCAM-1, CD34, sgp200 and PSGL-1 are sialomucins, containing serine- and threonine-rich areas capable of binding MAdCAM-1.

[h]MAdCAM-1 = Mucosal addressin cell adhesion molecule-1; L-selectin presumably recognizes a carbohydrate motif expressed on the mucin-like domain on a subset of MAdCAM-1 molecules synthesized in mesenteric lymph nodes (see refs. 8–10).

[i]P-Selectin mediates the adhesion of various leukocytes to endothelium during rolling; L- and P-selectin may be particularly important in the initial phases of leukocyte–endothelial interactions (capture), whereas the synergistic actions of L-, P-, and E-selectin may be required for optimal leukocyte rolling.

## C.   Immunoglobulin Superfamily

Members of this family of adhesion receptors (Table 3) are composed of variable numbers of globular, immunoglobulin-like, extracellular domains. The immunoglobulin superfamily is phylogenetically ancient, having been described
in insects, where they have been shown to function as adhesion receptors in
nervous system development. From an evolutionary standpoint, this implies
that the adhesive functions of the immunoglobulin superfamily members ante-
date their antigen recognition capacity. Only those immunoglobulin family
members associated with antigen recognition, namely immunoglobulin on B
cells and the T-cell receptor complex, undergo gene rearrangements and so-
matic mutation. Some adhesion molecules in the immunoglobulin superfamily,
for example CD31 and NCAM, are capable of mediating homotypic adhesion.
Others, such as intercellular adhesion molecule (ICAM)-1 vascular cell adhesion molecule (VCAM)-1, and mediate adhesion via interactions with integrins
(Table 3).

## D.   Other Adhesion Molecules

Several important adhesion molecules that cannot be classified into one of the
three families discussed above are shown in Table 4. Progress in the description
of the characteristics of many of these molecules has recently proceeded apace,
and their roles in allergic inflammation may be expected to be revealed in the not
too distant future.

## E.   Alternate Forms

An interesting consideration that has received increased attention of late is the
existence of alternative forms of certain adhesion molecules. For some adhesion molecules, such as CD44, different forms result from alternative gene
splicing. Isoforms of CD44 have been found to be of substantial prognostic importance as regards the metastatic potential of several malignancies (19). Posttranslational modifications, particularly glycosylation, also exert significant
effects on binding specificities. This variability may allow a greater degree of
specificity at the cellular or tissue level. For example, the binding specificity
and molecular weight of L-selectin differ on lymphocytes and neutrophils, presumably as the result of variable glycosylation. Variability in the glycosylation
of ICAM-1, which is also illustrated by different molecular weights, may underlie its inconstant affinity for ligand that has been demonstrated on a tissue
level.

**Table 3** Immunoglobulin Superfamily[a]

| Receptor | Counter-receptor | Distribution | MW (kD) | Regulation |
|---|---|---|---|---|
| ICAM-1 (CD54) [5 Ig domains] [domain 1 binds CD11a/CD18; domain 3 binds CD11b/CD18] binds rhinovirus | LFA-1 (CD11a/CD18), Mac-1 (CD11b/CD18), CD43 (leukosialin) | Widespread: EC, fibroblasts, epithelium, monocytes, lymphs, dendritic cells, chondrocytes; parenchymal cells (myocardiocytes, hepatocytes) after cytokine stimulation | 75–110 | Constitutive: expression increased(~40×) by IL-1, TNF-α, IFN-γ, LPS, substance P; functions optimally as a dimer |
| ICAM-2 (CD102) [2 Ig domains] [domain 1 binds CD11a/CD18] | LFA-1, [CD11b, through A domain] | EC (high expression); lymphocytes, monocytes, basophils, platelets (low expression) | 50 | Constitutive |
| ICAM-3 (CD50) [5 Ig domains] [~50% homology to CD54] [domain 1 binds CD11a/CD18; domains 3,4 bind α_D/CD18] | LFA-1, α_D/CD18 | Lymphocytes,[b] Monocytes, Neutrophils, Eosinophils, Basophils | 125 | Expression increased by activation |
| VCAM-1 (CD106) [6 or 7 Ig domains] [domains 1 and 4 bind VLA-4] | α4β1 (VLA-4, CD49d/CD29) α4β7 | EC, monocytes, fibroblasts, dendritic cells, bone marrow stromal cells, myoblasts | 90–110 | Expression increased by IL-1, TNF-α, IL-4, IL-13, LPS, oxidative stress |
| LFA-3 (CD58) [6 Ig domains] | CD2 | EC, leukocytes, epithelial cells | 50–75 | |

**Table 3** Continued

| Receptor | Counter-receptor | Distribution | MW (kD) | Regulation |
|---|---|---|---|---|
| PECAM-1 (CD31) [domains 1 and 2 mediate transendothelial migration, domain 6 mediates ECM migration] | CD31, heparin, other | EC (at EC–EC junctions), T-cell subsets, platelets, neutrophils, eosinophils, monocytes, smooth muscle cells, bone marrow stem cells | 130 | Polymorphic forms exist (? minor histocompatibility antigen) |
| NCAM (CD56 homologue) | NCAM, heparan $SO_4$ | Neural cells, glial cells, heart, muscle, kidney | 120–200 | Unknown |
| MAdCAM-1 (contains 4 domains: an ICAM-1/VCAM-1 like, a VCAM-1 like, and a $C\alpha2$-like) | $\alpha4\beta7$, L-selectin | HEV of Peyer's patch and mesenteric lymph nodes; also, mucosal EC (gut lamina propria, lactating mammary gland, exocrine pancreas); spleen sinus lining cells | | Expression increased by TNF-$\alpha$, IL-1, IFN-$\gamma$ |
| CD2 | CD58, CD59, CD48 | T lymphocytes | 40 | |

[a]Other immunoglobulin superfamily members include immunoglobulin, the CD3/T cell receptor complex, CD4, CD8, and MHC class I and II antigens.
[b]Transduces signal for T cells; cytoplasmic portion associates with p56[lck] and p59[fyn].

**Table 4** Other Adhesion Receptors

| Receptor | Counter-receptor | Distribution | MW | Regulation |
|---|---|---|---|---|
| CD44 (cartilage link protein family) | Hyaluronate, gp600 (serglycin; a proteoglycan stored in intracellular granules of lymphoid and myeloid cells), collagen type VI, osteopontin | Lymphocytes (role in homing, rolling on EC), monocytes, EC, fibroblasts, epithelial cells | 90 | Alternatively spliced (9 exons) isoforms exist (correlation with tumor metastases; inflammatory bowel disease) |
| VAP-1 (vascular adhesion protein-1) | Unknown (mediates lymphycyte–HEV binding); binding depends on sialic acid residues | HEV of lymph node; EC at chronic inflammatory sites (e.g., synovium, skin, etc.); dendritic cells | 170 | Expression increased at inflammatory sites |
| CD73; L-VAP-2 (lymphocyte vascular adhesion protein-2); ecto-5'-nucleotidase | Unknown | EC; subset (~20%) of lymphocytes (B cells, CD8 T cells) | 70 | Constitutive |
| Cadherins<br>• N-cadherin<br>• E-cadherin<br>• P-cadherin | Homophilic binding (also $\alpha^E\beta7$ is a counter-receptor for E-cadherin) | • Brain, muscle, lens<br>• Epithelium<br>• Placenta, epithelium | 120–135 | $Ca^{++}$-dependent |
| CD36 (platelet gpIV, gpIIIa) | Thrombospondin (also, EC receptor for Plasmodium falciparum–infected erythrocytes) | Platelets, monocytes, EC | 88 | Unknown |
| CD26 (dipeptidyl peptidase IV) adenosine deaminase binding protein | Fn, Co | EC, lymphocytes | 120 | Unknown |

## IV. CONCEPTS RELATED TO ADHESION MOLECULE UTILIZATION

Perhaps of greater clinical relevance than the progress in the delineation of the characteristics of individual adhesion molecules has been the appreciation of the integrated utilization of these molecules. Indeed, it is probably most appropriate that adhesion molecules be considered a cascade, much like the complement and coagulation systems. Thus, although cells possess a large repertoire of adhesion molecules on their surface, they are not utilized randomly or *en masse*. Rather, they function in an hierarchal, sequential manner (Table 5 and Fig. 2). In the initial phases of an inflammatory response, perturbations in the endothelium allow predominantly selectin-based interactions to slow the velocity of the circulating cells. Consequent to activation mediated by chemokines and adhesion molecule interactions, some of these rolling cells become flatter and more tightly adherent. Subsequently, these cells may exit the vessel by migrating between endothelial cells, a process mediated to a large extent by activated integrins and immunoglobulin-family receptors. Migration of the cell through the ECM and further activation of cells within the inflammatory site are mediated by integrins as well as chemokines and cytokines.

Early in the investigation of adhesion molecules it had been hypothesized that the intricacies of tissue localization of circulating cells might be entirely explained by specific adhesion receptors. Although there may be preferential expression of certain molecules at defined tissue sites (e.g., addressins and homing receptors), adhesion molecules exhibit substantial redundancy. It appears that the specific function of adhesion molecules, rather than depending on unique tissue expression, may be explained by other factors, such as tissue-specific posttranslational modification. Perhaps more importantly, the synergistic and sequential use of combinations of adhesion molecules may impart specificity, with certain com-

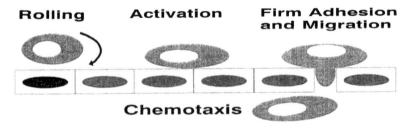

**Figure 2** Sequential adhesion molecule interactions at different stages of a developing inflammatory response. Components related to leukocytes and endothelium/tissue sites at different stages are indicated in Table 5.

**Table 5** The Adhesion Cascade

| | Inflammatory stimulus | Rolling | Activation | Firm adhesion and transmigration | Haptotaxis/chemotaxis; tissue retention |
|---|---|---|---|---|---|
| Leukocyte component | | sLe$^x$, L-selectin, PSGL-1, VLA-4, CD44 CD31 | Cytokine and chemokine receptors; CD31 | β2 integrins (also β1 and β7 integrins, CD31) | Chemokine receptors; β1 integrins |
| Endothelial and tissue component | Histamine, thrombin, bradykinin, LPS, IL-1, TNF-α; various lipid (e.g., leukotriene) and peptide (e.g., substance P) mediators; oxidative and shear stress | P-selectin, E-selectin, GlyCAM-1, MAdCAM-1, PAF; VCAM-1 | Chemokines, cytokines, CD31, PAF, C5a, fMLP | ICAM-1, 2 (also VCAM-1, CD31) | Chemokines (provide a directional gradient), lipid mediators, ECM molecules |

binations functioning like telephone area codes or postal zip codes (9). Specificity may also be engendered by other components of the immune response such as the chemokines. This diverse and expanding group of inflammatory mediator molecules play a central role in allergic and other immunologic reactions by mediating chemoattractant and activating functions (20). By activating certain adhesion receptors and up-regulating their avidity for ligand, chemokines liberated in a particular inflammatory milieu may help impart a degree of specificity. Moreover, because they can exhibit selectivity for target cells (e.g., RANTES is a specific chemoattractant for memory T cells and eosinophils, as is eotaxin for eosinophils), the repertoire of chemokines in the local milieu may contribute significantly to the specific composition of cells at that site. In addition, other factors that up-regulate the avidity of various adhesion receptors for their ligands, such as pro-inflammatory cytokines, and peptide and lipid mediators, may also contribute to specificity at inflammatory sites.

## V.  ROLE OF ADHESION MOLECULES IN ALLERGIC DISEASES

The potential utility of adhesion molecules as therapeutic targets in allergic diseases is predicated on the significant role these molecules play in the initiation and propagation of these diseases. As will be discussed further in subsequent chapters, several lines of evidence support such a role. Thus, adhesion molecules have been shown to be expressed on various cell types integral to the development of allergic inflammation. Moreover, the expression of many adhesion molecules has been demonstrated to be quantitatively as well as qualitatively up-regulated in vivo at sites of allergic inflammation (21–27). The importance of adhesion molecules to allergic disease is further supported by animal models in which adhesion receptor–directed therapies effectively attenuate allergic inflammation.

## VI.  ADHESION RECEPTORS AS THERAPEUTIC TARGETS IN ALLERGIC DISEASES

Although there may be multiple potential targets of antiadhesion therapy, there are two adhesion molecule/counter-receptor pairs that appear to play particularly critical roles in the pathogenesis of allergic diseases, and thus may be the most appealing therapeutic targets. These pairs, LFA-1/ICAM-1 and VLA-4/VCAM-1, are central to various intercellular interactions of T cells, eosinophils, and other cells, including their accrual at inflammatory sites and their activation. Nevertheless, as more is learned of the complex interactions of other adhesion molecules relevant to allergic disease, additional attractive targets will no doubt emerge.

## A. Approaches to Antiadhesion Therapy

There are several methods by which adhesion molecule function might be inhibited (Fig. 3). The most direct strategy involves specific inhibition of the adhesion molecule or its counter-receptor. This has been the method used most widely in both animal models of inflammation as well as in the few human studies using antiadhesion therapy (8,28). Most studies have utilized monoclonal antibodies (mAb) as the therapeutic agent. Monoclonal antibodies possess several desirable characteristics, including exquisite target specificity, availability in large quantities, and the ability to execute various effector functions via their Fc receptors. However, as most mAb produced to date have been generated in mice, there are potential limitations to their therapeutic use for human disease. Because they are foreign, murine mAb will elicit human anti-mouse antibody (HAMA) responses. On repeated administration, these HAMA may not only decrease the serum half-life and thereby the therapeutic utility of the mAb but may also cause potentially serious adverse effects. To circumvent these problems, several methods to reduce the immunogenicity of therapeutic mAb have been developed. Utilizing the techniques of molecular biology, researchers can substitute parts of human antibodies for the murine, yielding chimeric and humanized mAbs (28). The most eagerly awaited development is the ability to produce human mAbs directed against targets such as adhesion receptors.

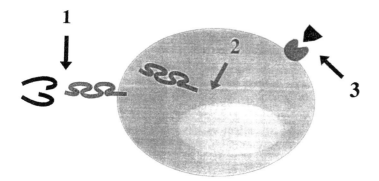

**Figure 3** Inhibition of adhesion receptors. Potential mechanisms by which the utilization of adhesion receptors might be inhibited include: (*1*); direct targeting of the adhesion receptor or its counter-receptor, (*2*); inhibition of the transcription and/or translation of DNA or RNA encoding for the adhesion receptor (*3*); inhibition of pro-inflammatory chemokines or cytokines, or their cell surface receptors, that up-regulate the adhesion receptor (alternatively, another cytokine that directly down-regulates an adhesion receptor or inhibits the activity of pro-inflammatory cytokines could be utilized) (*3*).

The function of adhesion receptors can be directly inhibited by agents other than mAb. For example, several adhesion molecules have been detected in soluble form. In various in vitro assays, soluble adhesion receptors have been effectively used as competitive inhibitors of adhesion receptor interactions. Moreover, therapeutically administered forms of soluble adhesion receptors have effectively abrogated inflammation in animal models (29). Some constructs of soluble adhesion receptors may ultimately be of use in allergic disease.

Traditionally, desirable characteristics of pharmacologic compounds have included low molecular weight, oral availability, and relatively low production cost. In the near future, agents possessing these characteristics and capable of inhibiting specific adhesion receptor interactions may become available. Perhaps the most notable progress in this field has been for adhesion receptors whose binding is mediated via the RGD (arginine–glycine–aspartic acid) motif. For example, several peptides and peptidomimetics capable of inhibiting the binding of the platelet integrin gpIIb/IIIa to the RGD sequence on fibrinogen have been discovered or synthesized (30). Several such agents are currently under investigation as inhibitors of platelet adhesion in human studies. There is a tremendous effort within the pharmaceutical industry to develop such inhibitors for other adhesion receptors, including those relevant to inflammatory diseases. Interestingly, some RGD-based inhibitors have been found to inhibit the adhesive interactions of ligands previously thought not to be mediated by the RGD domain. For example, adhesive interactions of VLA-4 with its ligands VCAM-1 and fibronectin (Fn) have been shown to be inhibited by a cyclic RGD peptide (31). Because VLA-4/VCAM-1 and VLA-4/Fn interactions presumably play a central role in allergic diseases, such inhibitors might be expected to be of therapeutic value. Peptide fragments of Fn have been used to attenuate inflammation in a rat model of arthritis (32). Potential small soluble inhibitors of interactions between integrins and immunoglobulin superfamily adhesion receptors, such as those mediated by CD11a/CD18, CD11b/CD18, ICAM-1, and ICAM-2, are an area of intense development (33,34). Interestingly, some investigators have turned to nature, uncovering prokaryotic and eukaryotic products capable of inhibiting specific adhesion receptor interactions (35).

Substantial progress in the development of small soluble adhesion receptor inhibitors has been made for the selectins. An intriguing early observation, indeed one that antedated the field of adhesion receptors *per se*, was that infusion of simple sugars was able to alter leukocyte recirculation (36). With the discovery that the adhesive interactions of the selectins are mediated via specific carbohydrate moieties, the potential utility of oligosaccharides as inhibitors of adhesion resurfaced. In both *in vitro* experiments as well as an *in vivo* animal model of inflammation, soluble carbohydrates have been effective inhibitors (36,37). These glycomimetics are another area of intense development in the pharmaceutical industry. As with the integrins, investigators have also discovered bacterial products capable of inhibiting selectins (38).

In addition to direct inhibition of adhesion receptors at the cell surface, there are other means by which these molecules can be inhibited (Fig. 3). Both the cell surface expression as well as the avidity for ligand of many adhesion receptors are modulated by the effects of cytokines (5–8). Pro-inflammatory cytokines such as interleukin-1 (IL-1) and tumor necrosis factor–$\alpha$ (TNF-$\alpha$) have also been suggested to play an important role in the immunopathogenesis of allergic disease, in part related to their ability to up-regulate various adhesion receptors. Therefore, inhibition of such cytokines, as can be accomplished by diverse methods (28), may ultimately exert an immunomodulatory effect by inhibiting adhesion receptor function. There is support for this concept. For example, the primary immunomodulatory mechanism of action of corticosteroids has been demonstrated to be the inhibition of the production of various cytokines, including IL-1 and IL-6 (39). Presumably via this mechanism, treatment with steroids has been shown to attenuate adhesion receptor–mediated accumulation of leukocytes at inflammatory sites (40). Also, therapy with cyclosporin, which acts predominantly via inhibition of IL-2, has been shown to decrease expression of ICAM-1 in psoriatic skin lesions (41). Recently, it has been shown that treatment of rheumatoid arthritis (RA) patients with a mAb directed against TNF-$\alpha$ resulted in decreases in serum levels of soluble E-selectin and ICAM-1. The decreases correlated with alterations in circulating lymphocyte counts and clinical response, suggesting that part of the mechanism of action of this agent relates to modulation of adhesion receptor function in the synovial endothelium (42).

Another novel approach to antiadhesion therapy involves direct inhibition of adhesion receptor synthesis. Inhibition of the transcription or translation of the DNA or RNA encoding adhesion receptor proteins can be achieved with specific antisense oligonucleotides (Fig. 3) (43,44). The efficacy of this approach has been proven in animal models (44). Another molecular approach involves the use of specific inhibitors of regulatory factors required for the processing of DNA encoding particular adhesion molecules (45). Such an approach could offer pharmacologic benefits such as oral availability. These elegant molecular biologic approaches will no doubt emerge as potential antiadhesion therapies for human disease in the near future.

## B.   Mechanisms of Action of Antiadhesion Therapy

Because adhesion receptors mediate interactions at various stages in the immunopathogenesis of allergic disease, targeting them might be expected to yield heterogeneous effects. Interestingly, in both animal models and human studies, it appears that several mechanisms of action might indeed contribute to the efficacy of antiadhesion therapy.

The most straightforward mechanism of action of antiadhesion therapy would be alterations in cellular trafficking. Inhibiting an adhesion receptor critical to the entry of a particular cell into an inflammatory site should block that cell's accrual at the site. While there is substantial evidence supporting this mechanism of action, it appears that the in vivo situation is even more complex. As noted, adhesion molecules function synergistically and sequentially, in a cascade fashion. Therefore, inhibition of a given adhesion receptor/counter-receptor pair may ultimately affect the utilization of other adhesion receptors. Consequently, it might be possible to block the function of a given adhesion receptor or a part of the immune response by targeting an interaction more proximal in the cascade. For example, the selectins have their primary role in the initial interactions between circulating leukocytes and the endothelium. However, inhibition of E-selectin function has been shown to attenuate the late phase airway response in an animal model of reactive airway disease (46). Another indirect mechanism by which blocking of adhesion receptors might produce clinical benefit is by attenuating the local damage secondary to leukocyte infiltration. During the course of transendothelial migration, circulating leukocytes release a host of products such as proteases and oxidants that are capable of injuring the endothelium, altering endothelial function, and degrading the extracellular matrix constituents. The sequelae of such injury, for example, increased vascular permeability, may have a profound potentiating effect on the underlying inflammatory response. This impairment may be attenuated by antiadhesion therapy (47).

It has been shown that certain adhesion receptors are capable of providing activation signals to immunocompetent cells. Thus, inhibition of the activation of cells may be another important mechanism of antiadhesion therapy. Indeed, in some animal models it appears that decreased activation of cells in the inflammatory site may have been an important mechanism of action of antiadhesive therapy. For example, treatment with an anti–VLA-4 antibody has been shown to exert significant effects on airway hyperresponsiveness, an indication of decreased eosinophil activity. Because the clinical benefit occurred without substantial alterations in the number of eosinophils recovered from bronchoalveolar lavage fluid, this suggests that blocking the activation of eosinophils at the inflammatory site—rather than simply impeding their access to the site—was a contributory mechanism (48). For immunologically driven diseases, the preeminent goal of antiadhesion receptor therapy may indeed be modulation of cellular activation. As noted, LFA-1 and VLA-4 have both been shown to function as costimulatory molecules on T cells (16,17). Inhibition of costimulatory molecules in the context of antigen presentation may facilitate the establishment of immunologic tolerance. Indeed, in animal models, antigen-specific tolerance of solid organ allografts has been achieved utilizing antiadhesion receptor therapies (44,49).

A relevant consideration in antiadhesion therapy is the potential for adverse effects, particularly increased susceptibility to infection. This is supported by two human immunodeficiencies, leukocyte adhesion deficiency (LAD) types 1 and 2, that are characterized, respectively, by deficiencies of CD11/CD18 and selectin ligand. Moreover, in some animal studies, an increased proclivity for infection has been noted as a sequela of antiadhesion therapy (50). The risk of infection may be affected by the choice of target. For example, it might be expected that targeting ICAM-1, which is up-regulated at inflammatory sites in comparison to normal tissue, might be less immunosuppressive than targeting CD18, which is present on all leukocytes. Support for this comes from knockout animals, where it has been shown that ICAM-1 deficient mice do not appear excessively susceptible to infection. Similarly, targeting VLA-4, which is expressed on all leukocytes with the exception of neutrophils, might conceivably engender less global immune suppression by leaving neutrophil function intact. Nevertheless, heightened vigilance for sequelae of immunosuppression is required for all studies using antiadhesion therapies.

## C. Studies Using Antiadhesion Therapy

As will be discussed in greater detail in subsequent chapters, several studies assessing the efficacy of targeting adhesion molecules in animal models have been undertaken. Targets in these trials have included VLA-4, ICAM-1, and E-selectin. As has been seen in other animal models of inflammatory disease, temporal associations can be germane. Generally, antiadhesion therapy has been most efficacious when administered either prior to or closely following antigen exposure.

To date, there have been no published reports of antiadhesion therapies used in patients with allergic disease. However, trials of antiadhesion receptor therapy have been undertaken in solid organ transplant as well as in rheumatoid arthritis (RA), an autoimmune disease with certain immunopathologic characteristics similar to allergic diseases. Thus, RA is considered to be a disease orchestrated by autoreactive T cells that are responding to antigens that have not yet been identified. Its manifestations relate to the accrual of various leukocytes at the inflammatory and the effects of their secreted products, similar to the case in allergic inflammation.

The greatest experience with antiadhesion therapy in RA has been with a murine anti–ICAM-1 mAb (51–55). The safety and efficacy of the anti–ICAM-1 mAb have been analyzed in heterogeneous populations of patients with active RA (51–55). Although the studies were open and noncontrolled, substantial efficacy was noted for a large subset of treated patients. In some cases, patients experienced dramatic improvement in several parameters of disease activity that persisted for more than 6 months after a single treatment course. Notably, patients with RA of shorter disease duration achieved greater clinical responses than those with longstanding disease. Importantly, no infectious complications related to therapy were noted.

Several analyses were conducted to help delineate the mechanism of action of the anti–ICAM-1 mAb. After therapy, anti–ICAM-1 mAb was readily detected on the surface of circulating leukocytes as well as on the vascular endothelium and perivascular leukocytes from skin biopsy specimens. As a correlate of this observation, serial delayed type hypersensitivity (DTH) testing revealed that anti–ICAM-1 induced transient cutaneous anergy for a number of patients. Analysis of the numbers of circulating leukocytes revealed a significant increase in the number of lymphocytes during therapy, without significant changes in neutrophils or monocytes. Phenotypic analysis revealed that the increase in lymphocytes consisted predominantly of CD3+ CD4+ T cells. Of note, in this population there was an increase in the numbers of activated cells, as evidenced by the increased expression of HLA-DR and the IL-2 receptor (CD25). These results suggest that treatment with anti–ICAM-1 altered the recirculation of T cells and may have caused a temporary redistribution of T cells out of the rheumatoid synovium. Further analysis revealed that there was an elevation of mRNA for interferon gamma (IFN-$\gamma$) from circulating mononuclear cells during therapy, suggesting an alteration in the circulatory pattern of cells with a Th1-like phenotype (53). The increase in IFN-$\gamma$ also correlated with clinical efficacy in treated patients, implying mechanistic relevance.

As noted, interference with adhesion receptor function might facilitate the induction of immunologic tolerance. In these studies, treatment with the anti–ICAM-1 mAb appeared to induce a state of T-cell hyporesponsiveness. Thus, T-cell proliferative responses to mitogens were impaired in some patients after therapy, whereas proliferative responses to recall antigens were preserved (54). Of note, there was a correlation between this T-cell hyporesponsiveness and clinical outcome. Therefore, in this study a form of peripheral T-cell anergy may have been induced. As noted, the ability to engender immunologic tolerance decreases with increasing chronicity of disease. Thus, it may be quite difficult for any immunomodulatory agent to achieve a longstanding remission of disease activity when used as monotherapy. Therefore, the clinical results seen in the cohort of patients with refractory RA who received anti–ICAM-1 mAb are quite encouraging. In addition, the extended clinical benefit noted from the treatment of some patients with relatively early disease are promising.

Eight patients received a second course of the murine anti–ICAM-1 mAb (55). Of note, six patients developed serum sickness–like symptoms several days after the second course of treatment. Although the precise underlying mechanisms have not been defined, these symptoms are presumed to relate to the formation of immune complexes, possibly consisting of HAMA/anti–ICAM-1 mAb/circulating ICAM-1. In support of this, all patients developed detectable HAMA after the first treatment. In addition, there was evidence of transient depletion of complement proteins during the second course of therapy that was not observed in the initial course. Moreover, the clinical benefit associated with the second course of therapy was far inferior to that of the first course, both in the number of patients responding as well as the duration of response. In summary, these observations indicate that although ICAM-1 appears to be an appropriate target for antiadhesion therapy in RA, a murine mAb is not a suitable agent on account of its immunogenicity.

Another agent with potential antiadhesion effects that has been used in patients with RA is the oligosaccharide dimeric sLe$^x$ (56). In a 6-month open study of 27 RA patients, 17 showed some improvement in clinical status related to intradermal therapy with this agent. As adverse effects were relatively minor, future therapies based on this approach seem to be warranted.

## D. Future Directions in Antiadhesion Therapy

There are great expectations that developments in antiadhesion therapy in the near future will yield important and useful therapeutic agents (3,4,5–8,57 and see Table 6). Substantial progress in several facets of antiadhesion therapy should be forthcoming. Prominent among these may be novel types of agents. The ultimate goal is the development of agents that are not only effective but also easy to administer, relatively inexpensive to produce, and nonimmunogenic. This is particularly germane, as the costs and hence the cost-efficacy of novel therapeutic

**Table 6**  Future Directions in Antiadhesion Therapy

Agents
> Human antibodies directed against adhesion receptors
> Peptide and peptidomimetic inhibitors of adhesion receptors
> Oligonucleotide-based therapies/gene-based therapies
> Orally available inhibitors
> Combination therapy (e.g., antiadhesion therapy + cytokine-directed therapy; combination of therapies directed against adhesion receptors and their counter-receptors)

Patients
> Define the most appropriate therapeutic subsets (e.g., patients most likely to derive benefit from antiadhesion therapy)

Targets
> Novel adhesion receptors
> Cytokines/chemokines that qualitatively or quantitatively up-regulate adhesion molecule

Goals
> Induction of tolerance to etiologic antigens

agents will have a substantial impact on their ultimate clinical utility (58). The combination of antiadhesion therapy with agents possessing distinct mechanisms of action, for example cytokine or chemokine-directed therapies, might be complementary or synergistic. Alternatively, combinations of antiadhesion therapy with more traditional therapeutic agents might produce increased efficacy for some patients. As it has been observed that there is substantial heterogeneity in patient response to immunomodulatory therapies, identification of the subsets of patients most likely to respond to antiadhesion therapy would be a significant advance. Finally, the goals of antiadhesion therapy might be altered as our therapeutic armamentarium expands and our experience broadens. The ultimate goal would be true disease modification, as might be expected to result from the establishment of immunologic tolerance. Further advancements in this exciting discipline are eagerly awaited.

## REFERENCES

1. T. A. Springer, *Nature, 346*:426–434 (1990).
2. S. M. Albelda and C. A. Buck, *FASEB J., 4*:2868 (1990).
3. E. C. Butcher, *Cell, 67*:1033 (1991).
4. R. Pardi, L. Inverardi, J. R. Bender, *Immunol. Today, 13*:224 (1992).
5. C. W. Smith, *Semin. Hematol., 30*:45–55 (1993).
6. C. R. Mackay and B. A. Imhof, Cell adhesion in the immune system, *Immunol. Today, 14*:99 (1993).
7. D. H. Adams and S. Shaw, *Lancet, 343*:831–836 (1994).
8. T. M. Carlos and J. M. Harlan, *Blood, 84*:2068 (1994).
9. T. A. Springer, *Cell, 76*:301 (1994).
10. N. Hogg and C. Berlin, *Immunol. Today, 16*:327 (1995).
11. Fishman AP, ed., *Annals of the New York Academy of Sciences*, Vol. 401. New York. pp 1–274 (1982).
12. E. R. Clark and E. L. Clark, *Am. J. Anat., 57*:385 (1935).
13. J. L. Gowans and E. J. Knight, *Proc. R. Soc. Lond., 159*:257 (1964).
14. W. M. Gallatin, I. L. Weissman, and E. C. Butcher, A *Nature, 304*:30 (1983).
15. M. Lub, Y. van Kooyk, and C. G. Figdor, *Immunol. Today, 16*:47 (1995).
16. G. A. van Seventer, Y. Shimizu, and S. Shaw, *Curr. Opinion Immunol., 3*:294 (1991).
17. N. K. Damle, K. Klussman, P. S. Linsley, and A. Aruffo, *J. Immunol., 148*:1985 (1992).
18. B. S. Bochner, S. A. Sterbinsky, C. A. Bickel, S. Werfel, M. Wein, and W. Newman, *J. Immunol., 152*:774 (1994).
19. Y. Matsumura and D. Tarin, *Lancet, 340*:1053 (1992).
20. K. B. Bacon and T. J. Schall, *Int. Arch. Allergy Immunol., 109*:97 (1996).
21. A. M. Bentley, S. R. Durham, D. S. Robinson, G. Menz, C. Storz, O. Cromwell, A. B. Kay, and A. J. Wardlaw, *J. Allergy Clin. Immunol., 92*:857 (1993).
22. A. L. Lazaar, S. M. Albelda, J. M. Pilewski, B. Brennan, E. Pure, and R. A. Panettieri, *J. Exp. Med., 180*:807 (1994).

23. B.-J. Lee, R. M. Naclerio, B. S. Bochner, R. M. Taylor, M. C. Lim, and F. M. Barrody, *J. Allergy Clin. Immunol.*, *94*:1006 (1994).
24. M. M. Hamawy, S. E. Mergenhagen, and R. P. Siraganian, *Immunol. Today*, *15*:62 (1994).
25. B. S. Bochner and R. P. Schleimer, *J. Allergy Clin. Immunol.*, *94*:427 (1994).
26. L. Hakansson, E. Björnsson, C. Janson, and B. Schmekel, *J. Allergy Clin. Immunol.*, *96*:941 (1995).
27. M. S. Weinberger, T. M. Davidson, and D. H. Broide, *J. Allergy Clin. Immunol.*, *97*:662 (1996).
28. J. J. Cush and A. F. Kavanaugh, *Rheum. Dis. Clin. North Am.*, *21*:797 (1995).
29. S. R. Watson, C. Fennie, and L. Lasky, Neutrophil influx into an inflammatory site inhibited by a soluble homing receptor-IgG chimera. *Nature*, *349*:164 (1991).
30. A. M. Krezel, G. Wagner, J. Seymour-Ulmer, and R. A. Lazarus, *Science*, *264*:1944 (1994).
31. P. M. Cardarelli, R. R. Cobb, D. M. Nowlin, W. Scholz, F. Gorcsan, M. Moscinski, M. Yasuhara, A.-L. Chiang, and T. J. Lobl, *J. Biol. Chem.*, *269*:18668 (1994).
32. S. M. Wahl, J. B. Allen, K. L. Hines, T. Imamichi, A. M. Wahl, L. T. Furcht, and J. B. McCarthy, *J. Clin. Invest.*, *94*:655 (1994).
33. L. Ross, F. Hassman, and L. Molony, *J. Biol. Chem.*, *267*:8537 (1992).
34. R. Li, P. Nortamo, L. Valmu, M. Tolvanen, J. Huuskonen, C. Kantor, and C. G. Gahmberg, *J. Biol. Chem.*, *23*:17513 (1993).
35. E. Rozdzinski, J. Sandros, A. van der Flier, A. Young, B. Spellerberg, C. Bhattacharyya, J. Straub, G. Musso, S. Putney, R. Starzyk, and E. Tuomanen, *J. Clin. Invest.*, *95*:1078 (1995).
36. E. Tuomanen, A *J. Clin. Invest.*, *93*:917 (1994).
37. A. Prokopova, V. Kery, M. Stancikova, J. Grimova, J. Capek, J. Sandula, and E. Orvisky, *J. Rheumatol.*, *20*:673 (1993).
38. E. Rozdzinski, W. N. Burnette, T. Jones, V. Mar, and E. Tuomanen, *J. Exp. Med.*, *178*:917 (1993).
39. W. Y. Almayi, M. L. Lipman, A. C. Stevens, B. Zanker, E. T. Hadro, and T. B. Strom, *J. Immunol.*, *146*:3523 (1991).
40. J. H. Wang, C. J. Trigg, J. L. Devalia, S. Jordan, R. J. Davies, *J. Allergy Clin. Immunol.*, *94*:1025 (1994).
41. P. Petzelbauer, G. Stingl, K. Wolff, and B. Volc-Platzer, *J. Invest. Dermatol.*, *96*:363 (1991).
42. E. M. Paleolog, M. Hunt, M. J. Elliott, J. N. Woody, M. Feldmann, and R. N. Maini, *Arthritis Rheum.*, (1996).
43. C. F. Bennett, T. P. Condon, S. Grimm, H. Chan, and M-Y. Chiang, *J. Immunol.*, *152*:3530 (1994).
44. S. M. Stepkowski, Y. Tu, T. P. Condon, and C. F. Bennett, *J. Immunol.*, *153*:5336 (1994).
45. C. C. Chen, C. L. Rosenbloom, D. C. Anderson, and A. M. Manning, *J. Immunol.*, *155*:3538 (1995).
46. R. H. Gundel, C. D. Wedner, C. A. Torcellini, et al., *J. Clin. Invest.*, *88*:1407 (1991).
47. X.-L. Ma, P. S. Tsao, and A. M. Lefer, *J. Clin. Invest.*, *88*:1237 (1991).
48. W. M. Abraham, M. W. Sielczak, and A. Ahmed, et al., *J. Clin. Invest.*, *93*:776 (1994).

49. M. Isobe, H. Yagita, K. Okumara, and A. Ihara, *Science, 255*: 1125 (1992).
50. S. R. Sharar, R. K. Winn, C. E. Murry, J. M. Harlan, and C. L. Rice, *Surgery, 110*:213 (1991).
51. A. F. Kavanaugh, L. S. Davis, L. A. Nichols, S. H. Norris, R. Rothlein, L. A. Scharschmidt, and P. E. Lipksy, *Arthritis Rheum., 37*:992 (1994).
52. A. F. Kavanaugh, L. S. Davis, R. I. Jain, L. A. Nichols, S. H. Norris, and P. E. Lipsky, *J Rheumatol.*, (1996).
53. H. Schulze-Koops, P. E. Lipsky, A. F. Kavanaugh, and L. S. Davis, *J. Immunol., 155*:5029 (1995).
54. L. S. Davis, A. F. Kavanaugh, L. A. Nichols, and P. E. Lipsky, *J. Immunol., 154*:3525 (1995).
55. A. F. Kavanaugh, L. S. Davis, L. A. Nichols, and P. E. Lipsky, *Arthritis Rheum., 38*(suppl):S280 (1995).
56. T. Ochi, S.-L. Hakomori, M. Fujimoto, M. Okamura, H. Owaki, S. Wakitani, Y. Shimaoka, K. Hayashida, T. Tomita, S. Kawamura, and K. Ono, *J. Rheumatol., 20*:2038 (1993).
57. M. Wein and B. S. Bochner, *Eur. Respir. J., 6*:1239 (1993).
58. A. F. Kavanaugh, G. Heudebert, J. J. Cush, and R. Jain, *Semin. Arthritis Rheum., 25*:297 (1996).

# 2

# The Endothelium in Leukocyte Recruitment

**Francis W. Luscinskas** *Brigham and Women's Hospital, and Harvard Medical School, Boston, Massachusetts*

## I.  INTRODUCTION

The vascular endothelium lines the entire cardiovascular system and acts as a nonthrombogenic and selectively permeable boundary between the bloodstream and extravascular space. The vascular endothelium actively participates in regulation and maintenance of vessel integrity. Endothelial cells are polar cells that are continuously subjected to hemodynamic forces of flowing blood on their luminal surface; their abluminal surface contacts basement membrane and extracellular matrix. A third surface is the lateral interface between adjacent endothelial cells, which forms intercellular junctions and regulates permeability. Thus, the endothelium by virtue of its anatomical location possesses a unique and influential position during many disease processes. This chapter highlights the role of the vascular endothelium in leukocyte trafficking.

Peripheral blood leukocytes interact with the vascular endothelium in a wide range of physiological and pathophysiological processes. During the inflammatory response in the nonlymphoid vasculature, a well-supported concept is that "activation" of the normally quiescent vascular endothelium is a critical event to initiate and sustain leukocyte emigration, conferring spatial and temporal localization, and leukocyte-type selectivity to the recruitment process. This chapter focuses on some of the characteristics of the endothelium that lines vessels in nonlymphoid tissues at sites of inflammation. It discusses the molecular organization of the endothelial surfaces, such as induction and apical expression of leukocyte adhesion molecules, endothelial cell-to-cell lateral junctions, and their relationship to blood leukocyte trafficking through these surfaces to gain passage into the extravascular space.

## II. THE PHENOTYPE OF ACTIVATED ENDOTHELIUM: MECHANISMS AND MOLECULES

Localized leukocyte accumulation is the cellular hallmark of inflammation. Although this has been recognized for more than a century, it is only in the past decade that the role of the vascular endothelial cell in leukocyte recruitment has been recognized. The notion that the endothelium actively participates in leukocyte recruitment initially gained support from in vitro studies demonstrating that treatment of cultured human umbilical vein endothelial (HUVEC) monolayers with inflammatory cytokines, namely interleukin-1 (IL-1) and tumor necrosis factor–α (TNF-α), and certain gram-negative bacterial endotoxin (LPS, lipid A), could "activate" the endothelium to become adhesive for peripheral blood leukocytes and leukocyte cell lines (1–4). Subsequently, many investigators have contributed to the identification and molecular cloning of several such endothelial surface adhesion molecules and their cognate ligands on leukocytes that support blood leukocyte adhesion to endothelium in a selective fashion (see Table 1). The primary examples are members of at least four different gene families: 1.) the selectins, E-, P-, and L-selectin (CD62E, CD62P, and CD62L, respectively) [reviewed in (5)]; 2.) the immunoglobulin (Ig) family (intercellular adhesion molecule-1 [ICAM-1] ICAM-2, and ICAM-3); vascular cell adhesion molecule-1 (VCAM-1); platelet-endothelial cell adhesion molecule-1 (PECAM-1, CD31); 3.) the heterodimeric integrins, principally leukocyte α4β1 (CD49d/CD29, VLA-4), α4β7, and β2-integrins (CD11a, b, c/CD18, respectively) [reviewed in (6–9)] and 4.) mucin-like adhesion molecules, P-selectin glycoprotein ligand-1 (PSGL-1), CD34, MAdCAM [see (6,11) and references therein]. The molecular structure of many of these members are examined in detail in other chapters of this book and have also been the subject of several comprehensive review articles (6,7,9–11). The members of a new family of closely related molecules that are chemotactic for leukocytes, chemokines, have also been shown to be intimately involved in leukocyte–endothelial interactions (12,13). Members of the chemokine family can act to amplify (12,14–16) or to reduce recruitment (17–21), depending on their local concentration at inflammatory sites (e.g., interstitial compartment versus luminal compartment). Several chapters in this book contain an in-depth discussion of the chemoattractants relevant to basophil and eosinophil recruitment.

### A. Endothelial–Leukocyte Adhesion: The Multistep Adhesion Cascade

Experiments performed in vivo in the microcirculation using intravital microscopy (22–25) and in in vitro models using two-dimensional laminar flow designed to simulate blood flow in the microcirculation (26–30) have provided keen insight into the cellular and molecular processes that occur during the extravasation of leukocytes at sites of inflammation. The original paradigm, ini-

**Table 1** Endothelial–Leukocyte Adhesion Molecules (nonlymphoid)

| Family | Individual member(s) | | Cellular/tissue distribution | Target counter-receptor/ligand |
|---|---|---|---|---|
| Integrins | β2 | CD11a/CD18 (LFA-1) | Leukocytes | ICAM-1, -2, -3 |
| | | CD11b/CD18 (Mac-1) | Baso, Eos, Neu, Mono | ICAM-1 |
| | | CD11c/CD18 (p150) | Baso, Eos, Mono, Neu | ? |
| | β1 | α4β1 (VLA-4) | Baso, Eos, Mono, Lymphs | VCAM-1, CS1 fibronectin |
| | β7 | α4β7 | Baso, Eos, B- and some T-lymphs | VCAM-1 |
| Ig | | ICAM-1 (CD54) | Endothelium and cell lines | β2-integrins (LFA-1, Mac-1) |
| | | ICAM-2 (CD102) | Endothelium | LFA-1 |
| | | ICAM-3 (CD50) | Leukocytes, platelets | LFA-1, ? other |
| | | VCAM-1 (CD106) | Endothelium, smooth muscle | |
| | | PECAM-1 (CD31) leukocytes | Endothelium, platelets, | PECAM-, $\alpha v\beta 3$, glycosaminoglycans |
| Selectins | | L-selectin (CD62L) | Leukocytes | sialyl-Lewis[x] and[a] |
| | | E-selectin (ELAM-1, CD62E) | Endothelium | " |
| | | P-selectin (CD62P, PADGEM) | Endothelium, platelets | " |
| Mucin-like | | PSGL-1 | Leukocytes | P-, L-, E-selectin; sialyl-Lewis[x,a] |

tially published by Butcher and colleagues (31), was that multiple receptor–ligand pairs function in an orchestrated fashion to support the adhesive interactions of leukocytes with the endothelial lining. Subsequently, the work of many investigators have confirmed and extended this adhesion cascade model to show that most leukocytes utilize multiple steps to emigrate from the blood, each step requiring multiple and overlapping function of different sets of molecules. The cascade is divided into at least three steps: 1.) initial attachment (tethering, rolling), 2.) subsequent firm adhesion (arrest, stable adhesion), and 3.) migration through endothelial junctions to the extravascular space (transmigration). Figure 1 illustrates this process for a human blood neutrophil interacting with a cytokine-activated HUVEC monolayer under defined laminar fluid flow conditions using an in vitro flow chamber designed and used in our laboratory (9,29,30,32). The neutrophil initially attaches to the apical surface of the endothelial cell (Fig. 1a, arrow depicts neutrophil that has just attached). Initial attachment is primarily dependent on both E- and L-selectin (33) and possibly P-selectin (34), depending on the experimental model and tissue(s) [see (25) for review and references therein]. Subsequently, the neutrophil rolls on the apical surface of the endothelium at low velocity (ca. ~10 µm/sec), relative to freely flowing cells, for a short distance. The cell then stops (Fig. 1b) and stably adheres on the apical surface (Fig. 1c, arrest). The arrest step is mediated by CD11/CD18 integris (β2-integrins) (22). The neutrophil crawls toward intercellular junctional borders between adjacent endothelial cells and ultimately migrates between intercellular junctions (Fig. 1d) The arrowheads in Fig. 1d identify pseudopods that appear to project through the lateral junctions, beneath the endothelial monolayer. The neutrophil rapidly migrates into the abluminal space (Fig. 1e) using a combination of molecules including CD11/CD18 integrins and CD31 (35) (see section below). Amazingly, the entire process resulting in transmigration, as illustrated in Fig. 1, occurs within minutes. Despite the frequency of leukocyte trafficking observed, rarely is there evidence of frank damage to the endothelial cell monolayer in this system as assessed by video phase microscopy (36) or in experimental models in vivo (37). For the most part, this paradigm appears to hold up for other leukocytes including T lymphocytes (32), monocytes (9,29), basophils, and eosinophils although few T-lymphs transmigrated in this 10 min assay (32). The precise molecular mechanisms and molecules involved in eosinophil and basophil interactions with endothelium are detailed in subsequent chapters of this book.

## B.  Inducible and Selective Expression of Adhesion Molecules and Chemoattractants

As mentioned above, activation of the endothelium with inflammatory cytokines or certain endotoxin results in a striking increase in leukocyte adhesion under static or fluid flow conditions in vitro. This increase in adhesion is correlated with

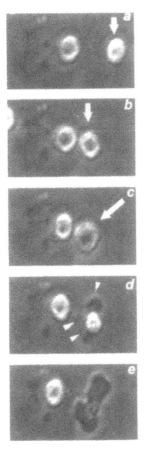

**Figure 1** Neutrophil rolling, arrest, spreading and transmigration across TNF-α–activated endothelium under defined flow conditions. Sequential video frames depicting each event listed above were digitized and photographed to generate the composite image as previously detailed (32,40).

the expression of multiple adhesion molecules and chemotactic factors by the endothelium (reviewed in (9,11,12)). It is hypothesized that the leukocyte selectivity and temporal pattern of recruitment is the result of regulated expression of specific adhesion molecules and chemoattractants by the endothelium at sites of inflammation. Studies of the promotor regions of certain of these genes indicated that cytokines can induce synergistic interactions among a small group of trans-activating factors, such as NF-κB and ATF-2/c-Jun, to form specialized complexes that activate transcription of certain genes [(38) and references therein]. In particular, incubation of endothelium with inflammatory cytokines (TNF-α, IL-

1β) and certain gram-negative bacterial endotoxin dramatically induced the expression of E-selectin and VCAM-1, while that of ICAM-1 increased severalfold Similarly, the genes encoding leukocyte subset specific chemoattractants, such as interleukin-8 (IL-8) and monocyte-chemoattractant protein-1 (MCP-1), which are members of the chemokine gene family [reviewed in (13)], are also dramatically induced by the above inflammatory cytokines (11,12). Subsequent chapters examine members of the chemokine family involved in both basophil and eosinophil recruitment in vivo and in vitro.

## C.  Regulation of Endothelial Cell Activation and Leukocyte Recruitment

Recent reports have shown that the proteasome pathway is involved in activation of NF-κB, which, as mentioned above, is a transcription factor necessary for activation of endothelial cell gene transcription of E-selectin, ICAM-1, and VCAM-1, as well as a regulator in immune responses (38). The proteasome is a large intracellular macromolecular complex localized in the cytosol and nuclear compartments, and contains a 20S proteolytic component [reviewed by Ciechanover (39) and references therein]. This nonlysosomal pathway for protein degradation removes abnormal proteins and biologically active proteins in many cell types. Small peptide aldehyde inhibitors (MG132, MG115) of the proteasome have been shown to dramatically reduce TNF-α induced cell surface expression of E-selectin, VCAM-1, and ICAM-1 in human umbilical vein endothelial cell monolayers (40). The inhibitors almost completely prevented induction of mRNA transcripts for both E-selectin and VCAM-1, whereas the mRNA levels for ICAM-1 remained at basal levels. This inhibition of transcriptional activation was largely due to a reduction in TNF-α induced NF-κB activation. The primary mechanism is correlated to prevention of NF-κB activation. Thus, minutes after addition of cytokines to HUVEC, the NF-κB complex disassociates, releasing the inhibitory subunit (I-κB) and allowing active p50 and p65 subunits to gain access to the nucleus and promote gene activation and transcription. Concomitantly, the level of I-κB subunits plummets via proteasomal degradation and does not recover for hours. Functionally, these inhibitors had a profound and striking effect on leukocyte (neutrophils and peripheral blood lymphocytes) interactions with endothelium. Neutrophil adhesion under static or flow conditions (1.8 dynes/cm$^2$) was significantly reduced, and under flow, lymphocyte adhesion was completely blated. A striking feature was that adherent neutrophils failed to undergo transendothelial migration. Fewer than 10% of adherent cells transmigrated, whereas upwards of 50% of adherent neutrophils migrated beneath the monolayers within 10 min in the presence of structurally inactive peptides or vehicle-treated HUVEC monolayers. This difference was clearly revealed by live-time video microscopic analysis of neutrophil interactions with endothelial monolayers pretreated with proteasome inhibitor MG132 prior to activation for 4 hr with TNF-$\alpha$ (40). After several minutes of perfusion, many adherent neutrophils had flattened and extended pseudopods into the junctional borders between endothelial cells but were unable to penetrate the junctions and gain access to the abluminal space. This behavior was never observed in media- or vehicle-treated TNF-$\alpha$–activated monolayers. Inhibition of transmigration raises the possibility that the proteasome participates in transendothelial passage of the leukocyte, either through a proteolytic degradation event(s) or indirectly through processing of a component(s) of a signaling cascade. This concept is supported by a recent report that demonstrated signal transduction through Notch 1 receptor molecules occurred by a proteasome-dependent proteolytic cleavage at an intracellular surface of Notch 1 (41). The next section will review the current picture of leukocyte migration and the structure of endothelial cell-to-cell junctions, and then address how the process of leukocyte transendothelial migration may occur.

## III.  LEUKOCYTE TRANSENDOTHELIAL MIGRATION

### A.  Molecules Implicated in Transmigration

Much of the current information on leukocyte transmigration is based on findings from patients with rare, heritable diseases that result in a general defect in the inflammatory response. Two such diseases, leukocyte adhesion deficiency syndrome type I and type II (LAD I and II), clearly implicate a critical role for integrins and selectins, respectively, in leukocyte recruitment. A critical role for $\beta$2-integrins is illustrated by the LAD I patients (42,43). Affected individuals have defective or absent $\beta$2-integrins on their circulating leukocytes and exhibit severe adhesion and motility defects, manifested in lack of pus at sites of inflammation and susceptibility to recurrent bacterial infections; however, their humoral and most immune responses appear normal (42). The importance of selectins in leukocyte recruitment is revealed by a second inherited disease, LAD type II (44). The afflicted patient had a poorly characterized genetic defect(s) in fucose metabolism resulting in failure to synthesize selectin ligands such as sialyl-Lewis[x] and related carbohydrate determinants. This global defect presumably interferes with efficient leukocyte adhesive interactions (rolling or initial attachments) with activated endothelium in vivo, thus accounting for increased incidence of bacterial infections. Consistent with this, leukocytes from LAD II patients have impaired rolling interactions with activated endothelial monolayers in vitro (45).

Two additional adhesion molecules implicated are members of the Ig gene family, ICAM-1 (CD54) and PECAM-1 (CD31), and a third is the recently described CD47 molecule, originally referred to as integrin associated protein (IAP). ICAM-1, which is the best-characterized endothelial counter-receptor and ligand for β2-integrins, LFA-1 and Mac-1 (46,47), is expressed at high levels on activated endothelium (48,49). Its requirement has been revealed through in vivo studies in ICAM-1 deficient mice (gene "knockout") (50,51) and by in vitro studies using function-blocking monoclonal antibodies (mAb) directed to ICAM-1 (52–54). Recent in vitro studies suggest that ICAM-1 expressed by stimulated adenocarcinoma cells (A549 cell line) exists in a dimeric form, and that the conformation of a dimeric form correlated with better binding of one of its ligands, LFA-1 (55,56). Further study of the native structure of ICAM-1 and its distribution, in the context of resting and activated endothelium, is needed to ascertain if the dimeric form is indeed a better ligand for leukocytes. A role for PECAM-1 (CD31) has come from both in vivo and in vitro experimental models. Studies by Muller and colleagues using cultured HUVEC and blocking mAb have revealed that PECAM-1, which was originally described as an endothelial-expressed "externally disposed plasmalemmal protein enriched in intercellular junctions" (57), is involved in leukocyte transmigration (35,58,59). Specifically, PECAM-1 has been proposed to mediate adhesive interactions distal to firm adhesion and was required for neutrophil and monocyte transendothelial migration under static (35) or flow conditions (36). One intriguing hypothesis for PECAM-1 regulation of migration is supported by the recent observation that ligand occupancy of PECAM-1 activates the affinity of neutrophil β2-integrins, principally Mac-1 (60). Thus, PECAM-1 engagement, concomitant with Mac-1 activation, may be a feedback loop to control Mac-1 interaction with its endothelial ligands, such as ICAM-1, thus regulating the leukocyte migratory behavior on endothelium at the apical surface and/or at intercellular junctions. One caveat, however, is that PECAM-1 expression on leukocytes per se does not appear to be necessary for transmigration, because memory T lymphocytes (CD45RO+), which are recruited during acute and chronic immune reactions, do not express PECAM-1 (61). The involvement of CD47 has been reported in multiple experimental models using endothelium and epithelium (62,63). First molecularly cloned as an ovarian tumor marker (64), it now appears that CD47 may act in a regulatory fashion during leukocyte transmigration, although the precise role of this molecule in the process of migration across endothelial (63,65) or epithelial monolayers is not clear (62).

## B.  Endothelial-Dependent Mechanisms in Leukocyte Transmigration

An important issue is how the endothelium coordinates the migratory event. This no doubt involves intracellular signaling events in either or both the adherent leukocyte and endothelial cell. The next section will focus on the signaling prospects in endothelial cells. Which intracellular signaling pathways in endothelial cells are involved in regulating the leukocyte transmigration? Recent findings by Huang, Silverstein, and colleagues (66) have revealed that intracellular cationic gradients in endothelium are intimately coupled to leukocyte transendothelial migration. Neutrophil-endothelial adhesive interactions induced a transient sevenfold increase in endothelial cell cytosolic free $Ca^{2+}$ ($[Ca^{2+}]_i$), which temporally mirrored the time course of neutrophil transmigration across 4-hr interleukin-1 (IL-1) treated endothelium. Pharmacologic clamping of endothelial cell $[Ca^{2+}]_i$ to resting levels using an intracellular $Ca^{2+}$ chelator, MAPTAM, prevented most (90%) neutrophil transendothelial migration but had no effect on neutrophil adhesion to the apical surface of the endothelial cell monolayers. These studies did not report on the spatial resolution of the $[Ca^{2+}]_i$ flux in individual endothelial cells, so it remains to be determined whether the cation flux correlates with the actual adhesion event or migratory event. One scenario would be that the rise in $[Ca^{2+}]_i$ in endothelium that accompanies leukocyte adhesion may aid in loosening of the intercellular junctions at very localized regions to allow subsequent transmigration. In effect, the junctional protein complexes deform and, concomitantly, actin filaments reorganize (perhaps via stimulation of depolymerize–polymerize cycles) at adjacent endothelial borders and passively "open" lateral junctions (see section C. below). Pfau and colleagues, using lymphocytes and endothelial cells, also report adhesion-dependent $[Ca^{2+}]_i$ alterations in both cell types (67). Neither study wasable to distinguish whether adhesion or migration triggered the $[Ca^{2+}]_i$ changes. That the leukocyte exhibited increases in $[Ca^{2+}]_i$ during adhesion has been previously described. Studies by Marks et al. using human neutrophils have shown that transient increases in intracellular $Ca^{2+}$ were necessary for their migration on glass-coated surfaces (68). Based on the data, Marks et al. inferred that localized increases in $[Ca^{2+}]_i$ mediate the process of release from previous sites of attachment.

Another event triggered in endothelium during leukocyte adhesion has recently been reported by Yoshida, Gimbrone, and colleagues (69). Binding of leukocytes to 4-hr cytokine-treated endothelial monolayers induced E-selectin linkage at its cytoplasmic domain to the endothelial cell actin cytoskeleton. The cytoskeletal association of E-selectin may supply cues, both physical and mechanical, to the endothelial cell that influence adhesion molecules on the apical surface (e.g., ICAM-1, VCAM-1) or molecules preferentially expressed at cell–cell lateral junctions (e.g., PECAM-1, VE-cadherin, p120; see Fig. 2), and thus provide guidance for adherent and migrating leukocytes.

## C.  The Process of Leukocyte Transendothelial Migration

Leukocyte transmigration, which is also termed extravasation, emigration, or diapedesis, across nonlymphoid, peripheral vascular endothelium during inflammation usually occurs at the level of postcapillary and collecting venules, where the endothelial junctions are least specialized [reviewed in (70)]. In situ and in vivo

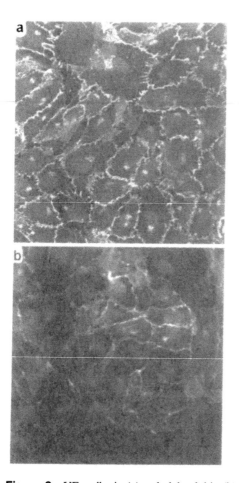

**Figure 2**  VE-cadherin (a) and plakoglobin (b) are preferentially localized to the lateral endothelial junction on cultured human endothelium. Confluent human umbilical vein endothelial monolayers (post confluent for 3 days) were washed and fixed in ice-cold methanol for 5 min on ice. Monoclonal antibodies to VE-cadherin (7 µg/ml, mAb TEA1/31; Immunotech Inc, Westbrook, ME) and plakoglobin (7 µg/ml, mAb PG5.1; Biodesign International, Kennebunk, ME) were incubated with monolayers for 3 hr and detected by a three-step sandwich immunofluorescence assay (84). Micrographs of selected fields were digitized using a scanning laser confocal inverted microscope and appropriate image analysis equipment (32,40).

studies using electron microscopy (70–73), and in vitro studies using cultured endothelium (e.g., human umbilical vein or bovine aortic endothelial monolayers) and live-time videomicroscopy (Fig. 1) (29,32,74) have reported that leukocyte migration occurs at intercellular junctions. However, certain experimental systems also have reported that leukocytes can gain passage through the endothelial lining at points other than intercellular junctions. The original reports of Marchesi and Florey (72,73) and recent studies by Granger and colleagues [(71) and original references therein] have elegantly described neutrophil movement through intercellular junctions at venules. These findings suggest that leukocytes extend pseudopods into the interendothelial junctions, and then use the mechanical leverage gained by pseudopods to squeeze through the now widened junctions. One can speculate that concomitantly, collapse of endothelial tight adherence junctional complexes may occur to permit leukocyte passage. Close inspection of electron micrographs (71) showed the presence of stricture points in migrating leukocytes, suggesting leukocytes can loosen the endothelial junctional interconnections to a limited extent. The leukocyte and endothelial cell membranes appear in close apposition and revealed prominent cytoskeletal structures and membrane vesicles. These features preceded the apparent movement of endothelial-derived membrane over the trailing end of the leukocyte, which ultimately enveloped the lumenal portion of the leukocyte and resealed the apical surface junctional barrier. Their observations, together with the above-mentioned studies by Huang and Silverstein, and Yoshida and coworkers, demonstrate active participation by the endothelium in the process of leukocyte migration. But how does the endothelium accommodate passage of one or many leukocytes through their lateral junctions without exhibiting frank signs of damage or gross changes in vessel permeability? Answers to such questions remain elusive, but significant progress has been made in determining the individual molecules that preferentially localize to the endothelial lateral junctions and several investigators have provided a conceptual framework in which to understand how the molecules work together to form junctional complexes.

## D. Molecular Organization of Endothelial Cell-to-Cell Lateral Junctions

As noted above, a great deal of information is available that details the activation-dependent, surface expression of adhesion molecules on endothelium, the relevant surface adhesion molecules on leukocytes, and the processes (i.e., adhesion molecule cascade) that are involved in leukocyte–endothelial adhesive interactions under flow conditions in vivo and in vitro. However, much less is known about the communication(s) between the migrating leukocyte and endothelium during transendothelial migration. Specifically, the identity and role of molecules expressed at the lateral inter-endothelial junctions that are involved in leukocyte transendothelial passage have yet to be clearly defined. Nor is it clear how the inter-endothelial cell junctions "retract" and "reseal" during leukocyte transmigration. However, the efforts of many investigators have identified certain of the components at endothelial cell–cell lateral junctions and provided insight into their regulation during physiologic and pathophysiologic situations.

The molecular structure and organization of the lateral junctions has been reviewed recently in detail by Dejana and colleagues (75). A working model depicting the structural organization of the cell-to-cell lateral junctions has been published by Dejana et al. (75). Endothelial cell–cell lateral junctions appear to contain at least four distinct zones based on morphological and functional characterization: starting from the apical surface and moving toward the basolateral surface, 1.) tight junctions (zona occludens), 2.) adherence junctions, 3.) gap junctions, and 4.) syndesmos (75). These zones are thought to be dynamic in nature and are composed of transmembrane and cytoplasmic molecules that associate with the cytoskeleton in a highly organized fashion. Together, these adhesion zones form a continuous belt at the boundary between the apical and lateral surfaces of adjacent endothelial cells. There appear to be many similarities, as well as obvious differences, between the structure of lateral junctions in endothelium and in epithelium (75,76).

The tight junctions in endothelium are composed of several different proteins, principally the transmembrane protein, occludin (77). The actual number of tight junctions in the vasculature decreases as vessel size narrows such that few if any are present in true capillaries (71) where junctional permeability is higher. Other cytoplasmic molecules such as ZO-1 (zonula occludens-1), ZO-2, cingulin, 7H6, rab13 are present in the intracellular region of tight junctions (75). Gap junctions (also termed connexons) are important for inter-endothelial cell communication. Three connexins have been described in endothelium: Co43, Co40, and Co37. Adherence junctions may serve as a focal point for the connections between cell plasma membrane and the underlying actin–cytoskeleton complex. The adherence junctions are the cellular membrane "joints" or "intercellular bonds" between individual endothelial cells, and they contain cadherins [e.g., E-, N-, or A-CAM, P-, L-CAM (78)], a family of single-span transmembrane glycoproteins that directly associate with structural components of the cytoskeleton and mediate $Ca^{2+}$-dependent cell–cell adhesion in a homotypic fashion. Cadherin-5, also termed VE-cadherin, is specific to vascular endothelium and localizes exclusively to lateral junctions of intact and confluent endothelium (Fig. 2a) as recently reported (79,80). The recent elucidation of the crystal structure of the amino-terminal domain of murine N-cadherin (81) has afforded insight into its structure–function relationship and suggests that cadherins, in general, form a "cell-adhesion zipper," which resembles the linear structure of the intracellular filaments where cadherins localize. This cell-adhesion zipper may allow for a mechanism to organize individual molecular interactions into strong intercellular bonds between cells (81). This conceptual framework for cadherin function in endothelium is reinforced by the recent finding of several investigators. Immunofluorscence microscopic studies in tightly confluent endothelial monolayers showed that VE-cadherin colocalized at lateral cell borders with α-catenin, β-catenin, γ-catenin [also termed plakoglobin (80)], and p110/p120 (82) (see Fig. 2a and b, VE-cadherin and plakoglobin staining pattern). Further biochemical analyses using immunoprecipitation and Western blotting revealed that VE-cadherin associates directly with α-, β-, and γ-catenin, and p100/120 (82). Although the precise mechanisms that mediate lateral junction formation and subsequently regulate their function are not well understood, recent studies have shed some light. Lampugnani and coworkers (80) have recently reported that VE-cadherin and α- and β-catenins form a complex and readily organize at nascent endothelial cell-to-cell contacts. Plakoglobin (γ-catenin) associates with cell to cell contacts only as endothelial cells approach confluence through an as yet unknown mechanism(s). Wounding (80) or $Ca^{2+}$ depletion (75) strategies further suggested that VE-cadherin and plakoglobin can rapidly and reversibly retract from the lateral junctions. Whether such events actually occur during leukocyte migration is not yet known, but the time course for cadherin cycling back and forth from the lateral junction is of the same scale. Recent studies also suggest that members of the catenin family may participate in the development of tight junctions in epithelial cells (83) through a rather complex series of steps. How the actin is regulated as cells near confluence has been carefully examined; less information is available about the process of leukocyte migration [reviewed in (70)]. Agents and drugs (e.g., phalloidin, antamanide) that disrupt actin function inhibit leukocyte migration in vivo and in vitro [reviewed in (70)]. Lastly, endothelial cells do not appear to have syndesmos zones and components of this zone (e.g., desmogleins and desmocollins) have not been observed in endothelium. For a more in-depth analysis of the adhesion zones the reader is referred to other excellent reviews and references therein (75,76).

## IV. CONCLUSIONS

The preceding sections have highlighted the contributions of the vascular endothelium in leukocyte recruitment to inflammatory sites. The chapter started with the adhesion molecule cascade, which brought the leukocytes into close apposition with the vessel wall, and proceeded to discuss leukocyte navigation of the lateral cell-to-cell junctions. Aside from important structural information that has been gained by interpretation of "static end point" electron microscopic studies of transmigration in experimental models, little information is available on the process of leukocyte passage through lateral junctions. One scenario is that leukocyte contact with the endothelial apical surface signals the endothelial cell to prepare for leukocyte passage. The minimal requirements for contact could be studied using an in vitro flow model and measurements of $[Ca^{2+}]_i$ to distinguish whether rolling or stably adherent leukocytes trigger responses. For example, leukocyte engagement of E-selectin (rolling cell) and/or ICAM-1 (stably adherent) would have multiple consequences: first, to provide the leukocyte secure footing (i.e., mechanical) in order to resist the shear force exerted by blood flow, and second, to alert the endothelial cell that a leukocyte(s) has bound via localized rises in $[Ca^{2+}]_i$ in endothelial cells. Leukocyte–endothelial homotypic PECAM-1 engagement may also achieve similar results via a direct up-regulation of leukocyte $\beta_2$-integrin affinity for ligand (60). Alternatively, other surface-expressed proteins that are not involved in adhesion may "sense" the presence of adherent leukocyte and transmit an activating signal. This "early warning signal" initiates endothelial responses necessary for lateral junction reorganization and actin depolymerization–repolymerization. To date, studies have not been devised to examine the composition of lateral junctions or to evaluate actin staining patterns in endothelium in vessels at sites of inflammation. Understanding how the components of lateral junctions accommodate migrating leukocytes and how they coordinate this function with the leukocyte is an essential question that will require more fundamental knowledge.

## ACKNOWLEDGEMENTS

The author wishes to thank Drs. Jennifer Allport, Han Ding, Doug Goetz, Michael Gimbrone, Masayuki Yoshida, and other members of the Vascular Research Division, Brigham and Women's Hospital, for helpful discussions, and Dr. Mary E. Gerritsen, Bayer Inc., for fruitful discussions and assistance in performing immunofluorescence studies reported here. The author is indebted to the support of National Institutes of Health (NHLBI) through grants HL-36028, HL-47646, and HL-53993.

## REFERENCES

1. M. P. Bevilacqua, J. S. Pober, M. E. Wheeler, R. S. Cotran, and M. A. Gimbrone, Jr., *J. Clin. Invest.*, *76*:2003 (1985).
2. R. P. Schleimer and R. K. Rutledge, *J. Immunol.*, *113*:6649 (1986).
3. J. R. Gamble, J. M. Harlan, S. J. Klebanoff, A. F. Lopez, and M. A. Vadas, *Proc. Natl. Acad. Sci. USA.*, *82*:8667 (1985).
4. M. A. Gimbrone, Jr., M. P. Bevilacqua, and M. I. Cybulsky, *Ann. NY Acad. Sci.*, *598*:77 (1990).
5. M. P. Bevilqua and R. M. Nelson, *J. Clin. Invest.*, *91*:379 (1993).
6. T. M. Carlos and J. M. Harlan, *Blood*, *84*:2068 (1994).
7. M. A. Arnaout, *Blood*, *75*:1037 (1990).
8. R. O. Hynes, *Cell*, *69*:11 (1992).
9. F. W. Luscinskas and J. Lawler, *FASEB. J.*, *8*:929 (1994).

10. M. P. Bevilacqua, *Ann. Rev. Immunol.*, *11*:767 (1993).
11. T. A. Springer, *Cell*, *76*:301 (1994).
12. M. B. Furie and G. J. Randolph, *Am. J. Pathol.*, *146*:1287 (1995).
13. I. Clark-Lewis, K.-S. Kim, K. Rajarathnam, J.-H. Gong, B. Dewald, B. Moser, M. Baggiolini, and B. D. Sykes, *J. Leuko. Biol.*, *57*:703 (1995).
14. A. Harada, N. Sekido, T. Akahoshi, T. Wada, N. Mukaida, and K. Matsushima, *J. Leuk. Biol.*, *56*:559 (1994).
15. A. R. Huber, S. L. Hunkel, R. F. Todd, III, and S. J. Wiess, *Science*, *254*:99 (1991).
16. T. W. Kuijpers, B. C. Hakkert, M. H. L. Hart, and D. Roos, *J. Cell. Biol.*, *117*:565 (1992).
17. M. A. Gimbrone, Jr., M. S. Obin, A. F. Brock, E. A. Luis, P. E. Hass, C. A. Hébert, Y. K. Yip, D. W. Leung, D. G. Lowe, and W. J. Kohr, *Science*, *246*:1601 (1989).
18. C. A. Hébert, F. W. Luscinskas, J.-M. Kiely, E. A. Luis, W. C. Darbonne, G. L. Bennett, C. C. Liu, J. B. Baker, M. S. Obin, and M. A. Gimbrone Jr., *J. Immunol.*, *145*:3033 (1990).
19. K. Ley, J. B. Baker, M. I. Cybulsky, M. A. Gimbrone, Jr., and F. W. Luscinskas, *J. Immunol.*, *151*:6357 (1993).
20. D. H. Hechtman, M. I. Cybulsky, H. J. Fuchs, J. B. Baker, and M. A. Gimbrone, Jr. *J. Immunol.*, *147*:883 (1991).
21. W. F. Westlin, J. M. Kiely, and M. A. Gimbrone, Jr., *J. Leuko. Biol.*, *52*:43 (1992).
22. U. H. von Andrian, J. D. Chambers, L. M. McEvoy, R. F. Bargatze, K. E. Arfors, and E. C. Butcher, *Proc. Natl. Acad. Sci. USA.*, *88*:7538 (1991).
23. S. D. House and H. H. Lipowsky, *Microvasc. Res.*, *34*:363 (1987).
24. K. Ley, D. C. Bullard, M. L. Arbones, R. Bosse, D. Vestweber, T. F. Tedder, and A. L. Beaudet *J. Exp. Med.*, *181*:669 (1995).
25. K. Ley, *Microcirculation*, *2*:141 (1995).
26. M. B. Lawrence and T. A. Springer, *Cell*, *65*:1(1991).
27. M. B. Lawrence, L. V. Mcintire, and S. G. Eskin, *Blood*, *70*:1284 (1987).
28. M. B. Lawrence, C. W. Smith, S. G. Eskin, and L. V. McIntire, *Blood*, *75*:227 (1990).
29. F. W. Luscinskas, G. S. Kansas, H. Ding, P. Pizcueta, B. E. Schleiffenbaum, T. F. Tedder, and M. A. Gimbrone, Jr., *J. Cell. Biol.*, *125*:1417 (1994).
30. J. Shen, F. W. Luscinskas, A. Connolly, C. F. Dewey, Jr., and M. A. Gimbrone Jr., *Am. J. Physiol.*, *262*:C384 (1992).
31. E. C. Butcher, *Cell*, *67*:1033 (1991).
32. F. W. Luscinskas, H. Ding, A. H. Lichtman, *J. Exp. Med.*, *181*:1179 (1995).
33. O. Abbassi, T. K. Kishimoto, L. V. Mcintire, D. C. Anderson, C. W. Smith, *J. Clin. Invest.*, *92*:2719 (1993).
34. T. N. Mayadas, R. C. Johnson, H. Rayburn, R. O. Hynes, and D. D. Wagner, *Cell*, *74*:541 (1993).
35. W. A. Muller, S. A. Weigl, X. Deng, and D. M. Phillips, *J. Exp. Med.*, *17*:449 (1993).
36. F. W. Luscinskas, H. Ding, P. Tan, D. Cumming, T. F. Tedder, and M. E. Gerritsen, *J. Immunol.*, *156*:326 (1996).
37. A. Thurston-Klein, P. Hedqvist, and L. Lindbom, *Tissue and Cell*, *1*:1 (1986).
38. T. Collins, M. A. Read, A. S. Neish, M. Z. Whitley, D. Thanos, and T. Maniatis, *FASEB J.*, *9*:899 (1995).
39. A. Ciechanover, *Cell*, *79*:13 (1994).

40. M. A. Read, A. S. Neish, F. W. Luscinskas, V. J. Palombella, T. Maniatis, and T. Collins, *Immunity*, *2*:493 (1995).
41. R. Kopan, E. H. Schroeter, H. Weintraub, and J. S. Nye, *Proc. Natl. Acad. Sci. USA.*, *93*:1683 (1996).
42. M. A. Arnaout, *Immunol. Rev.*, *11*:4147 (1990).
43. T. M. Carlos and J. M. Harlan, *Blood*, *84*:2068 (1994).
44. A. Etzioni, M. Frydman, S. Pollack, I. Avidor, M. L. Phillips, J. C. Paulson, and R. Gershoni-Baruch, *N. Eng. J. Med.*, *327*:1789 (1992).
45. T. H. Price, H. D. Ochs, R. Gershoni-Baruch, J. M. Harlan, and A. Etzioni, *Blood*, *84*:1635 (1994).
46. M. S. Diamond, D. E. Staunton, A. R. de Fougerolles, S. A. Stacker, J. Garcia-Aguilar, M. L. Hibbs, and T. A. Springer, *J. Cell Biol.*, *111*:3129 (1990).
47. M. S. Diamond, J. Garcia-Aguilar, J. K. Bickford, A. L. Corbi, and T. A. Springer, *J. Cell. Biol.*, *120*:1031 (1993).
48. J. M. Munro, J. S. Pober, and R. S. Cotran, *Am. J. Pathol.*, *135*:121 (1989).
49. R. Rothlein, M. L. Dustin, S. D. Marlin, and T. A. Springer, *J. Immunol.*, *137*:1270 (1986).
50. H. Xu, J. A. Gonzalo, Y. St. Pierre, I. R. Williams, T. S. Kupper, R. S. Cotran, T. A. Springer, and J.-C. Gutierrez-Ramos, *J. Exp. Med.*, *180*:95 (1994).
51. J. E. Sligh, Jr., C. M. Ballantyne, S. S. Rich, H. K. Hawkins, C. W. Smith, A. Bradley, and A. L. Beaudet, *Proc. Natl. Acad. Sci. USA*, *90*:8529 (1993).
52. C. W. Smith, S. D. Marlin, R. Rothlein, C. Toman, and D. C. Anderson, *J. Clin. Invest.*, *83*:2008 (1989).
53. B. C. Hakkert, T. W. Kuijpers, J. F. M. Leeuwenberg, J. A. van Mourik, and D. Roos, *Blood*, *78*:2721 (1991).
54. M. B. Furie, M. C. A. Tancino, and C. W. Smith, *Blood*, *78*:2089 (1991).
55. P. L. Reilly, J. R. Waska, D. D. Jeanfavre, E. McNally, R. Rothlein, and B. J. Bormann, *J. Immunol.*, *155*:529 (1995).
56. J. Miller, R. Knorr, M. Ferrone, R. Houdei, C. P. Carron, and M. L. Dustin, *J. Exp. Med.*, *172*:1231 (1995).
57. W. A. Muller, C. M. Ratti, S. L. McDonnell, and Z. A. Cohn, *J. Exp. Med.*, *170*:399 (1989).
58. W. A. Muller and S. A. Weigl. *J. Exp. Med.*, *176*:819 (1992).
59. S. Bogen, J. Pak, M. Garifallou, X. Deng, and W. A. Muller, *J. Exp. Med.*, *179*:1059 (1994).
60. M. E. Berman and W. A. Muller, *J. Immunol.*, *154*:299 (1995).
61. Y. Tanaka, S. M. Albelda, K. J. Horgan, G. A. van Seventer, Y. Shimizu, W. Newman, J. Hallam, P. J. Newman, C. A. Buck, and S. Shaw, *J. Exp. Med.*, *176*:245 (1992).
62. C. A. Parkos, S. P. Colgan, T. W. Liang, A. Nusrat, A. E. Bacarra, D. K. Carnes, J. L. Madara, *J. Cell. Biol.*, *132*:437 (1996).
63. D. Cooper, F. P. Lindberg, J. R. Gamble, E. J. Brown, and M. A. Vadas, *Proc. Natl. Acad. Sci. USA.*, *92*:3978 (1995).
64. I. G. Campbell, P. S. Freemont, W. Foulkes, and J. Trowsdale, *Cancer. Res.*, *52*:5416 (1992).
65. F. P. Lindberg, H. D. Gresham, E. Schwarz, and E. J. Brown, *J. Cell. Biol.*, *123:*485 (1993).

66. A. J. Huang, J. E. Manning, T. M. Bandak, M. C. Ratau, K. R. Hanser, and S. C. Silverstein, *J. Cell Biol.*, *120*:1371 (1993).
67. S. Pfau, D. Leitenberg, H. Rinder, B. R. Smith, R. Pardi, and J. R. Bender, *J. Cell. Biol.*, *128*:969 (1995).
68. P. W. Marks and F. R. Maxfield, *J. Cell. Biol.*, *110*:43 (1990).
69. M. Yoshida, W. F. Westlin, N. Wang, D. E. Ingber, A. Rosenzweig, N. Resnick, and M. A. Gimbrone, Jr., *J. Cell. Biol. 133*:445 (1996).
70. D. Shepro, *Physiology and pathophysiology of leukocyte adhesion.* (D. N. Granger, G. W. Schmid-Schöenbien, eds.), Oxford University Press, New York, p. 196 (1995).
71. H. J. Granger, Y. Yuan, and D. C. Zawieja, *Physiology and pathophysiology of leukocyte adhesion.* (D. N. Granger, G. W. Schmid-Schöenbien, eds.), Oxford University Press, New York, p. 185 (1995).
72. V. T. Marchesi, *Quart. J. Exp. Physiol.*, *46*:115 (1961).
73. V. T. Marchesi and H. W. Florey, *Quart. J. Exp. Physiol.*, *45*:343 (1860).
74. F. W. Luscinskas, M. I. Cybulsky, J.-M. Kiely, C. S. Peckins, V. D. Davis, and M. A. Gimbrone, Jr., *J. Immunol.*, *146*:1617 (1991).
75. E. Dejana, M. Corada, and G. Lampugnani, *FASEB J.*, *9*:910 (1995).
76. J. M. Anderson, M. S. Balda, and A. S. Fanning, *Curr. Opin. Cell. Biol.*, *5*:772 (1993).
77. M. Furuse, T. Hirase, M. Itoh, A. Nagafuchi, S. S. Yonemura, S. Tsukita, and S. Tsukita, *J. Cell. Biol.*, *123*:1777 (1993).
78. M. Takeichi, *Science*, *251*:1451 (1991).
79. M. G. Lampugnani, M. Resnati, M. Raiteri, R. Pigott, A. Pisacane, G. Houen, L. P. Ruco, and E. Dejana, *J. Cell. Biol.*, *118*:1511 (1992).
80. M. G. Lampugnani, M. Corada, L. Caveda, F. Breviario, O. Ayalon, B. Geiger, and E. Dejana, *J. Cell. Biol.*, *129*:203 (1995).
81. L. Shapiro, A. M. Fannon, P. D. Kwong, A. Thompson, M. S. Lehmann, G. Grubel, J.-F. Legrand, J. Als-Nielsen, D. R. Colman, and W. A. Hendrickson, *Nature*, *374*:327 (1995).
82. J. M. Staddon, C. Smales, C. Schulze, F. S. Esch, and L. L. Rubin, *J. Cell. Biol.*, *130*:369 (1995).
83. A. K. Rajasekaran, M. Hojo, T. Huima, and E. Rodriguez-Boulan, *J. Cell. Biol.*, *132*:451 (1996).
84. M. E. Gerritsen, C.-P. Shen, M. C. McHugh, W. J. Atkinson, J.-M. Kiely, D. S. Milstone, F. W. Luscinskas, and M. A. Gimbrone, Jr., *Microcirculation*, *2*:151 (1995).

# 3

# Epithelial Cells

*Phenotype, Substratum, and Mediator Production*

**Albert J. Polito and David Proud**   *Johns Hopkins Asthma and Allergy Center, Johns Hopkins University School of Medicine, Baltimore, Maryland*

## I. INTRODUCTION

The respiratory epithelium is classically viewed as a simple interface between the host and its environment. In fact, it is a remarkably developed tissue comprising a variety of cell types with specialized functions. Adhesion molecules are expressed on the surface of epithelial cells and are crucial to their interactions with the surrounding microenvironment. Moreover, epithelial cells play an active role in the inflammatory response by secreting an array of mediators, including cytokines, lipid and peptide products, and reactive oxygen species. The broad range of phenotypes, surface marker expression, and mediators secreted by the epithelium of the lung will be discussed in this chapter.

## II. EPITHELIAL PHENOTYPE

A continuous layer of epithelial cells lines the entire respiratory tract. This oversimplification, however, does not do justice to the enormous complexity of the epithelium, which has been found by electron microscopy to comprise at least eight different cell types, with many more intermediate or differentiating cells (1). A pseudostratified columnar morphology characterizes the larger proximal airways, with the bulk of the epithelium made up of ciliated cells, goblet cells, and basal cells. Ciliated cells decrease in number in the distal airways, as the epithelium becomes more cuboidal and basal cells and Clara cells become more prominent (2). The alveoli are lined largely by a much thinner layer of type I alveolar cells interspersed with type II alveolar cells (3).

Ciliated cells are a major part of the body's defense against inhaled pathogens and noxious particles. They function to propel the mucus of the tracheobronchial tree proximally, thus clearing the airway of trapped debris. In the upper respiratory tract, as much as 50% to 80% of the luminal surface is occupied by these cells (4), which do not form a continuous lining but occur in groups separated by fields of nonciliated cells and submucosal gland openings (5). Approximately 250 cilia and numerous smaller microvilli cover the apical surface of each cell (1,6). The effectiveness of these cells in clearing the airway hinges on the coordinated beating of each field of cilia in the same direction in metachronal waves (7). Two layers make up the airway secretions that bathe the luminal surface of the epithelium. The cilia beat primarily in the low viscosity sol phase, whereas the clawlike projections on their tips move in the thicker, more viscous gel phase (8). These projections "grab" the mucus and propel it in the cephalad direction toward the pharynx. Impaired mucociliary clearance results either from impaired ion and water transport processes that lead to a dehydrated sol phase, such as occurs in cystic fibrosis (9), or from dysfunctional cilia. This latter condition can arise as the result of inherited ciliary disorders such as primary ciliary dyskinesia, as well as from acquired diseases such as bronchiectasis and chronic bronchitis, in which there is damage to the ciliated epithelium (10). Ultrastructural abnormalities of cilia have been observed with increased frequency in chronic smokers and may account for the associated impairment of tracheobronchial mucus clearance (11).

Goblet cells are less prevalent in the respiratory tract than ciliated cells, there being an approximate ratio of 1:5 (1). They are metabolically active secretory cells, as evidenced by their mucous granules, numerous ribosomes, prominent Golgi apparatus, and extensive rough endoplasmic reticulum (5). Morphologically, these cells taper as they extend basally and thus acquire their goblet-like appearance (12). The content of the secretory granules is variable but predominantly consists of neutral mucins, sulfomucins, and sialomucins (13). Exposure to irritants can dramatically affect the proportion of goblet cells. For example, the airway epithelium of pathogen-free rats typically exhibits <1% goblet cells (8), whereas airways of rats exposed to tobacco smoke show a severalfold increase in goblet cell numbers (14). Similar effects of smoking have been noted in the human host (15).

Serous cells are another class of secretory cells that contribute to the fluid bathing the surface of the epithelium. In contrast to goblet cells, however, they produce low-viscosity secretions rich in lysozyme, neutral glycoprotein, and the epithelial transfer component of IgA (16). Electron microscopy has demonstrated that the secretory granules of serous cells are electron-dense, whereas those of goblet cells are electron-lucent (1). Serous and mucus cells are also the two cell types that line the submucosal glands that are found throughout the larger airways.

At the bronchiolar level, Clara cells constitute the majority of nonciliated epithelial cells. They are columnar in shape and typically project into the airway lumen (1). Their ultrastructure shows abundant rough endoplasmic reticulum, a prominent Golgi apparatus, and electron-dense granules, all indicative of a secretory role (17,18). Clara cell 16-kD protein (CC16kD, previously referred to as Clara cell 10-kD protein), an inhibitor of elastase (19) and phospholipase $A_2$ (20), is believed to be the major product of these cells (21,22). In addition, they may secrete surfactant-associated glycoprotein (23,24). Clara cells also serve as a site of extrahepatic oxidative metabolism of foreign substances through cytochrome P450–dependent mono-oxygenase reactions (25,26).

Basal cells, as their name indicates, are found adjacent to the basement membrane of the airways. They are small cells with a broad range of size and shape (27), but generally have a small cytoplasmic/nuclear ratio, few mitochondria, scant rough endoplasmic reticulum, and numerous keratin filaments (28). Columnar cells cannot form junctional attachments with the basement membrane (hemidesmosomes) and instead rely on basal cells for indirect anchoring. The basal cells act as a bridge by forming desmosomal attachments with columnar cells and hemidesmosomal attachments with the basement membrane (see below) (27).

Classically, basal cells have been viewed as the primary progenitor cell of the airway epithelium, but this has been disputed in recent years. In a series of articles, Keenan et al. (29–32) found that the secretory cell is primarily responsible for the proliferative response following mechanical injury to hamster tracheal epithelium. Studies on rat tracheal epithelium have also concluded that secretory cells form the major progenitorial compartment in the airway (28,33), whereas investigations on a rabbit model supported the traditional depiction of the basal cell as the germinal cell (34,35). There may indeed be species differences, and studies on human epithelial repair are, as yet, lacking.

Neuroendocrine cells also reside in the basal region of the epithelium and typically have thin cytoplasmic extensions to the luminal surface (36,37). Their numbers increase as the caliber of airways decreases to the bronchiolar level, but more distally, in the terminal bronchioles, they are very sparse (1,36). Collections of neuroendocrine cells, called neuroepithelial bodies, tend to be concentrated near the bifurcation points of intrapulmonary airways (38,39). The main ultrastructural feature of these cells is the presence of secretory granules, called dense-cored vesicles, which are characterized by an electron-dense core separated by a clear space from the outer membrane (37). They are localized to the basal portion of the cells and contain a broad spectrum of bioactive secretory products, including serotonin, calcitonin, calcitonin gene-related peptide, gastrin-releasing peptide (bombesin), enkephalin, somatostatin, cholecystokinin, and substance P (37).

The epithelial cells of the alveoli are designated alveolar type I and type II cells. Type I cells cover almost 95% of the alveolar surface and have a flattened morphology that allows for efficient gas exchange (3). In contrast, alveolar type II cells are cuboidal and occur at the junction of alveolar septae (5). Although they constitute only about 5% of the alveolar surface, type II cells represent 60% of alveolar cells by number (40). Their exposed surface is covered with microvilli, and their cytoplasm contains characteristic lamellated inclusion bodies. Phospholipids and proteins are found in these organelles and are released to form the surfactant of the lung (41,42). Type II cells also produce several proteins of the classical and alternative complement pathways (43). The type II cell is the stem cell of the alveolar epithelium. When the alveolar surface is damaged by exposure to oxidants or other injurious agents, the type II cells rapidly divide, repopulate the area, and then differentiate into type I cells (44).

## III. ADHESIVE STRUCTURES

Epithelial cell–cell adhesion is essential for the airway epithelium to function as a physical barrier to potentially harmful inhaled agents. Three ultrastructural components that are important in cell–cell adhesion have been identified: the desmosome (macula adherens); the intermediate junction (zonula adherens); and the tight junction (zonula occludens).

The desmosome is sometimes referred to as a "spot-weld" between cells (45). Within the pseudostratified morphology of the airway epithelium, desmosomes are present along the lateral aspects of the columnar cells, particularly towards the cell apex, and at the junction between the columnar and basal cells. Ultrastructurally, desmosomes measure between 0.1 and 1.5 μm in diameter and are delineated by an electron-dense plaque and electron-dense filaments that span the intercellular space (46). Anchoring tonofilaments fan out from the plaques into the adjacent cytoplasm. The tonofilaments loop repeatedly from the cytoplasm to the plaque and then back into the cytoplasm (47).

A variety of proteins and glycoproteins make up the different portions of the desmosome. Membrane glycoproteins, known as desmogleins or desmocollins, project into the intercellular space to form an intercellular adhesive structure (45). Three desmogleins have been identified (48): desmoglein I has a molecular weight of 150 kD, desmoglein II is a 110–120 kD protein, and desmoglein III is considerably smaller (22 kD). Although desmoglein I is also found in the desmosomal plaque (49), the cytoplasmic side of the desmosome is primarily composed of four nonglycosylated proteins called desmoplakins. Desmoplakins I and II (250 kD and 215 kD, respectively) are somewhat larger proteins than desmoplakins III and IV (83 kD and 78 kD, respectively) (50). Desmin, vimentin, and particularly cytokeratin, make up the tonofilaments (45). Cytokeratin expression is often used as a characteristic feature of epithelial cells.

Hemidesmosomes occur between basal cells and the underlying basement membrane. The few columnar cells that reach down to the basement membrane do not form adhesive links and instead are anchored in place through their desmosomal attachments to the basal cell layer (27). A line of cleavage is thus formed between the columnar and basal cell layers, such that desquamated epithelial cells from atopic asthmatics consist almost entirely of columnar cells (51). Specific members of the integrin family of adhesion molecules have been localized to the hemidesmosome and will be discussed below.

Intermediate junctions, or zonula adherens, are adhesive structures that encircle the cell and divide its lateral aspects into apical and basolateral regions. At the level of the intermediate junction, the intercellular space is narrowed to 25 to 35 nm in width (46) and contains filaments to which a calcium-dependent adhesion protein called E-cadherin (see below) has been localized. Electron-dense intracellular plaques anchor numerous actin filaments to the plasma membrane in a belt-like pattern that lies parallel to the cell membrane (45). Immunocytochemical analysis has also localized the actin-binding proteins $\alpha$-actinin and vinculin to the plaque region of intermediate junctions.

Tight junctions seal the apical aspects of the epithelial cell, eliminating the intercellular space to create a regulated paracellular barrier to the movement of water, solutes, and leukocytes (52). Ultrastructurally, freeze-fracture studies show that tight junctions resemble a complex network of linearly arranged strands (46). Biochemically, several proteins have been associated with tight junctions. These include E-cadherin (see below), ZO-1, a 220-kD protein that requires the availability of calcium and cell contact in order to be present in the plasma membrane (53), and ZO-2, a 160-kD protein (54). Within the cytoplasmic plaque of the tight junction, ZO-1 and ZO-2 are bound to each other (55), and to an as yet unidentified 130-kD protein (52). Cingulin, a dimeric protein that consists of two 108-kD polypeptide chains, also is found at the cytoplasmic side of the cell membrane (56). Furuse et al. (57) have recently identified a transmembrane protein called occludin, which may function as the sealing protein; this work was done, however, in cell membranes derived from chicken liver, and occludin has not yet been shown in respiratory tissues. Occludin is bound on the cytoplasmic surface to the plaque proteins ZO-1 and ZO-2, which in turn are linked to cytoskeletal filaments.

## IV. EXPRESSION OF ADHESION MOLECULES

A broad range of adhesion molecules is expressed on the surface of epithelial cells. They mediate adhesion of the epithelium to the underlying substratum as well as cell–cell interactions. The latter include both structural-type adhesion between epithelial cells and inflammatory-type adhesion between epithelial cells and leukocytes. The spectrum of expressed adhesion molecules is by no means fully defined, and ongoing investigations are rapidly expanding the current descriptions.

## A. Epithelial Cell–Substratum Adhesion

The principal adhesion molecules involved in the binding of epithelial cells to basement membrane are the integrins. Integrins are a family of at least 22 heterodimeric glycoproteins comprising noncovalently linked $\alpha$ and $\beta$ subunits, with subfamilies traditionally grouped on the basis of a common $\beta$ subunit (58). As the number of integrins identified has grown, it has become clear that this grouping is no longer entirely accurate, because certain $\alpha$ subunits can associate with more than one $\beta$ subunit (59). Albelda (60) has developed a simplified schema that categorizes integrins into two groups: those that bind primarily to constituents of the basement membrane (e.g., collagen, laminin) and those that bind primarily to matrix proteins found during inflammation, wound repair, and development (e.g., fibronectin, tenascin, fibrinogen, vitronectin, thrombospondin). Integrins associated with the respiratory epithelium are summarized in Table 1.

The $\beta_1$-integrins (the "very late activation"—VLA—subfamily) are the most important group of integrins that mediate adhesion to the cell substratum. At least nine different $\alpha$ subunits are known to associate with the $\beta_1$ subunit, and of these, $\alpha_2$, $\alpha_3$, and $\alpha_6$ are involved in the binding of epithelial cells to collagen and laminin (61–64). In the epithelium of the bronchi (45), cornea (65), and skin (66), the $\alpha_6$ subunit has been found in association with a different $\beta$ subunit, $\beta_4$. The $\alpha_6\beta_4$ heterodimer has been localized to the hemidesmosome complex (66), suggesting an important role in basal cell attachment to the basement membrane. Very

**Table 1**  Epithelial Integrins

Integrins binding to normal components of basement membrane

| Integrin | Ligand |
| --- | --- |
| $\alpha_2\beta_1$ | Collagen/laminin |
| $\alpha_3\beta_1$ | Laminin/collagen/(fibronectin) |
| $\alpha_6\beta_1$ | Laminin |
| $\alpha_6\beta_4$ | Laminin |

Integrins binding to ligands not normally present in basement membrane

| Integrin | Ligand |
| --- | --- |
| $\alpha_5\beta_1$ | Fibronectin (RGD site) |
| $\alpha_v\beta_1$ | Fibronectin/vitronectin |
| $\alpha_v\beta_5$ | Vitronectin |
| $\alpha_9\beta_1$ | Tenascin |
| $\alpha_v\beta_6$ | Fibronectin/tenascin(RGD sites) |

few fibronectin-binding integrins are normally present on airway epithelium. Among these is $\alpha_v\beta_1$, which uses vitronectin and fibronectin as its ligands (59). The $\alpha_v$ subunit can also associate with $\beta_5$ on the epithelium (67), and there is some binding of fibronectin by $\alpha_3\beta_1$. Two other integrin subunits, $\alpha_9$ and $\beta_6$, also have been recently identified on epithelial cells. The $\alpha_9$ subunit forms a heterodimer with the $\beta_1$ integrin subunit, and $\alpha_9\beta_1$ has been shown to mediate cell attachment to the matrix protein, tenascin (68). By contrast, the $\beta_6$ subunit forms a heterodimer with $\alpha_v$ to produce an integrin that has been shown to bind to fibronectin (69) and to mediate cell attachment to fibronectin (70) and tenascin (71).

Hyaluronic acid (HA) is a major component of the extracellular matrix, and both bronchial and alveolar epithelial cells interact with HA via a cell surface glycoprotein referred to as CD44 (72,73). CD44, which enjoys a wide tissue distribution, is a transmembrane protein comprising a central polypeptide core that is posttranslationally modified by addition of multiple S and O sulfated glycosaminoglycan side chains. The so-called "standard form" of the molecule, CD44s, functions as a receptor for collagen and fibronectin as well as for HA (74). In addition to CD44s, however, multiple isoforms of CD44 can be generated by alternative splicing of 10 variant exons (75). Immunofluorescence techniques have demonstrated that bronchial epithelial cells from normal adult human lungs show focal expression of the CD44v6 and CD44v9 isoforms; CD44s is observed only on basal cells (76). Positive staining for CD44v6 and CD44v9 is also observed on type II alveolar cells. The expression of these epithelial isoforms is reduced in lungs affected by pulmonary fibrosis, suggesting that they may be involved in epithelial-mesenchymal interactions and maintenance of the pulmonary histoarchitecture (76). Other recent work has focused on the role of CD44 in growth and metastasis of a number of malignancies, including epithelial-derived lung neoplasms (77).

Other non-integrin adhesion molecules have been identified on extrapulmonary epithelial tissues, but only a limited number have been demonstrated on the bronchial epithelium. The 67-kD high-affinity laminin receptor is present on human lung adenocarcinomas and may also function in recognition of $\gamma/\delta$ tumor-infiltrating T lymphocytes (78). Syndecan, a proteoglycan that binds collagen, fibronectin, and thrombospondin, is found on the basolateral surface of many epithelia but has not yet been shown on lung epithelium (79,80).

## B. Epithelial Cell–Cell Adhesion

Adhesion between cells within the epithelium is critical to the formation of a tight permeability barrier. Junctional structures, including desmosomes, intermediate junctions, and tight junctions, are a key feature of cell–cell contact. Cytoplasmic junctional proteins, discussed earlier, make up these structures and interact with specific cell adhesion molecules.

The cadherins are a group of calcium-dependent cell-surface adhesion molecules that mediate cell–cell adhesion. They are made up of a single polypeptide chain and engage in homophilic adhesion of one cadherin molecule to another on an opposing cell membrane (60). Epithelial cadherin (E-cadherin, also called uvomorulin, Arc-1, liver cell adhesion molecule, L-CAM, and cell CAM 120/80) has been identified as a constituent of intermediate junctions in the bronchial epithelium. It localizes by immunofluorescence to the lateral cell membrane close to the luminal surface (45). Recent determination of the solution structure of E-cadherin revealed unexpected structural similarities to the immunoglobulin fold. The molecule has seven β-strands arranged in a "β-barrel" topology, and two short α-helices, one of which provides the calcium-binding pocket. The putative adhesion interface is centered around the F-strand of the β-sheet and contains a conserved His-Ala-Val sequence (81). E-cadherin has been implicated in the suppression of tumor invasiveness, as its expression is decreased in the invasive phase of epithelial malignancies (82–84).

Some $\beta_1$-integrins also function in cell–cell adhesion. The $\alpha_2$ and $\alpha_3$ subunits have been identified in the region between cells, but their specific roles are not yet known (60).

## C.  Epithelial Cell–Leukocyte Adhesion

Interactions between leukocytes and the vascular endothelium have been extensively characterized over the past ten years (85), but it is only recently that research has targeted the mechanisms by which leukocytes interact with, and migrate through, the respiratory epithelium. Intercellular adhesion molecule–1 (ICAM-1), a member of the immunoglobulin gene superfamily also referred to as CD54, was first shown to be important in these interactions in a nonhuman primate model of asthma, in which antigen inhalation resulted in an increase in ICAM-1 expression on both endothelium and airway epithelium (86). Vignola et al. (87) have since demonstrated that bronchial epithelial cell ICAM-1 expression is enhanced in patients with asthma compared to normal subjects or to patients with chronic bronchitis; moreover, the level of expression correlates with disease severity. Specific inflammatory mediators, including histamine, interleukin (IL)-1β, tumor necrosis factor (TNF)-α, and interferon (IFN)-γ, have also been shown to increase ICAM-1 expression in both primary human airway epithelium and transformed epithelial cell lines (88–91). The effects on epithelial ICAM-1 expression of TNF-α, and IFN-γ are additive (Fig. 1).

Leukocytes form strong adhesive interactions with ICAM-1 through $\beta_2$-integrins (CD11/CD18) expressed on their surfaces (92). Interestingly, while preincubation of leukocytes with monoclonal antibodies to CD18 almost totally blocks adhesion to cultured epithelial cells, several studies have demonstrated that blockade of epithelial ICAM-1 is much less effective in reducing leukocyte

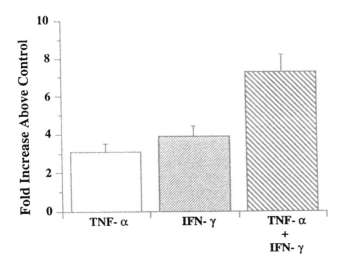

**Figure 1**  The effects of interferon-γ and tumor necrosis factor-α on expression of ICAM-1 on the human bronchial epithelial cell line BEAS-2B are additive. Cells were exposed for 18 hours to medium control, 30 U/mL of IFN-γ, 10 U/mL of TNF-α, or the combination of the two cytokines. Cells were then resuspended non-enzymatically and subjected to flow cytometry. Data are expressed as fold increase above medium control (n = 4).

adhesion (88,89,93,94). This has led to the suggestion that CD18-dependent, ICAM-1–independent adhesion mechanisms must also exist. Indeed, there is evidence that exposure of epithelial cells to ozone enhances ICAM-1–independent leukocyte adhesion (93), and infection of cells with parainfluenza has been reported to both increase expression of ICAM-1 and enhance ICAM-1–independent leukocyte adhesion (94). In contrast to the effects of parainfluenza virus infection on epithelial ICAM-1 expression, infection of an epithelial cell line with rhinovirus causes no alteration in the level of ICAM-1 expression nor does it alter leukocyte adhesion (95), even though ICAM-1 is actually the receptor for the majority of rhinovirus serotypes (96,97). Further support for the concept of ICAM-1-independent mechanisms of leukocyte adherence to, and transmigration through, the epithelium are provided by in vivo studies of double knockout mice, in which the genes for both P-selectin and ICAM-1 were mutated. Despite the deletion of both of these adhesion molecules, the mice demonstrated normal neutrophil emigration into the alveolar space during *Streptococcus pneumoniae* infection (98). Although preliminary studies have indicated that vascular cell adhesion molecule–1 (VCAM-1, CD106) can be induced on an epithelial cell line upon exposure to cytokines (99), it is unlikely that ICAM-1–independent leukocyte adhesion to epithelium involves either VCAM-1 or E-selectin, as these molecules are not expressed on cultured primary human epithelial cells (91). The delineation of additional adhesion molecues that play a role in leukocyte adherence to, and transmigration through, the epithelium is currently a major research focus.

## V. IMMUNOREGULATION

Epithelial cells may function as immune accessory cells through their expression of class II major histocompatibility (MHC) antigens. Human leukocyte antigen (HLA)-DR is expressed at all levels of the bronchial tree (100), and it is up-regulated on the surface of bronchial epithelial cells of patients with asthma (87,101) and on nasal epithelial cells of patients with nasal polyps (102). IFN-γ–treated epithelial cells show increased expression of HLA-DR (103,104) and are capable of presenting antigens to autologous T lymphocytes (105). Their ability to stimulate proliferation of CD4 and CD8 T cells in mixed lymphocyte cultures lends further credence to an immunoregulatory role (106).

## VI. EXTRACELLULAR MATRIX DEPOSITION

Repair of inflammatory injury to the airway is an essential feature of the normal immune response, and epithelial cells can contribute to such processes in several ways. Not only can the epithelial cell influence other parenchymal cells in the lung, but epithelial cells are now known to be capable of synthesizing several extracellular matrix proteins. Fibronectin, an integral matrix component to which basal cells adhere, has been identified as a product of epithelial cells (107,108) and has been reported to function as a chemoattractant stimulus for both fibroblasts (107) and epithelial cells themselves (109). Fibronectin can exist in several isoforms, generated as a result of alternative RNA splicing (110). Three regions of variability—EDA, EDB, and IIICS (also known as the oncofetal domain)— produce isoforms containing extra domains that are present less frequently in the plasma fibronectin generated by hepatocytes. Recent studies have shown that respiratory epithelial cells deposit only the EDA and oncofetal variants of fibronectin into the extracellular matrix (111).

Tenascin is another secreted glycoprotein that exists in several isoforms and has structural homology to fibronectin. The protein is composed of six polypeptides and exists as two molecular weight forms of 280 kD and 190 kD. Expression of tenascin is increased in the bronchial mucosa of patients with asthma, and the amount of tenascin correlates with disease severity (112). It has recently been shown that epithelial cells produce both the 280-kD and 190-kD isoforms of tenascin (111). Transforming growth factor (TGF)-β has been shown to stimulate production of both fibronectin and tenascin from epithelial cells (111, 113), while TNF-α and IFN-γ enhance production of tenascin but have minimal effects on epithelial elaboration of fibronectin (114).

Recently, entactin, a 150-kD glycoprotein component of basement membrane, was demonstrated to be produced by rat alveolar epithelial cells (115); it has the presumed function of facilitating basement membrane assembly by binding laminin and type IV collagen, as well as proteoglycans and fibronectin (116).

Finally, the epithelium influences extracellular matrix synthesis by other, more indirect, mechanisms. Many epithelial-derived mediators (see below), including IL-1, IL-6, and prostaglandin (PG)E$_2$, directly affect matrix production (117). Epithelial proteases and protease inhibitors can affect tissue remodeling (see below). Epithelial cells also exert several levels of control over fibroblasts and influence their migration (107,118), synthesis of DNA (119), and production of matrix components (120).

## VII. MEDIATOR RELEASE

Epithelial cells are often described as functioning as both "target" cells and "effector" cells (121). Airway secretions, mucociliary clearance, and ion transport can each be altered when a specific stimulus "targets" the cell. In contrast, the epithelial cell can act in an "effector" role as a central part of the inflammatory response by releasing a variety of mediators, including cytokines, arachidonic acid metabolites, peptide products, proteases and protease inhibitors, and reactive oxygen species.

## A. Cytokines

Epithelial cells are now known to produce a wide variety of cytokines. These can be categorized into several main groups: colony-stimulating factors; pleiotropic cytokines; chemoattractant cytokines; and growth factors (Table 2).

Among the epithelial-derived colony stimulating factors, granulocyte-macrophage colony-stimulating factor (GM-CSF) is the most extensively studied. It has been proposed as a central inflammatory mediator in the pathogenesis of asthma, in light of its ability to prolong eosinophil survival and enhance the cytotoxic activity, mediator release, and phagocytic ability of neutrophils, eosinophils, and macrophages (122–125). GM-CSF is constitutively synthesized and released by airway epithelial cells in vitro (126,127). This finding stemmed from the observation that supernatants from human nasal epithelial cell cultures promoted the formation of granulocyte-macrophage and metachromatic cell colonies from progenitor cells (128). In this study, there was a clearly defined difference between cells obtained from atopic individuals versus those obtained from non-atopics, with a higher level of activity noted in the former. Glucocorticoids decrease the in vitro production of GM-CSF (126), whereas histamine (126), IL-1 (126,129), and common respiratory viruses (95,130) increase its production.

**Table 2**  Epithelial Cytokines

| Colony-stimulating factors | Pleiotropic cytokines | Growth factors |
|---|---|---|
| GM-CSF | IL-6 | TGF-β |
| G-CSF | IL-11 | |
| CSF-1 | IL-10 | |
| M-CSF(?) | IL-1 | |
| | TNF-α (?) | |

*Chemoattractant cytokines*

| C-X-C/α chemokines | C-C/β chemokines | Other |
|---|---|---|
| IL-8 | RANTES | IL-16 |
| GROα | eotaxin | |
| GROγ | MCP-1 | |
| | MCP-4 | |

Increased GM-CSF gene expression and protein production have been shown in bronchial biopsy specimens from asthmatic patients (131–133). Not surprisingly, the levels of GM-CSF are also increased in the bronchoalveolar lavage (BAL) fluid of such patients. IL-1 has been proposed as the cause of enhanced GM-CSF expression in asthma, because IL-1 levels are increased in the airways of asthmatic subjects (134). Glucocorticoids down-regulate GM-CSF expression in bronchial epithelial cells from asthmatics both in vitro (131,132) and after inhalation therapy in vivo (133,135,136). Another class of anti-inflammatory drug, nedocromil sodium, has been noted to attenuate the IL-1–induced increase in epithelial-derived GM-CSF by more than 40% without any effect on constitutive production (137); these data, however, have not been confirmed by others.

Other colony-stimulating factors, including G-CSF (138,139) and CSF-1 (140), are also released by epithelial cells. M-CSF may be produced, but definitive data are still lacking (141,142).

Among the pleiotropic cytokines, it is clear that IL-6 is synthesized by human bronchial epithelial cells (95,127,138,140,143), and that increased gene expression of IL-6 is seen in bronchial biopsies of asthmatic subjects (131,132). In vitro, IL-6 protein has been reported to be released primarily in the apical direction from epithelial monolayers exposed to histamine (144), a finding that correlates with the in vivo observation of increased IL-6 levels in the BAL fluids from symptomatic asthmatics (134). Exposure to IL-1 or TNF-α up-regulates epithelial cell expression of IL-6 (127, 138), as does viral infection (95, 130). Production of IL-6 is decreased in epithelial cells obtained from asthmatic subjects receiving inhaled glucocorticoids (132), but exposure of cultured cells to glucocorticoids in vitro does not affect IL-6 production (138). IL-6—in addition to being a growth factor for several types of cells and to inducing production of acute phase proteins from hepatocytes—has several properties of relevance to allergic airway diseases. This cytokine induces B-cell differentiation, T-cell activation and proliferation, mucosal IgA production, and neural differentiation (145). Consideration must also be given, however, to the concept that IL-6 production may actually be a normal host response to an abnormally reactive airway, because transgenic mice that overexpress IL-6 in airway epithelial cells have diminished responses to methacholine, despite the development of a chronic peribronchial lymphocytic infiltrate (146).

Another cytokine with some similarities to IL-6, namely IL-11, is also released by epithelial cells (147). In vitro, exposure of epithelial cells to IL-1 or TGF-β enhances IL-11 production, while retinoic acid synergistically increases TGF-β–stimulated IL-11 release and inhibits IL-1-stimulated release (147). The role of IL-11 in the pathogenesis of airway diseases is not well understood, but it is an extremely cationic protein that can regulate neural phenotype (148) and can synergize with both IL-4 (149) and stem cell factor (150). In addition, IL-11 can activate B cells via a mechanism that is T cell dependent (151). In vivo, elevated IL-11 levels are present in the nasal secretions of children with upper respiratory symptoms, with the highest levels of IL-11 being found in subjects with audible wheezing (152). Further support for a potential role of IL-11 in asthma was provided by bronchial challenge studies in mice, showing that IL-11 provocation induced marked airways hyperresponsiveness and mononuclear cell infiltration (152).

Data regarding epithelial production of other pleiotropic cytokines is more limited. Epithelial cells obtained from normal, healthy subjects by bronchial brushings have been reported to constitutively produce IL-10, a potent immunoregulatory cytokine originally called "cytokine synthesis inhibitory factor." Interestingly, IL-10 production is down-regulated in cells from patients with cystic fibrosis (153), and is not detectable from immortalized cell lines (D. Proud, unpublished data). Several studies have demonstrated that epithelial cells express mRNA for IL-1 but, to date, IL-1 protein product been detected primarily upon cell lysis (154) or upon exposure to cytotoxic stimuli, such as toluene diisocyante (155). Similarly, although it has been reported that epithelial cells can produce low levels of TNF-α (156), this is still somewhat controversial and additional studies are needed to confirm this finding.

In recent years, considerable attention has focused on the epithelial cell as a source of potent chemoattractant cytokines (chemokines) that could play an important role in recruitment of inflammatory cells to the airway during allergic diseases. It is now known that epithelial cells produce members both of the C-X-C, or α, chemokine family, and of the C-C, or β, chemokine group.

The archetypal C-X-C chemokine that is produced in striking amounts by epithelial cells is IL-8 (127,138,140,157,158). Increased epithelial production of

IL-8 in vitro can be induced by a variety of stimuli including cell deformation (159), cytokines, such as IL-1 or TNF-α (127, 138, 157), toxic agents including nitrogen dioxide, ozone or asbestos (156,160,161), bacterial products (162), and respiratory viruses (95,130,163). Moreover, an increase in IL-8 gene expression is seen in bronchial epithelium of asthmatic subjects obtained by biopsies (131,132,135), a clinically important finding because IL-8 has been postulated to contribute to the pathophysiology of asthma via its abilities to be chemotactic for neutrophils (164), some T-lymphocyte subsets (165,166), and eosinophils that have been pre-incubated with GM-CSF or IL-3 (167). Production of IL-8 is decreased in epithelial cells obtained from asthmatic subjects receiving inhaled glucocorticoids (132, 135), but exposure of cultured cells to glucocorticoids in vitro does not effect IL-8 release (138).

Other C-X-C chemokines are also produced by epithelial cells, most notably GROα and GROγ (168). These chemokines could contribute to neutrophil recruitment during airway inflammation and have also been reported to be chemotactic for basophils (169).

RANTES (regulated on activation, normal T-cell expressed and presumably secreted), a member of the C-C chemokine family, is a chemoattractant for eosinophils, basophils, memory T-lymphocytes, and monocytes (170–172). Alam et al. (173) suggested a role for RANTES in the etiology of asthma when they found elevated levels of the chemokine in the BAL fluid of asthmatics compared with that of normal subjects; Venge and coworkers have shown that RANTES and IL-5 are the major eosinophil chemoattractants in the asthmatic lung (174). Marked epithelial expression of RANTES has also been observed in nasal polyps (175). In vitro, unstimulated human bronchial epithelial cells produce little or no RANTES but production is increased in response to IL-1, TNF-α, and IFN-γ (175–177). Glucocorticoids inhibit RANTES synthesis both in vitro (175,176) and in vivo (177).

Given that eosinophilia is characteristic of allergic diseases of the airways, great interest has focused on stimuli that can induce selective eosinophil recruitment. Although RANTES induces eosinophil recruitment both in vitro (170–172) and in vivo (178,179), this chemokine also recruits other cell types. By contrast, the recently discovered C-C chemokine, eotaxin, appears to be a selective chemoattractant for eosinophils. Eotaxin was initially described as the major eosinophil chemotactic factor in the bronchoalveolar lavage fluid of allergen-challenged guinea pigs (180); both the guinea pig (181) and human (182,183) genes have been cloned. Eotaxin is expressed by nasal polyp epithelium (182) and by human epithelial cells stimulated in vitro with TNF-α, and IFN-γ (183). The human protein is not chemotactic for monocytes, lymphocytes, or neutrophils (182,183).

Cultured epithelial cells also produce the C-C chemokine, monocyte chemoattractant protein (MCP)-1 (168) and MCP-4 (formerly known as CKβ10) (184).

Levels of MCP-1 are elevated in the BAL fluid of asthmatics compared to normal subjects (173), and immunohistochemical studies on bronchial biopsies of atopic asthmatics have shown increased epithelial expression of MCP-1 (185). MCP-1 is chemotactic for monocytes, basophils, and memory T-lymphocytes (186,187). Patients with idiopathic pulmonary fibrosis also have increased levels of MCP-1 in their bronchial epithelium (188).

The final chemoattractant cytokine to be discussed does not belong to either the C-X-C or C-C chemokine families. IL-16 is a selective chemoattractant for CD4$^+$ cells, such as lymphocytes and eosinophils; and is a growth factor for the former cell type. This cytokine, also referred to as "lymphocyte chemoattractant factor," is produced by epithelial cells (189,190) and is detected in increased amounts in bronchoalveolar lavage fluids of antigen-challenged asthmatics (191), raising the possibility that it may contribute to the recruitment of the eosinophils and CD4$^+$ T cells observed in the airway mucosa during asthma.

The data regarding the elaboration of cytokine growth factors by lung epithelium is somewhat confusing. The production of TGF-$\beta_1$ and TGF-$\beta_2$ by cultured epithelial cells has been reported (192,193). Moreover, immunohistochemical staining of human bronchial epithelium has revealed the presence of TGF-$\beta$ within individual cells (194,195). Pelton et al. (196) have suggested, however, that this cytokine is actually produced by adjacent cells (smooth muscle, connective tissue cells) and internalized by the epithelial cell. A recent report on lung specimens from patients with pulmonary fibrosis showed ubiquitous expression of TGF-$\beta_2$ and TGF-$\beta_3$ throughout the epithelium of both healthy and diseased lungs; in contrast, TGF-$\beta_1$ was only present in those areas where there was advanced pulmonary fibrosis and honeycombing, irrespective of the etiology (197). Additional research is needed to clarify these issues.

## B. Lipid Mediators

Arachidonic acid metabolites are a large group of biologically active mediators that are generated within the epithelium by lipoxygenase and cyclooxygenase pathways. Regulation of arachidonic acid metabolism is dependent on the available substrate concentration (198), with maximal epithelial cyclooxygenase activity occurring at lower concentrations of arachidonic acid than for lipoxygenase (199,200).

Mammalian respiratory epithelia are characterized by the 5-, 12-, and 15-lipoxygenase pathways (198), the last of which predominates in the human host (201). The 15-lipoxygenase enzyme converts arachidonic acid to the metabolically important 15-hydroxyeicosatetraenoic acid (15-HETE) and a broad range of hydroperoxy-, epoxyhydroxy-, keto-, and dihydroxy-acids (199,201). Murray et al. (202) were the first to report significant levels of 15-HETE in the BAL fluids of chronic stable asthmatics subjected to antigen provocation. It was not clear, however, that airway epithelial cells were the source of this metabolite because eosinophils also contain high levels of 15-lipoxygenase (203). Activity of 15-lipoxygenase has since been shown to be increased in asthmatic bronchial epithelium by both metabolic assay (204) and direct immunohistochemical staining (205).

The metabolites of 15-lipoxygenase induce a variety of inflammatory responses on the molecular and cellular levels. For example, 15-HETE activates the 5-lipoxygenase pathway in mast cells and ultimately causes the release of leukotriene (LT)$C_4$ (206). It also acts as an active secretagogue of mucus glycoprotein in cultured human airways (207) and augments the early asthmatic response to allergen challenge (208). The 8S,15S-diHETE metabolite is a chemotactic factor for neutrophils both in vitro (209) and in vivo (210), and, once recruited, those neutrophils generate the trihydroxy acid lipoxin A from 15-HETE via the 5-lipoxygenase pathway (211). In turn, lipoxin A has a number of effects on the inflammatory response, including activation of protein kinase C (212), generation of superoxide radicals by neutrophils (211), inhibition of natural killer (NK) cell activity (213), and contraction of guinea pig lung strips (211,214) and human bronchi (215).

Arachidonic acid may also be metabolized via the cyclooxygenase pathway in airway epithelial cells (199,216,217). The predominant products are the counterregulatory prostaglandins $PGE_2$ and $PGF_{2\alpha}$, which are generated in approximately equal amounts (199,218). $PGE_2$ inhibits histamine-induced airway smooth muscle contraction (219,220) and mediates mucus glycoprotein secretion (221). It has also been observed to inhibit mast cell degranulation (222) and generation of LTB4 by alveolar macrophages (223). The in vivo effects of $PGE_2$ include enhancement of cough sensitivity (224), blockade of the early and late bronchoconstrictor responses to inhaled allergen (225), and attenuation of exercise-induced bronchospasm (226). $PGF_{2\alpha}$ has the opposing effect of inducing marked bronchoconstriction (198). Stimulators of epithelial prostaglandin production include bradykinin, histamine, and platelet-activating factor (PAF) (218,219,227).

Platelet-activating factor itself is another lipid mediator produced in small amounts by human bronchial epithelial cells (228). It is a potent chemoattractant for neutrophils and eosinophils and also causes increased vascular permeability and airway hyperreactivity (229). Calcium ionophore A23187 (228) and ozone (230) each augment the release of PAF by primary bronchial epithelium. Phorbol myristate acetate (PMA), a protein kinase C activator, has a similar effect in the human lung epithelial cell line ATC-CCL-185 (231).

## C.  Peptide Mediators

The endothelin family of peptides consists of three closely related mediators, designated endothelin-1, -2, and -3 (232), that act as potent constrictors of both vascular and airway smooth muscle (233). Canine, porcine, and human bronchial epithelial cells all release these mediators (234,235). Elevated levels of endothelin-1 and endothelin-3 have been detected in the BAL fluids from asthmatic patients (236,237), and a correlation has been suggested between the level of endothelin activity and the severity of symptoms (238). Increased gene expression of the precursor peptide, preproendothelin, has been demonstrated in asthmatic bronchial epithelium, as has increased endothelin-1 product by immunohistochemistry (239). Release of biologically active endothelin has also been reported, with the amount of peptide released decreasing significantly after treatment with hydrocortisone (240). Laporte et al. (241) recently demonstrated secretion of endothelin-1 by cultured guinea pig Clara cells. Multiple stimuli, including thrombin (242), TNF (243,244), lipopolysaccharide (243–245), IL-1 (244,246), IL-2, IL-6, IL-8, and TGF-β (246), enhance endothelin release. In contrast, relatively few inhibitors of endothelin secretion (IFN-γ, platelet-derived growth factor [PDGF]) have been reported (244). Endothelin's contribution to the asthmatic disease process may extend beyond the regulation of airway smooth muscle tone, for it also induces release of lipid mediators from the epithelium (a possible autocrine role) (244,247,248) and stimulates smooth muscle proliferation (249).

Sporadic reports of other peptide mediators produced by the epithelium have appeared in recent years. Immunohistochemical studies of rabbit tracheal epithelial cells have revealed the presence of vasopressin and substance P (250). Thrombin induces release of vasopressin but has no effect on substance P release (242). Another bronchoconstrictor neuropeptide, calcitonin gene-related peptide (CGRP), has been demonstrated in the secretory granules of serous cells in rat tracheal epithelium (251).

## D.  Proteases and Protease Inhibitors

The cysteine protease, cathepsin B, has been detected in bronchoalveolar lavage fluids and sputum from patients with chronic bronchitis, bronchiectasis, and emphysema (252–254). Because this protease can hydrolyze extracellular matrix proteins (255), it could contribute to connective tissue damage in the airway. Cathepsin B is usually stored as an inactive proenzyme in the lysosomes of cells, and the proenzyme is activated upon exposure to alkaline pH conditions, probably by proteases in lung secretions. It has recently been reported that bronchial epithelial cells synthesize and secrete procathepsin B that can be activated by neutrophil elastase to the form of the active enzyme that is found in airway secretions (256).

Epithelial cells can also regulate tissue destruction, however, by virtue of their ability to release protease inhibitors. Cystatin C, the major inhibitor of cathepsin B, is produced by epithelial cells (256). Although this inhibitor can be inactivated by neutrophil elastase, several types of epithelial cells can also produce potent elastase inhibitors. Both Clara cells and type II alveolar epithelial cells have been shown to produce secretory leukocyte protease inhibitor (SLPI), and elafin (257). SLPI, which is thought to be the major elastase inhibitor in upper bronchial secretions (258,259) is also capable of inhibiting neutrophil cathepsin G, chymase, chymotrypsin, and trypsin (260). By contrast, elafin, also known as elastase-specific inhibitor (ESI), inhibits elastase and proteinase 3 but not trypsin, chymotrypsin, or cathepsin G (261). Epithelial production of both SLPI and elafin is enhanced upon stimulation with IL-1$\beta$ or TNF (260).

## E.  Reactive Oxygen Species

Reactive oxygen species are important determinants of the pathologic changes associated with airway inflammation, and it is now recognized that, in addition to leukocytes (262, 263), epithelial cells may serve as a source of these mediators. Guinea pig tracheal and bovine bronchial epithelial cells release hydrogen peroxide in response to PAF or PMA (264–266), and rat tracheal explants generate hydrogen peroxide and superoxide anion at the apical cell membrane when exposed to cigarette smoke (267). Although the amount of hydrogen peroxide released by individual epithelial cells is small (approximately 100-fold less than that derived from macrophages), the respiratory tract taken as a whole can be a significant contributor to this aspect of airway inflammation (268). Hydrogen peroxide also has secondary inflammatory effects, in that it stimulates epithelial production of IL-8 (269).

A large amount of research in the past few years has focused on the role of nitric oxide (NO) in airway diseases. NO is generated from L-arginine upon conversion to L-citrulline by the enzyme NO synthase (NOS). NOS is a family of enzymes consisting of at least three isoforms, which are referred to as the constitutive form (cNOS), the neuronal form (nNOS), and the inducible form (iNOS). Both cNOS and iNOS have been detected in primary and transformed respiratory epithelial cell lines (270,271). Lipopolysaccharide, IL-1, and zymosan-activated serum each result in a marked increase in NO production, but similar effects of TNF-$\alpha$ and IFN-$\gamma$ have not been consistently demonstrated (270,272).

The effect of NO in asthma remains unclear, for it may function either in a therapeutic role as a bronchodilator (273,274) or in a deleterious role as an inflammatory contributor (275). Expression of iNOS is clearly increased in the epithelium of transbronchial biopsy specimens from patients with asthma compared with controls (276), and exhaled NO levels are also elevated in such patients (277,278). Treatment of asthmatic subjects with systemic (278) or inhaled (279) glucocorticoids leads to a reduction in levels of exhaled NO. Production of NO can lead to tissue damage in the airways of asthmatics if NO reacts with superoxide anion to produce the highly toxic peroxynitrite radical (280). It may also favor expansion of the Th2 lymphocyte phenotype and thus increase production of IL-4, which causes local production of IgE, and IL-5, which recruits eosinophils to the region (281).

Both primary and immortalized epithelial cells also have protective enzymes that detoxify potentially harmful reactive oxygen species and provide an important airway defensive barrier. Consistent with this protective role, epithelial scavenging of hydrogen peroxide has been reported to be faster at the apical surface (282). Antioxidant mechanisms include Mn superoxide dismutase (SOD), Cu/Zn SOD, catalase, and the glutathione-redox cycle (283,284). Inducers of antioxidant genes include TNF-$\alpha$ (284), hyperoxia (285), and influenza virus (286).

## VIII. SUMMARY

The epithelium is a complex collection of many different cell types with specialized functions. In the setting of injury to the airway, the epithelium is capable of regenerating itself through differentiation of specific cell subpopulations and generation of extracellular matrix components. Individual cells interact with each other, with leukocytes, and with the surrounding matrix through specific adhesion molecules expressed on the cell membrane. Several different classes of mediators are secreted by the epithelium, and they may contribute to the pathophysiology of airway diseases such as asthma. Ongoing investigations into the dynamic properties of respiratory epithelium may ultimately yield insight not only into the mechanisms of disease, but also into potential targets for therapeutic intervention.

## REFERENCES

1. R. G. Breeze and E. B. Wheeldon, *Am. Rev. Respir. Dis.*, *116*:705 (1977).
2. C. G. Plopper, A. T. Mariassy, D. W. Wilson, J. L. Alley, S. J. Nishio, and P. Nettesheim, *Exp. Lung Res.*, *5*:281 (1983).
3. J. D. Crapo, B. E. Barry, P. Gehr, M. Bachofen, and E. R. Weibel, *Am. Rev. Respir. Dis.*, *125*:332 (1982).
4. D. Romberger, A. A. Floreani, R. A. Robbins, A. B. Thompson, J. H. Sisson, J. R. Spurzem, S. Von Essen, S. I. Rennard, and I. Rubinstein, *Textbook of Pulmonary Diseases* (G. L. Baum and E. Wolinsky, eds.), Little & Brown, Boston, p. 23 (1994).
5. D. B. Gail and C. J. M. Lenfant, *Am. Rev. Respir. Dis.*, *127*:366 (1983).
6. M. A. Sleigh, *Respiratory Defense Mechanisms, Part I* (J. D. Brain, D. F. Proctor, and L. M. Reid, eds.), Marcel Dekker, New York, p. 247 (1977).
7. M. J. Sanderson and M. A. Sleigh, *J. Cell Sci.*, *47*:331 (1981).
8. P. K. Jeffery and L. Reid, *J. Anat.*, *120*:295 (1975).
9. M. J. Welsh, *Physiol. Reviews*, *67*:1143 (1975).

10. M. A. Sleigh, J. R. Blake, and N. Liron, *Am. Rev. Respir. Dis.*, *137*:726 (1988).

11. F. Verra, E. Escudier, F. Lebargy, J. F. Bernaudin, H. de Crémoux, and J. Bignon, *Am. J. Respir. Crit. Care Med.*, *151*:630 (1995).

12. A. T. Mariassy and C. G. Plopper, *Anat. Rec.*, *209*:523 (1984).

13. A. B. Thompson, R. A. Robbins, D. J. Romberger, J. H. Sisson, J. R. Spurzem, H. Teschler, and S. I. Rennard, *Eur. Respir. J.*, *8*:127 (1995).

14. D. Lamb and L. Reid, *Br. Med. J.*, *1*:33 (1969).

15. R. V. Ebert and M. J. Terracio, *Am. Rev. Respir. Dis.*, *111*:4 (1975).

16. R. J. Phipps, *International Review of Physiology, Respiratory Physiology III* (J. G. Widdicombe, ed.), University Park Press, Baltimore, p. 213 (1981).

17. M. N. Smith, S. D. Greenberg, and H. J. Spjut, *Am. J. Anat.*, *155*:15 (1979).

18. C. G. Plopper, L. H. Hill, and A. T. Mariassy, *Exp. Lung Res.*, *1*:171 (1980).

19. R. P. Gupta, S. E. Patton, A. M. Jetten, and G. E. R. Hook, *Biochem. J.*, *248*:337 (1987).

20. G. Singh, S. L. Katyal, W. E. Brown, S. Phillips, A. L. Kennedy, J. Anthony, and N. Squeglia, *Biochim. Biophys. Acta*, *950*:329 (1988).

21. G. Singh, J. Singh, S. L. Katyal, W. E. Brown, J. A. Kramps, I. L. Paradis, J. H. Dauber, T. A. Macpherson, and N. Squeglia, *J. Histochem. Cytochem.*, *36*:73 (1988).

22. A. Bernard, X. Dumont, H. Roels, R. Lauwerys, I. Dierynck, M. De Ley, V. Stroobant, and E. de Hoffman, *Clin. Chim. Acta*, *223*:189 (1993).

23. J. U. Balis, J. F. Paterson, J. E. Paciga, E. M. Haller, and S. A. Shelley, *Lab. Invest.*, *52*:657 (1985).

24. D. S. Phelps and J. Floros, *Exp. Lung Res.*, *17*:985 (1991).

25. M. R. Boyd, *Nature*, *269*:713 (1977).

26. J. D. Beckmann, J. R. Spurzem, and S. I. Rennard, *Cell Tissue Res.*, *274*:475 (1993).

27. M. J. Evans, R. A. Cox, S. G. Shami, B. Wilson, and C. G. Plopper, *Am. J. Respir. Cell Mol. Biol.*, *1*:463 (1989).

28. N. F. Johnson and A. F. Hubbs, *Am. J. Respir. Cell Mol. Biol.*, *3*:579 (1990).

29. K. P. Keenan, J. W. Combs, and E. M. McDowell, *Virchows Arch. [Cell Pathol.]*, *41*:193 (1982).

30. K. P. Keenan, J. W. Combs, and E. M. McDowell, *Virchows Arch. [Cell Pathol.]*, *41*:215 (1982).

31. K. P. Keenan, J. W. Combs, and E. M. McDowell, *Virchows Arch. [Cell Pathol.]*, *41*:231 (1982).

32. K. P. Keenan, T. S. Wilson, and E. M. McDowell, *Virchows Arch. [Cell Pathol.]*, *43*:213 (1983).

33. M. J. Evans, S. G. Shami, L. J. Cabral-Anderson, and N. P. Dekker, *Am. J. Pathol.*, *123*:126 (1986).

34. Y. Inayama, G. E. R. Hook, A. R. Brody, G. S. Cameron, A. M. Jetten, L. B. Gilmore, T. Gray, and P. Nettesheim, *Lab. Invest.*, *58*:706 (1988).

35. Y. Inayama, G. E. Hook, A. R. Brody, A. M. Jetten, T. Gray, J. Mahler, and P. Nettesheim, *Am. J. Pathol.*, *134*:539 (1989).

36. R. Tateishi, *Arch. Pathol.*, *96*:198 (1973).

37. D. Adriaensen and S. W. Scheuermann, *Anat. Rec.*, *263*:70 (1993).
38. N. A. Edmondson and D. J. Lewis, *Thorax*, *35*:371 (1980).
39. K. Wasano and T. Yamamoto, *Cell Tissue Res.*, *216*:481 (1981).
40. J. D. Crapo, S. L. Young, E. K. Fram, K. E. Pinkerton, B. E. Barry, and R. O. Crapo, *Am. Rev. Respir. Dis.*, *128*:S42 (1983).
41. H. Hamm, H. Fabel, and W. Bartsch, *Clin. Investig.*, *70*:637 (1992).
42. J. Johansson, T. Curstedt, and B. Robertson, *Eur. Respir. J.*, *7*:372 (1994).
43. R. C. Strunk, D. M. Eidlen, and R. J. Mason, *J. Clin. Invest.*, *81*:1419 (1988).
44. D. H. Bowden, *Thorax*, *36*:801 (1981).
45. S. Montefort, C. A. Herbert, C. Robinson, and S. T. Holgate, *Clin. Exper. Allergy*, *22*:511 (1992).
46. E. E. Schneeberger, *The Lung: Scientific Foundations* (R. G. Crystal, J. B. West, P. J. Barnes, N. S. Cherniak, and E. R. Weibel, eds.), Raven Press, New York, p. 205 (1991).
47. B. R. Stevenson and D. L. Paul, *Curr. Opin. Cell Biol.*, *1*:884 (1989).
48. D. R. Garrod, *Curr. Opin. Cell Biol.*, *5*:30 (1993).
49. P. Cowin, W. W. Franke, C. Grund, H.-P. Kapprell, and J. Kartenbeck, *The Cell in Contact: Adhesions and Junctions as Morphogenetic Determinants* (G. M. Edelman and J.-P. Thiery, eds.), John Wiley, New York, p. 427 (1985).
50. H. Mueller and W. W. Franke, *J. Mol. Biol.*, *163*:647 (1983).
51. S. Montefort, J. A. Roberts, R. Beasley, S. T. Holgate, and W. R. Roche, *Thorax*, *47*:499 (1992).
52. J. M. Anderson and C. M. Van Itallie, *Am. J. Physiol.*, *269*:G467 (1995).
53. E. M. Willott, S. Balda, A. S. Fanning, B. Jameson, C. Van Itallie, and J. M. Anderson, *Proc. Natl. Acad. Sci. USA*, *90*:7834 (1993).
54. L. A. Jesaitis and D. A. Goodenough, *J. Cell Biol.*, *124*:949 (1994).
55. B. Gumbiner, T. Lowenkopf, and D. Apatira, *Proc. Natl. Acad. Sci. USA*, *88*:3460 (1991).
56. S. Citi, H. Sabanay, J. Kendrick-Jones, and B. Geiger, *J. Cell Sci.*, *93*:107 (1989).
57. M. Furuse, T. Hirase, M. Itoh, A. Nagafuchi, S. Yonemura, S. Tsukita, and S. Tsukita, *J. Cell Biol.*, *123*:1777 (1993).
58. R. O. Hynes, *Cell*, *48*:549 (1987).
59. J. M. Pilewski and S. M. Albelda, *Am. Rev. Respir. Dis.*, *148*:S31 (1993).
60. S. M. Albelda, *Am. J. Respir. Cell Mol. Biol.*, *4*:195 (1991).
61. R. J. Sapsford, J. L. Devalia, A. E. McAulay, A. J. D'Ardenne, and R. J. Davies, *J. Allergy Clin. Immunol.*, *87*:A303 (1991).
62. L. Damjanovich, S. M. Albelda, S. A. Mette, and C. A. Buck, *Am. J. Respir. Cell Mol. Biol.*, *6*:197 (1992).
63. S. A. Mette, J. Pilewski, C. A. Buck, and S. M. Albelda, *Am. J. Respir. Cell Mol. Biol.*, *8*:562 (1993).
64. N. D. Manolitsas, C. J. Trigg, A. E. McAulay, J. H. Wang, S. E. Jordan, A. J. D'Ardenne, and R. J. Davies, *Eur. Respir. J.*, *7*:1439 (1994).
65. M. A. Stepp, S. Spurr-Michaud, A. Tisdale, J. Elwell, and I. K. Gipson, *Proc. Natl. Acad. Sci. USA*, *87*:8970 (1990).
66. A. Sonnenberg, J. Calafat, H. Janssen, H. Daams, L. M. H. van der Raaij-Helmer,

R. Falcioni, S. J. Kennel, J. D. Aplin, J. Baker, M. Loizidou, and D. Garrod, *J. Cell Biol.*, *113*:907 (1991).

67. J. W. Smith, D. J. Vestal, S. V. Irwin, T. A. Burke, and D. A. Cheresh, *J. Biol. Chem.*, *265*:11008 (1990).

68. Y. Yokosaki, E. L. Palmer, A. L. Prieto, K. L. Crossin, M. A. Bourdon, R. Pytela, and D. Sheppard, *J. Biol Chem*, *269*:26691 (1994).

69. M. Busk, R. Pytela, and D. Sheppard, *J. Biol Chem*, *267*:5790 (1992).

70. A. Weinacker, A. Chen, M. Agrez, R. I. Cone, S. Nishimura, and E. Wayner, *J. Biol Chem*, *269*:6940 (1994).

71. A. L. Prieto, G. M. Edelman, and K. L. Crossin, *Proc. Natl. Acad. Sci. USA*, *90*:10154 (1993).

72. A. Aruffo, I. Stamenkovic, M. Melnick, C. B. Underhill, and B. Seed, *Cell*, *61*:1303 (1990).

73. L. J. Picker, M. Nakache, and E. C. Butcher, *J. Cell Biol.*, *109*:927 (1989).

74. J. Lesley, R. Hyman, and P. W. Kincade, *Adv. Immunol.*, *54*:271 (1993).

75. U. Günthert, *Curr. Topics Microbiol. Immunol.*, *184*:47 (1993).

76. M. Kasper, U. Günthert, P. Dall, K. Kayser, D. Schuh, G. Haroske, and M. Müller, *Am. J. Respir. Cell Mol. Biol.*, *13*:648 (1995).

77. M. B. Penno, J. T. August, S. B. Baylin, M. Mabry, I. Linnoila, V. S. Lee, D. Croteau, X. L. Yang, and C. Rosada, *Cancer Res.*, *54*:1381 (1994).

78. M. Ferrarini, S. M. Pupa, M. R. Zocchi, C. Rugarli, and S. Ménard, *Int. J. Cancer*, *57*:486 (1994).

79. S. Saunders, M. Jalkanen, S. O'Farrell, and M. Bernfield, *J. Cell Biol.*, *108*:1547 (1989).

80. T. E. Hardingham and A. J. Fosang, *FASEB J.*, *6*:861 (1992).

81. M. Overduin, T. S. Harvey, S. Bagby, K. I. Tong, P. Yau, M. Takeichi, and M. Ikura, *Science*, *267*:386 (1995).

82. K. Vleminckx, L. Vakaet, Jr., M. Mareel, W. Fiers, and F. Van Roy, *Cell*, *66*:107 (1991).

83. S. M. Albelda, *Lab. Invest.*, *68*:4 (1993).

84. W. Birchmeier, J. Hulsken, and J. Behrens, *Ciba Found. Symp.*, *189*:124 (1995).

85. D. H. Adams and S. Shaw, *Lancet*, *343*:831 (1994).

86. C. D. Wegner, R. H. Gundel, P. Reilly, N. Haynes, L. G. Letts, and R. Rothlein, *Science*, *247*:456 (1990).

87. A. M. Vignola, A. M. Campbell, P. Chanez, J. Bousquet, P. Paul-Lacoste, F. B. Michel, and P. Godard, *Am. Rev. Respir. Dis.*, *148*:689 (1993).

88. M. F. Tosi, J. M. Stark, C. W. Smith, A. Hamedani, D. C. Gruenert, and M. D. Infeld, *Am. J. Respir. Cell Mol. Biol.*, *7*:214 (1992).

89. D. C. Look, S. R. Rapp, B. T. Keller, and M. J. Holtzman, *Am. J. Physiol.*, *263*:L79 (1992).

90. A. M. Vignola, A. M. Campbell, P. Chanez, P. Lacoste, F. B. Michel, P. Godard, and J. Bousquet, *Am. J. Respir. Cell Mol. Biol.*, *9*:411 (1993).

91. P. G. M. Bloemen, M. C. van den Tweel, P. A. J. Henricks, F. Engels, S. S. Wagenaar, A. A. J. J. L. Rutten, and F. P. Nijkamp, *Am. J. Respir. Cell Mol. Biol.*, *9*:586 (1993).

92. T. K. Kishimoto, R. S. Larson, A. L. Corbi, M. L. Dustin, D. E. Staunton, and T. A. Springer, *Adv. Immunol.*, *46*:149 (1989).

93. M. F. Tosi, A. Hamedani, J. Brosovich, and S. E. Alpert, *J. Immunol.*, *152*:1935 (1994).

94. M. F. Tosi, J. M. Stark, A. Hamedani, C. W. Smith, D. C. Gruenert, and Y. T. Huang, *J. Immunol.*, *149*:3345 (1992).

95. M. C. Subauste, D. B. Jacoby, S. M. Richards, and D. Proud, *J. Clin. Invest.*, *96*:549 (1995).

96. J. M. Greve, G. Davis, A. M. Meyer, C. P. Forte, S. C. Yost, C. W. Marlor, M. E. Kamarck, and A. McClelland, *Cell*, *56*:839 (1989).

97. D. E. Staunton, V. J. Merluzzi, R. Rothlein, R. Barton, S. D. Marlin, and T. A. Springer, *Cell*, *56*:849 (1989).

98. D. C. Bullard, L. Qin, I. Lorenzo, W. M. Quinlin, N. A. Doyle, R. Bosse, D. Vestweber, C. M. Doerschuk, and A. L. Beaudet, *J. Clin. Invest.*, *95*:1782 (1995).

99. J. Atsuta, B. S. Bochner, S. A. Sterbinsky, and R. P. Schleimer, *J. Allergy Clin. Immunol.*, *97*:292 (abstr.) (1996).

100. A. R. Glanville, H. D. Tazelaar, J. Theodore, E. Imoto, R. V. Rouse, J. C. Baldwin, and E. D. Robin, *Am. Rev. Respir. Dis.*, *139*:330 (1989).

101. A. M. Vignola, P. Chanez, A. M. Campbell, A. M. Pinel, J. Bousquet, F. B. Michel, and P. Godard, *Clin. Exp. Immunol.*, *96*:104 (1994).

102. A. E. Stoop, D. M. H. Hameleers, P. E. M. v Run, J. Biewenga, and S. van der Baan, *J. Allergy Clin. Immunol.*, *84*:734 (1989).

103. G. A. Rossi, O. Sacco, B. Balbi, S. Oddera, T. Mattioni, G. Corte, C. Ravazzoni, and L. Allegra, *Am. J. Respir. Cell Mol. Biol.*, *3*:431 (1990).

104. J. R. Spurzem, O. Sacco, G. A. Rossi, J. D. Beckmann, and S. E. Rennard, *J. Lab. Clin. Med.*, *120*:94 (1992).

105. M. Mezzetti, M. Soloperto, A. Fasoli, and S. Mattoli, *J. Allergy Clin. Immunol.*, *87*:930 (1991).

106. T. H. Kalb, M. T. Chuang, Z. Marom, and L. Mayer, *Am. J. Respir. Cell Mol. Biol.*, *4*:320 (1991).

107. S. Shoji, K. A. Rickard, R. F. Ertl, R. A. Robbins, J. Linder, and S. I. Rennard, *Am. J. Respir. Cell Mol. Biol.*, *1*:13 (1989).

108. G. Stoner, J.-M. Katoch, B. Foidart, P. Trump, P. Steinert, and C. Harris, *In Vitro*, *17*:577 (1981).

109. S. Shoji, R. F. Ertl, J. Linder, D. J. Romberger, and S. I. Rennard, *Am. Rev. Respir. Dis.*, *141*:218 (1990).

110. E. Ruoslahti, *Annu. Rev. Biochem.*, *57*:375 (1989).

111. A. Linnala, V. Kinnula, L. A. Laitinen, V-P. Lehto, and I. Virtanen, *Am. J. Respir. Cell Mol. Biol.*, *13*:578 (1995).

112. A. Laitinen, A. Altraja, M. Kämpe, M. Linden, G. Stalenheim, P. Venge, L. Hakansson, I. Virtanen, and L. A. Laitinen, *Am. J. Respir. Crit. Care Med.*, *149*:A942 (1994).

113. D. J. Romberger, J. D. Beckmann, L. Claassen, R. F. Ertl, and S. I. Rennard, *Am. J. Respir. Cell Mol. Biol.*, *7*:149 (1992).

114. E. Härkönen, I. Virtanen, A. Linnala, L. L. Laitinen, and V. L. Kinnula, *Am. J. Respir. Cell Mol. Biol.*, *13*:109 (1995).

115. R. M. Senior, G. L. Griffin, M. S. Mudd, M. A. Moxley, W. J. Longmore, and R. A. Pierce, *Am. J. Respir. Cell Mol. Biol.*, *14*:239 (1996).

116. A. E. Chung and M. E. Durkin, *Am. J. Respir. Cell Mol. Biol.*, *3*:275 (1990).
117. S. E. McGowan, *FASEB J.*, *6*:2895 (1992).
118. M. D. Infeld, J. A. Brennan, and P. B. Davis, *Am. J. Physiol.*, *262*:L535 (1992).
119. Y. Nakamura, L. Tate, R. Ertl, M. Kawamoto, T. Mio, Y. Adachi, D. Romberger, S. Koizumi, G. Gossman, R. A. Robbins, J. R. Spurzem, and S. I. Rennard, *Am. Rev. Respir. Dis.*, *147*:A278 (1993).
120. M. Kawamoto, Y. Nakamura, L. Tate, Jr., R. F. Ertl, D. J. Romberger, and S. I. Rennard, *Am. Rev. Respir. Dis.*, *145*:A842 (1992).
121. D. Proud, *Asthma: Physiology, Immunopharmacology, and Treatment, Fourth International Symposium* (S. T. Holgate, K. F. Austen, L. M. Lichtenstein, and A. B. Kay, eds.), Academic Press, New York, p. 199 (1993).
122. A. F. Lopez, D. J. Williamson, J. R. Gamble, C. G. Begley, J. M. Harlan, S. J. Klebanoff, A. Waltersdorph, G. Wong, S. C. Clark, and M. A. Vadas, *J. Clin. Invest.*, *78*:1220 (1986).
123. S. Heidenreich, J.-H. Gong, A. Schmidt, M. Nain, and D. Gemsa, *J. Immunol.*, *143*:1198 (1989).
124. C. Ruef and D. L. Coleman, *Rev. Infect. Dis.*, *12*:41 (1990).
125. L. A. Burke, M. P. Hallsworth, T. M. Litchfield, R. Davidson, and T. H. Lee, *J. Allergy Clin. Immunol.*, *88*:226 (1991).
126. L. Churchill, B. Friedman, R. P. Schleimer, and D. Proud, *Immunology*, *75*:189 (1992).
127. O. Cromwell, Q. Hamid, C. J. Corrigan, J. Barkans, Q. Meng, P. D. Collins, and A. B. Kay, *Immunology*, *77*:330 (1992).
128. H. Otsuka, J. Dolovich, M. Richardson, J. Bienenstock, and J. A. Denburg, *Am. Rev. Respir. Dis.*, *136*:710 (1987).
129. M. Marini, M. Soloperto, M. Mezzetti, A. Fasoli, and S. Mattoli, *Am. J. Respir. Cell Mol. Biol.*, *4*:519 (1991).
130. T. L. Noah and S. Becker, *Am. J. Physiol.*, *265*:L472 (1993).
131. S. Mattoli, M. Marini, and A. Fasoli, *Chest*, *101*:27S (1992).
132. M. Marini, E. Vittori, J. Hollemborg, and S. Mattoli, *J. Allergy Clin. Immunol.*, *89*:1001 (1992).
133. A. R. Sousa, R. N. Poston, S. J. Lane, J. A. Nakhosteen, and T. H. Lee, *Am. Rev. Respir. Dis.*, *147*:1557 (1993).
134. S. Mattoli, V. L. Mattoso, M. Soloperto, L. Allegra, and A. Fasoli, *J. Allergy Clin. Immunol.*, *87*:794 (1991).
135. J. H. Wang, C. J. Trigg, J. L. Devalia, S. Jordan, and R. J. Davies, *J. Allergy Clin. Immunol.*, *94*:1025 (1994).
136. C. J. Trigg, N. D. Manolitsas, J. Wang, M. A. Calderón, A. McAulay, S. E. Jordan, M. J. Herdamn, N. Jhalli, J. M. Duddle, S. A. Hamilton, J. L. Devalia, and R. J. Davies, *Am. J. Respir. Crit. Care Med.*, *150*:17 (1994).
137. M. Marini, M. Soloperto, Y. Zheng, M. Mezzetti, and S. Mattoli, *Pulm. Pharmacol.*, *5*:61 (1992).
138. S. J. Levine, P. Larivée, C. Logun, C. W. Angus, and J. H. Shelhamer, *Am. J. Physiol.*, *265*:L360 (1993).
139. G. Cox, J. Gauldie, and M. Jordana, *Am. J. Respir. Cell Mol. Biol.*, *7*:507 (1992).

140. M. Bédard, C. D. McClure, N. L. Schiller, C. Francoeur, A. Cantin, and M. Denis, *Am. J. Respir. Cell Mol. Biol.*, *9*:455 (1993).
141. T. Ohtoshi, C. Vancheri, G. Cox, J. Gauldie, J. Dolovich, J. A. Denburg, and M. Jordana, *Am. J. Respir. Cell Mol. Biol.*, *4*:255 (1991).
142. Z. Xing, T. Ohtoshi, P. Ralph, J. Gauldie, and M. Jordana, *Am. J. Respir. Cell Mol. Biol.*, *6*:212 (1992).
143. H. Takizawa, T. Ohtoshi, K. Ohta, S. Hirohata, M. Yamaguchi, N. Suzuki, T. Ueda, A. Ishii, G. Shindoh, T. Oka, K. Hiramatsu, and K. Ito, *Biochem. Biophys. Res. Commun.*, *187*:596 (1992).
144. T. L. Noah, A. M. Paradiso, M. C. Madden, K. P. McKinnon, and R. B. Devlin, *Am. J. Respir. Cell Mol. Biol.*, *5*:484 (1991).
145. T. Kishimoto, *Blood*, *74*:1 (1989).
146. B. F. DiCosmo, G. P. Geba, D. Picarella, J. A. Elias, J. A. Rankin, B. R. Stripp, J. A. Whitsett, and R. A. Flavell, *J. Clin. Invest.*, *94*:2028 (1994).
147. J. A. Elias, T. Zheng, O. Einarsson, M. Landry, T. Trow, N. Rebert, and J. Panuska, *J. Biol. Chem.*, *269*:22261 (1994).
148. P. H. Patterson and H. Nawa, *Cell*, *72(suppl)*:123 (1993).
149. M. Musashi, S. C. Clark, T. Sudo, D. L. Urdal, and M. Ogawa, *Blood*, *78*:1448 (1991).
150. R. M. Lemoli, M. Fogli, A. Fortuna, M. R. Motta, S. Rizzi, C. Benini, and S. Tura, *Exp. Hematol.*, *21*:1668 (1993).
151. T. Yin, P. Schendel, and Y-C. Yang, *J. Exp. Med.*, *175*:211 (1992).
152. O. Einarsson, G. P. Geba, Z. Zhu, M. Landry, and J. A. Elias, *J. Clin. Invest.*, *97*:915 (1996).
153. T. L. Bonfield, M. W. Constan, P. Burfeind, J. R. Panuska, J. B. Hilliard, and M. Berger, *Am. J. Respir. Cell Mol. Biol.*, *13*:257 (1995).
154. J. S. Kenney, C. Baker, M. R. Welch, and L. C. Altman, *J. Allergy Clin. Immunol.*, *93*:1060 (1994).
155. S. Mattoli, S. Miante, F. Calabrò, M. Mezzetti, A. Fasoli, and L. Allegra, *Am. J. Physiol.*, *259*:L320 (1990).
156. J. L. Devalia, A. M. Campbell, R. J. Sapsford, C. Ruszak, D. Quint, P. Godard, J. Bousquet, and R. J. Davies, *Am. J. Respir. Cell Mol. Biol.*, *9*:271 (1993).
157. T. J. Standiford, S. L. Kunkel, M. A. Basha, S. W. Chensue, J. P. Lynch III, G. B. Toews, J. Westwick, and R. M. Strieter, *J. Clin. Invest.*, *86*:1945 (1990).
158. H. Nakamura, K. Yoshimura, H. A. Jaffe, and R. G. Crystal, *J. Biol. Chem.*, *266*:19611 (1991).
159. Y. Shibata, H. Nakamura, S. Kato, and H. Tomoike, *J. Immunol.*, *156*:772 (1996).
160. R. B. Devlin, K. P. McKinnon, T. Noah, S. Becker, and H. S. Koren, *Am. J. Physiol.*, *266*:L612 (1994).
161. G. J. Rosenthal, D. R. Germolec, M. E. Blazka, E. Corsini, P. Simeonova, P. Pollack, L-Y. Kong, J. Kwon, and M. I. Luster, *J. Immunol.*, *153*:3237 (1994).
162. P. P. Massion, H. Inoue, J. Reichman-Eisenstat, D. Grunberger, P. G. Jorens, B. Housset, J. F. Pittet, J. P. Weiner-Kronish, and J. A. Nadel, *J. Clin. Invest.*, *93*:26 (1994).
163. A. M. K. Choi and D. B. Jacoby, *FEBS Lett.*, *309*:327 (1992).
164. E. J. Leonard and T. Yoshimura, *Am. J. Respir. Cell Mol. Biol.*, *2*:479 (1990).

165. C. G. Larsen, A. O. Anderson, E. Appella, J. J. Oppenheim, and K. Matsushima, *Science*, *243*:1464 (1989).

166. D. D. Taub, M. Anver, J. J. Oppenheim, D. L. Longo, and W. J. Murphy, *J. Clin. Invest.*, *97*:1931 (1996).

167. R. A. J. Warringa, L. Koenderman, P. T. M. Kok, J. Kreukniet, and P. L. B. Bruijnzeel, *Blood*, *77*:2694 (1991).

168. S. Becker, J. Quay, H. S. Koren, and J. S. Haskill, *Am. J. Physiol.*, *266*:L278 (1994).

169. M. Baggiolini, B. Dewald, and B. Moser, *Adv. Immunol.*, *55*:97 (1994).

170. T. J. Schall, K. Bacon, K. J. Toy, and D. V. Goeddel, *Nature*, *347*:669 (1990).

171. T. J. Schall, *Cytokine*, *3*:165 (1991).

172. A. Rot, M. Krieger, T. Brunner, S. C. Bischoff, T. J. Schall, and C. A. Dahinden, *J. Exp. Med.*, *176*:1489 (1992).

173. R. Alam, J. York, M. Boyars, S. Stafford, J. A. Grant, J. Lee, P. Forsythe, T. Sim, and N. Ida, *Am. J. Respir. Crit. Care Med.*, *153*:1398 (1996).

174. J. Venge, M. Lampinen, L. Hakansson, S. Rak, and P. Venge, *J. Allergy Clin. Immunol.*, *97*:1110 (1996).

175. C. Stellato, L. A. Beck, G. A. Gorgogne, D. Proud, T. J. Schall, S. J. Ono, L. M. Lichtenstein, and R. P. Schleimer, *J. Immunol.*, *155*:410 (1995).

176. O. J. Kwon, P. J. Jose, R. A. Robbins, T. J. Schall, T. J. Williams, and P. J. Barnes, *Am. J. Respir. Cell Mol. Biol.*, *12*:488 (1995).

177. J. H. Wang, J. L. Devalia, C. Xia, R. J. Sapsford, and R. J. Davies, *Am. J. Respir. Cell Mol. Biol.*, *14*:27 (1996).

178. R. Meurer, G. Van Riper, W. Feeney, P. Cunningham, D. Hora, M. S. Springer, D. E. MacIntyre, and H. Rosen, *J. Exp. Med.*, *178*:1913 (1993).

179. L. Beck, C. Bickel, S. Sterbinsky, C. Stellato, R. Hamilton, H. Rosen, B. Bochner, and R. P. Schleimer, *FASEB J.*, *9*:A804 (1995).

180. P. J. Jose, D. A. Griffiths-Johnson, P. D. Collins, D. T. Walsh, R. Moqbel, N. F. Totty, O. Truong, J. J. Hsuan, and T. J. Williams, *J. Exp. Med.*, *179*:881 (1994).

181. P. J. Jose, I. M. Adcock, D. A. Griffiths-Johnson, N. Berkman, T. N. Wells, T. J. Williams, and C. A. Power, *Biochem. Biophys. Res. Commun.*, *205*:788 (1994).

182. P. D. Ponath, S. Qin, D. J. Ringler, I. Clark-Lewis, J. Wang, N. Kassam, H. Smith, X. Shi, J-A. Gonzalo, W. Newman, J-C. Gutierrez-Ramos, and C. R. Mackay, *J. Clin. Invest.*, *97*:604 (1996).

183. E. A. Garcia-Zepeda, M. E. Rothenberg, R. T. Ownbey, J. Celestin, P. Leder, and A. D. Luster, *Nature Medicine*, *2*:449 (1996).

184. C. Stellato, L. Beck, L. Schwiebert, H. Li, J. White, and R. P. Schleimer, *J. Allergy Clin. Immunol.*, *97*:304 (abstr.) (1996).

185. A. R. Sousa, S. J. Lane, J. A. Nakhosteen, T. Yoshimura, T. H. Lee, and R. N. Poston, *Am. J. Respir. Cell Mol. Biol.*, *10*:142 (1994).

186. E. J. Leonard, M. Takeya, A. Skeel, and T. Yoshimura, *Adv. Exp. Med. Biol.*, *305*:57 (1991).

187. M. W. Carr, S. J. Roth, E. Luther, S. S. Rose, and T. A. Springer, *Proc. Natl. Acad. Sci. USA*, *91*:3652 (1994).

188. H. N. Antoniades, J. Neville-Golden, T. Galanopoulos, R. L. Kradin, A. J. Valente, and D. T. Graves, *Proc. Natl. Acad. Sci. USA*, *89*:5371 (1992).

189. A. Bellini, H. Yoshimura, E. Vitori, M. Marini, and S. Mattoli, *J. Allergy Clin. Immunol.*, *92*:412 (1993).

190. S. Laberge, P. Ernst, O. Ghaffar, W. W. Cruikshank, H. Kornfeld, D. M. Center, and Q. Hamid, *Am. J. Respir. Crit. Care Med.*, *153*:A880 (1996).

191. W. W. Cruikshank, A. Long, R. E. Tarpy, H. Kornfeld, M. P. Carroll, L. Teran, S. T. Holgate, and D. M. Center, *Am. J. Respir. Cell Mol. Biol.*, *13*:738 (1995).

192. V. Cazals, B. Mouhieddine, B. Maitre, Y. Le Bouc, K. Chadelat, J. S. Brody, and A. Clement, *J. Biol. Chem.*, *269*:14111 (1994).

193. O. Sacco, D. Romberger, A. Rizzino, J. D. Beckmann, S. I. Rennard, and J. R. Spurzem, *J. Clin. Invest.*, *90*:1379 (1992).

194. A. Magnan, I. Frachon, B. Rain, M. Peuchmaur, G. Monti, B. Lenot, M. Fattal, G. Simonneau, P. Galanaud, and D. Emilie, *Thorax*, *49*:789 (1994).

195. N. Khalil, R. N. O'Connor, H. W. Unruh, P. W. Warren, K. C. Flanders, A. Kemp, O. H. Bereznay, and A. H. Greenberg, *Am. J. Respir. Cell Mol. Biol.*, *5*:155 (1991).

196. R. W. Pelton, M. D. Johnson, E. A. Perkett, L. I. Gould, and H. L. Moses, *Am. J. Respir. Cell Mol. Biol.*, *5*:522 (1991).

197. N. Khalil, R. N. O'Connor, K. C. Flanders, and H. Unruh, *Am. J. Respir. Cell Mol. Biol.*, *14*:131 (1996).

198. M. J. Holtzman, *Annu. Rev. Physiol.*, *54*:303 (1992).

199. M. J. Holtzman, J. R. Hansbrough, G. D. Rosen, and J. Turk, *Biochim. Biophys. Acta*, *963*:401 (1988).

200. J. R. Hansbrough, A. B. Atlas, J. Turk, and M. J. Holtzman, *Am. J. Respir. Cell Mol. Biol.*, *1*:237 (1989).

201. J. A. Hunter, W. E. Finkbeiner, J. A. Nadel, E. J. Goetzl, and M. J. Holtzman, *Proc. Natl. Acad. Sci. USA*, *82*:4633 (1985).

202. J. J. Murray, A. B. Tonnel, A. R. Brash, L. J. Roberts II, P. Gosset, R. Workman, A. Capron, and J. A. Oates, *N. Engl. J. Med.*, *315*:800 (1986).

203. M. J. Holtzman, A. Pentland, N. L. Baenziger, and J. R. Hansbrough, *Biochim. Biophys. Acta*, *1003*:204 (1989).

204. M. Kumlin, M. Hamberg, E. Granström, T. Björck, B. Dahlén, H. Matsuda, O. Zetterström, and S.-E. Dahlén, *Arch. Biochem. Biophys.*, *282*:254 (1990).

205. V. R. Shannon, P. Chanez, J. Bousquet, and M. J. Holtzman, *Am. Rev. Respir. Dis.*, *147*:1024 (1993).

206. E. J. Goetzl, M. J. Phillips, and W. M. Gold, *J. Exp. Med.*, *158*:731 (1983).

207. Z. Marom, J. H. Shelhamer, F. Sun, and M. Kaliner, *J. Clin. Invest.*, *72*:122 (1983).

208. C. K. W. Lai, R. Polosa, and S. T. Holgate, *Am. Rev. Respir. Dis.*, *141*:1423 (1990).

209. S. Shak, H. D. Perez, and I. M. Goldstein, *J. Biol. Chem.*, *258*:14948 (1983).

210. C. M. Kirsch, E. Sigal, T. D. Djokic, P. D. Graf, and J. A. Nadel, *J. Appl. Physiol.*, *64*:1792 (1988).

211. C. N. Serhan, K. C. Nicolaou, S. E. Webber, C. A. Veale, S.-E. Dahlén, T. J. Puustinen, and B. Samuelsson, *J. Biol. Chem.*, *261*:16340 (1986).

212. A. Hansson, C. N. Serhan, J. Haeggström, M. Ingelman-Sundberg, and B. Samuelsson, *Biochem. Biophys. Res. Commun.*, *134*:1215 (1986).

213. U. Ramstedt, J. Ng, H. Wigzell, C. N. Serhan, and B. Samuelsson, *J. Immunol.*, *135*:3434 (1985).

214. S.-E. Dahlén, J. Raud, C. N. Serhan, J. Björk, and B. Samuelsson, *Acta Physiol. Scand.*, *130*:643 (1987).

215. S.-E. Dahlén, L. Franzén, J. Raud, C. N. Serhan, P. Westlund, E. Wikström, T. Björck, H. Matsuda, S. E. Webber, C. A. Veale, T. Puustinen, J. Haeggström, K. C. Nicolaou, and B. Samuelsson, *Lipoxins: Biosynthesis, Chemistry, and Biological Activities* (P. Y.-K. Wong and C. N. Serhan, eds.), Plenum Press, New York, p. 107 (1988).

216. T. E. Eling, R. M. Danilowicz, D. C. Henke, K. Sivarajah, J. R. Yankaskas, and R. C. Boucher, *J. Biol. Chem.*, *261*:12841 (1986).

217. Z. M. Duniec, T. E. Eling, A. M. Jetten, T. E. Gray, and P. Nettesheim, *Exp. Lung Res.*, *15*:391 (1989).

218. L. Churchill, F. H. Chilton, J. H. Resau, R. Bascom, W. C. Hubbard, and D. Proud, *Am. Rev. Respir. Dis.*, *140*:449 (1989).

219. D. A. Knight, G. A. Stewart, M. L. Lai, and P. J. Thompson, *Eur. J. Pharmacol.*, *272*:1 (1995).

220. D. A. Knight, G. A. Stewart, and P. J. Thompson, *Eur. J. Pharmacol.*, *272*:13 (1995).

221. Z. Marom, J. H. Shelhamer, and M. Kaliner, *J. Clin. Invest.*, *67*:1695 (1981).

222. S. P. Peters, E. S. Schulman, R. P. Schleimer, D. W. MacGlashan, Jr., H. H. Newball, and L. M. Lichtenstein, *Am. Rev. Respir. Dis.*, *126*:1034 (1982).

223. B. W. Christman and J. W. Christman, *Am. Rev. Respir. Dis.*, *141*:A393 (1990).

224. N. B. Choudry, R. W. Fuller, and N. B. Pride, *Am. Rev. Respir. Dis.*, *140*:137 (1989).

225. I. D. Pavord, C. S. Wong, J. Williams, and A. E. Tattersfield, *Am. Rev. Respir. Dis.*, *148*:87 (1990).

226. E. Melillo, K. L. Woolley, P. J. Manning, R. M. Watson, and P. M. O'Byrne, *Am. J. Respir. Crit. Care Med.*, *149*:1138 (1994).

227. L. Churchill, F. H. Chilton, and D. Proud, *Biochem. J.*, *276*:593 (1991).

228. M. J. Holtzman, B. Ferdman, A. Bohrer, and J. Turk, *Biochem. Biophys. Res. Commun.*, *177*:357 (1991).

229. L. M. McManus and S. I. Deavers, *Clin. Chest Med.*, *10*:107 (1989).

230. J. M. Samet, T. L. Noah, R. B. Devlin, J. R. Yankaskas, K. McKinnon, L. A. Dailey, and M. Friedman, *Am. J. Respir. Cell Mol. Biol.*, *7*:514 (1992).

231. H. Salari and A. Wong, *Eur. J. Pharmacol.*, *175*:253 (1990).

232. T. Sakurai, M. Yanagisawa, and T. Masaki, *Trends Pharmacol. Sci.*, *13*:103 (1992).

233. Y. Uchida, H. Ninomiya, M. Saotome, A. Nomura, M. Ohtsuka, M. Yanagisawa, K. Goto, T. Masaki, and S. Hasegawa, *Eur. J. Pharmacol.*, *154*:227 (1988).

234. P. N. Black, M. A. Ghatei, K. Takahashi, D. Bretherton-Watt, T. Krausz, C. T. Dollery, and S. R. Bloom, *FEBS Lett.*, *255*:129 (1989).

235. S. Mattoli, M. Mezzetti, G. Riva, L. Allegra, and A. Fasoli, *Am. J. Respir. Cell Mol. Biol.*, *3*:145 (1990).

236. S. Mattoli, M. Soloperto, M. Marini, and A. Fasoli, *J. Allergy Clin. Immunol.*, *88*:376 (1991).

237. M. Marini, A. Fasoli, and S. Mattoli, *Am. Rev. Respir. Dis.*, *143*:A158 (1991).

238. A. Nomura, Y. Uchida, M. Kameyama, M. Saotome, K. Oki, and S. Hasegawa, *Lancet*, *1*:747 (1989).

239. V. Ackerman, S. Carpi, A. Bellini, G. Vassalli, M. Marini, and S. Mattoli, *J. Allergy Clin. Immunol.*, *96*:618 (1995).

240. E. Vittori, M. Marini, A. Fasoli, R. de Franchis, and S. Mattoli, *Am. Rev. Respir. Dis.*, *146*:1320 (1992).
241. J. Laporte, P. D'Orléans-Juste, and P. Sirois, *Am. J. Respir. Cell Mol. Biol.*, *14*:356 (1996).
242. R. E. Rennick, P. Milner, and G. Burnstock, *Eur. J. Pharmacol.*, *230*:367 (1993).
243. J. Nakano, H. Takizawa, T. Ohtoshi, S. Shoji, M. Yamaguchi, A. Ishii, M. Yanagisawa, and K. Ito, *Clin. Exp. Allergy*, *24*:330 (1994).
244. B. A. Markewitz, D. E. Kohan, and J. R. Michael, *Am. J. Physiol.*, *268*:L192 (1995).
245. H. Ninomiya, Y. Uchida, Y. Ishii, A. Nomura, M. Kameyama, M. Saotome, T. Endo, and S. Hasegawa, *Eur. J. Pharmacol.*, *203*:299 (1991).
246. T. Endo, Y. Uchida, H. Matsumoto, N. Suzuki, A. Nomura, F. Hirata, and S. Hasegawa, *Biochem. Biophys. Res. Commun.*, *186*:1594 (1992).
247. T. Wu, R. D. Rieves, P. Larivée, C. Logun, M. G. Lawrence, and J. H. Shelhamer, *Am. J. Respir. Cell Mol. Biol.*, *8*:282 (1993).
248. T. Nagase, Y. Fukuchi, C. Jo, S. Teramoto, Y. Uejima, K. Ishida, T. Shimizu, and H. Orimo, *Biochem. Biophys. Res. Commun.*, *168*:485 (1990).
249. T. Nakaki, M. Nakayama, S. Yamamoto, and R. Kato, *Biochem. Biophys. Res. Commun.*, *158*:880 (1989).
250. R. E. Rennick, A. Loesch, and G. Burnstock, *Thorax*, *47*:1044 (1992).
251. P. Baluk, J. A. Nadel, and D. M. McDonald, *Am. J. Respir. Cell Mol. Biol.*, *8*:446 (1993).
252. M. Orlowski, J. Orlowski, M. Lesser, and K. H. Kilbuurn, *J. Lab. Clin. Med.*, *97*:467 (1981).
253. D. Burnett, J. Crocker, and R. A. Stockley, *Am. Rev. Respir. Dis.*, *128*:915 (1983).
254. D. Burnett and R. A. Stockley, *Clin. Sci.*, *68*:469 (1985).
255. M. C. A. Burleigh, A. J. Barrett, and G. L. Lazarus, *Biochem. J.*, *137*:387 (1974).
256. D. Burnett, M. Abrahamson, J. L. Devalia, R. J. Sapsford, R. J. Davies, and D. J. Buttle, *Arch. Biochem. Biophys.*, *317*:305 (1995).
257. J-M. Sallenave, A. Silva, M. E. Marsden, and A. P. Ryle, *Am. J. Respir. Cell Mol. Biol.*, *8*:126 (1993).
258. R. C. Thompson and K. Ohlsson, *Proc. Natl. Acad. Sci. USA*, *83*:6692 (1986).
259. H. Fritz, *Biol. Chem. Hoppe Seyler*, *369*:79 (1988).
260. J-M. Sallenave, J. Shulmann, J. Crossley, M. Jordana, and J. Gauldie, *Am. J. Respir. Cell Mol. Biol.*, *11*:733 (1994).
261. O. Wiedow, J. Lundemann, and B. Utecht, *Biochem. Biophys. Res. Commun.*, *174*:6 (1991).
262. L. R. DeChatelet, P. S. Shirley, L. C. McPhail, C. C. Huntley, H. B. Muss, and D. A. Bass, *Blood*, *50*:525 (1977).
263. F. Rossi, *Biochim. Biophys. Acta*, *853*:65 (1986).
264. A. Lopez, S. Shoji, J. Fujita, R. Robbins, and S. Rennard, *Am. Rev. Respir. Dis.*, *137*:A81 (1988).
265. K. B. Adler, V. L. Kinnula, N. Akley, J. Lee, L. A. Cohn, and J. D. Crapo, *Chest*, *101*:53S (1992).
266. V. L. Kinnula, K. B. Adler, N. J. Ackley, and J. D. Crapo, *Am. J. Physiol.*, *262*:L708 (1992).

267. J. Hobson, J. Wright, and A. Churg, *Am. J. Pathol.*, *139*:573 (1991).
268. V. L. Kinnula, J. I. Everitt, A. R. Whorton, and J. D. Crapo, *Am. J. Physiol.*, *261*:L84 (1991).
269. L. E. DeForge, A. M. Preston, E. Takeuchi, J. Kenney, L. A. Boxer, and D. G. Remick, *J. Biol. Chem.*, *268*:25568 (1993).
270. K. Asano, C. B. E. Chee, B. Gaston, C. M. Lilly, C. Gerard, J. M. Drazen, and J. S. Stamler, *Proc. Natl. Acad. Sci. USA*, *91*:10089 (1994).
271. P. W. Shaul, A. J. North, L. C. Wu, L. B. Wells, T. S. Brannon, K. S. Lau, T. Michel, L. R. Margraf, and R. A. Star, *J. Clin. Invest.*, *94*:2231 (1994).
272. H. H. Gutierrez, B. R. Pitt, M. Schwarz, S. C. Watkins, C. Lowenstein, I. Caniggia, P. Chumley, and B. A. Freeman, *Am. J. Physiol.*, *268*:L501 (1995).
273. P. M. Dupuy, S. A. Shore, J. M. Drazen, C. Frostell, W. A. Hill, and W. M. Zapol, *J. Clin. Invest.*, *90*:421 (1992).
274. M. Högman, C. G. Frostell, H. Hedenström, and G. Hedenstierna, *Am. Rev. Respir. Dis.*, *148*:1474 (1993).
275. P. J. Barnes and F. Y. Liew, *Immunol. Today*, *16*:128 (1995).
276. Q. Hamid, D. R. Springall, V. Riveros-Moreno, P. Chanez, P. Howarth, A. Redington, J. Bousquet, S. Godard, S. Holgate, and J. M. Polak, *Lancet*, *342*:1510 (1993).
277. S. A. Kharitonov, D. Yates, R. A. Robbins, R. Logan-Sinclair, E. A. Shinebourne, and P. J. Barnes, *Lancet*, *343*:133 (1994).
278. A. F. Massaro, B. Gaston, D. Kita, C. Fanta, J. S. Stamler, and J. M. Drazen, *Am. J. Respir. Crit. Care Med.*, *152*:800 (1995).
279. S. A. Kharitonov, D. Yates, and P. J. Barnes, *Am. J. Respir. Crit. Care Med.*, *153*:454 (1996).
280. J. S. Stamler, D. J. Singel, and J. Loscalzo, *Science*, *258*:1898 (1992).
281. A. W. Taylor-Robinson, F. Y. Liew, A. Severn, D. Xu, S. J. McSorley, P. Garside, J. Padron, and R. S. Phillips, *Eur. J. Immunol.*, *24*:980 (1994).
282. L. A. Cohn, V. L. Kinnula, and K. B. Adler, *Am. J. Physiol.*, *266*:L397 (1994).
283. V. L. Kinnula, J. R. Yankaskas, L. Chang, I. Virtanen, A. Linnala, B. H. Kang, and J. D. Crapo, *Am. J. Respir. Cell Mol. Biol.*, *11*:568 (1994).
284. V. L. Kinnula, L. Chang, J. I. Everitt, and J. D. Crapo, *Am. J. Physiol.*, *262*:L69 (1992).
285. B. A. Freeman, R. J. Mason, M. C. Williams, and J. D. Crapo, *Exp. Lung Res.*, *10*:203 (1986).
286. D. B. Jacoby and A. M. K. Choi, *Free Radical Biol. Med.*, *16*:821 (1994).

# 4

# Integrins, Integrin Regulators, and the Extracellular Matrix

## The Role of Signal Transduction and Leukocyte Migration

**Stephen W. Hunt III**   *Parke-Davis Pharmaceutical Research, Division of Warner-Lambert Company, Ann Arbor, Michigan*

**Sirid-Aimée Kellermann** and **Yoji Shimizu**   *Center for Immunology, University of Minnesota Medical School, Minneapolis, Minnesota*

## I.  INTRODUCTION

T lymphocytes require physical contact with the extracellular environment in order to develop, function, and migrate (1,2). These physical interactions involve transient adhesive contacts with other cells as well as with extracellular matrix (ECM) components (1,3). Consequently, the molecules that mediate T cell adhesion play an absolutely essential role in T cell recognition of foreign antigen and migration to specific anatomic sites in vivo. Integrins are a large family of αβ heterodimeric cell surface proteins that are important in a diverse array of biological processes that are dependent on cell–cell and cell–ECM interactions, including the immune response, embryogenesis, and tumor growth and metastasis (4,5). At least 21 different integrin heterodimers have been described and almost every cell type expresses at least one integrin receptor. It is clear that the integrins play a pivotal role in normal T-cell trafficking as well as recruitment of T cells to sites of inflammation and disease. Integrins were initially thought to be involved only in adhesion of T cells to target cells, endothelial cells, and ECM; it is now clear, however, that integrins, as well as other cell surface adhesion molecules, are capable of transducing signals from the extracellular to the intracellular compartments ("outside-in" signaling). Presumably these signals are integrated with other signals derived from the antigen-specific CD3/T cell receptor (CD3/TCR) complex and other accessory molecules to provide the cell with the signals to carry out its specific effector functions. It has also become clear that the ability of integrins to bind ligand is regulated, and that this regulation by activation through other cell surface receptors ("inside-out" signaling) plays an essential role in the trafficking and response of T cells in normal and disease situations.

In this chapter, we will discuss recent highlights of the role of integrins in T-cell trafficking and T-cell responses, with a particular emphasis on the signaling functions of integrins. We have attempted to focus on: 1.) adhesion molecules and mechanisms involved in the T-cell adhesion cascade and interaction with ECM; 2.) the concept of "inside-out" signaling and the components and pathways involved; 3.) the concept of "outside-in" signaling, the components and pathways involved, and the possible outcomes on this process. We apologize to our colleagues that limitations of space do not permit an exhaustive review of the field, nor complete citation of the literature. Instead, we have tried to describe recent reports that have provided particular insight and have given rise to general concepts in the field. Where possible we have cited relevant, recent reviews that the reader may acquire for additional information in this area.

## II.  STEPS IN THE ADHESION CASCADE

A distinguishing characteristic of leukocytes is their motility, a necessary feature for immune surveillance as well as migration to sites of tissue injury and inflammation. Lymphocyte trafficking, whether into lymphoid organs during the normal course of recirculation or into tissue during an immune response, involves a cascade of adhesion events during which leukocytes leave blood vessels and extravasate through the vascular endothelial lining into the underlying tissue. This process of arrest, diapedesis, and interstitial migration is dependent on the expression of specific adhesion and signaling molecules by leukocytes. We present a brief description of this process and the relevant molecules, and direct the reader to several recent reviews that provide a more in-depth description of the adhesion cascade and the spectrum of adhesion molecules implicated in leukocyte interaction with endothelium, epithelium, and ECM (2,6–8).

Lymphocyte trafficking can be divided into four discrete stages summarized in the widely accepted multi-step paradigm: 1.) tethering and rolling, 2.) activation, 3.) stable adherence, and 4.) extravasation (2,8).

In both lymphocytes and neutrophils, the initial phase of the adhesion cascade, tethering and rolling, is mediated to a large extent by the selectins. The selectins are a group of cell surface glycoproteins found on endothelium, platelets, and leukocytes (reviewed in 9–11). The three-member family includes E-selectin (CD62E, ELAM-1), P-selectin (CD62P, GMP-140), and L-selectin (CD62L). L-selectin is the key adhesion molecule involved in lymphocyte tethering and rolling on high endothelial venules (HEV), specialized postcapillary venules of lymphoid tissue, and is implicated as a homing receptor involved in lymphocyte migration into peripheral lymph nodes (PLN) (reviewed in 10). L-selectin is expressed on most peripheral blood leukocytes and binds to mucin-like glycoproteins, including GlyCAM-1 and CD34 expressed on PLN HEV. L-selectin also binds to mucosal addressin cell adhesion molecule-1 (MAdCAM-1) (12), which is expressed primarily on Peyer's patch HEV, mesenteric lymph node (LN) HEV, and mucosal lamina propria venules (13). MAdCAM-1 is a unique protein in that it contains both a mucin-like domain capable of binding L-selectin, and an immunoglobulin superfamily-like domain that can bind $\alpha4\beta7$ (12). Thus, MAdCAM-1 can support both lymphocyte rolling and integrin-mediated firm adhesion (see below). Additional members of the mucin family that are important in lymphocyte-endothelial interactions include the P-selectin glycoprotein ligand, PSGL-1, found on the surface of myeloid cells and activated lymphocytes, and the E-selectin ligand, CLA, which is selectively expressed on memory T cells that migrate to the skin (13,14).

The importance of L-selectin in T-cell tethering and rolling has been demonstrated in numerous in vitro and in vivo studies (15). However, studies have shown that the integrins $\alpha4\beta1$ and $\alpha4\beta7$ (15,16), as well as P- and E-selectin (17), can play a role in this initial adhesion step. Mice deficient in P-selectin display no leukocyte rolling (18,19). E-selectin deficiency alone does not affect rolling (20), but when both P- and E-selectin functions are blocked by either gene knockout (21) or a combination of gene knockout and blocking antibodies (20), the observed rolling impairment confirms a role for E-selectin in combination with P-selectin.

The second phase of the adhesion and transmigration process involves activation of T-cell adhesiveness by soluble factors, such as the chemokines. Chemokines such as monocyte chemoattractant proteins 1, 2, and 3 (MCP-1, 2, 3), macrophage inhibitory protein-1$\alpha$ and $\beta$ (MIP-1$\alpha$ and $\beta$), RANTES (regulated on activation normal T expressed and secreted), and eotaxin act primarily on monocytes, lymphocytes, eosinophils, and basophils. In addition to inducing increased adhesion, these peptide chemoattractants serve to direct the T cell to the site of tissue destruction or disease. The ECM may serve as a reservoir for such chemoattractants (22). Soluble factors such as chemokines may be immobilized and presented by proteoglycans in the ECM as well as on cell surfaces. This may be critical for providing a sufficiently high local concentration to activate a response by leukocytes (23).

Although chemokines have been proposed to be the major activating stimulus in the adhesion cascade, other receptor–ligand interactions up-regulate integrin activity and thus may play a role in this second step of the adhesion cascade. L-selectin engagement results in increased functional activity of $\beta2$ integrins on neutrophils (24). Engagement of CD31 on T cells, perhaps via CD31 or the $\alpha v\beta3$ integrin on endothelium, may play a role in activating $\beta1$ integrin receptors expressed on T cells and natural killer (NK) cells that are weakly tethered to the endothelial surface (25,26).

Stable arrest, the third step in the multistep paradigm, is mediated by activated integrins on the surface of the lymphocyte and immunoglobulin superfamily (IgSF) adhesion molecules on the endothelium. The increase in integrin-mediated leukocyte adhesiveness that is critical to the transition from tethering and rolling to stable arrest is accomplished without an increase in the numbers of integrin molecules on the surface of the lymphocyte.

The final step in the successful exit of lymphocytes out of the bloodstream into lymphoid organs or nonlymphoid tissue sites involves leukocyte penetration through the underlying basement membrane and into the surrounding interstitial stroma (27). Although the precise mechanism by which T cells carry out this process of diapedesis remains poorly defined, leukocytes express several functionally active receptors that can interact with major components of the ECM. It is clear that integrin:IgSF interactions, as well as CD31 homophilic interactions, are involved (2,26). Diapedesis toward a region of tissue damage is thought to be unidirectional. Thus, chemokines, or perhaps tethered cytokines and other soluble factors, may provide a gradient for lymphocytes to follow. Recent studies have clearly revealed that the ECM provides a reservoir for cytokines that modulate adhesion (22). It is possible that the chemokines involved in activation are used for establishing the chemotactic gradient.

## III. THE EXTRACELLULAR MATRIX AND LYMPHOCYTE MIGRATION

The ECM is a complex and highly organized mixture of fibrous proteins, such as collagens, laminin, fibronectin (Fn), and proteoglycans. Adhesion of leukocytes to the ECM is mediated by the interaction of leukocyte integrins and other adhesion molecules with ECM components (see Chap. 1). Resting peripheral T cells and monocytes express the $\alpha 4\beta 1$ and $\alpha 5\beta 1$ integrins, which have been shown to mediate activation-dependent T-cell adhesion to Fn (28). These two integrins bind to distinct sites on the Fn molecule, the type IIICS in the major cell adhesion domain (for $\alpha 4\beta 1$) and the arg-gly-asp (RGD) sequence (for $\alpha 5\beta 1$). The $\alpha 6\beta 1$ integrin is expressed on monocytes, platelets, eosinophils, and resting T cells, and mediates cell adhesion to laminin (28), a major component of the basement membrane. Although resting peripheral T cells do not express $\alpha 2\beta 1$, activation of T cells in vitro induces $\alpha 2\beta 1$ (5) and subsequent $\alpha 2\beta 1$-mediated binding to collagen (29). Other important integrins involved in leukocyte adhesion include the $\beta 3$ integrins $\alpha v\beta 3$ and $\alpha IIb\beta 3$, which are expressed on platelets and resting monocytes and mediate adhesion to vitronectin and fibrinogen, respectively. Finally, CD44 is expressed at high levels on a variety of leukocytes (30) and binds to hyaluronic acid (HA). This binding appears to be tightly regulated on resting T cells, for these cells express CD44 but do not bind to soluble or immobilized HA (30). ICAM-1, which is expressed at low levels on resting T cells as well as NK cells and resting monocytes, has also been reported to be a HA receptor (31). The physiological significance of ICAM-1-mediated cell adhesion to HA remains to be established.

Although numerous *in vitro* studies have shown that leukocytes express multiple functional ECM receptors, the physiological relevance of leukocyte adhesion molecule–ECM interactions during processes such as diapedesis is not as well established. Inhibition of T cell adhesion to, and migration through, cultured high EC can be achieved *in vitro* using specific antibodies and peptides containing the relevant $\alpha4\beta1$ and $\alpha5\beta1$ binding sequences on Fn (CS-1 and RGD respectively) (32–34). Similar treatments can inhibit T cell–mediated immune responses *in vivo* (35–38). These blocking reagents appear to inhibit T cell migration rather than T cell activation, although the precise mechanism of trafficking inhibition remains difficult to establish in these models.

Extracellular matrix components may promote leukocyte migration not only by presenting soluble chemoattractants, as described above, but also by serving as chemoattractants themselves. Studies with several T-cell lines and phorbol myristate acetate (PMA)-stimulated T cells have shown both haptotactic and chemotactic migration of these cells on Fn, collagen type IV, and laminin (39). Locomotion of T cells in a three-dimensional collagen matrix results in the generation of a novel T-cell population that expresses $\alpha2\beta1$, an integrin that is not typically expressed on human peripheral T cells (40). These results indicate that T-cell interactions with the ECM affect differentiation-induced expression of adhesion molecules.

## IV. "INSIDE-OUT" SIGNALING

The multistep adhesion cascade model predicts that successful movement of leukocytes from the bloodstream into tissue sites requires a signaling event that rapidly and transiently up-regulates the functional activity of integrin receptors. This activation-dependent up-regulation of integrin activity is an important regulatory event in leukocyte function, because integrins on circulating leukocytes are normally not able to bind effectively to endothelial cells or ECM. The rapidity of this "inside-out" signal ensures that leukocytes can quickly and appropriately modify their adhesiveness in response to stimuli in the external environment. For example, stimulation of T cells with monoclonal antibodies (mAbs) specific for the CD3/TCR complex results in a rapid up-regulation of $\beta2$ integrin–mediated adhesion to ICAMs, as well as $\beta1$ integrin–mediated adhesion to ECM components (28,41,42). Such antigen-specific up-regulation of integrin function is necessary in order to enhance the bidirectional delivery of signals between T cells and antigen-presenting cells that occurs during a T cell–mediated immune response. Furthermore, activation-induced increases in T-cell adhesion to the ECM ensures that these cells are retained at the site of an ongoing immune response.

The phenomenon of inside-out signaling involves rapid changes in integrin functional activity that do not require changes in integrin expression on the cell surface. The signaling pathways and individual components involved in inside-out signaling remain poorly characterized. In this section, we will review recent progress in understanding inside-out signaling, beginning with the concept of integrin regulators and signaling pathways, and ending with changes in the integrin itself that might lead to increased functional activity.

## A.  Integrin Regulators

The complexity of inside-out signaling is illustrated by the identification of multiple activation stimuli that can up-regulate integrin functional activity. These activation signals can be grouped as pharmacologic agents or receptor-mediated signals. Treatment of T cells with either the phorbol ester PMA or the $Ca^{2+}$ ionophore A23187 has been shown to up-regulate integrin-mediated T-cell adhesion (28,41,43,44). This suggests that both protein kinase C (PKC) activation and changes in intracellular $Ca^{2+}$ are involved in the intracellular signaling events that upregulate integrin activity. More significantly, mAb-mediated cross-linking of a litany of cell surface receptors can induce an inside-out signal. Such receptors are designated "integrin regulators" in this review. Leukocytes express a wide array of integrin regulators on the cell surface. Using T cells as an example, ligation of a number of cell surface receptors can result in up-regulation of $\beta 1$ and $\beta 2$ integrin functional activity. T-cell integrin regulators include the CD3/TCR complex, as well as the accessory molecules CD2, CD28, and CD7 (41–43). CD2, CD7, and CD28 are all IgSF members. The CD2 antigen is a 45–55 kD protein that was initially identified as an important signaling molecule on human T cells and NK cells (45,46). Subsequently, CD2 was shown to mediate T-cell adhesion by binding to its counter-receptor, LFA-3 (47). The CD28 antigen is a 44 kD molecule that has been extensively studied because of its importance in delivering a signal to T cells that is believed to be critical to the prevention of T-cell tolerance upon encounter with antigen (48). At least two distinct CD28 counter-receptors, B7-1 and B7-2, have been identified and there has been extensive analysis of CD28-mediated signaling as it relates to T-cell proliferation and cytokine production (48). CD7 is a 40 kD antigen that is expressed on T cells and NK cells. The function of CD7 is unknown at present. All three of these integrin regulators facilitate CD3-mediated T cell proliferation, suggesting that inside-out signaling mediated by these molecules is particularly important in enhancing T cell responses to foreign antigen.

As discussed above, receptors for chemokines and other soluble factors are believed to be a particularly critical class of integrin regulators for leukocyte interactions with endothelium (2). The chemokines MCP-1, MIP-1$\alpha$, MIP-1$\beta$, RANTES, and interferon (IFN)-inducible protein-10 (IP-10), as well as hepatocyte growth factor, have all been shown to induce up-regulation of $\beta 1$ and/or $\beta 2$ integrin-mediated adhesion of T cells to relevant counter-receptors and ECM ligands (49–52). Chemokine receptors belong to the seven-membrane-spanning family of G protein–coupled receptors (53). Chemokine-mediated inside-out signaling is believed to be particularly rapid in nature, inducing integrin activity within minutes of receptor engagement. Studies of mast cells suggest that mitogenic growth factors, such as stem cell factor and platelet-derived growth factor (PDGF), can also be classified as integrin regulators (54–57). Thus, other growth factor receptors that initiate the same downstream signaling pathways as the PDGF receptor may also function as integrin regulators.

Increases in integrin activity can also be induced by altering the divalent cation milieu, as well as with certain "activating" antibodies that recognize integrin α and β subunits (58). These two modes of activating integrins are believed to bypass the requirement for inside-out signaling by directly altering the affinity of the integrin for ligand (see further discussion below).

## B. Inside-Out Signal Transduction Pathways

Although there has been considerable progress in defining the various ways in which T-cell integrin activity can be modulated, the biochemical mechanisms by which activation signals up-regulate integrin activity remain poorly defined. Studies with primary T cells have utilized various pharmacological inhibitors to implicate tyrosine kinases, PKC, protein kinase A, and cytoskeletal changes in integrin up-regulation mediated by integrin regulators (41,43,59,60). The activation-dependent conversion from rolling to stable, integrin-mediated adhesion under conditions of shear flow can be inhibited by pertussis toxin, which inhibits G protein–mediated signal transduction (61). This finding implicates a G protein–coupled seven-membrane-spanning receptor as the critical integrin regulator in the multistep adhesion cascade. Chemokines bind to this class of cell surface receptor (53), thereby indirectly implicating these soluble factors as the critical activation "triggers" in the adhesion cascade. Recent studies with C3 transferase, an exoenzyme that ADP ribosylates the rho subfamily of Ras-related GTP-binding proteins, have also implicated rho in inside-out signaling mediated by the fMLP and IL-8 receptors (62). Inhibition of PMA-mediated up-regulation of β1 integrin function in HL-60 myelomonocytic cells by a recently described inhibitor of MAP kinase kinase or MEK, suggests a role for the Ras/Raf/MEK/MAP kinase signaling cascade in inside-out signaling (63). Interestingly, this same inhibitor did not block CD2-mediated inside-out signaling in the same cell type, suggesting heterogeneity in integrin regulator signaling pathways (63).

Recent studies of inside-out signaling in several different immune cell systems have implicated the lipid kinase phosphatidylinositol 3-kinase (PI 3-K) as a pivotal mediator of the inside-out signal generated by several different integrin regulators. PI 3-K is an intracellular enzyme that phosphorylates the D-3 position of the inositol ring of PI, PI-4-phosphate, and PI-4,5-bisphosphate (Fig. 1). PI 3-K is activated by a wide range of growth factors, oncogenes, and other nonmitogenic stimuli and has been implicated in a number of cellular functions, including growth factor–dependent mitogenesis, membrane ruffling, and glucose uptake (64). Structurally, PI 3-K consists of two subunits, p110 and p85. The p110 subunit is the catalytic subunit that contains a lipid kinase domain (65). The p85 subunit lacks kinase activity but contains several amino acid motifs (SH2, SH3, and proline-rich) that allow for the association of PI 3-K with cytoplasmic receptors to form signaling complexes (66). Different isoforms of both p110 and p85 have been identified, suggesting further complexity and specificity in PI 3-K–depen-

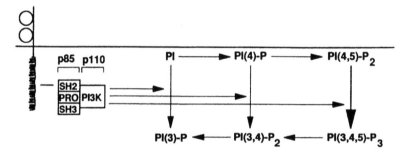

**Figure 1** Production of 3-phosphorylated lipids by PI 3-K. The PI 3-K complex illustrated consists of an 85-kD regulatory subunit (p85) and a 110-kD catalytic subunit (p110). The p85 subunit contains an amino-terminal SH3 domain, two SH2 domains, two proline-rich sequences, and a region homologous to the C-terminal part of the Breakcluster region (Bcr) gene product. The SH2 and SH3 domains of p85 have been implicated in mediating protein–protein interactions. Association of p85 with tyrosine phosphorylated sequences in the cytoplasmic domains of several integrin regulators, including CD28, CD7, and the platelet derived growth factor (PDGF) receptor, has been demonstrated. The p85 subunit has also been shown to associate with CD2, although the complete lack of tyrosine residues in the CD2 cytoplasmic domain suggests that this interaction is not mediated by the SH2 domains in the p85 subunit. The p110 subunit contains enzymatic activity and phosphorylates the D-3 position of the inositol ring to generate 3-phosphorylated lipids. $PI(3,4,5)$-$P_3$ is believed to be the most physiologically relevant lipid product generated by PI 3-K. The 3-phosphorylated lipids can act directly as effectors to initiate downstream signaling cascade (see text for details). In addition, the presence of protein binding motifs in the p85 subunit suggest that the p85–p110 complex is capable of associating with and recruiting additional signaling molecules that would initiate signaling cascades independent of the lipid kinase activity of PI 3-K. Multiple isoforms of both p85 and p110 have been isolated, although the significance of this structural heterogeneity remains to be elucidated. A novel p110 subunit that does not associate with p85 and is sensitive to stimulation by G proteins has also been identified and is not shown in the illustration.

dent signaling. However, the true functional significance of the heterogeneity in the PI 3-K family has not been established.

There are several lines of evidence implicating PI 3-K in inside-out signaling events that up-regulate integrin function. First, the PI 3-K signaling pathway has been implicated in several cellular responses that would be expected to have an impact on integrin function, notably agonist-induced actin polymerization and membrane ruffling (64). Second, several integrin regulators associate with and activate PI 3-K. The cytoplasmic domains of both the CD7 and CD28 antigens contain the SH2 binding motif YXXM, which can mediate protein–protein interactions when the tyrosine residue is phosphorylated (66). SH2-dependent association of PI 3-K with the YMNM sequence in the cytoplasmic domain of CD28 has been demonstrated by several groups (67). CD7 stimulation also results in the association and activation of PI 3-K in the Jurkat T cell line (68). Treatment of mast cells with stem cell factor results in increased $\beta$1 integrin activity (54,56,57). The stem cell factor receptor, the c-kit receptor, has also been shown to associate with and activate PI 3-K. Although the CD2 antigen lacks cytoplasmic tyrosine residues, CD2 stimulation has been reported to result in the accumulation of D-3 phosphorylated lipids (69). In addition, constitutive association of CD2 with PI 3-K has been reported in both human T cells and in HL-60 cell transfectants expressing human CD2 (70). Although PI 3-K is constitutively associated with CD2, PI 3-K becomes enzymatically active only on CD2 stimulation. The mode of association and activation of PI 3-K with CD2 clearly differs from the more conventional tyrosine phosphorylation–dependent association via SH2 domains.

Third, the fungal metabolite wortmannin, a relatively specific inhibitor of PI 3-K, can inhibit inside-out signaling mediated by several different integrin regulators. Wortmannin was initially found to inhibit $\alpha$IIb$\beta$3-dependent aggregation of platelets induced by thrombin (71). In recent studies, human CD2 and a CD2 chimeric receptor containing the CD28 cytoplasmic domain (CD2/28) were expressed in HL-60 cells, which express $\beta$1 integrins that can be rapidly up-regulated by PMA treatment. Stimulation of CD2 or CD2/28 resulted in rapid increases in $\beta$1 integrin–mediated adhesion that could be completely inhibited by wortmannin (70,72). The relevance of these findings for our understanding of integrin regulator function on human T cells is suggested by the fact that wortmannin can also inhibit CD2- and CD28-dependent up-regulation of $\beta$1 integrin function in peripheral T cells (70 and T. Zell, J. L. Mobley, and Y. Shimizu, unpublished data).

Fourth, the wortmannin inhibition studies have been complemented by studies demonstrating that abrogation of integrin regulator association with PI 3-K by site-directed mutagenesis of cytoplasmic tails results in a loss of inside-out signaling. These are important extensions of the wortmannin studies, because recent evidence suggests that wortmannin may inhibit PI 4-kinase and phospholipase A$_2$

as well as PI 3-K (67). In HL-60 cell transfectants, cross-linking of a CD2/28 chimera containing a tyrosine-to-phenylalanine substitution in the YMNM motif does not result in PI 3-K association and also does not induce increases in β1 integrin-mediated adhesion to FN (72). Coupled with the inhibitory effects of wortmannin, these results clearly suggest that CD28-mediated inside-out signaling in this system requires activation of PI 3-K. However, because the Grb2 adapter protein has also been reported to associate with the YMNM motif in the CD28 cytoplasmic domain (73), a role for a PI 3-K-independent pathway of integrin regulation by CD28 cannot be formally excluded.

Inside-out signaling mediated by PI 3-K can also be demonstrated upon expression of the PDGF receptor in mast cells (55). Treatment of these transfectants with PDGF resulted in increases in mast cell adhesion to Fn that were comparable to that observed with stem cell factor treatment. Analysis of PDGF receptors mutated in specific autophosphorylation sites that mediate protein–protein binding demonstrated that induction of adhesion by this receptor is mediated by two independent signaling pathways, one involving PI 3-K and another involving phospholipase Cγ1 and PKC. Cytoplasmic domain mutations that abrogated association of the PDGF receptor with either PI 3-K or phospholipase Cγ1 were insufficient to inhibit PDGF-induced integrin function. However, the effect of these mutations on integrin regulation could be demonstrated by inhibiting the other pathway, such as with prolonged PMA treatment to down-regulate PKC or with the use of wortmannin to inhibit PI 3-K. PDGF was not able to up-regulate adhesion of mast cells expressing PDGF receptors with mutations in both the PI 3-K and phospholipase Cγ1 association sites. Thus, while studies of CD2 and CD28 suggest that these integrin regulators rely exclusively on PI 3-K to mediate inside-out signaling (70,72), the PDGF receptor utilizes two independent pathways, either of which by itself is sufficient to mediate the inside-out signal. A role for phospholipase Cγ1 in receptor-mediated integrin regulation is not surprising, given the potent effects of phorbol esters on up-regulating integrin activity. Mutation of the PI 3-K binding site in the c-kit receptor also affects the ability of this receptor to up-regulate β1 integrin function upon stem cell factor treatment (56). However, unlike the PDGF receptor studies, there was a partial inhibition of agonist-induced inside-out signaling when the PI 3-K binding site was eliminated. It is not clear from this study whether PI 3-K can associate with other sites in the c-kit cytoplasmic domain (and therefore still transmit an inside-out signal) or whether other signaling effector molecules are involved in c-kit-mediated integrin regulation. Nevertheless, these studies again illustrate a role for PI 3-K in inside-out signaling mediated by a cell surface receptor.

The role of PI 3-K in inside-out signaling mediated by seven-membrane-spanning chemokine receptors is not well established. However, several indirect lines of evidence suggest that PI 3-K may play a role in G protein–sensitive inside out-signaling. First, wortmannin has been shown to inhibit T-cell migration induced by the chemokine RANTES (74). Second, activation via the G protein–coupled receptors for fMLP and thrombin, both of which up-regulate integrin activity, also activates PI 3-K (75,76). Finally, both the α and βγ subunits of G proteins have been shown to induce PI 3-K activity (77–79). Some studies suggest that G proteins activate "conventional" p85/p110 PI 3-K (78). However, the potential unique nature of G protein–sensitive induction of PI 3-K activity is suggested by the identification of a novel PI 3-K isotype, designated p110γ, that does not interact with the p85 subunit and can be activated by G proteins (79).

These studies suggest that the PI 3-K signaling cascade is likely to be a common mechanism by which different cell surface receptors can converge to regulate β1 integrin activity upon stimulation. Although the signaling pathway(s) initiated by PI 3-K activation remain poorly defined, several downstream molecular targets of PI 3-K have recently been identified. PI 3-K activation mediated by platelet-derived growth factor (PDGF), insulin, and the T-cell growth factor IL-2 results in the activation of the 70/85K S6 kinases (pp70$^{S6k}$)(80,81). This kinase has been implicated in the regulation of progression through the G1 phase of the cell cycle (82). Activation of pp70$^{S6k}$ can be regulated by a PI 3-K–dependent pathway, as well as a PKC-dependent pathway. The immunosuppressant rapamycin is a specific inhibitor of pp70$^{S6k}$. Although the role of this kinase in PI 3-K–dependent regulation of integrin regulation requires further analysis, rapamycin has not been found to inhibit CD28-mediated inside-out signaling in HL-60 cells (T. Zell and Y. Shimizu, unpublished observations).

Protein kinase C has also been implicated in the PI 3-K signaling cascade. The lipid products generated by PI 3-K have been shown to activate the PKC isozymes δ, ε, ζ, and λ *in vitro* (83–86). Activation by D-3 phosphorylated lipids of conventional PKC isozymes, such as α and β, was not observed. These results suggest that PKC may play a role in PI 3-K–dependent activation of integrin activity. This model is consistent with studies demonstrating that PKC inhibitors can partially inhibit CD2- and CD28-mediated inside-out signaling (70,72). However, a direct role for PKC as an effector downstream of PI 3-K in the integrin regulatory signaling pathway has yet to be established.

Activation of PI 3-K has also been reported to activate the serine-threonine kinase encoded by akt, the cellular homologue of the viral oncogene v-akt (87,88). Although the function of Akt has yet to be determined, this kinase does have structural homology with PKC. A role for Akt in PI 3-K–dependent inside-out signaling that regulates integrin-mediated adhesion is unknown.

Analysis of inside-out signaling in several cultured human cell lines has also revealed currently undefined defects in this signaling event. Although the Jurkat T cell line expresses high levels of the LFA-1 integrin, adhesion assays reveal no adhesion of Jurkat cells to ICAM-1 (89,90). Furthermore, PMA stimulation is incapable of increasing LFA-1 activity, even though PMA stimulation up-regulates the functional activity of β1 integrins on the same cell (89). However, divalent cation conditions that increase the affinity of LFA-1 are capable of increasing LFA-1–mediated adhesion of Jurkat cells to ICAM-1 (89). This suggests that there is not a structural defect in the LFA-1 integrin that prevents it from mediating adhesion. The ability of PMA to selectively up-regulate $\beta1$ integrin activity on Jurkat cells also implies some heterogeneity in inside-out signaling to $\beta1$ versus $\beta2$ integrins. The nature of the defect in LFA-1 function in Jurkat cells remains undefined, although there are similar defects in LFA-1 activity in other T-cell lines as well (89).

A genetic approach has also been employed to elucidate the mechanisms by which activation upregulates $\beta1$ integrin activity (91). Mutants of the Jurkat T cell line were isolated by $\gamma$-irradiation and selection for cells that were unable to bind to Fn-coated plates following stimulation with PMA or anti-CD3 mAbs. These mutant cell lines expressed normal levels of $\beta1$ integrins. Furthermore, the ability of these mutants to bind to Fn and VCAM-1 following direct stimulation of $\beta1$ integrins with an activating $\beta1$-specific mAb suggested that the mutation induced in these cells did not alter the structural integrity of the ligand-binding domain. Genetic complementation studies indicated that at least three genetically distinct mutant types were isolated. The defect in $\beta1$ integrin function in one of these mutant types was also associated with the expression of an altered form of the MAPK isoform ERK-1 and defective production of IL-2 following CD3 stimulation. While the precise nature of the defects in $\beta1$ integrin function in these mutant cells remains to be elucidated, early signaling events that impact integrin functional activity, such as tyrosine kinase activity, PKC activity, and intracellular $Ca^{2+}$ flux, appear normal.

## C.  Inside-out Signaling and the Integrin

Recent data indicate that inside-out signals generated by integrin regulators stimulate integrin adhesiveness via at least two mechanisms: an alteration in individual integrins that leads to increased affinity (as detected by binding studies with soluble integrin ligands), and redistribution of integrin receptors on the cell surface that result in increased avidity.

"Integrin activating" antibodies, as well as divalent cations such as $Mg^{2+}$ and $Mn^{2+}$, are thought to augment adhesion by altering the affinity of the integrin. Faull et al. (92) showed that the $\beta1$-specific activating antibody 8A2 induces T-cell adhesion to immobilized Fn, but did not induce cell spreading. 8A2 treatment resulted in increased binding of soluble fibronectin to the T cells, indicating that this antibody caused a change in integrin affinity. That 8A2 operates by increasing $\alpha4\beta1$ affinity is supported by observations that cytochalasin D, which disrupts actin filaments, did not inhibit 8A2-stimulated adhesion, suggesting that the cytoskeleton is not involved in this process. In another study that investigated divalent cation-mediated up-regulation of integrin activity (93), $Mg^{2+}$ induced increased LFA-1 adhesion to both soluble and immobilized ICAM-1 as well as increased binding of mAb 24, which recognizes an activation epitope on LFA-1 and is an indicator of the high affinity state. $Mg^{2+}$ increased the affinity of LFA-1 because the appearance of the mAb 24 epitope was not inhibited by cytochalasin D, and $Mg^{2+}$ mediated LFA-1 activation did not lead to cell spreading.

In contrast, PMA also induces adhesion of T cells to Fn, but does so by increasing integrin avidity, as evidenced by increased cytochalasin D–sensitive cell spreading and adhesion to immobilized Fn following PMA stimulation (92). Activation of LFA-1 adhesion to ICAM-1 on T cells by another phorbol ester, phorbol dibutyrate (PdBu), had similar characteristics (93). Furthermore, this increased adhesion was not accompanied by altered binding of mAb 24, suggesting that LFA-1 remained in a low affinity state. No increase in integrin affinity, as measured by binding of soluble ligand, was detectable in either of these studies following phorbol ester treatment (92,93). Taken together, these studies indicate that activation through $Mg^{2+}$ or an activating antibody, both of which act on extracellular portions of the integrin, augment adhesion to ligand by increasing integrin affinity. On the other hand, phorbol esters increase adhesion by altering integrin avidity. In contrast to these findings, Lollo et al. (94) have shown that PMA is capable of inducing high affinity in 15% to 30% of murine LFA-1 molecules.

Stimulation through the CD3/TCR complex by anti-TCR mAb cross-linking in vitro has been shown to lead to increased $\alpha4\beta1$- and $\alpha5\beta1$-mediated adhesion to soluble or immobilized Fn, without any evidence of cell spreading (92). These data suggest that activation of the TCR leads to an increase in integrin affinity, and not avidity. On the other hand, Jakubowski et al. showed that cross-linking CD3 did not result in the generation of high affinity $\alpha4\beta1$ on T cell blasts (95). Further evidence that CD3 activation may affect the avidity and not the affinity of integrins was presented by Stewart et al.(93). In their hands, cross-linking the of the CD3/TCR complex led to increased LFA-1 adherence to immobilized but not soluble ICAM-1, as well as cell spreading. This indicates that LFA-1 remains in a low-affinity state upon TCR-mediated signaling. The different conclusions of these studies have not been resolved.

Carr et al. (52) recently examined the effects of chemokine stimulation on integrin activity in T cells. Stimulation with the chemokine MCP-1 resulted in increased cytochalasin D–sensitive adherence to Fn as well as endothelial-secreted ECM under shear flow conditions. Interestingly, unstimulated T cells exhibited a high level of basal binding to the purified endothelial ligand VCAM-1 or to activated human umbilical vein endothelial cells, which was not augmented by MCP-1 stimulation. These results suggest that $\alpha4\beta1$ is in an activated state with respect to VCAM-1 binding. These data support a model in which $\alpha4\beta1$ exists in at least three different activation states: an inactive state, a partially active state that binds VCAM-1 but not Fn, and a fully active state that binds both VCAM-1 and Fn (95,96). α4β1 on peripheral blood T cells may be in the partially active state and thus can bind VCAM-1, which (in addition to selectin:mucin interactions) could be important for initial binding to vascular endothelium (2). Following stable adhesion and diapedesis, MCP-1 could activate the T cells, resulting in increased adherence to ECM via α4β1 and α5β1, allowing migration through tissue.

Interestingly, MCP-1 stimulation did not increase LFA-1 binding of T cells to immobilized ICAM-1, even though PMA was able to induce increased binding to ICAM-1 (52). Therefore, β1 and β2 integrins may require distinct activation signals to interact efficiently with their ligands, or may have different sensitivities to chemokine induction. Furthermore, the data suggest that MCP-1 does not play a role in stable adherence or diapedesis of T cells. In contrast, Lloyd et al. (51) have shown that MIP-1α, MIP-1β, RANTES, and IP-10 induced rapid β1 and β2 integrin-mediated adhesion to immobilized recombinant human ICAM-1 and VCAM-1, as well as ECM in a rapid and dose-dependent manner.

The signal transduction cascades generated by integrin regulators lead, ultimately, to the physical association of signaling and cytoskeletal elements with the integrin subunits, thereby inducing either a conformational change in the integrin (resulting in increased affinity) or integrin clustering (increased avidity). Analysis of this final step in integrin regulation has concentrated on investigating sequence requirements within the α and β cytoplasmic tails.

The β subunit cytoplasmic tail is important in determining ligand binding function (97) (Fig. 2). Most β subunits contain an NPXY motif, which O'Toole et al. showed to be important in generating the high-affinity binding state of β1 and β3 integrins expressed in CHO cells (98). In particular, the asparagine and proline residues appeared to be critical. Conservative mutations of the tyrosine in this motif, as well as other potential carboxy-terminal phosphorylation sites, had little or no effect on adhesion, arguing that phosphorylation of β cytoplasmic tails does not play a role in converting integrins to an active state.

The importance of the NPXY motif was further suggested by LaFlamme et al. (99), who showed that when the extracellular portion of a nonintegrin protein (CD25) was joined to either the β1, β3, or β5 cytoplasmic domains, these chimeras localized to focal adhesion complexes (FACs) independent of any α subunit association (99). However, a chimera containing the β3 splice variant β3b, which lacks the NPXY and NPIY motifs, did not localize to FACs. In contrast to the results presented by O'Toole et al. (98), these data seem to indicate that NPXY motifs are relevant for integrin clustering and hence increasing avidity (Fig. 2). The expression of CD25/β1 or CD25/β3 chimeras in CHO cells reduces endogenous integrin avidity, arguing that these chimeras compete for limiting β subunit–binding factors (100).

Additional domains of the β cytoplasmic tail may be relevant in modulating the avidity of integrins. Mutations of the β2 cytoplasmic tail that delete or destroy the putative α-helical structure of the sequence 734–741 completely abrogate cytoskeletal association (97) (Fig. 2). This domain includes most of the sequence REYRRFEKEK, which is highly conserved among various integrin β subunits. Mutants that retain this motif but delete the C-terminal sequence, which includes two NPXF motifs, are impaired in their ability to associate with cytoskeletal elements. Thus, cytoskeletal associations appear to involve several domains throughout the β cytoplasmic tail. In addition, an earlier study showed that β2-deficient lymphoid cells transfected with a β2 mutant truncated at residue 732 were unable to bind to ICAM-1. An even shorter truncation removing only the 5 carboxy-terminal residues was similarly defective (101) (Fig. 2). Cumulatively, these studies indicate a correlation between β cytoplasmic domains necessary for 1.) cytoskeletal association and 2.) increase in integrin binding avidity.

Cytoskeletal elements may also affect the affinity of an integrin. The association of the LFA-1 β2 subunit with vinculin and α-actinin has been reported to correlate with conversion to the high-affinity state (59), an observation that in theory might apply to other β2 integrins as well. Results from another study (97) confirmed the association of these two cytoskeletal elements with LFA-1 expressed in COS-7 cells, but suggested that the interaction was constitutive, as it was not augmented by phorbol ester stimulation (Fig. 2).

The α subunit is thought to have a regulatory function in inside-out signaling (reviewed in 58). Integrin α subunits share little sequence homology beyond a highly conserved membrane proximal GFFKR motif, which has been implicated in stabilizing the αβ heterodimer. The GFFKR motif may be responsible for transducing intracellular signals into conformational changes in the extracellular domains of α and β subunits, leading to changes in affinity for ECM ligands. Deletion of the GFFKR motif in αIIb locks αIIbβ3 into a high-affinity state (102). Furthermore, deletion of the β3 membrane-proximal residues 717–723 (which are presumably juxtaposed with the α subunit sequence GFFKR) has the same effect (103). The significance of α cytoplasmic sequence diversity is emphasized by observations that different α cytoplasmic tails confer different levels of affinity upon a chimeric integrin (102).

Few α subunit-associated proteins have been characterized to date. Recent work found that activation of α2β1 integrin with phorbol ester or stimulatory mAbs led to the association of calreticulin with the α2 GFFKR sequence; calreticulin may serve to stabilize the high affinity state of the integrin (104).

Additional molecules are no doubt involved in up-regulating integrin activity by interaction with integrins. For example, a novel lipid, identified as integrin modulating factor (IMF), has been shown to enhance the binding of Mac-1 (105). This lipid might interact directly with Mac-1, as it is also able to increase the binding affinity of purified Mac-1.

Another protein that associates with β2 integrins is CD87, the GPI-anchored

urokinase plasminogen activator-receptor (106). CD87 is expressed on monocytes, activated T cells, and granulocytes and plays a role in ECM migration by
activating bound uPA, which in turn digests ECM plasminogen. However, CD87
also appears to be capable of generating intracellular signals important in adhesion and chemotactic migration. In monocytes, this protein was found in a multi-
molecular complex that also included β2 integrins (LFA-1 or Mac-1) and
members of the Src family of tyrosine kinases such as Fyn, Lck, Hck, and Fgr
(106). The intimate association of CD87 with integrins was demonstrated by the
codistribution of CD87 with Mac-1 upon Mac-1 capping, and the comodulation
of CD87 with β2 integrins.

Members of the 4-transmembrane protein superfamily (TM4), such as CD9,
CD63, and CD81, have been colocalized with certain β1 and β3 integrins
(107–109). TM4:integrin interaction appears to involve extracellular regions
(107). Expression of CD9 on an early B-cell line enhances motility, but not adhesion, of this cell line on Fn and laminin (109). This activity was blocked by antibodies against either CD9 or the β1 integrin subunit, as well as the tyrosine
kinase inhibitor herbimycin A. Potential biochemical functions of TM4 proteins
might include ion transport or as adapter proteins that can couple to G proteins
(107).

## V. INTEGRIN-MEDIATED SIGNAL TRANSDUCTION: THE "OUTSIDE-IN" SIGNAL

As described above, adhesion molecules play a key role in cell trafficking
through the body. It is clear that different families of adhesion molecules are involved in primary and secondary adhesion events and that engagement of the
CD3/TCR complex and other cell surface receptors can lead to T-cell activation
and increased integrin-mediated adhesion. Concomitant with receptor cross-linking is the rapid phosphorylation of a number of cellular targets and activation of
several protein tyrosine kinases. Although adhesion molecules were initially
thought to be involved only in cell–cell and cell–ECM interaction, evidence for
adhesion molecule–mediated signal transduction in lymphocytes was suggested
by the fact that engagement of α4β1 and α5β1 integrins could provide costimulatory signals to human T cells (110–112). It is now clear that adhesion molecules
are capable of providing additional signals that synergize with the signals induced by TCR ligation and amplify downstream effector functions in T cells. Indeed, recent reports have shown that integrin engagement alone is capable of
transducing signals and that phosphorylation is involved in this signaling cascade
(see below).

A significant amount of work has been accomplished in studies of integrin signaling in non-T cells, especially platelets and fibroblasts. Many of the proteins
and pathways involved in this process have been elucidated, although the molec-

|  | *1a* ———  *1b* ——————  *2b* | *2a* —————— |  |  |  |
|---|---|---|---|---|---|
| β1 | KLLMI IHDRREFAKFEKEKMNAKWDTGE | NPIY | KSAVTTVV | NPKY | EGK |
|  | *4a* ———————— |  |  |  |  |
|  | *3.4b* ———————— |  |  |  | *5* |
| β2 | KALIHLSDLREYRRFEKEKLKSQWNND | NPLF | KSATTTVM | NPKF | AES |
|  | 7* * * | 6 | 8* | * |  |
| β3 | KLLITIHDRKEFAKFEEERARAKWDTAN | NPLY | KEATSTFT | NITY | RGT |
| β7 | RLSVEIYDRREYSRFEKEQQQLNWKQDS | NPLY | KSAITTTI | NPRF | QEADSPT |

1. Binding of FAK and paxillin to β1 localizes to the 13 membrane-proximal amino acids (1a). Substitution of alanine residues for D and E in this region, which are conserved in all β subunits shown, eliminates association of FAK and paxillin. α-actinin binds a distinct region (1b) (119).

2. β1 mutants lacking the 13 C-terminal amino acids fail to colocalize talin, actin, or FAK (2a), while α-actinin colocalization depends on a different region (2b) (121).

3. Deletion of the β2 amino acids $E^{734}$-$E^{741}$, or substitution of glycines for the arginine and glutamic acid residues contained in this sequence, eliminates cytoskeletal association: deletion of the cytoplasmic tail distal to residue $E^{741}$ greatly impairs cytoskeletal association (97).

4. Filamin associates with the N-terminal portion of the β2 cytoplasmic domain (4a), but binds to a region distinct from α-actinin (4b) (126).

5. Deletion of even the five most C-terminal residues of β2 abrogates ICAM-1 binding of LFA-1 in β2-deficient lymphoid cells (101).

6. Substitution of A for $N^{744}$ or $Y^{747}$ in β3 abrogates αvβ3-mediated binding to immobilized vitronectin; deletion of residues C-terminal to $T^{741}$ results in loss of chemo-and haptotaxis (124).

7. Residues marked with an asterisk affect binding of a mAb specific for the active conformation of β3 (98)

8. Binding of β3-endonexin is β3-specific; mutation of $S^{752}$ to P greatly diminishes this interaction (127).

**Figure 2** The sequences of the integrin β subunit cytoplasmic tails are highly conserved. The β subunits relevant to leukocyte adhesion are shown. The NXX(Y/F) motifs are boxed. Overlined sequences or residues marked with an asterisk (*) have been shown to be important in association of signaling or cytoskeletal elements. The numbers refer to the notes listed below, which describe the nature of the interaction and the relevant citations.

ular mechanisms remain ill-defined. In this section, we will briefly summarize the known mechanisms of adhesion molecule signaling in non–T cells and then focus on data from studies on T cells.

## A. The Role of Integrin Cytoplasmic Domains in Outside-In Signaling

Outside-in signaling is initiated by the integrin β subunit cytoplasmic domain-de-pendent rearrangement of cytoskeletal components and actin into organized structures, known as focal adhesion complexes (FACs), that are found in the area of the cell membrane interacting with the ECM (Fig. 3). Integrins are the major

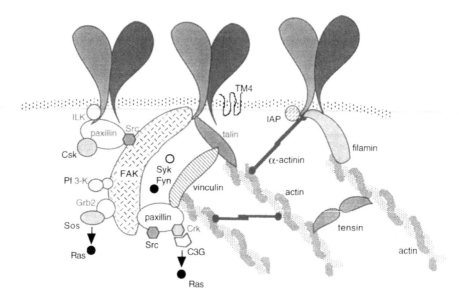

**Figure 3** Numerous signaling and cytoskeletal elements associate with integrins. Most of these elements are recruited upon initiation of the outside-in signal to sites of integrin ligation and clustering. The molecular interactions are described in more detail in the text. The figure highlights the most relevant elements that have been shown to interact directly or indirectly with integrins and is not meant to imply that all interactions illustrated occur in every cell type at all times, or in every sort of adhesion complex. FAK, focal adhesion kinase; IAP, integrin-associated protein; ILK, integrin-linked kinase; PI 3-K, phosphatidylinositol 3-kinase; TM4, member of the 4-transmembrane protein superfamily.

adhesive elements associated with focal adhesions, although other cell surface membrane proteins may be involved. Focal adhesions also include several cytoskeletal components, including talin, vinculin, α-actinin, tensin, and paxillin. These cytoskeletal proteins are involved in forming the framework of the FAC (113,114). Focal adhesion complexes serve as the major site of actin filament attachment at the contact surface. Formation of FACs in adherent cells is likely associated with cell spreading. Interestingly, cells that remain motile, nonadherent, or transiently adherent, such as lymphocytes, lack FACs. It is presumed that these cells must have similar cytoskeletal structures, but they are likely to be smaller, less distinctive, and transient.

Upon recruitment, a number of integrin-associated proteins subsequently become phosphorylated, although phosphorylation of integrins themselves does not appear to be a prerequisite for localizing to FACs (115). In fibroblasts and several other nonlymphoid cell lines, a predominant FAC component is the tyrosine kinase, pp 125^FAK, or focal adhesion kinase (FAK) (116,117). FAK has been shown to physically interact with the cytoplasmic domain of integrin β subunits via its noncatalytic amino-terminal domain (118). Several studies have mapped this interaction to the membrane-proximal segment of the β1, β2, and β3 cytoplasmic tails (119) (Fig. 2). Deletions or mutations in these sequences abrogate binding of both FAK and paxillin, a cytoskeletal signaling protein found in FACs (120). However, other studies have mapped the interaction of FAK with β1 to the carboxy-terminal end of β1 (121) (Fig. 2). In turn, FAK is thought to recruit several signaling molecules to FACs (see below). In endothelial cells, signaling molecules, e.g., Src, PI 3-K, phospholipase C-γ, as well as other proteins such as the Na⁺-H⁺ antiporter localize to FACs (122).

Besides FAK, the β1 integrin cytoplasmic tail interacts directly or indirectly with cytoskeletal proteins, such as talin, vinculin, α-actinin, tensin, and paxillin, which in turn organize F-actin into stress fibers (117,123). Studies using tyrosine kinase inhibitors have found that the accumulation of these proteins is phosphorylation-independent, with the exception of paxillin (reviewed in 123), which is phosphorylated in a FAK-dependent manner (117). The binding site for paxillin on the β1 cytoplasmic tail overlaps the binding site for FAK, although the interactions of FAK and paxillin with the β1 tail appear to be independent events. The β1 cytoplasmic tail can also bind α-actinin at a site that is distinct from the region that interacts with FAK and paxillin (119) (Fig. 2).

Recently, Lewis and Schwartz (121) mapped association of cytoskeletal elements with β1 *in vivo*. Colocalization of FAK, talin, and actin with a transfected chicken β1 protein expressed in mouse NIH 3T3 cells was dependent on a common region of the carboxy-terminus of β1, while α-actinin colocalization appeared to require a more membrane-proximal sequence (Fig. 2). These results, in combination with earlier work, led the authors to propose that the carboxy-terminal region, containing two NPXY motifs, is necessary for talin association. Filardo and coworkers (124) showed that similar motifs in the β3 cytoplasmic tail were important for adhesion, spreading, and migration of melanoma cells (Fig. 2). Similarly, several mutations in the NXXY motif of the β3 cytoplasmic tail abolished cell spreading and FAC formation in COS cell transfectants (125). These data assign cytoskeletal interaction roles to these two motifs in the β3 cytoplasmic tail. In addition, mutations in two other regions of β3 (727–733, 752) inhibited β3 integrin recruitment to preformed FACs. This region is homologous to that found in β1 to be important in interactions with α-actinin (119) (Fig. 2).

Much work has been done in elucidating the components of FACs in nonleukocyte cell types, although it is not clear whether the cytoskeletal and signaling components found in FACs associate with integrins in leukocytes. FACs are structures generally associated with adhesion of stationary cells to a substratum, and therefore may not accurately represent the molecules and events involved in leukocyte locomotion during extravasation and migration through ECM. Using affinity chromatography, several proteins, including talin, α-actinin, gelsolin, vinculin, and filamin, were found to elute from a column containing peptides derived from the β2 integrin cytoplasmic domain (126) (Fig. 2). However, only filamin and α-actinin bound directly to β2 peptide. Filamin, an actin-binding protein that stabilizes the cortical cytoskeleton, and α-actinin bound β2 via distinct but overlapping sites in the membrane-proximal half of the β2 cytoplasmic tail.

Using a two-hybrid system, Shattil et al. (127) have identified a novel 111 aa protein termed β3-endonexin that interacts with the β3 cytoplasmic tail (Fig. 2). This interaction was specific and functional, for a point mutation that disrupts integrin signaling also reduced interaction of the β3 integrin with β3-endonexin. β3-endonexin bound to β3 in extracts from platelets and fibroblasts and could be identified by immunoblotting in cell lysates from platelets and peripheral blood mononuclear cells. The function of this protein is not known, although its presence in a lymphocyte library suggests that it may play a role in lymphocyte adhesion.

Another protein, the integrin-associated protein (IAP), has been shown to associate with β3 integrins on neutrophils and other myeloid cells (reviewed in 128) and is necessary for at least certain β2- and β3-mediated signal transduction events (129–131). Because IAP is also found in erythrocytes independent of integrins, it may have a more general signaling function.

Hannigan et al. (132) have recently described a novel serine/threonine kinase that appears to associate with the β1–integrin cytoplasmic domain. This 59-kD protein, called integrin-linked kinase (ILK), was identified in a two-hybrid screen and shares similarity with other tyrosine and serine/threonine protein kinases. A GST-ILK fusion protein autophosphorylated itself, but more importantly, phosphorylated a peptide representing the β1-integrin cytoplasmic domain. Anti-ILK antibodies showed that ILK colocalizes with β1-integrin in focal plaques, and analysis of immunoprecipitates using anti-Fn receptor (α5/α3β1), anti-vitronectin receptor (αvβ3/β5), or anti-β1 antibodies identified ILK. Over-expression of ILK resulted in reduced adherence of these cells to laminin, Fn, and vitronectin.

Although integrin-mediated outside-in signaling is typically considered to lead to positive signals, such as cell growth and the prevention of apoptosis, an interesting exception has recently been found in β1c, an alternatively spliced form of β1. Expression of the β1c isoform inhibits cell cycle progression in a ligand-independent manner. Two independent studies have mapped the domain responsible for this inhibition to slightly different regions, although both regions lie within a β1c-unique sequence (133,134). The signal transducer(s) that interact with this critical domain of β1c are unknown.

Outside-in signaling is also regulated by α cytoplasmic tails. In a study comparing the functional properties of α2, α4, and α5 cytoplasmic domains expressed in a chimeric context, chimeras containing either the α2 or α5 cytoplasmic domains localized predominantly to FACs and showed increased spreading on ECM compared to those containing α4 cytoplasmic domains (135). In contrast, elevated chemo- and haptotactic migration correlated with the expression of the α4 cytoplasmic domain. These results suggest that the α4 cytoplasmic domain, in contrast to the α2 and α5 cytoplasmic domains, is responsible for weaker integrin:cytoskeletal interactions. Thus, the α4β1 integrin, which is predominantly expressed in leukocytes in adults, might endow these cells with their unique motility. Further evidence that α cytoplasmic tails can participate in outside-in signaling comes from a study that showed that a peptide corresponding to the α2 cytoplasmic tail interacted directly with F-actin, whereas an α1 cytoplasmic domain peptide did not (136). In addition, work on αIIb indicated that the carboxy-terminal portion of its cytoplasmic tail may play a role in inhibiting FAK phosphorylation in nonadherent cells (120). Thus, in contrast to inside-out signaling, where the importance of α cytoplasmic domains seems to be primarily restricted to a regulatory role of the conserved GFFKR motif, the role of α cytoplasmic tails in outside-in signaling appears to be more subunit-specific and diverse (137). It has been proposed that the α subunit sequence divergence distal to the GFFKR motif may result in the interaction of different, cell type spcific elements to the α cytoplasmic tail (102).

## B.  Integrin Signaling Pathways

Clustering of integrins within FACs results in the tyrosine phosphorylation of a series of proteins in the 120–130 kD range (138–140). FAK is the major phosphorylated protein found in FACs (139,141). Numerous studies have now shown that FAK is autophosphorylated on a tyrosine residue following integrin engagement in a variety of cell types (118) and predict a role for FAK in linking integrin activation to downstream effector functions. However, the precise role that FAK plays is not yet clear. Although leukocytes are nonadherent cells and do not form classical FACs when interacting with ECM proteins, β1 integrin–mediated phosphorylation of FAK in a T-cell line has been reported (142,143), and mAb cross-linking of the CD3/TCR complex has also been reported to result in tyrosine phosphorylation of a novel substrate with homology to FAK, named fakB (144).

In addition to FAK, integrin engagement leads to the activation of Src and several related Src family tyrosine kinases (117,145). Association of Src with integrin-mediated cytoskeletal complexes may result through interaction of SH3 domains with proline-rich regions of paxillin, whereas localization of Src to FACs may be mediated through the interaction of Src SH2 domains with phosphorylated peptide motifs on FAK. Other protein kinases that are found in focal adhesions include Syk and Fyn, which may regulate both kinase activation and localization of Src to focal adhesions (146,147). These other kinases may become associated with integrins through cytoskeletal complexes induced by integrin ligation and may interact with FAK and paxillin through their SH2 domains. Clustering of these proteins in FACs may result in activation of protein kinases. PI-3K has also been found to interact with FAK following integrin engagement (148) through interaction of the SH2 domain of the p85 subunit of PI-3K and a phosphorylated motif of FAK. Finally, Csk (C-terminal Src kinase) was likewise found to be associated with FAK and paxillin in FACs and may function in the regulation of Src activity (149).

In addition to interaction with kinases, FAK may transduce signals through integrin-dependent interaction with adapter proteins such as Grb2 (localized to FAK residue $Y^{925}$), SOS, and p130$^{Cas}$ (Cas) (150–152). Adapter proteins, which are small proteins composed almost exclusively of SH2 and SH3 domains (66) and which have no intrinsic kinase activity, may link FAK to the Ras pathway. SOS is a guanine nucleotide exchange factor (GNEF) that functions by converting inactive Ras-GDP to active Ras-GTP (153). Cas (Crk-associated tyrosine kinase substrate) is found in FACs and is likely phosphorylated by FAK and Src. Cas may serve as a docking protein that recruits additional signaling molecules, including Crk, to focal adhesions following integrin activation. Crk, which contains both SH2 and SH3 domains and may bind to C3G, a putative GNEF for Ras (152,154), has also been shown to associate with tyrosine-phosphorylated paxillin through its SH2 domain (155).

Several reports indicate that integrin activation results in activation of the Ras/MAP kinase pathway (118,156–159). MAP kinase is a downstream target of Ras (118,156). Activation of the Ras-MAP kinase pathway may regulate proliferation and differentiation as a function of integrin engagement, through phosphorylation and activation of transcription factors (160). MAP kinase is also known to phosphorylate and activate phospholipase $A_2$, which leads to production of arachidonic acid and lysophospholipid (161). Finally, MAP kinase may regulate cytoskeletal changes necessary for cell spreading. Growth factors also stimulate the Ras pathway and may synergize with integrin-mediated signaling to enhance Ras-MAP kinase activity. Integrin engagement has also been shown to result in increased intracellular $Ca^{2+}$ concentrations (162,163), and increases in intracellular pH due to activation of the $Na^+/H^+$ antiporter (164,165).

By analogy to studies of adherent cells, numerous studies have attempted to identify how signals are transduced following integrin occupancy in T cells. Several studies have identified FAK in both human T and B cells (143,144,166). In a recent report, Maguire et al. (143) showed that cross-linking the CD3/TCR gave rise to low levels of FAK phosphorylation in human CD4$^+$ T cell blasts and the Jurkat T-cell line, whereas incubation on immobilized Fn or VCAM-1 together with CD3/TCR cross-linking led to a synergistic increase in phosphorylation of FAK. This report confirmed the role of the $\alpha5\beta1$ integrin, and identified a previously unknown role of the $\alpha4\beta1$ integrin, in FAK regulation. The kinetics of FAK phosphorylation in T cells following TCR or integrin engagement imply that FAK may play a role in several downstream events. The increase in phosphorylation is maximal within 5–15 minutes, suggesting that FAK phosphorylation may be involved in immediate effector functions, such as T-cell killing or extravasation. However, because the phosphorylated state is maintained for greater than 2 hours, FAK may also play a role in additional downstream functions, such as T-cell costimulation, which require prolonged interaction of the T cell with a target cell.

In addition to FAK, a variety of other proteins become phosphorylated in T cells following integrin engagement. Nojima et al. (167) have demonstrated increased phosphorylation of a 105-kD protein following cross-linking of the $\alpha4\beta1$ integrin in H9 cells as well as resting peripheral T cells. This 105-kD protein appears to be antigenically distinct from FAK (142). Phosphorylation of this protein could also be induced with the CS-1 containing fragment of Fn. Gismondi et al. (168) have shown increased tyrosine phosphorylation of two proteins (105 and 115 kD) following $\alpha4\beta1$ or $\alpha5\beta1$ integrin cross-linking in freshly isolated NK cells. Brando and Shevach (169) have also shown increased phosphorylation of a 115-kD protein following engagement of $\alpha v\beta3$ or TCR on a murine T-cell hybridoma. This phosphorylated protein was not FAK, and did not have intrinsic kinase activity. Similarly, Kapron-Bras et al. (159) have shown that ligation of the $\alpha2\beta1$ integrin on Jurkat T cells with antibodies to either the $\alpha2$ or $\beta1$ chains resulted in the accumulation of Ras in the active GTP-bound state, as well as increased phosphorylation of intracellular targets in the 47–52 kD range. Finally, LFA-1 engagement on T and B cells results in the increased phosphorylation of a number of as yet unidentified proteins (170,171).

In addition to the integrins, members of several other cell surface adhesion molecule families have been shown to be involved in outside-in signaling. Several recent reports have demonstrated a role for L-selectin in leukocyte signaling (172,173). Cross-linking of L-selectin on neutrophils led to increased intracellular $Ca^{2+}$ and potentiation of the oxidative burst. Increased phosphorylation of a number of intracellular targets was observed, including MAP kinase. Incubation on sulfatides and sulfated glycolipids, the suspected natural ligands of L-selectin, also resulted in a dose-dependent increase in tyrosine phosphorylation and increased MAP kinase activity. It is not known whether L-selectin expressed on T cells also functions as a signaling molecule. L-selectin can also be classified as an integrin regulator because cross-linking of L-selectin on neutrophils results in increased $\beta2$ integrin activity (24).

Recent data suggest that IgSF adhesion molecules are also involved in T-cell signaling. Chirathaworn et al. (174) showed that tyrosine phosphorylation of a 34-kD cdc2 protein kinase was observed after cross-linking ICAM-1 in a T-cell leukemic cell line and in human peripheral T cells. Phosphorylation of cdc2 through ICAM-1 engagement led to inactivation of cdc2. ICAM-1 cross-linking in peripheral T cells also led to increased phosphorylation of several other targets of 45, 115, and 145 kD. Campanero et al. (175) have demonstrated that ICAM-3 is involved in the regulation of homotypic aggregation mediated by LFA-1 and ICAM-1. In addition, ICAM-3 engagement could induce T cell activation and proliferation, as well as induce expression of the T-cell activation antigens CD25 and CD69 (175,176). It has also been shown that some anti–ICAM-3 antibodies can induce tyrosine phosphorylation of intracellular targets of 125, 70, and 38 kD (170). Immunofluorescence studies indicated that these tyrosine phosphorylated proteins were preferentially located at the regions of cell–cell interaction, suggesting their involvement in the adhesion process (177).

## VI. NEGATIVE REGULATION OF INTEGRIN FUNCTION: TURNING OFF ADHESION

By necessity, adhesion of a migrating leukocyte to the surrounding ECM is transient and cycling, which might suggest that any integrin-associated intracellular complexes are equally transient. Alternatively, such complexes are constitutively present in a complete or semi-formed state, but require activation signals to be functional. There has been some recent progress in identifying negative regulators of integrin-mediated outside-in signaling.

There are several possible means of regulating the transient organization of cytoskeletal and signaling elements, for example 1.) transience of the activating signal, 2.) restricted localization of the activating signal (i.e., at the leading edge of a migrating cell) and/or 3.) activation of a negative signal.

Recent work by Rovere et al. (178) implies feedback regulation of β2-mediated adhesion in T cells. These investigators showed that costimulation of T cells via CD3 and LFA-1 elicited the PKC-dependent appearance of an adenylyl cyclase isoform, which in turn led to the elevation of cAMP. Subsequent activation of cAMP-dependent kinase led to the dissociation of LFA-1 from cytoskeletal elements and de-adhesion.

Lawson and Maxfield (179) showed that αvβ3 integrins that bind vitronectin at the leading edge of migrating neutrophils regularly detached from the ECM ligand and are endocytosed and recycled as the cell extends forward. This release and endocytosis required the $Ca^{2+}$-dependent serine-threonine phosphatase calcineurin, whose activity cycled in response to fluctuating intracellular $Ca^{2+}$ concentrations. When $Ca^{2+}$ levels were buffered, neutrophils became stuck to vitronectin at their trailing edge, where αvβ3 colocalized with talin and α-actinin. The authors proposed that calcineurin mediated integrin release from the substratum by dephosphorylating a component associated with the αvβ3 integrin adhesion complex. It is conceivable that different integrins, or different forms of contact (FACs, point contacts, etc.) are sensitive to different types and extents of negative regulation. For example, using the same experimental system, the authors were also able to show that release of β2 attachments to ECM substrate was not dependent on $Ca^{2+}$ signaling (179). Regen and Horwitz observed that motile chick skeletal fibroblasts attached to Fn via FAC-like macroaggregates containing clustered α5β1 integrin (180). As these cells moved forward, the macroaggregates at the trailing edge were not detached so much as the membrane (including a large portion of the integrins) remained behind on the substratum in a ripping process. However, cytoskeletal elements such as vinculin were not left behind. In this experimental system, negative regulators may dismantle adhesion complexes without necessarily downregulating the affinity of the integrin for its ligand. In addition, recycling of integrins is probably not essential. Finally, an overt negative regulator was identified in platelets in the form of calpain, a $Ca^{2+}$-dependent protease. In addition to αIIbβ3 activation, platelet activation leads to increased calpain levels; calpain in turn cleaves β3 at sites flanking the NPXY and NXXY motifs (181). Furthermore, calpain cleaves talin, protein-tyrosine phosphatase 1B, Src, and PKC. These results suggest coordinate negative regulation of αIIbβ3 integrin and associated cytoskeletal and signaling elements. Future work will address the issue of negative regulation of leukocyte integrins during inside-out as well as outside-in signaling.

## REFERENCES

1. T. A. Springer, *Nature*, *346*:425 (1990).
2. T. A. Springer, *Cell*, *76*:301 (1994).
3. Y. Shimizu and S. Shaw, *FASEB J.*, *5*:2292 (1991).
4. R. O. Hynes, *Cell*, *69*:11 (1992).
5. M. E. Hemler, *Annu. Rev. Immunol.*, *8*:365 (1990).
6. C. Rosales and R. L. Juliano, *J. Leukocyte Biol.*, *57*:189 (1995).
7. U. Dianzani and F. Malavasi, *Crit. Rev. Immunol.*, *15*:167 (1995).
8. S. W. Hunt, III, E. S. Harris, S.-A. Kellermann, and Y. Shimizu, *Crit. Rev. Oral Biol.*, *In press.* (1996).
9. T. M. Carlos and J. M. Harlan, *Blood*, *84*:2068 (1994).
10. L. A. Lasky, *Annu. Rev. Biochem.*, *64*:113 (1995).
11. T. F. Tedder, D. A. Steeber, A. Chen, and P. Engel, *FASEB J.*, *9*:866 (1995).
12. E. L. Berg, L. M. McEvoy, C. Berlin, R. F. Bargatze, and E. C. Butcher, *Nature*, *366*:695 (1993).
13. L. J. Picker, *Curr. Opin. Immunol.*, *6*:394 (1994).
14. Y. Shimizu and S. Shaw, *Nature*, *366*:630 (1993).
15. M. L. Arbonès, D. C. Ord, K. Ley, H. Ratech, C. Maynard-Curry, G. Otten, D. J. Capon, and T. F. Tedder, *Immunity*, *1*:247 (1994).
16. C. Berlin, E. L. Berg, M. J. Briskin, D. P. Andrew, P. J. Kilshaw, B. Holzmann, I. L. Weissman, A. Hamann, and E. C. Butcher, *Cell*, *74*:185 (1993).
17. D. A. Jones, L. V. McIntire, C. W. Smith, and L. J. Picker, *J. Clin. Invest.*, *94*:2443 (1994).

18. T. N. Mayadas, R. C. Johnson, H. Rayburn, R. O. Hynes, and D. D. Wagner, *Cell*, 74:541 (1993).
19. M. Subramaniam, S. Saffaripour, S. R. Watson, T. N. Mayadas, R. O. Hynes, and D. D. Wagner, *J. Exp. Med.*, 181:2277 (1995).
20. M. A. Labow, C. R. Norton, J. M. Rumberger, et al., *Immunity*, 1:709 (1994).
21. P. S. Frenette, T. Mayadas, H. Rayburn, R. O. Hynes, and D. Wagner, *Cell*, 84:563 (1996).
22. D. Gilat, L. Cahalon, R. Hershkoviz, and O. Lider, *Immunol. Today*, 17:16 (1996).
23. Y. Tanaka, D. H. Adams, and S. Shaw, *Immunol. Today*, 14:111 (1993).
24. S. I. Simon, A. R. Burns, A. D. Taylor, P. K. Gopalan, E. B. Lynam, L. A. Sklar, and C. W. Smith, *J. Immunol.*, 155:1502 (1995).
25. Y. Tanaka, S. M. Albelda, K. J. Horgan, G. A. van Seventer, Y. Shimizu, W. Newman, J. Hallam, P. J. Newman, C. A. Buck, and S. Shaw, *J. Exp. Med.*, 176:245 (1992).
26. L. Piali, P. Hammel, C. Uherek, F. Bachmann, R. H. Gisler, D. Dunon, and B. A. Imhof, *J. Cell Biol.*, 130:451 (1995).
27. A. Ager, *Trends Cell Biol.*, 4:326 (1994).
28. Y. Shimizu, G. A. van Seventer, K. J. Horgan, and S. Shaw, *Nature*, 345:250 (1990).
29. R. Goldman, J. Harvey, and N. Hogg, *Eur. J. Immunol.*, 22:1109 (1992).
30. J. Lesley, R. Hyman, and P. W. Kincade, *Adv. Immunol.*, 54:271 (1993).
31. P. A. G. McCourt, B. Ek, N. Forsberg, and S. Gustafson, *J. Biol. Chem.*, 269:30081 (1994).
32. Z. Szekanecz, M. J. Humphries, and A. Ager, *J. Cell Sci.*, 101:885 (1992).
33. A. Ager and M. J. Humphries, *Int. Immunol.*, 2:921 (1990).
34. H. Hourihan, T. D. Allen, and A. Ager, *J. Cell Sci.*, 104:1049 (1993).
35. T. A. Ferguson, H. Mizutani, and T. S. Kupper, *Proc. Natl. Acad. Sci. USA*, 88:8072 (1991).
36. T. A. Ferguson and T. S. Kupper, *J. Immunol.*, 150:1172 (1993).
37. R. Hershkoviz, N. Greenspoon, Y. A. Mekori, R. Hadari, R. Alon, G. Kapustina, and O. Lider, *Clin. Exp. Immunol.*, 95:270 (1994).
38. K. L. Hines, A. B. Kulkarni, J. B. McCarthy, H. Tian, J. M. Ward, M. Christ, N. L. McCartney-Francis, L. T. Furcht, S. Karlsson, and S. M. Wahl, *Proc. Natl. Acad. Sci. USA*, 91:5187 (1994).
39. D. Hauzenberger, J. Klominek, and K.-G. Sundqvist, *J. Immunol.*, 153:960 (1994).
40. P. Friedl, P. B. Noble, and K. S. Zanker, *J. Immunol.*, 154:4973 (1995).
41. M. L. Dustin and T. A. Springer, *Nature*, 341:619 (1989).
42. Y. van Kooyk, P. van de Wiel–van Kemenade, P. Weder, T. W. Kuijpers, and C. G. Figdor, *Nature*, 342:811 (1989).
43. Y. Shimizu, G. A. van Seventer, E. Ennis, W. Newman, K. J. Horgan, and S. Shaw, *J. Exp. Med.*, 175:577 (1992).
44. B. M. C. Chan, J. G. P. Wong, A. Rao, and M. E. Hemlet, *J. Immunol.*, 147:398 (1991).
45. S. C. Meuer, R. E. Hussey, M. Fabbi, D. Fox, O. Acuto, K. A. Fitzgerald, J. C. Hodgdon, J. P. Protentis, S. F. Schlossman, and E. L. Reinherz, *Cell*, 36:897 (1984).
46. B. E. Bierer and W. C. Hahn, *Semin. Immunol.*, 5:249 (1993).

47. P. Selvaraj, M. L. Plunkett, M. Dustin, M. E. Sanders, S. Shaw, and T. A. Springer, *Nature*, *326*:400 (1987).
48. C. H. June, J. A. Bluestone, L. M. Nadler, and C. B. Thompson, *Immunol. Today*, *15*:321 (1994).
49. Y. Tanaka, D. H. Adams, S. Hubscher, H. Hirano, U. Siebenlist, and S. Shaw, *Nature*, *361*:79 (1993).
50. D. H. Adams, L. Harvath, D. P. Bottaro, R. Interrante, G. Catalano, Y. Tanaka, A. Strain, S. G. Hubscher, and S. Shaw, *Proc. Natl. Acad. Sci. USA*, *91*:7144 (1994).
51. A. R. Lloyd, J. J. Oppenheim, D. J. Kelvin, and D. D. Taub, *J. Immunol.*, *156*:932 (1996).
52. M. W. Carr, R. Alon, and T. A. Springer, *Immunity*, *4*:179 (1996).
53. T. J. Schall and K. B. Bacon, *Curr. Opin. Immunol.*, *6*:865 (1994).
54. J. Dastych and D. D. Metcalfe, *J. Immunol.*, *152*:213 (1994).
55. T. Kinashi, J. A. Escobedo, L. T. Williams, K. Takatsu, and T. A. Springer, *Blood*, *86*:2086 (1995).
56. H. Serve, N. S. Yee, G. Stella, L. Sepp-Lorenzino, J. C. Tan, and P. Besmer, *EMBO J.*, *14*:473 (1995).
57. N. L. Kovach, N. Lin, T. Yednock, J. M. Harlan, and V. C. Broudy, *Blood*, *85*:159 (1995).
58. M. S. Diamond and T. A. Springer, *Curr. Biol.*, *4*:506 (1994).
59. R. Pardi, L. Inverardi, C. Rugarli, and J. R. Bender, *J. Cell Biol.*, *116*:1211 (1992).
60. A. S. H. Chan, P. J. Reynolds, and Y. Shimizu, *Eur. J. Immunol.*, *24*:2602 (1994).
61. R. F. Bargatze and E. C. Butcher, *J. Exp. Med.*, *178*:367 (1993).
62. C. Laudanna, J. J. Campbell, and E. C. Butcher, *Science*, *271*:981 (1996).
63. J. L. Mobley, T. Zell, and Y. Shimizu, *Submitted* (1996).
64. R. Kapeller and L. C. Cantley, *BioEssays*, *16*:565 (1994).
65. I. D. Hiles, M. Otsu, S. Volinia, M. J. Fry, I. Gout, R. Dhand, G. Panayotou, F. Ruiz-Larrea, A. Thompson, N. F. Totty, J. J. Hsuan, S. A. Courtneidge, P. J. Parker, and M. J. Waterfield, *Cell*, *70*:419 (1992).
66. G. B. Cohen, R. Ren, and D. Baltimore, *Cell*, *80*:237 (1995).
67. S. G. Ward, C. H. June, and D. Olive, *Immunol. Today*, *17*:187 (1996).
68. S. G. Ward, R. Parry, C. LeFeuvre, D. M. Sansom, J. Westwick, and A. I. Lazarovits, *Eur. J. Immunol.*, *25*:502 (1995).
69. S. G. Ward, S. C. Ley, C. MacPhee, and D. A. Cantrell, *Eur. J. Immunol.*, *22*:45 (1992).
70. Y. Shimizu, J. L. Mobley, L. D. Finkelstein, and A. S. H. Chan, *J. Cell Biol.*, *131*:1867 (1995).
71. T. J. Kovacsovics, C. Bachelot, A. Toker, C. J. Vlahos, B. Duckworth, L. C. Cantley, and J. H. Hartwig, *J. Biol. Chem.*, *270*:11358 (1995).
72. T. Zell, S. W. Hunt,III, L. D. Finkelstein, and Y. Shimizu, *J. Immunol.*, *156*:883 (1996).
73. Y.-C. Cai, D. Cefai, H. Schneider, M. Raab, N. Nabavi, and C. E. Rudd, *Immunity*, *3*:417 (1995).
74. L. Turner, S. G. Ward, and J. Westwick, *J. Immunol.*, *155*:2437 (1995).
75. M. Eberle, A. E. Traynor-Kaplan, L. A. Sklar, and J. Norgauer, *J. Biol. Chem.*, *265*:16725 (1990).

76. J. Zhang, M. J. Fry, M. D. Waterfield, S. Jaken, L. Liao, J. E. B. Fox, and S. E. Rittenhouse, *J. Biol. Chem.*, *267*:4886 (1992).

77. L. Stephens, A. Smrcka, F. T. Cooke, T. R. Jackson, P. C. Sternweis, and P. T. Hawkins, *Cell*, *77*:83 (1994).

78. P. A. Thomasen, S. R. James, P. J. Casey, and C. P. Downes, *J. Biol. Chem.*, *269*:16525 (1994).

79. B. Stoyanov, S. Volinia, T. Hanck, et al. *Science*, *269*:690 (1995).

80. J. Chung, T. C. Grammer, K. P. Lemon, A. Kazlauskas, and J. Blenis, *Nature*, *370*:71 (1994).

81. M. Monfar, K. P. Lemon, T. C. Grammer, L. Cheatham, J. Chung, C. J. Vlahos, and J. Blenis, *Mol. Cell. Biol.*, *15*:326 (1995).

82. M. M. Chou and J. Blenis, *Curr. Opin. Cell Biol.*, *7*:806 (1995).

83. S. Moriya, A. Kazlauskas, K. Akimoto, S. Hirai, K. Mizuno, T. Takenawa, Y. Fukui, Y. Watanabe, S. Ozaki, and S. Ohno, *Proc. Natl. Acad. Sci. USA*, *93*:151 (1996).

84. H. Nakanishi, K. A. Brewer, and J. H. Exton, *J. Biol. Chem.*, *268*:13 (1993).

85. A. Toker, M. Meyer, K. K. Reddy, J. R. Falck, R. Aneja, S. Aneja, A. Parra, D. J. Burns, L. M. Ballas, and L. C. Cantley, *J. Biol. Chem.*, *269*:32358 (1994).

86. K. Akimoto, R. Takahashi, S. Moriya, N. Nishioka, J. Takayanagi, K. Kimura, Y. Fukui, S. Osada, K. Mizuno, S. Hirai, A. Kazlauskas, and S. Ohno, *EMBO J.*, *15*:788 (1996).

87. T. F. Franke, S.-I. Yang, T. O. Chan, K. Datta, A. Kazlauskas, D. K. Morrison, D. R. Kaplan, and P. N. Tsichlis, *Cell*, *81*:727 (1995).

88. B. M. T. Burgering and P. J. Coffer, *Nature*, *376*:599 (1995).

89. J. L. Mobley, E. Ennis, and Y. Shimizu, *Blood*, *83*:1039 (1994).

90. M. Lub, Y. van Kooyk, and C. G. Figdor, *Immunol. Today*, *16*:479 (1995).

91. J. L. Mobley, E. Ennis, and Y. Shimizu, *J. Immunol.*, *156*:948 (1996).

92. R. J. Faull, N. L. Kovach, J. M. Harlan, and M. H. Ginsberg, *J. Exp. Med.*, *179*:1307 (1994).

93. M. P. Stewart, C. Cabañas, and N. Hogg, *J. Immunol.*, *156*:1810 (1996).

94. B. A. Lollo, K. W. H. Chan, E. M. Hanson, V. T. Moy, and A. A. Brian, *J. Biol. Chem.*, *268*:21693 (1993).

95. A. Jakubowski, B. Newman Ehrenfels, R. B. Pepinsky, and L. C. Burkly, *J. Immunol.*, *155*:938 (1995).

96. A. Masumoto and M. E. Hemler, *J. Biol. Chem.*, *268*:228 (1993).

97. R. Pardi, G. Bossi, L. Inverardi, E. Rovida, and J. R. Bender, *J. Immunol.*, *155*:1252 (1995).

98. T. E. O'Toole, J. Ylanne, and B. M. Culley, *J. Biol. Chem.*, *270*:8553 (1995).

99. S. E. LaFlamme, L. A. Thomas, S. S. Yamada, and K. M. Yamada, *J. Cell Biol.*, *126*:1287 (1994).

100. Y.-P. Chen, T. E. O'Toole, T. Shipley, J. Forsyth, S. E. LaFlamme, K. M. Yamada, S. J. Shattil, and M. H. Ginsberg, *J. Biol. Chem.*, *269*:18307 (1994).

101. M. L. Hibbs, H. Xu, S. A. Stacker, and T. A. Springer, *Science*, *251*:1611 (1991).

102. T. E. O'Toole, Y. Katagiri, R. J. Faull, K. Peter, R. Tamura, V. Quaranta, J. C. Loftus, S. J. Shattil, and M. H. Ginsberg, *J. Cell Biol.*, *124*:1047 (1994).

103. P. E. Hughes, T. E. O'Toole, J. Ylänne, S. J. Shattil, and M. H. Ginsberg, *J. Biol. Chem.*, *270*:12411 (1995).

104. M. Coppolino, C. Leung-Hagesteijn, S. Dedhar, and J. Wilkins, *J. Biol. Chem.*, *270*:23132 (1995).

105. A. Hermanowski-Vosatka, J. A. G. van Strijp, W. J. Swiggard, and S. D. Wright, *Cell*, *68*:341 (1992).

106. J. Bohuslav, V. Horejsí, C. Hansmann, J. Stöckl, U. H. Weidle, O. Majdic, I. Bartke, W. Knapp, and H. Stockinger, *J. Exp. Med.*, *181*:1381 (1995).

107. F. Berditchevski, M. M. Zutter, and M. E. Hemler, *Mol. Biol. Cell*, *7*:193 (1996).

108. E. Rubinstein, F. Le Naour, M. Billard, M. Prenant, and C. Boucheix, *Eur. J. Immunol.*, *24*:3005 (1994).

109. A. R. E. Shaw, A. Domanska, A. Mak, A. Gilchrist, K. Dobler, L. Visser, S. Poppema, L. Fliegel, M. Letarte, and B. J. Willett, *J. Biol. Chem.*, *270*:24092 (1995).

110. Y. Shimizu, G. A. van Seventer, K. J. Horgan, and S. Shaw, *J. Immunol.*, *145*:59 (1990).

111. T. Matsuyama, A. Yamada, J. Kay, K. M. Yamada, S. K. Akiyama, S. F. Schlossman, and C. Morimoto, *J. Exp. Med.*, *170*:1133 (1989).

112. G. A. van Seventer, Y. Shimizu, K. J. Horgan, and S. Shaw, *J. Immunol.*, *144*:4579 (1990).

113. C. E. Turner and K. Burridge, *Curr. Opin. Cell Biol.*, *3*:849 (1991).

114. B. M. Gumbiner, *Cell*, *84*:345 (1996).

115. M. F. Barreuther and L. B. Grabel, *Exp. Cell Res.*, *222*:10 (1996).

116. A. Richardson and J. T. Parsons, *BioEssays*, *17*:229 (1995).

117. E. A. Clark and J. S. Brugge, *Science*, *268*:233 (1995).

118. M. D. Schaller and J. T. Parsons, *Curr. Opin. Cell Biol.*, *6*:705 (1994).

119. M. D. Schaller, C. A. Otey, J. D. Hildebrand, and J. T. Parsons, *J. Cell Biol.*, *130*:1181 (1995).

120. L. Leong, P. E. Hughes, M. A. Schwartz, M. H. Ginsberg, and S. J. Shattil, *J. Cell Sci.*, *108*:3817 (1995).

121. J. M. Lewis and M. A. Schwartz, *Mol. Biol. Cell*, *6*:151 (1995).

122. G. E. Plopper, H. P. McNamee, L. E. Dike, K. Bojanowski, and D. E. Ingber, *Mol. Biol. Cell*, *6*:1349 (1995).

123. K. M. Yamada and S. Miyamoto, *Curr. Opin. Cell Biol.*, *7*:681 (1995).

124. E. J. Filardo, P. C. Brooks, S. L. Deming, C. Damsky, and D. A. Cheresh, *J. Cell Biol.*, *130*:441 (1995).

125. J. Ylänne, J. Huuskonen, T. E. O'Toole, M. H. Ginsberg, I. Virtanen, and C. G. Gahmberg, *J. Biol. Chem.*, *270*:9550 (1995).

126. C. P. Sharma, R. M. Ezzell, and M. A. Arnaout, *J. Immunol.*, *154*:3461 (1995).

127. S. J. Shattil, T. O'Toole, M. Eigenthaler, V. Thon, M. Williams, B. M. Babior, and M. H. Ginsberg, *J. Cell, Biol.*, *131*:807 (1995).

128. S. W. Edwards, *Trends Biochem. Sci.*, *20*:362 (1995).

129. S. D. Blystone, I. L. Graham, F. P. Lindberg, and E. J. Brown, *J. Cell Biol.*, *127*:1129 (1994).

130. S. D. Blystone, F. P. Lindberg, S. E. LaFlamme, and E. J. Brown, *J. Cell Biol.*, *130*:745 (1995).

131. D. Cooper, F. P. Lindberg, J. R. Gamble, E. J. Brown, and M. A. Vadas, *Proc. Natl. Acad. Sci. USA*, *92*:3978 (1995).

132.  G. E. Hannigan, C. Leung-Hagesteijn, L. Fitz-Gibbon, M. G. Coppolino, G. Radeva, J. Filmus, J. C. Bell, and S. Dedhar, *Nature*, *379*:91 (1996).

133.  J. Meredith, Jr., Y. Takada, M. Fornaro, L. R. Languino, and M. A. Schwartz, *Science*, *269*:1570 (1995).

134.  M. Fornaro, D. Q. Zheng, and L. R. Languino, *J. Biol. Chem.*, *270*:24666 (1995).

135.  P. D. Kassner, R. Alon, T. A. Springer, and M. E. Hemler, *Mol. Biol. Cell*, *6*:661 (1995).

136.  J. D. Kieffer, G. Plopper, D. E. Ingber, J. H. Hartwig, and T. S. Kupper, *Biochem. Biophys. Res. Commun.*, *217*:466 (1995).

137.  P. D. Kassner, S. Kawaguchi, and M. E. Hemler, *J. Biol. Chem.*, *269*:19859 (1994).

138.  L. J. Kornberg, H. S. Earp, C. E. Turner, C. Prockop, and R. L. Juliano, *Proc. Natl. Acad. Sci. USA*, *88*:8392 (1991).

139.  L. Kornberg, H. S. Earp, J. T. Parsons, M. Schaller, and R. L. Juliano, *J. Biol. Chem.*, *267*:23439 (1992).

140.  J.-L. Guan, J. E. Trevithick, and R. O. Hynes, *Cell Regul.*, *2*:951 (1991).

141.  S. K. Hanks, M. B. Calalb, M. C. Harper, and S. K. Patel, *Proc. Natl. Acad. Sci. USA*, *89*:8487 (1992).

142.  Y. Nojima, K. Tachibana, T. Sato, S. F. Schlossman, and C. Morimoto, *Cell Immunol.*, *161*:8 (1995).

143.  J. E. Maguire, K. M. Danahey, L. C. Burkly, and G. A. van Seventer, *J. Exp. Med.*, *182*:2079 (1995).

144.  S. B. Kanner, A. Aruffo, and P.-Y. Chan, *Proc. Natl. Acad. Sci. USA*, *91*:10484 (1994).

145.  S. J. Shattil, M. H. Ginsberg, and J. S. Brugge, *Curr. Opin. Cell Biol.*, *6*:695 (1994).

146.  E. A. Clark, S. J. Shattil, M. H. Ginsberg, J. Bolen, and J. S. Brugge, *J. Biol. Chem*, *269*:28859 (1994).

147.  B. S. Cobb, M. D. Schaller, T. H. Leu, and J. T. Parsons, *Mol. Cell Biol.*, *14*:147 (1994).

148.  H.-C. Chen and J.-L. Guan, *Proc. Natl. Acad. Sci. USA*, *91*:10148 (1994).

149.  H. Sabe, A. Hata, M. Okada, H. Nakagawa, and H. Hanafusa, *Proc. Natl. Acad. Sci., USA*, *91*:3984 (1994).

150.  S. Miyamoto, S. K. Akiyama, and K. M. Yamada, *Science*, *267*:883 (1995).

151.  S. Miyamoto, H. Teramoto, O. A. Coso, J. S. Gutkind, P. D. Burbelo, S. K. Akiyama, and K. M. Yamada, *J. Cell Biol.*, *131*:791 (1995).

152.  T. R. Polte and S. K. Hanks, *Proc. Natl. Acad. Sci. USA*, *92*:10678 (1995).

153.  P. van der Geer, T. Hunter, and R. A. Lindberg, *Annu. Rev. Cell Biol.*, *10*:251 (1994).

154.  S. Tanaka, T. Morishita, Y. Hashimoto, S. Hattori, S. Nakamura, M. Shibuya, K. Matuoka, T. Takenawa, T. Kurata, K. Nagashima, and M. Matsuda. *Proc. Natl. Acad. Sci. USA*, *91*:3443 (1994).

155.  R. B. Birge, J. E. Fajardo, C. Reichman, S. E. Shoelson, Z. Songyang, L. C. Cantley, and H. Hanafusa, *Mol. Cell Biol.*, *13*:4648 (1993).

156.  Q. Chen, M. S. Kinch, T. H. Lin, K. Burridge, and R. L. Juliano, *J. Biol. Chem*, *269*:26602 (1994).

157.  N. Morino, T. Mimura, K. Hamasaki, K. Tobe, K. Ueki, K. Kikuchi, K. Takahara, T. Kadowaki, Y. Yazaki, and Y. Nojima, *J. Biol. Chem.*, *270*:269 (1995).

158. X. Zhu and R. K. Assoian, *Mol. Biol. Cell*, *6*:273 (1995).
159. C. Kapron-Bras, L. Fitz-Gibbon, P. Jeevaratnam, J. Wilkins, and S. Dedhar, *J. Biol. Chem*, *268*:20701 (1993).
160. C. S. Hill and R. Treisman, *Cell*, *80*:199 (1995).
161. D. Piomelli, *Curr. Opin. Cell Biol.*, *5*:274 (1993).
162. M. C. Wacholtz, S. S. Patel, and P. E. Lipsky, *J. Exp. Med.*, *170*:431 (1989).
163. J. Ng-Sikorski, R. Andersson, M. Patarroyo, and T. Andersson, *Exp. Cell Res.*, *195*:504 (1991).
164. M. A. Schwartz and C. Lechene, *Proc. Natl. Acad. Sci. USA*, *89*:6138 (1992).
165. M. A. Schwartz, D. E. Ingber, M. Lawrence, T. A. Springer, and C. Lechene, *Exp. Cell Res.*, *195*:533 (1991).
166. G. S. Whitney, P.-Y. Chan, J. Blake, W. L. Cosand, M. G. Neubauer, A. Aruffo, and S. B. Kanner, *DNA Cell Biol.*, *12*:823 (1993).
167. Y. Nojima, D. M. Rothstein, K. Sugita, S. F. Schlossman, and C. Morimoto, *J. Exp. Med.*, *175*:1045 (1992).
168. A. Gismondi, M. Milella, G. Palmieri, M. Piccoli, L. Frati, and A. Santoni, *J. Immunol.*, *154*:3128 (1995).
169. C. Brando and E. M. Shevach, *J. Immunol.*, *154*:2005 (1995).
170. A. G. Arroyo, M. R. Campanero, P. Sánchez-Mateos, J. M. Zapata, M. A. Ursa, M. A. Del Pozo, and F. Sánchez-Madrid, *J. Cell Biol.*, *126*:1277 (1994).
171. S. C.-T. Wang, S. B. Kanner, J. A. Ledbetter, S. Gupta, G. Kumar, and A. E. Nel, *J. Leuk. Biol.*, *57*:343 (1995).
172. T. K. Waddell, L. Fialkow, C. K. Chan, T. K. Kishimoto, and G. P. Downey, *J. Biol. Chem.*, *269*:18485 (1994).
173. T. K. Waddell, L. Fialkow, C. K. Chan, T. K. Kishimoto, and G. P. Downey, *J. Biol. Chem.*, *270*:15403 (1995).
174. C. Chirathaworn, S. A. Tibbetts, M. A. Chan, and S. H. Benedict, *J. Immunol.*, *155*:5479 (1995).
175. M. R. Campanero, M. A. Del Pozo, A. G. Arroyo, P. Sánchez-Mateos, T. Hernández-Caselles, A. Craig, R, Pulido, and F. Sánchez-Madrid, *J. Cell Biol.*, *123*:1007 (1993).
176. T. Hernandez-Caselles, G. Rubio, M. R. Campanero, M. A. Del Pozo, M. Muro, F. Sanchez-Madrid, and P. Aparicio, *Eur. J. Immunol.*, *23*:2799 (1993).
177. M. Angel del Pozo, P. Sanchez-Mateos, and F. Sanchez-Madrid, *Immunol. Today*, *17*:127 (1996).
178. P. Rovere, L. Inverardi, J. R. Bender, and R. Pardi, *J. Immunol.*, *156*:2273 (1996).
179. M. A. Lawson and F. R. Maxfield, *Nature*, *377*:75 (1995).
180. C. M. Regen and A. F. Horwitz, *J. Cell Biol.*, *119*:1347 (1992).
181. X. P. Du, T. C. Saido, S. Tsubuki, F. E. Indig, M. J. Williams, and M. H. Ginsberg, *J. Biol. Chem.*, *270*:26146 (1995).

# 5

# Pathways of Cell Recruitment to Mucosal Surfaces

**Michael J. Briskin**   *LeukoSite Inc., Cambridge, Massachusetts*

## I. INTRODUCTION

The mucosal immune system, consisting of mucosal tissues associated with the lacrimal, salivary, gastrointestinal, respiratory, and urogenital tracts and lactating breasts, qualitatively makes up the majority of the lymphoid tissue in the body. The gastrointestinal mucosal immune system has a number of features: it contains specialized structures, such as the Peyer's patches (PPs), where primary immune responses are though to be initiated, followed by patterns of specific recirculation of lymphocyte to mucosal tissues. Subsets of lymphoid cells, primarily IgA-positive B cells and memory T cells predominate at mucosal surfaces, and the predominant mucosal immunoglobulin, secretory IgA, is particularly well adapted to host defenses at mucosal surfaces. This combination of factors in the gastrointestinal immune system thus serves two specific purposes—to protect the host from harmful pathogens while at the same time being tolerant of ubiquitous dietary antigens and microbial flora. In this chapter we emphasize the mechanisms by which leukocytes (primarily lymphocytes) acquire a phenotype to selectively localize from the circulation to mucosal sites.

Lymphocytes are unique among all leukocyte classes: they possess the ability to migrate and recirculate thousands of times during their life history. All other, leukocyte classes—including neutrophils, monocytes, eosinophils, and basophils—cannot recirculate, although the molecular mechanisms governing their homing capacity are remarkably similar to those of lymphocytes. A description of the mechanisms by which all leukocyte classes can migrate beyond the vasculature to different mucosal sites is beyond the scope of this chapter, and many aspects of this topic will be covered in various other chapters of this book. We will

primarily address the molecular mechanisms controlling organ-specific recruitment of lymphocytes to mucosal surfaces, with special emphasis on the gastrointestinal tract.

## II.  DEVELOPMENT OF A HOMING PHENOTYPE

The capacity of lymphocytes to home to tissues in an organ-specific manner is a
developmental process, resulting in specific homing phenotypes in mature B and
T cells. Early B-lymphocyte development occurs in the bone marrow, whereby
hematopoeitic stem cells give rise to lymphoid precursors, B-lineage precursors,
and then mature virgin B cells (1,2). Differentiation at this stage is independent
of exogenous stimuli; the tempo of this process might be regulated by the peripheral immune system. T-cell development, like B cells, also originates in the bone
marrow; limited development occurs here but continues in the thymic microenvironment, a stage that requires specific homing to the thymus. These precursors
are characterized by surface expression of CD7, CD2, cytoplasmic CD3 and are
still CD34$^+$ (3). Mechanisms of selective homing of thymocytes from the bone
marrow to the thymus is not well understood, as the population of this subset is
very small and difficult to directly observe.

Mature B and T lymphocytes leaving primary, lymphoid organs have a full
complement of homing receptors that enable them to circulate to secondary
lymphoid tissues, such as lymph nodes and PPs. These cells migrate poorly, if at
all, to tertiary tissue, even during inflammation. Within lymph nodes and PPs,
lymphocytes cross endothelial barriers though morphologically distinct postcapillary venules (4,5) characterized by the presence of large cuboidal endothelial
cells. These specialized structures, termed high endothelial venules (HEV), are a
distinct feature of lymph nodes and mucosal lymphoid organs (Peyer's patches
and appendix). High endothelial venules are not present in spleen, but this organ
has specialized endothelial cells lining sinuses of the white pulp that may also
participate in recruitment to this site. While normally not found in extralymphoid sites, HEV-like vessels can also be observed in extralymphoid tissues in
sites of chronic inflammation. The study of lymphocyte interactions with HEV
has therefore been of paramount importance in our understanding of the mechanisms regulating lymphocyte traffic, both to lymphoid tissue and to sites of inflammation.

Trafficking of lymphocytes to secondary lymphoid tissues is regulated both by
the expression of receptors in lymphocytes, termed homing receptors, and their
cognate ligands on endothelial cells, termed vascular addressins (6–9). Although
naive B and T cells both have multiple (albeit low levels) of homing receptors
that allow traffic in and out of lymph nodes, Peyer's patches, and the spleen,
there appear to be fundamental differences in the capability of B and T cells to
traffic to peripheral and mucosal sites. This is reflected in the relative numbers of

B and T cells in these organs: B cells appear to preferentially localize to the PPs whereas T cells are the predominant cell population in peripheral organs. In vitro binding assays and short-term homing experiments have demonstrated that B cells preferentially bind to Peyer's patch HEV and T cells to peripheral node HEV (10,11). The preponderance of B cells in the PPs is important for secretion of IgA and IgM into the mucosa (10). Tissue-specific differences between CD4[+] and CD8[+] T-lymphocyte subsets have also been reported, although this distinction appears to be a minor one (11).

As naive B and T cell express unique antigen receptors, it is logical that they possess random migratory patterns, so as to maximize their opportunity to interact with antigen in whatever lymphoid tissue it may exist. Cannulation experiments that show depletion of most B and T lymphocytes from the thoracic duct demonstrate that lymphocytes do continuously recirculate until they either expand into memory/effector populations or die (12). Lymphocytes that encounter antigen within secondary lymphoid tissues that become activated begin to express higher levels and/or activated forms of lymphocyte integrins, including β1 integrins (such as very late antigen [VLA]-4) and β2 integrins (e.g., lymphocyte function associated [LFA]-1) (6,8,9,13–16), which then facilitates interactions at extralymphoid sites of acute and chronic inflammation. These events are mediated by both Ig-like adhesion receptors and extracellular matrix components (6,8,9,15). Clonal expansion and differentiation of lymphocytes into memory and effector subsets results in cell populations that, unlike their naive precursors, now demonstrate preferential migratory pathways and show selective migration to nonlymphoid sites in the body. Evidence of this process comes from studies of sheep that demonstrate the existence of distinct skin-tropic and gut-tropic memory cells and corresponding specialization of their surface phenotype (17).

## III. ORGAN-SPECIFIC LYMPHOCYTE RECRUITMENT AS A MULTI-STEP PROCESS

The ability of leukocytes to migrate out the vasculature is a result of a complex series of interactions, involving multiple receptors and activating events, as illustrated in Fig. 1. Although lymphocytes differ from all other leukocyte classes in their ability to recirculate many times in their life span, the process of extravasation is mediated by very similar and sometimes identical sets of receptors. The initial step in this process is the attachment or tethering to the vessel wall, which is a transient adhesiveness characterized by rolling in the direction of flow (18–21); this is primarily mediated by selectin/carbohydrate interactions (with exceptions to be discussed below). Secondary signals are mediated by local concentrations of chemoattractants or chemokines in the vasculature and their cognate receptors on the leukocyte (22–24). Chemoattractants play a significant role in both the activation of integrin adhesiveness and in the direction of migration

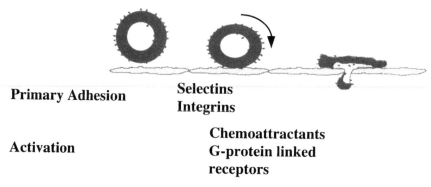

**Figure 1**  Multistep model of leukocyte extravasation. As explained in the text, the process whereby a leukocyte extravasates from the vasculature is a series of sequential steps involving 1.) initial contact mediated by selectins and, in some instances, by integrins; 2.) activation mediated by chemoattractants or chemokines, resulting in 3.) firm adhesion and diapedesis toward a chemoattractant gradient, mediated by integrins and Ig-like adhesion receptors and the extracellular matrix.

patterns of leukocytes through vascular walls. Chemokines can elicit changes in both the level of expression and the affinity of integrins for their cellular ligands. The best-characterized examples of such activation-dependent adhesion receptors are the $\beta1$ and $\beta2$ integrins, capable of mediating leukocyte attachments to vascular Ig-like receptors such as intercellular adhesion molecule (ICAM)-1, ICAM-2, ICAM-3, and vascular cell adhesion molecule (VCAM)-1 (6,9,15,25–32) and extracellular matrix (ECM) proteins such as fibronectin and laminin (14,25,33–35).

Integrin interactions with both Ig-like adhesion receptors and the extracellular matrix result in firm adhesion and subsequent diapedesis through the vessel wall. The selectivity by which leukocytes preferentially migrate to various tissues during both normal circulation and inflammation is due to a combinatorial process that requires each of the above-mentioned interactions (9,19,20,36). The initial transient rolling interaction provided by selectins and their carbohydrate ligands allows the cell to sample the microenvironment for the specific activating signals. A specific activating signal, possibly triggered by chemokines and their cognate

Figure 1 was kindly provided by Charles MacKay.

G protein–coupled receptors, would result in firm adhesion, and chemoattractant gradients would then signal subsequent transendothelial migration via integrins and their ligands.

The proposed multistep adhesion model implies that the specificity of leukocyte endothelial recognition may be determined at any one of these steps: primary adhesion, secondary activation, or firm adhesion/diapedesis. I shall discuss, in more detail, the roles these receptors/factors play in homing with particular respect to their contribution in localization to mucosal sites.

## IV. SELECTINS

To date, three selectin molecules have been identified—all with the shared molecular structure of a single N-terminal lectin domain followed by a single epidermal growth factor (EGF)-like repeat, followed by different numbers of consensus repeats similar to those found in complement-binding proteins (9,37,38). L-selectin is expressed on all circulating leukocytes, except certain memory cell subsets; P-selectin is stored in Weibel-Palade bodies of platelets or $\alpha$-granules in endothelial cells (9,37,38). E-selectin can be induced in vascular endothelial cells by inflammatory mediators such as interleukin (IL)-1, LPS, or tumor necrosis factor (TNF)-$\alpha$.

L-selectin was originally termed the peripheral lymph node (PLN) homing receptor and consistent with this early observation, high-affinity carbohydrate ligands for this receptor are constitutively expressed in PLN HEV (9,37,38). L-selectin has, however, been shown in short-term recruitment assays to play a role in homing to both peripheral and mucosal lymphoid tissues (39–42). Lymphocytes from mice possessing a targeted disruption of the L-selectin gene are completely impaired in their ability to bind and localize to peripheral lymph nodes (43); the role of the L-selectin gene in localization to PPs was also shown in this knockout as a 50% reduction in migration to PP HEVs (43). Studies with an L-selectin Ig chimera demonstrated inhibition of neutrophil recruitment to the peritoneum in response to inflammatory stimuli (44). L-selectin, therefore, appears to facilitate multiple roles in homing to lymphoid tissues and sites of inflammation. The widespread distribution of L-selectin implies that expression alone is not a sufficient indication of the homing capacity of a particular type of leukocyte subset. The role of L-selectin in mediating homing to mucosal sites is likely to be more significant for migration of naive lymphocytes through lymphoid tissue (e.g., PPs), as a mechanism of allowing their sampling of several microenvironments prior to differentiation into memory/effector cells. L-selectin expression on memory/effector cells is heterogeneous, as is the cells ability to extravasate to peripheral vs. mucosal sites, and it is likely that homing to tertiary mucosal sites occurs independent of contributions from L-selectin (discussed in detail below).

## V. CHEMOATTRACTANTS/CHEMOKINES

Although chemoattractants and chemokines are highly implicated in homing, little information exists regarding their capability to direct specificity of leukocyte homing. It is known, however, that certain classes of chemokines prefer to act on certain cell types. Classic chemoattractants such as FMLP and C5a are known to act on a wide range of leukocytes such as neutrophils, basophils, monocytes, and eosinophils (22,24,45,46). Chemokines are small (70–80 residues) polypeptides that are suggested to have greater selectivity for specific leukocyte subsets (22,45). Two subfamilies of chemokines have been described that have homologous sequences differentiated by two cysteine residues–either separated (C-X-C branch) or not separated (C-C branch) by a single amino acid. C-X-C chemokines such as IL-8 and gro/MGSA are chemoattractants for neutrophils and basophils and on fibroblasts in wound healing (22,23,45). C-C chemokines are known to act on several types of leukocytes, including subsets of lymphocytes (22,24,47).

A recently cloned C-C chemokine, eotaxin, has been shown to be highly selective for binding and chemotaxis of purified preparation of eosinophils (48). Analysis of RNA distribution shows that eotaxin is contstitutively expressed in the intestine; immunohistochemical analysis also shows dramatic staining of eotaxin in inflamed respiratory epithelium. The selective expression of this chemokine may contribute to eosinophil migration to these mucosal sites and may help explain the migration of eosinophils that has been suggested to play a role in pathogenesis of asthma and other allergic diseases (49).

With respect to lymphocyte homing, little is known about the role of chemokines in directing specific migratory patterns, although data suggests that regional expression of chemokines might add to the selective recruitment of leukocytes to specific organs in the body. Additionally, it has been shown that pertussis toxin can inhibit adhesion and subsequent diapedesis of lymphocytes into inflammatory sites in vivo (50,51). As pertussis toxin inhibits G protein–coupled signaling events, these data implicate chemokine receptors, as they are all known to couple to G proteins (46) Chemokines can induce adhesiveness for Ig family adhesion receptors and components of the extracellular matrix for both lymphocytes and neutrophils. FMLP has been shown to induce MAC-1 binding on neutrophils (20) to ICAM-1 and in a transfected cell line, the FMLP receptor can also induce VLA-4 mediated adhesiveness for VCAM-1 (52). With respect to lymphocytes, MIP1-$\alpha$, MIP1-$\beta$, RANTES and most recently MCP-1 have all been shown to be chemotactic for lymphocytes. MIP1-$\alpha$ can preferentially act on activated CD8$^+$ T cells; MIP1-$\beta$ preferentially acts on CD4$^+$ T cells in both lymphocyte binding and migration assays (9,53). Additionally, both MIP-1-$\beta$ and RANTES can, in severe combined immunodeficiency (SCID) mice, augment trafficking of human T cells to peripheral lymphoid tissues (54). Direct demonstration of a chemokine that selectively recruits lymphocytes to mucosal sites, however, has yet to be observed.

## VI.  INTEGRINS

Integrins are heterodimeric glycoproteins that play a prominent role in diverse immune cell responses, as well as regulation of lymphocyte trafficking (9,14,15,55). While not a sole determinant of homing capacity, integrin expression can be used to partially classify both cellular distribution and migratory potential of all leukocytes. Among the β2 family of integrins, LFA-1 is more widely distributed on lymphocytes, monocytes, and neutrophils; MAC-1 is expressed in monocytoid lines and neutrophils.

The integrin α4β7 is expressed on subsets of B and T lymphocytes and is also highly expressed on eosinophils, basophils, mast cells, and natural killer (NK) cells (56–59). An early suggestion of the role of the β7 integrin in specific binding to mucosal sites was suggested by studies of lymphocyte adhesion to frozen sections of murine PPs, which demonstrated that an anti-α4 mAb, R1-2, could inhibit binding of the mucosal specific lymphoma TK1 to PP HEV (36,60). It was subsequently shown that a novel β chain, termed βp, associated with α4, and this integrin was named LPAM for lymphocyte Peyer's patch adhesion molecule-1.

Additional mAb were generated; the β chain was cloned and subsequently identified as the β7 integrin, a novel chain restricted in expression to hematopoiteic lineages (61–63). Interestingly, the notion of α4β7 as the Peyer's patch homing receptor was subsequently dismissed after these initial experiments as the anti-α4 mAb R1-2 mAb failed to block adhesion of TK1 cells to affinity-purified mucosal addressin (see below). Only after additional mAbs were generated was the relationship of α4β7 to mucosal addressin cell adhesion molecule-1 (MAdCAM-I) reexamined and firmly established (64,65). The anti-α4 mAb PS/2 and anti-β7 mAbs FIB 22, DATK 32 (which sees a combinatorial epitope of α4β7), and anti-β7 mAb M301 all were shown to block binding of TK1 cells to affinity-purified MAdCAM-1 and CHO cells transfected with recently cloned MAdCAM-1 cDNA (64) (see below). These data confirmed that α4β7 is identical to the previously described Peyer's patch homing receptor LPAM 1 and is a ligand for MAdCAM-1. Studies with the anti-human α4β7 mAb Act-1 (57) have shown that for both CD4+ and CD8+ T cells and B-cell subsets, α4β7 represents a unique mucosal subpopulation (56,58,59)

Another β7 integrin is restricted to a subpopulation of lymphocytes known as intestinal intraepithelial lymphocytes (ilELs). ilELs are located on the basolateral surface of the basement membrane and are poised to respond to antigenic challenges from the intestinal lumen and to interact with epithelial cells. These cells are almost exclusively T lymphocytes and are > 80% CD4+CD8+; they have a restricted repertoire of T-cell receptor variable region gene segment usage relative to peripheral T lymphocytes (66,67). Intestinal intraepithelial lymphocytes are predominantly found in mucosal sites although the mechanism by which this subpopulation selectively migrates is unknown. As ilELs are found in fetal epithelium (68) and in the intestine of germ-free mice, the localization of ilELs is at least in part antigen-independent and may be developmentally regulated (68,69).

The alpha chain that associated with $\beta7$ in ilELS is a novel subunit referred to as $\alpha^E$, which is expressed on > 95% of ilELs in the breast and intestine but < 2% of peripheral T cells (70–72). Expression of $\alpha^E$ is increased by exposure to TGF-$\beta$ in cultured ilEL lines (73) In the mucosal T cell lymphoma TK1, $\alpha4\beta7$ expression is coordinately decreased while $\alpha^E\beta7$ increases in response to TGF-$\beta$, suggesting that $\alpha4\beta7$ positive lymphocytes that migrate to the intestine may (after exposure to TGF-$\beta$) differentiate into $\alpha^E\beta7$ ilELs (36). In vitro adhesion assays demonstrated that $\alpha^E\beta7$ interacted with a novel ligand expressed on epithelial cells and further molecular characterization demonstrated that this counter-receptor for $\alpha^E\beta7$ was E-cadherin (70–72). These results imply that both $\alpha4\beta7$ and $\alpha^E\beta7$ are integrin receptors associated with populations of either resident or homing mucosal lymphocytes.

## VII.  Ig SUPERFAMILY MEMBERS

Interactions between immunoglobulin-related vascular adhesion receptors and integrins play a key role in firm adhesive interactions and possibly transendothelial migration of leukocytes as well. ICAM-1, ICAM-2, and ICAM-3 are all similar yet distinct receptors that can interact with LFA-1 (9,15,26,27,30–32,74). ICAM-1 can also bind to MAC-1 by virtue of an interaction that is distinct from its binding to LFA-1 (75). ICAM-1 expression is induced in vascular endothelium and endothelial cell lines (76–78) in response to pro-inflammatory cytokines, and increased expression plays a significant role in homing to inflammatory sites, albeit not in a tissue-specific manner. ICAM-2 is constitutively expressed and may play a role in trafficking of leukocytes in uninflamed tissues, as is the case for recirculating lymphocytes. All of the ICAMS contribute to antigen-specific interactions inhibition experiments show that mAbs to all three receptors are required to completely block LFA-1–dependent antigen-specific T-cell responses (9,37,74). LFA-1 binding to one of the ICAM receptors contributes to lymphocyte localization in mucosal tissues as shown by in vivo homing experiments and in situ videomicroscopy (see below) (39,41,42,79,80).

Vascular cell adhesion molecule 1 (VCAM-1) is, like ICAM-1, also induced on vascular endothelium by inflammatory stimuli and was originally identified as a six-domain Ig-like structure (29), but several groups subsequently identified a predominant form containing seven domains. The fourth, alternatively sliced domain is a duplicate of domain 1, and both are involved in binding VLA-4 (29,83–85). VCAM-1 also participates in binding to $\alpha4\beta7$ (64,86,87) and recent studies have indicated that distinctions exist for amino acid residues critical for binding VLA-4 vs. $\alpha4\beta7$ (88).

VCAM-1 is expressed on several nonvascular cells (but is more restricted than ICAM-1 ) such as follicular dendritic cells, bone marrow stromal cells, and synovial cells in inflamed joints (89–91). VCAM-1 interacts with mononuclear cells (monocytes and lymphocytes) and eosinophils and basophils primarily by binding VLA-4 ($\alpha4\beta1$) (37,82,92–96,97). Leukocytes expressing the $\alpha4\beta7$ integrin also bind VCAM-1 but apparently with much lower affinity, as certain activating stimuli (phorbol esters or $Mn^{2+}$) are required to achieve significant adhesive interactions (64,86,87). Studies have shown that induction of VCAM-1 is correlated with increased mononuclear cell infiltrates, consistent with its ability to support adhesion of monocytes and lymphocytes but not neutrophils. This interaction is a component of diverse inflammatory events such as atherosclerosis, delayed-type hypersensitivity (101), and rheumatoid arthritis (37,98–101,102). While VCAM-1 is highly induced in these diverse sites of inflammation (37,102), it is rarely observed except weakly on venules of mucosal sites, even in inflammatory bowel disease. VCAM-1 expression may play a role in homing to non-gut associated mucosal sites such as the lung, as VCAM-1 expression has been seen in the endothelium of the trachea of mice that were sensitized by antigen inhalation (103).

The mucosal vascular addressin MAdCAM-1 was first identified in the mouse by the monoclonal antibody MECA-367 as a 60-kd glycoprotein specifically expressed in Peyer's patches and not peripheral lymph nodes (104,105). MAdCAM-1 is also expressed in the lamina propria of the small and large intestine, the lactating mammary gland, and in peripheral lymph node HEV in newborn mice (7). MAdCAM-1 is highly expressed on HEV-like vessels in the chronically inflamed pancreas of the nonobese diabetic mouse (106) and up-regulation of expression in lamina propria venules has recently been observed in two murine models of inflammatory bowel disease, IL-2 and IL-10 knockout mice (107,108; D. Ringler and D. Picarrella, personal communication). MECA-367, like the anti-$\beta7$ and anti-$\alpha4$ antibodies described above, was shown to block binding of the mucosal homing T-cell lymphoma, TK1, to PP HEV in the Stamper-Woodruff assay of lymphocyte binding to frozen tissue sections (104,105), to immuno-isolated MAdCAM-1, and to CHO cells expressing recombinant MAdCAM-1 (64). Additional adhesion assays have shown that MAdCAM-1, unlike its structural relative VCAM-1, fails to bind $\alpha4\beta1$ and binds only lymphocytes expressing $\alpha4\beta7$ (64,86).

## VIII. MOLECULAR CHARACTERIZATION OF MAdCAM-1/$\alpha4\beta7$ INTERACTIONS: MURINE MAdCAM-1

As deduced from cDNA cloning, murine MAdCAM-1 is a member of the immunoglobulin supergene family with three immunoglobulin domains (65). The two amino terminal domains are homologous to its closest relatives among Ig-like adhesion receptors, ICAM-1 and VCAM-1. The third (or membrane proximal) domain, although unrelated to adhesion receptors of this class, does in fact share homology with another mucosal-related Ig family member, namely IgA. In addition to the three immunoglobulin-like domains, MAdCAM-1 has a serine threonine-rich mucin-like domain between the second and third Ig loops. The unique combination of these structural elements has led us to propose that MAdCAM-1 may have the capacity to facilitate more than one function in cell adhesion cascades. Recent domain swapping experiments have shown that the two (ICAM-1 and VCAM-1-like) amino terminal Ig domains alone can facilitate MAdCAM-1/α4β7 interactions (109). Additionally, it has been shown that MAdCAM-1, when expressed in mesenteric lymph nodes, can present L-selectin binding carbohydrates associated with the peripheral node addressin epitope, MECA-79 (110). The addition of these carbohydrates is a likely modification of the mucin domain, due to its relationship to other selectin-binding mucin-like proteins such as GLYCAM-1, CD34, and PSGL-1 (111–113). This modified species of MAdCAM-1 is functional, as lymphocytes can roll on this substrate via L-selectin in laminar flow assays (110). L-selectin plays a role in the circulation of resting lymphocytes through the PPs, and it is likely that MAdCAM-1 is the predominant L-selectin ligand in these tissues (see below). Migration of memory/effector/ T cells and gut-derived thoracic duct blasts is, however, predominately dominated by the α4β7/MAdCAM-1 pathway (41). It is therefore likely that the α4β7 interactions with MAdCAM-1 are more significant than selectin-mediated events with respct to homing to extra-lymphoid tissues (see below).

## IX. HUMAN MAdCAM-1

Low-stringency cross-hybridization to zooblots, degenerate polymerase chain reaction PCR based on murine MAdCAM-1 cDNA sequences, and alignments with related family members indicated that MAdCAM-1 gene sequences were not evolutionarily well conserved. In vitro frozen-section assays, however, demonstrate that mucosal T-cell lines and gut-derived immunoblasts bind to human appendix HEV and that adhesion is blocked only by antibodies against the β7 integrin (and not α4β1) (58,114,115). Additional studies with human lymphoblastoid cell lines also demonstrated conservation of α4β7 binding to CHO cells expressing murine MAdCAM-1 (59). As α4β7[hi] memory T cells appear to define a gut-homing phenotype (56,58,59), these experiments indicated conservation of lymphocyte trafficking to mucosal tissues across species barriers. The conservation of α4β7-dependent interactions thus served as the basis of our design of a functional cloning approach similar to the one used to clone VCAM-1 (116), which we successfully employed to identify the human homologue to MAdCAM-1 (117).

Human MAdCAM-1 avidly binds RPMI 8866s, a human B-cell line that expresses α4β7, whereas VCAM-1 fails to bind this cell line (Fig. 2). Conversely the α4β1 high T cell line, Jurkat, only binds VCAM-1. These results are consistent with previous observations that α4β7/VCAM-1 interactions are activation

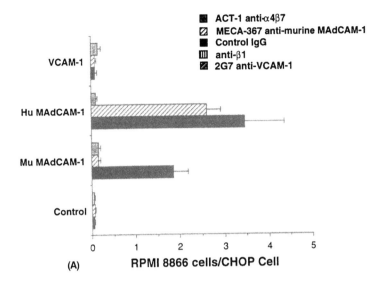

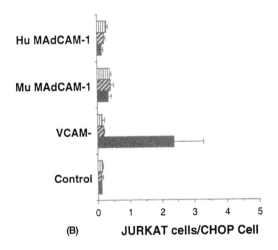

**Figure 2** Human MAdCAM-1 binds only lymphocytes expressing α4β7. (A) RPMI 8866 cells, which express high levels of α4β1, bind CHO/P cells expressing murine and human MAdCAM-1 but fail to bind VCAM-1 or control transfectants. (B) Jurkat cells, which express high levels of α4β1 but no α4β7, bind VCAM-1 transfectants but do not bind control, murine, and human MAdCAM-1 transfectants.

Figures 2 through 4 were reproduced from articles published with permission of the *Journal of Immunology* (refs. 109 and 117). Copyright 1995/1996, *The American Association of Immunologists.*

dependent (64,87) whereas MAdCAM-1 binds α4β7 in the absence of activation. Recently, we have used multicolor microfluorimetry to analyze differences in lymphocyte binding to murine MAdCAM-1 vs. VCAM-1 These studies also have shown that human lymphocyte subsets expressing high levels of α4β7 preferentially bind murine MAdCAM-1 (58). Although β7$^{hi}$ CD4$^+$ memory cells can also bind VCAM-1, adhesion of these cells is inhibited by over 90% by blockade with an anti-β1 MAb. These studies clearly confirmed that α4β7 is necessary for interaction with MAdCAM-1 and expression of this integrin can subdivide circulating B and T cell populations with respect to their ability to bind these vascular ligands. Conversely, α4β7$^-$β1$^{hi}$ T and B cells preferentially bind VCAM-1, which, as stated above, is a vascular ligand preferentially expressed in sites of inflammation other than the gut. This differential adhesive potential is likely a major point of delineation between mucosal and nonmucosal trafficking compartments, allowing separation of specialized immune responses. The cloning of this homologue demonstrates that this distinction is retained in humans.

Although human and murine MAdCAM-1 sequences are quite divergent, sequence comparisons demonstrate that the first two (N-terminal) Ig-like domains, at 57% identity (compared to 39% for the entire protein), are the most highly conserved regions of these receptors. The 2 Ig loops display several features conserved with murine MAdCAM-1, including 1.) double cysteine residues, separated by 3 amino acids in the first Ig domain; 2.) an identical nine amino acid stretch in Ig domain 1, containing the sequence LDTSL, which aligns with a consensus motif for integrin/IG family member interactions (Fig. 3) (116,118–120); and 3.) a uncharacteristically large second immunoglobulin domain, with approximately 70 amino acids between cysteine residues. This is a novel feature in comparison to other Ig-like adhesion receptors, which usually have 40 to 50 residues between the cysteines in each domain (121,122). Interestingly, the homology to IgA observed in murine MAdCAM-1 is not retained in human MAdCAM-1. The function of the IgA-like domain in murine MAdCAM-1 has yet to be determined.

Although the LDTSL motif in domain 1 has general conservation with respect to other Ig adhesion receptors such as ICAM-1, ICAM-2, ICAM-3, and VCAM-1 (29–32,118), this exact sequence has only been found in murine MAdCAM-1 (65). The functional significance of this motif has been demonstrated by the fact that a point mutation in amino acid 61 (illustrated in Fig. 3), which changed the first L to an R in this sequence, had a dramatic effect on α4β7 binding to murine MAdCAM-1 (109). Recent crystallographic analysis of the two N-terminal domains of VCAM-1 indicate that the analogous sequence, QIDSPL, resides in the highly exposed C-D loop of Ig domain 1 and is thus free to interact with its integrin ligand (123,124). The conservation of this sequence, in addition to studies showing that it can be exchanged with its homologous sequence in ICAM-1 with no change in function, suggests that this motif is a general integrin recognition el-

| Species | Protein | Sequence | | | | | | | | | |
|---|---|---|---|---|---|---|---|---|---|---|---|
|  | Ig consensus |  |  | G/Q | I/L | E/D | T/S | P/S |  |  |  |
| Hu | ICAM-1 D1 | L | L | G | I | E | T | P | L | P | K |
| Mu | ICAM-1 D1 | S | L | G | L | E | T | Q | W | L | K |
| Hu | ICAM-2 D1 | V | G | G | L | E | T | S | L | N | K |
| Mu | ICAM-2 D1 | M | G | G | L | E | T | P | T | N | K |
| Hu | ICAM-3 D1 | K | I | A | L | E | T | S | L | K | S |
| Hu/Ra | VCAM-1 D1 | R | T | Q | I | D | S | P | L | N | G |
| Mu | VCAM-1 D1 | R | T | Q | I | D | S | P | L | N | A |
| Hu | VCAM-1 D4 | R | T | Q | I | D | S | P | L | S | G |
| Mu/Ra | VCAM-1 D4 | R | T | Q | I | D | S | P | L | N | G |
| Mu | MAdCAM-1 D1 | W | R | G | L | D | T | S | L | G | S |
| Mu | " (L-R61 mutant) | W | R | G | R | D | T | S | L | G | S |
| Hu | MAdCAM-1 D1 | W | R | G | L | D | T | S | L | G | S |
| Mu | α4 | N | V | S | L | D | V | H | R | K | A |
| Hu | FN CS1 | P | E | I | L | D | V | P | S | T | V |

**Figure 3**  A common integrin binding motif in immunoglobulin-like adhesion receptors is also present in MAdCAM-1. An amino acid motif believed to be involved in binding to integrins is shown for several species of ICAM-1,2, and 3, VCAM-1, and MAdCAM-1. The wild-type sequence of MAdCAM-1 is compared to the leucine-to-arginine mutation that affects binding to α4β7. This sequence is identical in human MAdCAM-1 In addition, α4 integrin binding LDV sequences from the CS1 peptide of fibronectin and the integrin α4 chain are shown.

ement (116). Within domain 2 of VCAM-1 is a highly extended C′-E loop with a similar sequence that is even more accentuated in all species of MAdCAM-1 clones. This proposed human MAdCAM-1 C′E loop, which consists of nine negatively charged (D or E) residues is 19 amino acids in length compared to 10 to 12 amino acids in VCAM-1 (123,124). Although the C′E loops of the MAdCAM-1 clones are not identical, the length and the highly charged character of this region is well conserved. The crystal structure of VCAM-1 shows that this loop is also highly exposed and in close proximity to the CD loop of domain 1 and, along with its unique features, may also contribute to its specificity of α4β7 binding. Mutagenesis and domain-swap experiments are currently in progress to determine the relevant contributions of each of these elements in binding α4β7.

The mucin regions are the most divergent sequences in the MAdCAM-1 homologues. Extensive polymorphism and sequence divergence has been documented in other mucin-like sequences as well. Episialin, for example, contains a 20 amino acid repeat that can vary in copy number from 30 to 90 (125,126);

repetitive portions of intestinal mucins are not well conserved between rodents and humans (125,126). It is of interest to speculate on the capability of these divergent sequences to serve as substrates for L-selectin–binding carbohydrates. No specific sequence requirement has been identified with this modification: GLYCAM-1, CD34, and murine MAdCAM-1 all bind L-selectin but are only related by their high content of S/T/P residues (65,110–113). It is therefore likely that these variant sequences in the MAdCAM-1 homologues described are all capable of presentation of this carbohydrate structure. An alternative possibility, however, is that there is post-transcriptional processing to generate variants of the cDNAs isolated that might differ in their abilities to bind L-selectin. Along these lines, we have recently described a alternatively processed form of murine MAdCAM-1 that lacks the mucin/IgA domain (127) and is thus proposed to lack the capability to bind L-selectin. Monoclonal antibodies to the human receptor have been produced and are currently in use to ask about the carbohydrate modifications of human MAdCAM-1 and their contribution to binding L-selectin. The proposed structural differences between human and murine MAdCAM-1 are shown in Fig. 4.

## X. IN VITRO AND IN SITU STUDIES OF MUCOSAL LYMPHOCYTE HOMING

The selectivity of MAdCAM-1 for $\alpha4\beta7$ positive lymphocytes has been demonstrated in studies of short-term recruitment assays of labeled lymphocyte homing to peripheral vs. mucosal sites in the mouse (39,41). The anti-MAdCAM-1 mAb MECA-367 completely blocks homing to PPs and partially blocks homing to the mesenteric lymph nodes but has little effect on localization to peripheral lymph nodes. This would be consistent with the gradient of MAdCAM-1 expression from a peripheral tissue where it is almost absent to an intermediate site (mesenteric lymph nodes) to a mucosal lymphoid site where expression is extremely high (PPs). Although anti-MAdCAM-1 mAbs completely block homing to PPs, this event was only partially inhibited by pretreatment of lymphocytes with mAbs to the MAdCAM-1 ligands, whether it be anti-$\alpha4$, anti-$\beta7$, or anti-$\alpha4\beta7$ mAbs. Complete levels of inhibition were only observed when these mAbs were used in combination with the anti-L-selectin mAb MEL-14. In addition, antibodies to LFA-1 (CD18) also reduce lymphocyte homing to PPs, indicating that $\beta2$ integrins play a role in this pathway. When these experiments were repeated with isolated immunoblasts, however, anti-$\alpha4$ or anti-$\beta7$ mAbs completely blocked homing to the PPs (41) and to the intestine as well the anti-L-selectin mAb had little effect on homing to the intestine.

The discrepancy between the effect of anti-$\beta7$ and anti-$\alpha4$ mAbs and the effect of anti-MAdCAM-1 mAbs to differentially inhibit homing of lymph node cells (LNC) vs. immunoblasts to PPs has been partially resolved by in situ

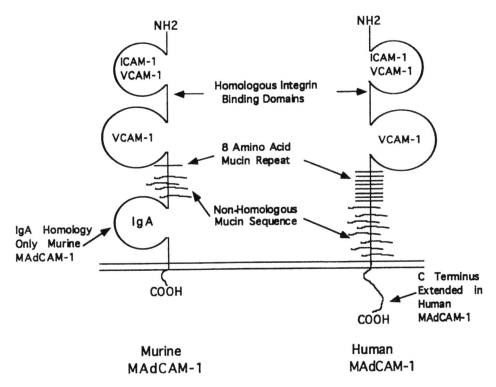

**Figure 4** Compared structures of murine and human MAdCAM-1. Highlighted are the similar integrin-binding amino terminal domains while the membrane proximal IgA-like domain in murine MAdCAM-1 is replaced by additional mucin sequences. The mucin sequences are believed to be the sites for addition of L-selectin binding carbohydrates; domain-swapping experiments with murine MAdCAM-1 have shown that both Ig domains 1 and 2 are required for binding $\alpha 4 \beta 7$.

videomicroscopy (79). Lymph node cells from young mice were shown to roll within PP HEV, and this rolling behavior was completely abrogated by pretreatment with anti–L-selectin mAb MEL 14. Without this rolling step, only a small subpopulation of these cells could firmly attach to the vessel wall. Interestingly, the anti-MAdCAM-1 mAb MECA-367 inhibited L-selectin mediated rolling to the same extent as anti–L-selectin mAbs. As MAdCAM-1 can be modified to present L-selectin–binding carbohydrates, it appears that the anti–MAdCAM-1 MAb disrupts L-selectin binding, even though MECA-367 maps to the amino terminal domain of MAdCAM-1(109), and it is probably the mucin region that bears the L-selectin binding glycotopes (109). It is likely that this effect is due to cross-linking, as Fab fragments fail to demonstrate the same effect as whole mAb. This modified species of MAdCAM-1 is therefore the major L-selectin ligand in the PP. As LFA-1 was shown to contribute to homing to the PP, this would explain the inability of an anti-α4 mAb to completely inhibit localization of mesenteric lymph node lymphocytes, as a combination of L-selectin–mediated rolling and LFA-1–mediated firm adhesion could recruit a subpopulation of these cells in the absence of a contribution from α4β7.

These data suggest that overlapping functions are provided by several different adhesion receptors in allowing their localization to lymphoid tissue. This might be considered advantageous to naive cells, as to maximize their chances of encountering antigen in the event that one adhesive interaction is mutant or missing.

In contrast to resting LNCs, primary interactions with preactivated LNC (treated with Mn²⁺ or PMA) or α4β7ʰⁱ TK1 cells could be inhibited by treatment with anti-α4 mAbs. The specificity of inhibition of preactivated LNC or α4β7ʰⁱ TK1 cells by anti-α4β7 mAbs might be considered representative of levels and types of adhesion receptors (and consequently homing phenotypes) expressed by memory vs. naive lymphocytes. It was recently shown that α4β7 can mediate L-selectin–independent primary interactions in vitro and in small venules in intestinal lamina propria as well (128). In the PPs where L-selectin interactions play a significant role with resting LNCs, lymphocytes with activated α4β7 can bypass the requirement for selectin-mediated interactions. Thus, lymphocyte activation can serve to regulate both the mechanism and sites of lymphocyte homing. The ability of α4β7 to serve as both a primary and secondary adhesion receptor may help explain the ability of mucosal immunoblasts to home efficiently to the intestine without an influence of a selectin interaction. It may also explain the gut-selective trafficking of subsets of memory T cells (17,129), many of which are α4β7ʰⁱ but lack L-selectin. The inefficiency of α4 integrins to solely mediate initial contact of resting lymphocytes in vessels would explain a selectivity of naive cells to traffic preferentially through L-selectin–binding lymphoid organ HEV, thus allowing naive cells to sample a full repertoire of antigens; only after exposure to antigen would a lymphocyte differentiate into a population committed in its homing potential. The differences in receptors that control homing of naive vs. immunoblasts (or memory cells ) are illustrated in Figure 5. For naive cells, L-selectin, α4β7, and LFA-1 all operate in a sequential yet overlapping function, whereas α4β7 alone can facilitate homing of memory cells or immunoblasts.

## XI. MUCOSAL HOMING OUTSIDE THE GASTROINTESTINAL TRACT; OTHER LEUKOCYTE CLASSES

The integrin α4β1 may play a role similar to α4β7 in regulation of leukocyte traffic to nonmucosal sites in the body. Studies have revealed that VLA-4 expressed on eosinophils, lymphocytes, and in a transfected cell line can, like

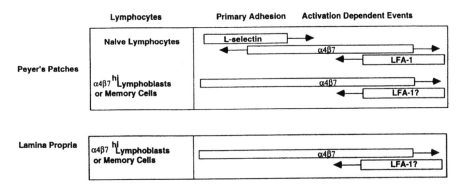

**Figure 5** The sequential and overlapping roles of multiple adhesion receptors are different for naive vs. memory lymphocytes. Naive cells enter the Peyer's patches using L-selectin and α4β7, both of which interact with MAdCAM-1. LFA-1 also participates in extravasation, presumably by interacting with either ICAM-1, ICAM-2, or ICAM-3. Lymphoblasts and memory cells, by virtue of having activated α4β7, do not require a selectin carbohydrate interaction and thus both primary adhesion and secondary events are mediated by α4β7; LFA-1 may still participate in later events.

α4β7, also support rolling via an interaction with VCAM-1 (130,131). In the rat, anti-α4 integrin inhibition of trafficking of T cells to cutaneous sites of inflammation and arthritic joints is likely mediated by VCAM-1/α4β1 interactions (132,133). It would thus appear that the subdivision of lymphocytes in mutually exclusive subsets with respect to levels of α4β1 vs. α4β7 may be the best indication of a cell's commitment to homing to a mucosal vs. a nonmucosal site.

Studies have shown that there is a common mucosal immune system; however, the expression of α4β7 may not be a sole prediction of homing to all mucosal tissues. Pulmonary T cells, for example, express low levels of α4β7 and L-selectin; this argues against the functions of these receptors in homing of T cells to the lung (134). Consistent with this observation, MAdCAM-1 RNA is not detected in lung and MAdCAM-1 is histologically negative in venules of lung parenchyma (65,104) in mice. We have yet to fully determine whether human MAdCAM-1 is significantly expressed in normal or asthmatic lung. In contrast to blood T cells, in which the predominant β7 integrin is α4β7, the integrin $\alpha^E\beta7$ is highly enriched on pulmonary T cells; this suggests that this integrin, more than α4β7, is involved in homing to the lung. It should be pointed out, however, that bronchiolar lavage fluid T cells also express high levels of the β1 integrin and half the T cells in the lung lack expression of $\alpha^E\beta7$. As $\alpha^E\beta7$ has been shown to bind to *epithelial* cells via its interaction with E-cadherin (135) it is possible that

---

the enrichment of this subset is independent of homing via $\alpha^E\beta7$. The mechanisms for enrichment of this subset may involve an uncharacterized receptor that mediates homing to the lung. Alternatively, local factors in lung epithelium (e.g., TGF-$\beta$) may induce expression of this integrin on subsets that were recruited by different mechanisms. Lastly, it should be mentioned that in vitro frozen-section assays have indicated that the lung may actually have a homing capacity between a peripheral tissue and mucosal lymphoid tissue (136). It would therefore be difficult to characterize a single lung-associated homing receptor, as in mesenteric lymph nodes, that co-express high levels of MAdCAM-1 and L-selectin binding carbohydrates with associated MECA-79 reactivity (7,8).

The overlapping functions contributing to tissue-specific homing of lymphocytes may very well apply to other leukocyte subsets such as monocytes, mast cells, basophils, and eosinophils, which also coordinately express selectins, $\beta2$ integrin, and $\alpha4$ integrin adhesion receptors. Eosinophil and basophil adhesion has been shown to be mediated in part by VCAM-1/VLA-4 interactions (93,96,97,136,137); basophils, eosinophils, and mast cells also express moderate levels of $\alpha4\beta7$, and in vitro assays have shown that all these leukocyte classes can bind MAdCAM-1 via $\alpha4\beta7$ (138 to 140, and M. Briskin, unpublished) although activation is required for eosinophil binding to MAdCAM-1. Previous in vivo studies of eosinophil migration in models of allergic inflammation have employed the HP1/2 mAb (141), which as an anti-$\alpha4$ mAb will inhibit $\alpha4\beta1$ and $\alpha4\beta7$ binding to VCAM-1 and $\alpha4\beta7$ binding to MAdCAM-1. Future studies will have to be employed using specific mAbs to $\alpha4\beta7$ (57) to fully understand the nature of these molecular interactions, along with histology to look at expression of human MAdCAM-1 in mucosal sites other than the gut. Lastly, it will be of interest to determine whether VCAM-1 interactions with $\alpha4\beta7$ are physiologically related to any inflammatory events in the mucosal tissues as well.

## XII.  CONCLUDING REMARKS

During B and T-cell ontogeny, lymphocytes acquire a capacity to differentially migrate to distinct tissues in the body. The ability to home to a particular site can be classified in part by the nature, activation state and levels of adhesion receptors expressed on each lymphocyte subset. With respect to homing to mucosal sites represented by the gastrointestinal system, the best-characterized receptor ligand interaction involves the integrin $\alpha4\beta7$ binding to the mucosal vascular addressin, MAdCAM-1. L-selectin also plays a role in migration of primarily naive cells to gut-associated lymphoid tissue (PPs), and LFA-1 can also facilitate migration of mucosal lymphocytes, but neither is specific to mucosal tissue. With respect to mucosal tissues other than the intestine (such as the lung), the selectivity of $\alpha4\beta7$ as a mucosal receptor has yet to be determined. $\alpha^E\beta7$ is also an integrin associated with mucosal lymphocytes, but it is not clear whether it is actually a homing receptor or derived from α4β7 expressing lymphocytes and/or expressed on resident mucosal lymphocytes. Furthermore, the ubiquitous expression of E-cadherin would argue against a role in homing to mucosal sites.

Activation of α4β7 plays a role in determination of specificity of interactions in mucosal tissues; it remains to be seen if specific chemokines play a role in this activation event and selective recruitment to tissue. Finally, the recent identification of human MAdCAM-1 now gives us the tools to further understand the role of this vascular addressin in both normal circulation and inflammation.

## ACKNOWLEDGMENTS

I would like to thank Eugene C. Butcher for the initial opportunity to contribute to what has been an exciting body of research over the past 5 years. I would also like to thank all those whom I have worked with and who have greatly contributed to this field, including Ellen Berg, Leslie McEvoy, Connie Berlin, Rob Bargatze, Lucia Rott, Sylvia Sampaio, and Anne Shyjan.

## REFERENCES

1. P. W. Kincade, G. Lee, C. E. Pietrangeli, S.-I. Hayashi, and J. M. Gimble, *Annu. Rev. Immunol.*, 7:11 (1989).
2. F. M. Uckum, *Blood*, 76:1908 (1990).
3. L. W. M. M. Terstappen, S. Huang, and L. J. Picker, *Blood*, 79:666 (1992).
4. J. L. Gowens and E. J. Knight, *Proc. R. Soc. London*, 159:257 (1964).
5. V. T. Marachis and J. L. Gowens, *Proc. R. Soc. London*, 159:283 (1964).
6. E. C. Butcher, *Cell*, 67:1033 (1991).
7. E. L. Berg, L. J. Picker, M. K. Robinson, P. R. Streeter, and E. C. Butcher, *Cellular and molecular mechanisms of inflammation*, 2:111 (1991).
8. L. J. Picker and E. C. Butcher, *Ann. Rev. Immunol.*, 10:561 (1992).
9. T. A. Springer, *Cell*, 76:301–314 (1994).
10. S. K. Stevens, I. L. Weissman, and E. C. Butcher, *J. Immunol.*, 128:844 (1982).
11. S. T. Pals, G. Kraal, E. Horst, A. de Groot, R. J. Scheper, and C. J. L. M. Meijer, *J. Immunol.*, 137:760 (1986)
12. J. Sprent, *Cell Immunol.*, 7:10 (1973).
13. B. M. C. Chan, J. G. P. Wong, A. Rao, and M. E. Hemler, *J. Immunol.*, 147:398 (1991).
14. R. O. Hynes, *Cell*, 69:11 (1992).
15. T. A. Springer, *Nature*, 346:426 (1990).
16. M. A. Aranaout, *Blood*, 75:037 (1990).
17. C. R. Mackay, W. L. Marston, L. Dudler, L. O. Spertini, T. F. Tedder, and W. R. Hein, *Eur J. Immunol.*, 22:887 (1992).
18. K. Ley, P. Gaehtgens, C. Fennie, M. S. Singer, L. A. Lasky, and S. D. Rosen, *Blood*, 77:2553 (1991).
19. U. H. vonAndrain, J. D. Chambers, L. M. McEvoy, R. F. Bargatze, K.-E. Arfors and E. C. Butcher, (1991) *Proc. Natl. Acad. Sci. USA*, 88:7538 (1991).

20. M. B. Lawrence and T. A. Springer, *Cell, 65*:859–873 (1991).
21. T. N. Mayadas, R. C. Johnson, H. Rayburn, R. O. Hynes, and D. D. Wagner, *Cell, 74*:541 (1993).
22. J. J. Oppenheim, C. O. C. Zachariae, N. Mukaida, and K. Matsushima, *Annu. Rev. Immunol., 9*:617 (1991).
23. M. D. Miller and M. S. Krangel, *CRC Crit. Rev. Immunol., 12*:17 (1992).
24. P. Kuna, S. R. Reddigari, T. J. Schall, D. Rucinski, M. Sadick, and A. P. Kaplan, *Immunol., 150*:1932 (1993).
25. M. E. Hemler, *Annu Rev. Immunol., 8*:365 (1990).
26. A. R. deFougerolles, S. A. Stacker, R. Schwarting, and T. A. Springer, *J. Exp. Med., 174*:253 (1991).
27. A. R. deFougerolles and T. A. Springer, *J. Exp. Med., 175*:185 (1992).
28. J. Fawcett, C. L. L. Holness, L. A. Needham, H. Turley, K. C. Gatter, D. Y. Mason, and D. L. Simmons, *Nature, 360*:481 (1992).
29. L. Osborn, C. Hession, R. Tizard, C. Vassallo, S. Luhowskyj, G. Chi-Rosso, and R. Lobb, *Cell, 59*:1203 (1989).
30. D. E. Staunton, S. D. Marlin, C. Stratowa, M. L. Dustin, and T. A. Springer, *Cell, 52*:925 (1988).
31. D. E. Staunton, M. L. Dustin, and T. A. Springer, *Nature, 339*:61 (1989).
32. R. Vazeux, P. A. Hoffman, J. K. Tornita, E. S. Dickinson, R. L. Jasmine, T. St. John, and W. M. Gallatin, *Nature, 360*:485 (1992).
33. E. Ruoslahti, *J. Clin. Invest., 87*:1 (1991).
34. A. L. Komoriya, L. J. Green, M. Mervic, S. S. Yamada, K. M. Yamada, and M. J. Humphries, *J. Biol. Chem., 265*:4020 (1991).
35. J. L. Guan and R. O. Hynes, *Cell, 60*:53 (1990).
36. E. C. Butcher, R. G. Scollay and I. L. Weissman, *Eur. J. Immunol., 10*:556 (1980).
37. M. P. Bevilacqua, *Annu. Rev. Immunol., 11*:767 (1993).
38. D. Vestweber, *Curr. Top. Micro. Immunol., 184*:65 (1993).
39. A. D. Hamaan, D. Jablonski-Westrich, P. Jonas and H.-G. Thiele, *J. Immunol., 147*:4178 (1991).
40. R. F. Bargatze, P. R. Streeter and E. C. Butcher, *J. Cell. Biochem., 42*:219 (1990).
41. A. Hamann, D. P. Andrew, D. Jablonski-Westrich, B. Holzmann, and E. C. Butcher, *J. Immunol., 152*:3282 (1994).
42. A. Hamaan and H.-G.Thiele, *Immunol. Rev., 108*:19 (1989).
43. M. L. Arbones, D. C. Ord, K. Ley, H. Ratech, C. Maynard Curry, G. Otten, D. J. Capon, and T. F. Tedder, *Immunity, 1*:247 (1994).
44. S. R. Watson, C. Fennie, and L. A. Laskey, *Nature, 349*:164 (1991).
45. T. J. Schall, *Cytokine, 3*:165 (1991).
46. C. Gerard and N. P. Gerard, *Molecular biology of human neutrophil chemotactic receptors. Immunopharmacology of Neutrophils* (P. J. Hallewell, ed.), 4, Academic Press, London (1994).
47. D. D. Taub, K. Conlon, A. R. Lloyd, J. J. Oppenheim, and D. J. Kelvin, *Science, 260*:355 (1993).
48. P. D. Ponath, S. Qin, D. J. Ringler, I. Clark-Lewis, J. Wang, N. Kassam, H. Smith, X. Shi, J.-A. Gonzalo, W. Newman, J.-C. Gutierrez-Ramos, and C. R. Mackay, *J. Clin. Invest., 97*:604 (1996).

49. P. F. Weller, *Curr. Opin. Immunol.*, *4*:782 (1992).
50. G. J. Sprangrude, F. Sacchi, H. R. Hill, D. E. Van Epps, and R. A. Daynes, *J. Immunol.*, *135*:4135 (1985).
51. R. F. Bargatze, and E. C. Butcher, *J. Exp. Med.*, *178*:367 (1993).
52. S. Honda, J. J. Campbell, D. P. Andrew, B. Englehardt, B. A. Butcher, R. A. Warnock, R. D. Ye, and E. C. Butcher, *J. Immunol.*, *152*:4026 (1994).
53. T. Tanaka, D. H. Adams, S. Hubscher, H. Hirano, U. Siebenlist, and S. Shaw, *Nature*, *361*:79 (1993).
54. W. J. Merphy, Z.-G. Tian, O. Asai, S. Funakoshi, P. Rotter, M. Henry, R. M. Streiter, S. Kunkel, D. L. Longo, and D. D. Taub, *J. Immunol.*, *156*:2104 (1996).
55. R. S. Larson and T. A. Springer, *Immunol. Rev.*, *144*:181 (1990).
56. T. Schweighoffer, Y. Tanaka, M. Tidswell, D. J. Erle, K. J. Horgan, G. E. G. Luce, A. I. Lazarovits, D. Buck, and S. Shaw, *J. Immunol.*, *151*:717 (1993).
57. A. I. Lazarovits, R. A. Moscicki, J. T. Kumick, D. Camerini, A. K. Bhan, L. G. Baird, M. Erikson, and R. B. Colvin, *J. Immunol.*, *133*:1857 (1984).
58. L. S. Rott, M. J. Briskin, D. P. Andrew, E. L. Berg, and E. C. Butcher, *J. Immunol.*, *156*:3727 (1996).
59. D. J. Erle, M. J. Briskin, E. C. Butcher, A. Garcia-Pardo, A. I. Lazarovits, and M. Tidswell, *J. Immunol.*, *153*:517 (1994).
60. B. Holzman, B. W. McIntyre, and I. L. Weissman, *Cell*, *56*:37 (1989).
61. Q. Yuan, W.-M. Jiang, E. Leung, D. Hollander, J. D. Watson, and G. W. Krissansen, *J. Biol. Chem.*, *267*:7352 (1992).
62. D. J. Erle, C. Ruegg, D. Sheppard, and R. Pytella, *J. Biol. Chem.*, *266*:11009 (1991).
63. M. C.-T. Hu, D. T. Crowe, I. L. Weissman, and B. Holzmann, *Proc. Natl. Acad. Sci. USA*, *89*:8254 (1992).
64. C. Berlin, E. L. Berg, M. J. Briskin, D. P. Andrew, P. J. Kilshaw, B. Holzmann, I. L. Weissman, A. Hamann, and E. C. Butcher, *Cell*, *74*:185 (1993).
65. M. J. Briskin, L. M. McEvoy, and E. C. Butcher, *Nature*, *363*:461 (1993).
66. D. M. Asarnow, T. Goodman, L. LeFrancois, and J. P. Allison, *Nature*, *341*:60 (1989).
67. C. Van Kerckhove, G. J. Russell, K. Deusch, K. Reich, A. K. Bhan, and M. B. Brenner, *J. Exp. Med.*, *175*:57 (1991).
68. L. Lefrancois, *Semin. Immunol.*, *3*:99 (1991).
69. A. Bandiera, T. Moto-Santos, S. Itohara, S. Degermann, C. Heusser, S. Tonegawa, and A. Coutinho, *J. Exp. Med.*, *172*:239 (1990).
70. S. K. Shaw, K. L. Cepek, E. A. Murphy, G. J. Russel, M. B. Brenner, and C. M. Parker, *J. Biol. Chem.*, 6016 (1994).
71. C. M. Parker, C. Pujades, M. B. Brenner, and M. E. Hemler, *J. Biol. Chem.*, *268*:7028 (1993).
72. C. M. Parker, K. Cepek, G. J. Russell, S. K. Shaw, D. Posnett, R. Schwarting, and M. B. Brenner, *Proc. Natl. Acad. Sci. USA*, *89*:1924 (1992).
73. K. L. Cepek, C. M. Parker, J. L. Madara, and M. B. Brenner, *J. Immunol.*, *150*:3459 (1993).
74. A. R. deFougerolles, X. Qin, and T. A. Springer, *J. Exp. Med.*, *179*:619 (1994).
75. M. S. Diamond, D. E. Staunton, S. D. Marlin, and T. A. Springer, *Cell*, *65*:961 (1991).

76. E. E. Sikorski, R. Hallmann, E. L. Berg, and E. C. Butcher, *J. Immunol.*, *151*:5239 (1993).
77. J. S. Pober, M. A. Gimbrone, L. A. Lapierre, D. L. Mendrick, W. Fiers, R. Rothlein, and T. A. Springer, *J. Immunol.*, *137*:1893 (1986).
78. M. L. Dustin, R. Rothlein, A. K. Bhan, C. A. Dinarello, and T. A. Springer, *J. Immunol.*, *137*:245 (1986).
79. R. F. Bargatze, M. A. Jutila, and E. C. Butcher, *Immunity*, *3*:99 (1995).
80. A. Hamann, D. Jablonski-Westrich, A. Duijvestijn, E. C. Butcher, H. Baisch, R. Harder, and H.-G. Thiele, *J. Immunol.*, *140*:693 (1988).
81. G. E. Rice, J. M. Munro, C. Corless, and M. P. Bevilacqua, *Am. J. Pathol.*, *138*:385 (1991).
82. G. E. Rice, J. M. Munro, and M. P. Bevilacqua, *J. Exp. Med.*, *171*:1369 (1990).
83. L. C. Osborn, C. Vassallo, and C. D. Benjamin, *J. Exp. Med.*, *176*:99 (1992).
84. T. Polte, W. Newman, and T. Venkat Gopal, *Nucleic Acids Res.*, *18*:5901 (1990).
85. C. Hession, R. Tizard, C. Vassallo, S. B. Schiffer, D. Goff, P. Moy, G. Cji-Rosso, S. Luhowskyj, R. Loss, and L. Osborn, *J. Biol. Chem.*, *266*:6682–6685 (1991).
86. U. G. Strauch, A. Lifka, U. GoBlar, P. J. Kilshaw, J. Clements, and B. Holzmann, *Int. Immunol.*, *6*:263 (1994).
87. C. Ruegg, A. A. Postigo, E. E. Sikorski, E. C. Butcher, R. Pytela, and D. J. Erle, *Cell Biol.*, *117*:179 (1992).
88. H. H. Chiu, D. T. Crowe, M. E. Renz, L. G. Presta, S. Jones, I. L. Weissman, and S. Fong, *J. Immunol.*, *155(11)*:5257 (1995).
89. A. S. Freedman, J. M. Munro, C. Morimoto, B. W. McIntyre, K. Rhynhardt, N. Lee, and L. M. Nadler, *Blood*, *79*:206 (1992).
90. K. Miyake, K. Medina, K. Ishihara, M. Kimoto, R. Auerbach, and P. W. Kincade, *J. Cell Biol.*, *114*:557 (1991).
91. D. H. Ryan, B. L. Nucci, C. N. Abboud, and J. M. Winslow, *J. Clin. Invest.*, *88*:996 (1991).
92. N. Graber, T. Venkat Gopal, D. Wilson, L. D. Beall, T. Polte, and W. Newman, *J. Immunol.*, *1145*:819 (1990).
93. B. S. Bochner, F. W. Luscinskas, M. A. Gimbrone, W. Newman, S. A. Sterbinsky, C. P. Derse-Anthony, D. Klunk, and R. P. Schleimer, *J. Exp. Med.*, *173*:1553 (1991).
94. B. R. Schwartz, E. A. Waymer, T. M. Carlos, H. D. Ochs, and J. M. Harlan, *J. Clin. Invest.*, *85*:2019 (1990).
95. N. Oppenheimer-Marks, L. S. Davis, D. Tompkins Bogue, J. Ramberg, and P. E. Lipsky, *J. Immunol.*, *147*:2913 (1991).
96. A. Dobrina, R. Menegazzi, T. M. Carlos, E. Nardon, R. Cramer, T. Zacchi, J. M. Harlan, and M. Patriarca, *J. Clin. Invest.*, *88*:20 (1991).
97. U. Kyan-Aung, D. O. Haskard, and T. H. Lee, *Am. J. Respir. Cell Mol. Biol.*, *5*,445 (1991).
98. M. I. Cybulsky and M. A. J. Gimbrone, *Science*, *251*:788 (1991).
99. D. M. Briscoe, F. J. Schoen, G. E. Rice, M. P. Bevilacqua, P. Ganz, and J. S. Pober, *Transplantation*, *51*:537 (1991).
100. A. E. Koch, J. C. Burrows, G. K. Haines, T. M. Carlos, J. M. Harlan, and S. J. Leibovich, *Lab. Invest.*, *64*:313 (1991).

101. P. Norris, R. N. Poston, D. S. Thomas, M. Thornhill, J. Hawk, and D. O. Haskard, *J. Invest. Dermatol.*, *763*:70 (1991).

102. A. A. Postigo, J. Teixido, and F. Sanchez-Madrid, *Res. Immunol.*, *144*:657 (1993).

103. H. Nakajima, H. Sano, T. Nishimura, S. Yoshida, and I. Iwamoto, *J. Exp. Med.*, *179*:1145 (1994).

104. P. R. Streeter, E. Lakey-Berg, B. T. N. Rouse, R. F. Bargatze, and E. C. Butcher, *Nature*, *331*:41 (1988).

105. M. Nakache, E. Lakey-Berg, P. R. Streeter, and E. C. Butcher, *Nature*, *337*:179 (1989).

106. A. Hanninen, C. Taylor, P. R. Streeter, L. S. Stark, J. M. Sarte, J. A. Shizuru, O. Simell, and S. A. Michie, *J. Clin. Invest.*, *92*:2590 (1993).

107. R. Kuhn, J. Lohler, D. Rennick, K. Rajewsky, and W. Muller, *Cell*, *7*:263 (1993).

108. B. Sadlack, H. Merz, H. Schorle, A. Schimpl, A. C. Feller, and I. Horak, *Cell*, *75*:253 (1993).

109. M. J. Briskin, L. S. Rott, and E. C. Butcher, *J. Immunol.*, *156*:719 (1996).

110. E. L. Berg, L. M. McEvoy, C. Berlin, R. F. Bargatze, and E. C. Butcher, *Nature*, *366*:695 (1993).

111. S. Baumhueter, M. S. Singer, W. Henzel, S. Hemmerich, M. Renz, S. D. Rosen, and L. A. Lasky, *Science*, *262*:436 (1993).

112. L. A. Lasky, M. S. Singer, D. Dowbenko, Y. Imai, W. J. Henzel, C. Grimley, C. Fennie, N. Gillett, S. R. Watson, and S. D. Rosen, *Cell*, *69*:927 (1992).

113. D. Sako, X.-J. Chang, K. M. Barone, G. Vachino, H. M. White, G. Shaw, G. M. Veldman, K. M. Bean, T. J. Ahern, B. Furie, D. A. Cumming, and G. R. Larsen, *Cell*, *75*:1179 (1993).

114. M. A. Salmi, D. P. Andrew, E. C. Butcher, and S. Jalkanen, *J. Exp. Med.*, *181*:137 (1995).

115. N. W. Wu, S. Jalkanen, P. R. Streeter, and E. C. Butcher, *J. Cell. Biol.*, *107*:845 (1988).

116. L. Osborn, C. Vassallo, B. G. Browning, R. Tizard, D. O. Haskard, C. D. Benjamin, I. Dougas, and T. Kirchhausen, *J. Cell Biol.*, *124*:601 (1994).

117. A. M. Shyjan, M. Bertognolli, C. J. Kenney, and M. J. Briskin, *J. Immunol.*, *156*:2851 (1996).

118. C. A. Holness, P. A. Bates, A. J. Littler, C. D. Buckley, A. L. McDowall, D. Bossey, N. Hogg, and D. L. Simmons, *J. Biol. Chem.*, *270*:877 (1995).

119. R. H. Vonderheide, T. F. Tedder, T. A. Springer, and D. Stuanton, *J. Cell Biol.*, *125*:215 (1994).

120. D. E. Staunton, M. L. Dustin, H. P. Erickson, and T. A. Springer, *Cell*, *61*:243 (1990).

121. T. Hunkapiller and L. Hood, *Adv. Immunol.*, *44*:1 (1989).

122. A. F. Williams and A. N. Barclay, *Annu. Rev. Immunol.*, *6*:381 (1988).

123. J.-H. Wang, R. B. Pepinsky, T. Steele, J.-H. Lui, M. Karpusas, B. Browning, and L. Osbom, *Proc. Natl. Acad. Sci. U.S.A.*, *92*:5714 (1995).

124. E. Y. Jones, K. Harlos, M. J. Bottomley, R. C. Robinson, P. C. Driscoll, R. M. Edwards, J. M. Clements, T. J. Dudgen, and D. I. Stewart, *Nature*, *373*:539 (1995).

125. N. Jentoft, *Trends Biochem. Sci.*, *15*:291 (1990).

126. J. M. Hilkens, J. L. Ligtenberg, H. L. Vos, and S. V. Litvinov, *Trends Biochem. Sci.*, *17*:859 (1992).

127. S. O. Sampaio, L. Xu, J. Luna, U. Francke, E. C. Butcher, and M. B. Briskin, *J. Immunol.*, *145*:2477 (1995).

128. C. Berlin, R. F. Bargatze, J. J. Campbell, U. H. von Andrian, M. C. Szabo, S. R. Hasslen, R. D. Nelson, E. L., Berg, S. L. Erlandsen, and E. C. Butcher, *Cell*, *80*:413 (1995).

129. C. R. Mackay, W. L. Marston, and L. Dudler, *J. Exp. Med.*, *171*:801 (1990).

130. P. Sriramarao, U. H. von Andrian, E. C. Butcher, M. A. Bourdon, and D. H. Broide, *J. Immunol.*, *153*:4238 (1994).

131. R. Alon, P. D. Kassner, M. W. Carr, E. B. Finger, M. E. Hemler, and T. A. Springer, *J. Cell Biol.*, *128*:1243 (1995).

132. T. B. Issekutz, *J. Immunol.*, *147*:4178 (1991).

133. T. B. Issekutz and A. C. Issekutz, *Clin. Immunol. Immunopathol.*, *61*:436 (1991).

134. D. J. Erle, T. Brown, D. Christian, and R. Aris, *Am. J. Respir. Cell Mol. Biol.*, *10*:237 (1994).

135. K. L. Cepek, S. K. Shaw, C. M. Parker, G. J. Russel, J. S. Morrow, D. L. Rimm, and M. B. Brenner, *Nature*, *372*:190 (1994).

136. G. J. V. D. Brugge-Gamelkoorn and G. Kraal, *J. Immunol.*, *134*:3746 (1985).

137. R. P. Schleimer, S. A. Sterbinsky, J. Kaiser, C. A. Bickel, D. A. Klunk, K. Tomika, W. Newman, F. W. Luscinskas, M. A. Gimbrone, B. W. McIntyre, and B. S. Bochner, *J. Immunol.*, *148*:1086 (1992).

138. B. S. Bochner, S. Sterbinsky, M. J. Briskin, S. S. Saini, and D. W. MacGlashan, *J. Immunol.* *157*:844 (1996).

139. G. M. Walsh, F. A. Symon, A. I. Lazarovits, and A. J. Wardlaw, *Eur. J. Immunol.*, (1996).

140. A. Palecanda, M. J. Briskin, R. Lobb, and T. B. Issekutz, *J. Immunol.*, (1996)

141. R. Pulido, M. J. Elices, L. Campanero, L. Osborn, S. Schiffer, A. Garcia-Pardo, M. E., Hemler, and F. Sanchez-Madrid, *J. Biol. Chem.*, *266*:10241 (1991).

# 6

# Phenotypic and Functional Characteristics of Adhesion Molecules on Human Basophils

**Sarbjit S. Saini, Kenji Matsumoto, and Bruce S. Bochner** *Johns Hopkins Asthma and Allergy Center, Johns Hopkins University School of Medicine, Baltimore, Maryland*

## I. INTRODUCTION

Human basophils are believed to be a major participatory leukocyte in human allergic diseases [reviewed in (1–3)]. Experimental models of allergen challenge in the skin, nose, and lung have all demonstrated the arrival of basophils at challenge sites in relatively large proportions compared with their numbers in the circulation (4–8). The reappearance of clinical symptoms 6 to 12 hours after initial allergen challenge corresponds to a second rise in inflammatory mediators such as histamine; the cellular source of histamine during this late-phase time period has been confirmed to be the basophil (6,9). In a variety of disorders, including allergic and contact dermatitis, rhinitis, and asthma, basophils and their progenitors are found at increased numbers in local tissue sites (10–22). Thus, basophils, like other granulocytes, can emigrate from the intravascular compartment into specific tissues. They can be selectively recruited to allergic and other inflammatory lesions, where they actively participate in the reaction as producers of mediators that are correlated with inflammation and disease activity. In this chapter, the expression and function of adhesion molecules in the migration of basophils from the circulation to specific tissue sites will be discussed; the chapter is written to supplement and update previous reviews of this topic (23–28).

## II.  ADHESION MOLECULES EXPRESSED BY HUMAN BASOPHILS AND BASOPHIL-LIKE CELLS

### A.  Peripheral Blood Basophils

With the availability of extensive panels of monoclonal antibodies, basophils have recently undergone careful phenotypic analysis (25,26,29,30). Using immunofluorescence and flow cytometry, as well as other techniques, investigators have found that basophils express a unique pattern of cell surface adhesion molecules. This pattern of constitutive expression includes molecules belonging to the integrin, selectin, and immunoglobulin gene superfamilies, as well as carbohydrate molecules that are ligands for selectins (Table 1). Expression of some of these molecules, such as β2 integrins and L-selectin, is altered during cell activation (31–33). Identical phenotypic changes occur in basophils that have been recruited into the lower airways or nose after allergen challenge in vivo (34,35). Other expressed surface molecules on basophils of potential relevance to inflammatory responses include the high-affinity IgE receptor (FcεRI), the C3b complement receptor (CD35), the hyaluronate receptor (CD44), intercellular adhesion molecule (ICAM)1, ICAM-2, ICAM-3 and one of the low-affinity IgG receptors (CD32) (3,26,36,37).

### B.  KU812 Cells

Basophilic cell lines, such as the KU812 human leukemia line, have been studied as surrogates for basophils given many similarities between the cell types. Like basophils, KU812 cells bear FcεRI (although in fewer numbers) and contain intracellular histamine (38,39). Common phenotypic markers with basophils include most if not all of the same adhesion molecules (Table 1) as well as CD13, CD17, CD25, CD33, CD38, CD40, CD43, and BSP-1 [(25,38) and unpublished observations].

### C.  Cord Blood and Bone Marrow–Derived Cells

Another type of basophilic cells include those that can be generated from progenitor cells in human bone marrow when grown in vitro with interleukin (IL)-3. These cells develop intracellular histamine and express many phenotypic features of peripheral blood basophils (e.g., CD9, CD13, CD33, FcεRI; see Table 1) but not mast cells (MAX1, MAX24, c-kit receptor) (40,41). Umbilical cord blood cells cultured with IL-3 (41) for 6–8 weeks also express surface markers common to blood basophils (Table 1) and use β1 integrins to bind to fibronectin [(42) and unpublished observations]. Cord blood–derived cells may also adopt hybrid cell features with simultaneous eosinophil and basophil granule expression under various cytokine-containing culture conditions and in patients with chronic myelogenous leukemia (43,44). Extensive surface phenotyping of these cells has not been reported.

**Table 1**　Adhesion Molecules Expressed by Mature Peripheral Blood Human Basophils and Other Basophil-like Cells

| Type | Name (CD designation) | Ligands | KU812 | CBB |
|---|---|---|---|---|
| Integrins | | | | |
| β1 family | α4β1 (CD49d/CD29) | VCAM-1, Fn | + | + |
| | α5β1 (CD49e/CD29) | Fn | + | + |
| β2 family | LFA-1 (CD11a/CD18) | ICAM-1,-2,-3 | + | + |
| | Mac-1 (CD11b/CD18) | C3bi, ICAM-1 | + | + |
| | p150,95 (CD11c/CD18) | C3bi, others | + | + |
| | αd/CD18 | ICAM-3 | ND | ND |
| β7 family | α4β7 | MAdCAM-1, VCAM-1, Fn | ND | ND |
| Selectins | L-selectin (CD62L) | Unknown endothelial ligand | + | + |
| Immunoglobulin gene superfamily | PECAM-1 (CD31) | CD31, αvβ3 | + | + |
| | ICAM-1 (CD54) | LFA-1, Mac-1 | + | + |
| | ICAM-2 (CD102) | LFA-1 | + | ND |
| | ICAM-3 (CD50) | LFA-1, αd/CD18 | + | + |
| Selectin ligands | Sialyl-Lewis$^x$ (CD15s) | E- and P-selectin | ND | + |
| | Sialyl-dimeric-Lewis$^x$ | E-selectin | ND | ND |
| Others | CD35 | C3b | + | ND |
| | CD44 | Hyaluronate | + | + |
| | FcεRI | IgE | + | + |
| | FcγRII (CD32) | Multimeric IgG | ND | ND |

Abbreviations: CBB—cord blood basophils, Fn—fibronectin, ND—not determined, +—present. Includes data and methods described elsewhere (26,30,42) as well as unpublished observations.

## III.　FUNCTION OF ADHESION MOLECULES ON HUMAN BASOPHILS

A series of steps is necessary for selective leukocyte migration into tissues (45). Circulating cells must first undergo processes termed "rolling" and "tethering," during which they marginate, attaching loosely to the endothelium via selectins and their carbohydrate counterligands—although the α4β1 integrin may also serve this function as well (46). Subsequent triggering of leukocyte adhesion and transendothelial migration may then occur, presumably under the influence of chemoattractants including chemokines. These migration events have not yet been formally studied for the basophil, but it is presumed that they are mediated by interactions of integrins with endothelial counterligands of the immunoglobulin gene superfamily including ICAM-1, ICAM-2, vascular cell adhesion molecule (VCAM)-1, and platelet endothelial cell adhesion molecule (PECAM)-1. It is also likely that during firm adhesion and transendothelial migration, changes occur in both the avidity and expression of integrins on the basophil surface. Therefore, the preferential recruitment of a basophil will ultimately occur as a result of many adhesion-dependent events, rather than the result of any one single cell–specific chemoattractant or adhesion molecule pathway.

In studying basophil adhesion, two general types of pro-adhesive stimuli have been identified: those that activate basophil adhesiveness directly, and those that induce expression of basophil adhesion ligands on other cell types (e.g., on endothelium). Results of studies examining the adhesiveness of basophils have identified several chemotactic factors (C5a, fMLP, PAF), cytokines (IL-3), and IgE-dependent secretagogues (allergen, anti-IgE) that can directly stimulate divalent cation-dependent attachment of basophils to resting endothelium in vitro; other nonselective leukocyte activators (phorbol esters, the calcium ionophore A23187) have similar effects (31,47–49). For both IL-3 and IgE-dependent secretagogues, however, these effects are selective for basophils, and they can occur at concentrations too low to induce histamine release. Therefore, exposure of circulating basophils to low levels of these agents locally in vivo may contribute to their selective adhesion and recruitment during certain inflammatory responses.

Recently, several C-C or β family chemokines have been shown to promote basophil chemotaxis and/or activation. These include RANTES, MIP-1α, MCP-1, MCP-3, and MCP-4 (50–53). Although none of these chemokines had any detectable effect on basophil adhesion molecule expression or adhesion to unstimulated endothelium in vitro [(30) and unpublished observations], studies of their effect on basophil transendothelial migration and integrin avidity should help to clarify their roles in basophil recruitment responses.

Studies from several laboratories have demonstrated that rapid changes in the avidity of integrins on leukocytes can occur as a result of cell activation or by manipulating the extracellular concentration of divalent cations [reviewed in (54,55)]. For example, specific monoclonal antibodies can be used to detect neoepitopes that become exposed when conformational changes occur in association with augmented integrin adhesiveness. Integrins on basophils have been shown to possess a calcium-sensitive epitope on CD11a (30), and a manganese-sensitive activation epitope on β1 integrin (Fig. 1). The findings are similar to those recently observed in human eosinophils (56–58) and serve to further highlight the importance of considering both the level of expression and the level of activation of integrins when examining the adhesive phenotype of a particular population of cells.

In the second type of experiment, induction on other cells of expression of ligands for basophil adhesion has been observed. The endothelium clearly can play an active role during basophil recruitment by expressing new or enhanced levels of adhesion molecules on its surface after being activated by specific

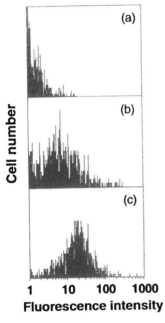

**Figure 1**   Flow cytometric analysis of levels of activated β1 integrins on platelet-free peripheral blood basophils detected by dual-color immunofluorescence and flow cytometry (37) using a mAb recognizing an activation-dependent epitope on β1 integrin (mAb 15/7) in the absence (a) or presence (b) of 1 mM $MnCl_2$ (58). Levels of total β1 integrin (defined in the absence of $MnCl_2$ using another mAb, 8A2 (72) are shown for comparison (c). In the absence of $MnCl_2$, expression of the 15/7 epitope is not above that seen with an irrelevant isotype-matched control mAb.

stimuli. The cytokines IL-1 and tumor necrosis factor (TNF), as well as phorbol esters and bacterial endotoxin (LPS), stimulate cultured endothelial cells to express ICAM-1, E-selectin, and VCAM-1, each of which is a ligand for basophils (47,59). Ligands on basophils for these endothelial adhesion molecules have been delineated. Like other granulocytes, human basophils utilize sialylated surface molecules to interact with endothelial E-selectin. However, the expression of these sialylated Le$^x$ structures on basophils is unusual in that it is bimodal, and panning experiments suggest that only the sialylated Le$^{x\ (high)}$ subset of basophils are capable of interacting with E-selectin (49). Whether this distinct distribution of sialylated Le$^x$-containing structures influences basophil adhesion to P-selectin has not yet been determined. As would be expected, β2 integrins function as ligands for ICAM-1. Both ICAM-1 and E-selectin are ligands for all granulocytes, and therefore do not appear, by themselves, to account for selective basophil recruitment responses in vivo. In contrast, basophils, unlike neutrophils, can adhere to the cytokine-inducible endothelial cell adhesion molecule VCAM-1 because they express counter-receptors for VCAM-1, namely the integrins VLA-4 ($\alpha4\beta1$, CD49d/CD29) and $\alpha4\beta7$ [(49,59,60) and see Table 1]. Therefore, conditions under which VCAM-1 is selectively induced, such as endothelial exposure to IL-4 or IL-13 (60,61), may favor basophil recruitment. It is not yet known whether these or other adhesion molecules participate in the process of basophil transendothelial migration. Besides its interaction with VCAM-1, VLA-4 on basophils can also be a ligand for the CS-1 region of fibronectin, as can VLA-5, both of which are expressed on these cells [Table 1 and (42)]. In addition, we have recently determined that $\alpha4\beta7$ expressed on basophils functions as a ligand for both VCAM-1 and MAdCAM-1 (mucosal addressin cell adhesion molecule-1) (49), an adhesion molecule that has structural homology to both VCAM-1 and ICAM-1 and is expressed in the gut mucosa (62–64). This is intriguing in view of the longstanding interest in the role of basophils in Crohn's disease and parasite immunity in various animal models (1,65), and suggests a possible role for the interaction of $\alpha4\beta7$ with MAdCAM-1 in trafficking of basophils to the gut during certain immune responses in vivo.

## IV.  SIGNALING THROUGH BASOPHIL ADHESION MOLECULES

Adherence via surface integrin molecules has been shown to activate a number of signaling pathways in other leukocytes (66); however, evidence in the basophil is more limited. Early studies with the rat basophilic cell line RBL-2H3 suggest that integrin-dependent adhesion to fibronectin does not lead to mediator release; however, it does augment release from IgE-dependent or IgE-independent secretagogues (67). Further, adherent RBL cells exhibit tyrosine phosphorylation of intracellular signaling proteins such as pp125[FAK] (focal adhesion kinase) which is further enhanced by FcεRI cross-linking, phorbol esters, or the calcium ionophore A23187 (68). In human cells, cross-linking of $\alpha4$ integrins on blood basophils from asthmatic donors (but not from normal donors) led to histamine release (69).

Signaling can also occur through other non-integrin adhesion molecules. On basophils, preliminary studies from our laboratory suggest that cross-linking of ICAM-3, a molecule that appears to function as a transmembrane signaling and activating structure in lymphocytes (70), potentiates IgE-dependent mediator release and adhesion to stimulated and unstimulated endothelial cell monolayers (71). This effect was not seen when other related immunoglobulin family members, ICAM-1 and ICAM-2, were cross-linked in a similar manner, lending support to the idea that ligation of some but not all surface immunoglobulin family adhesion molecules may activate pathways involved in basophil mediator release and adhesiveness. Further studies will be necessary to delineate these "outside-in" and "inside-out" signaling pathways and their roles in affecting basophil functions in vitro and in vivo.

## V. CONCLUSIONS

Human basophils express a number of cell adhesion molecules on their surface, including selectins, integrins, sialyl-Le$^x$–containing carbohydrates, and immunoglobulin adhesion molecule family members. The function of many of these molecules has now been established in a variety of static in vitro adhesion assays (Fig. 2), yet their role under physiologic conditions of vascular shear stress, during transendothelial migration, during interaction with extracellular matrix proteins, or in vivo remain essentially unexplored. Further, the activation of cellular signaling pathways through adhesion molecule binding is poorly understood. Defining the roles of adhesion molecule interactions between basophils, endothelial cells, epithelial cells and matrix proteins is imperative if we are to learn how these leukocytes are selectively recruited to the skin and airways of individuals suffering from asthma and other allergic diseases. With this information, it may be possible to selectively block the influx of basophils or modulate their inflammatory contributions to allergic inflammation in vivo.

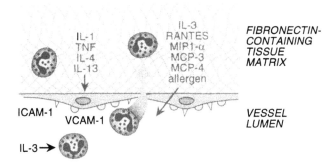

**Figure 2** Proposed pathways, influenced by cytokines and chemokines, by which adhesion molecule expression and/or function may contribute to selective migration of circulating human basophils into extravascular inflammatory tissue sites.

## REFERENCES

1. E. B. Mitchell and P. W. Askenase, *Clin. Rev. Allergy*, *1*:427 (1983).
2. L. M. Lichtenstein and B. S. Bochner, *Ann. N Y Acad. Sci.*, *629*:48 (1991).
3. B. S. Bochner, *Samter's Immunological Diseases* (M. M. Frank, K. F. Austen, H. N. Claman, and E. R. Unanue, eds), Little, Brown, and Company (5th ed.), Boston, p. 259 (1995).
4. A. B. Felarca and F. C. Lowell, *J. Allergy Clin. Immunol.*, *48*:125 (1971).
5. R. Bascom, M. Wachs, R. M. Naclerio, U. Pipkorn, S. J. Galli, and L. M. Lichtenstein, *J. Allergy Clin. Immunol.*, *81*:580 (1988).
6. E. N. Charlesworth, A. F. Hood, N. A. Soter, A. Kagey-Sobotka, P.S. Norman, and L. M. Lichtenstein, *J. Clin. Invest.*, *83*:1519 (1989).
7. O. Iliopoulos, F. M. Baroody, R. M. Naclerio, B. S. Bochner, A. Kagey-Sobotka, and L. M. Lichtenstein, *J. Immunol.*, *148*:2223 (1992).
8. C. B. Guo, M. C. Liu, S. J. Galli, B. S. Bochner, A. Kagey-Sobotka, and L. M. Lichtenstein, *Am. J. Respir. Cell Mol. Biol.*, *10*:384 (1994).
9. R. M. Naclerio, D. Proud, A. G. Togias, N. F. Adkinson Jr., D. A. Meyers, A. Kagey-Sobotka, M. Plaut, P. S. Norman, and L. M. Lichtenstein, *N. Engl. J. Med.*, *313*:65 (1985).
10. N. Aspegren, S. Fregert, and H. Rorsman, *Int. Arch. Allergy Immunol.*, *23*:150 (1963).
11. H. F. Dvorak and M. C. Mihm, *J. Exp. Med.*, *135*:235 (1972).
12. I. Kimura, Y. Moritani, and Y. Tanizaki, *Clin. Allergy*, *3*:195 (1973).
13. I. Kimura, Y. Tanizaki, K. Saito, K. Takahashi, N. Ueda, and S. Sato, *Clin. Allergy*, *5*:95 (1975).
14. R. Hastie, B. Chir, J. H. Heroy, and D. A. Levy, *Lab. Invest.*, *40*:554 (1979).
15. M. Okuda and H. Ohtsuka, *Arch. Otorhinolaryngol.*, *214*:283 (1977).
16. H. Otsuka, J. Dolovich, D. Befus, J. Bienenstock, and J. Denburg, *Am. Rev. Respir. Dis.*, *133*:757 (1986).
17. H. Otsuka, J. Dolovich, A. D. Befus, S. Telizyn, J. Bienenstock, and J. A. Denburg, *J. Allergy Clin. Immunol.*, *78*:365 (1986).
18. E. B. Mitchell, J. Crow, M. D. Chapman, S. S. Jouchal, F. M. Pope, and T. A. E. Platts-Mills, *Lancet*, *1*:127 (1982).
19. T. Koshino, S. Teshima, N. Fukushima, T. Takaishi, K. Hirai, Y. Miyamoto, Y. Arai, Y. Sano, K. Ito, and Y. Morita, *Clin. Exp. Allergy.*, *23*:919 (1993).
20. J. A. Denburg, M. Wooller, B. Leber, M. Linden, and P. O'Byrne, *J. Allergy Clin. Immunol.*, *94*:1135 (1994).
21. N. Maruyama, G. Tamura, T. Aizawa, T. Ohrui, S. Shimura, K. Shirato, and T. Takishima, *Am. J. Respir. Crit. Care Med.*, *150*:1086 (1994).
22. S. Molossi, M. Elices, T. Arrhenius, R. Diaz, C. Coulber, and M. Rabinovitch, *J. Clin. Invest.*, *95*:2601 (1995).
23. B. S. Bochner, R. P. Schleimer, E. N. Charlesworth, A. M. Lamas, and L. M. Lichtenstein, *Progress in Allergy and Clinical Immunology* (W. J. Pichler, B. M. Stadler, C. A. Dahinden, A. R. Pecoud, P. Frei, C. H. Schneider, and A. L. de Weck, eds.), Hogrefe and Huber Publishers, Toronto, p. 12 (1989).

24. B. S. Bochner and L. M. Lichtenstein, *Clin. Exp. Allergy*, 22:973 (1992).
25. P. Valent and P. Bettelheim, *Adv. Immunol.*, 52:333 (1992).
26. P. Valent, *J. Allergy Clin Immunol*, 94:1177 (1994).
27. B. S. Bochner and R. P. Schleimer, *J. Allergy Clin. Immunol.*, 94:427 (1994).
28. A. J. Wardlaw, G. M. Walsh, and F. A. Symon, *Allergy*, 49:797 (1994).
29. P. Valent and P. Bettelheim, *Crit. Rev. Oncol. Hematol.*, 10:327 (1990).
30. B. S. Bochner, S. A. Sterbinsky, E. F. Knol, B. J. Katz, L. M. Lichtenstein, D. W. MacGlashan Jr., and R. P. Schleimer, *J. Allergy Clin. Immunol.*, 94:1157 (1994).
31. B. S. Bochner, A. A. McKelvey, S. A. Sterbinsky, J. E. K. Hildreth, C. P. Derse, D. A. Klunk, L. M. Lichtenstein, and R. P. Schleimer, *J. Immunol.*, 145:1832 (1990).
32. B. S. Bochner and S. A. Sterbinsky, *J. Immunol.*, 146:2367 (1991).
33. E. F. Knol, F. P. J. Mul, H. Jansen, J. Calafat, and D. Roos, *J. Allergy Clin. Immunol.*, 88:328 (1991).
34. S. N. Georas, M. C. Liu, W. Newman, W. D. Beall, B. A. Stealey, and B. S. Bochner, *Am. J. Respir. Cell Mol. Biol.*, 7:261 (1992).
35. F. M. Baroody, B.-J. Lee, M. C. Lim, and B. S. Bochner, *Eur. Arch. Otorhinolaryngol.*, 252 *(suppl. 1)*:S50 (1995).
36. L. M. Anselmino, B. Perussia, and L. L. Thomas, *J. Allergy Clin. Immunol.*, 84:907 (1989).
37. B. S. Bochner, A. A. McKelvey, R. P. Schleimer, J. E. K. Hildreth, and D. W. MacGlashan Jr., *J. Immunol. Methods*, 125:265 (1989).
38. M. J. Elices, V. Tsai, D. Strahl, A. S. Goel, V. Tollefson, T. Arrhenius, E. A. Wayner, F. C. A. Gaeta, J. D. Fikes, and G. S. Firestein, *J. Clin. Invest.*, 93:405 (1994).
39. T. Blom, R. Y. Huang, M. Aveskogh, K. Nilsson, and L. Hellman, *Eur. J. Immunol*, 22:2025 (1992).
40. P. Valent, G. Schmidt, J. Besemer, P. Mayer, G. Zenke, E. Liehl, W. Hinterberger, K. Lechner, D. Maurer, and P. Bettelheim, *Blood,* 73:1763 (1989).
41. M. Ebisawa, H. Saito, D. Reason, K. Ohno, T. Sudo, K. Kurihara, T. Nagakura, and Y. Iikura, *Jpn. J. Allergol.*, 38:442 (1989).
42. K. Miura, M. Ebisawa, M. Shichijo, H. Tachimoto, T. Onda, H. Saito, Y. Iikura, K. Okumura, and C. Ra, *J. Allergy Clin. Immunol.*, 95:293 (1995).
43. J. A. Denburg, *Blood*, 79:846 (1992).
44. P. G. M. Bloemen, P. A. J. Henricks, L. Van Bloois, M. C. Van den Tweel, A. C. Bloem, F. P. Nijkamp, D. J. A. Crommelin, and G. Storm, *FEBS Lett.*, 357:140 (1995).
45. T. A. Springer, *Cell*, 76:301 (1994).
46. P. Sriramarao, U. H. von Andrian, E. C. Butcher, M. A. Bourdon, and D. H. Broide, *J. Immunol.*, 153:4238 (1994).
47. B. S. Bochner, P. T. Peachell, K. E. Brown, and R. P. Schleimer, *J. Clin. Invest.*, 81:1355 (1988).
48. B. S. Bochner, D. W. MacGlashan Jr., G. V. Marcotte, and R. P. Schleimer, *J. Immunol.*, 142:3180 (1989).
49. B. S. Bochner, S. A. Sterbinsky, S. S. Saini, M. Briskin, and D. W. MacGlashan Jr., *(submitted)*, (1996).
50. S. C. Bischoff, M. Krieger, T. Brunner, A. Rot, V. Vontscharner, M. Baggiolini, and C. A. Dahinden, *Eur. J. Immunol.*, 23:761 (1993).

51. C. A. Dahinden, T. Geiser, T. Brunner, V. von Tscharner, D. Caput, P. Ferrara, A. Minty, and M. Baggiolini, *J. Exp. Med.*, *179*:751 (1994).
52. M. Baggiolini and C. A. Dahinden, *Immunol. Today*, *15*:127 (1994).
53. C. Stellato, L. Beck, L. Schweibert, H. Li, J. White, and R. P. Schleimer, *J. Allergy Clin. Immunol.*, *97*:304 (1996).
54. M. S. Diamond and T. A. Springer, *Curr Biol*, *4*:506 (1994).
55. F. W. Luscinskas and J. Lawler, *FASEB J.*, *8*:929 (1994).
56. K. Matsumoto, R. P. Schleimer, and B. S. Bochner, *J. Allergy Clin. Immunol.*, *95*:338 (1995).
57. K. Matsumoto, D. Zhou, R. P. Schleimer, and B. S. Bochner, *FASEB J.*, *9*:A226 (1995).
58. S. Werfel, T. Yednock, K. Matsumoto, S. Sterbinsky, R. Schleimer, and B. Bochner, *Am. J. Respir. Cell Mol. Biol.*, *14*:45 (1996).
59. B. S. Bochner, F. W. Luscinskas, M. A. Gimbrone Jr., W. Newman, S. A. Sterbinsky, C. Derse-Anthony, D. Klunk, and R. P. Schleimer, *J. Exp. Med.*, *173*:1553 (1991).
60. R. P. Schleimer, S. A. Sterbinsky, J. Kaiser, C. A. Bickel, D. A. Klunk, K. Tomioka, W. Newman, F. W. Luscinskas, M. A. Gimbrone Jr., B. W. McIntyre, and B. S. Bochner, *J. Immunol.*, *148*:1086 (1992).
61. B. S. Bochner, D. A. Klunk, S. A. Sterbinsky, R. L. Coffman, and R. P. Schleimer, *J. Immunol.*, *154*:799 (1995).
62. M. J. Briskin, L. M. McEvoy, and E. C. Butcher, *Nature*, *363*:461 (1993).
63. C. Berlin, E. L. Berg, M. J. Briskin, D. P. Andrew, P. J. Kilshaw, B. Holzmann, I. L. Weissman, A. Hamann, and E. C. Butcher, *Cell*, *74*:185 (1993).
64. D. J. Erle, M. J. Briskin, E. C. Butcher, A. Garcia Pardo, A. I. Lazarovits, and M. Tidswell, *J. Immunol.*, *153*:517 (1994).
65. P. W. Askenase, *Prog. Allergy*, *23*:199 (1977).
66. E. A. Clark and J. S. Brugge, *Science*, *268*:233 (1995).
67. M. M. Hamawy, C. Oliver, S. E. Mergenhagen, and R. P. Siraganian, *J. Immunol.*, *149*:615 (1992).
68. M. M. Hamawy, S. E. Mergenhagen, and R. P. Siraganian, *J. Biol. Chem.*, *268*:5227 (1993).
69. S. E. Lavens, K. Goldring, L. H. Thomas, and J. A. Warner, *Am. J. Respir. Cell Mol. Biol.*, *14*:95 (1996).
70. M. C. Cid, J. Esparza, M. Juan, A. Miralles, J. Ordi, R. Vilella, A. Urbano-Marquez, A. Gaya, J. Vives, and J. Yague, *Eur. J. Immunol.*, *24*:1377 (1994).
71. S. Saini, J. White, W. M. Gallatin, P. A. Hoffman, L. M. Lichtenstein, and B. S. Bochner, *J. Allergy Clin. Immunol.*, *97*:264 (1996).
72. N. L. Kovach, T. M. Carlos, E. Yee, and J. M. Harlan, *J. Cell Biol.*, *116*:499 (1992).

# 7

# Integrin-Dependent Responses in Human Basophils

**Jane A. Warner, Kirsty Rich, and Kirstin Goldring**   *University of Southampton, Southampton, England*

## I.  INTRODUCTION

Basophils represent less than 1% of the circulating leukocytes, and identifying them in the tissues has proved difficult; therefore, defining their precise role in health and disease has been problematic. Their numerous granules within the cytoplasm contain histamine (1,2), and activation leads to the release of a range of lipid mediators and cytokines, including interleukins 4, 8, and 13 (3–6). Like the tissue mast cells, they possess numerous high-affinity IgE receptors that bind circulating IgE. Though other leukocytes are now known to possess the FcεRI (7,8), the levels on basophils are almost 250-fold higher than on monocytes (9), whereas eosinophil expression of FcεRI appears to be minimal except in hypereosinophilic patients. This combination of high levels of FcεRI and the presence of histamine and other inflammatory mediators has obviously focused attention on the role these cells play in the allergic response. Basophils have been identified in the skin and bronchoavelolar lavage of patients during the late-phase response to antigen (10,11), and the current paradigm suggests that they, like other leukocytes, are recruited to the tissues following mast cell activation. Once within the tissue, their activation and degranulation leads to a second wave of mediator release and a reoccurrence of symptoms. The migration of cells from the blood to the tissues involves a complex series of adhesive and de-adhesive interactions, and once within the tissues the cell must operate in a completely different environment. The alterations in cellular responses that occur as the basophil is recruited from the bloodstream and moves through the dense network of proteins and proteoglycans that make up the extracellular matrix may well lead to a clearer understanding of the role of the basophil in allergic and inflammatory responses.

## II. BASOPHIL ADHESION MOLECULES

Basophils possess a range of different adhesion molecules on their cell surface that facilitate migration and movement through the tissues(12–15 and see Chap. 6). Like most leukocytes, resting basophils express L-selectin and low levels of the β2 integrins, CD11a/CD18 and CD11b/CD18, though the latter are rapidly increased following stimulation (16). Successful migration into the tissues requires the interactions of very late activation antigen (VLA) 4 with its counter-ligand, vascular cell adhesion molecule (VCAM) 1, on the endothelial cell surface (17). This interaction may provide a mechanism for the selective accumulation of basophils (and eosinophils) during the allergic response, as neutrophils do not express VLA-4. Unlike the β2 integrins, the increased adhesion to VCAM-1 appears to involve changes in the affinity of the β1 integrin for VCAM-1 rather than increased expression of VLA-4. Once within the tissues, the β1 integrins are responsible for interactions with the extracellular matrix (18). Basophils express a range of β1 integrins including α5β1 (VLA-5), which binds to the central RGD-containing segment of fibronectin. Finally, in addition to binding to VCAM-1, VLA-4 also binds to the CS-1 region of fibronectin (18). Basophils and eosinophils also express substantial amounts of the marker CD9, which is thought to be involved in homotypic adhesion. However, CD9 has also been reported to function as a receptor for fibronectin (19), though there is evidence that it associates with VLA-4 and and may regulate its affinity (20,21). Basophils also express all three members of the intercellular adhesion molecule (ICAM) family, though expression of CD102 (intercellular adhesion molecule-2 [ICAM-2]) and CD50 (ICAM-3) is lower than CD 54 (ICAM-1). Finally, basophils express substantial levels of CD58 (LFA-3), CD31 (platelet endothelial cell adhesion molecule [PECAM), CD43 (leukosialin), and CD44.

## III. EFFECTS OF ADHESION MOLECULES ON CELL FUNCTION

In addition to their ability to regulate adhesion and migration, a substantial body of evidence has accumulated to suggest that adhesion molecules fulfill a much wider spectrum of roles. Interactions of the selectins or β2 integrins with their counter-ligands may cause conformational changes in the other adhesion proteins (22–25), facilitating firm adhesion to the endothelial surface. However, L-selectin is rapidly shed from the cell surface by proteolytic cleavage (26,27), and the function of β2 integrins are also down-regulated within a relatively short time frame (28,29), suggesting that their effects on cell function are likely to be short lived. This leaves the β1 integrins, which modulate firm adhesion to VCAM-1 and act as receptors for the extracellular matrix relaying signals from the external environment into the cell interior. These signals allow the cell to respond to its environment and modulate its responses accordingly; they have been shown to regulate many different aspects of cell function, including growth, differentiation, receptor expression, gene expression, and cell movement (30–34). Much of our knowledge of basophil integrins is derived from models developed in closely related cells; thus, it is relevant to summarize some of the effects of integrin ligation in mast cells and eosinophils.

The mast cell undergoes the final stages of maturation in the tissues, and cell–matrix interactions are likely to have profound effects on mast cell phenotype, distribution, activation state, and survival. These interactions may take place over extended periods, making investigations into their roles more complex. However, several matrix proteins, including laminin, fibronectin, and vitronectin, have been shown to have chemotactic activity for different rodent mast cell lines (35–38), which supports a role for cell–matrix interactions in regulating mast cell distribution. In addition, studies in the rodent mast cell line, rat basophilic leukemia (RBL)-2H3, have shown that there is a complex dialogue between adhesion and IgE-dependent responses (39–42). Mast cells that are allowed to adhere to a fibronectin matrix release more of their granule contents following IgE-dependent stimulation than cells maintained in solution (40). Conversely, cross-linking of IgE increases adhesion to fibronectin (41). This cross talk between IgE and adhesion appears to involve common signal transduction pathways, particularly increased tyrosine phosphorylation and the activation of the focal adhesion tyrosine kinase, P125[FAK] (42). More recently evidence has come to light suggesting that the effects of cell–matrix interactions on IgE signal transduction in RBL cells may be affected by cell age and the point in the cell cycle (43) and that this dialogue may be more complex than originally anticipated.

Cell–matrix interactions have also been shown to modulate cytokine-dependent responses in both human and rodent mast cell lines. Thus, the chemotactic activity of stem cell factor for the human mast cell line HMC-1 can be inhibited by RGD peptides (44); adherence to vitronectin increased the proliferation of mouse bone marrow derived mast cells (45). There is also evidence that cytokines modulate the nature of the cell–matrix interaction (46).

It is still not clear if the rodent mast cell lines provide an accurate model for human mast cell activation and migration. Adhesion molecules are intimately linked with cell survival (47) and so it is not surprising that immortalization of cell lines may alter expression of the integrins (48,49) or their regulation (50). Human mast cells express many of the same integrins that modulate adhesion in the rodent cell lines and have the capacity to interact with the same panel of matrix proteins; however, some discrepancies have been reported. Thus, in the mouse bone marrow derived mast cell, cross-linking of IgE leads to an increased adhesion to laminin (36), but in human skin mast cells there is a decrease in adhesion and increased cell motility (51). Biopsies of human lung tissue 6 hours after antigen challenge have shown alterations in mast cell distribution (52), providing indirect evidence that IgE cross-linking increases human mast cell migration, allowing cells to accumulate at an inflammatory focus rather than remaining firmly adhered in their original location.

The interactions of the eosinophil with the extracellular matrix again emphasize the complex nature of cell–matrix interactions and their effects on cell function. Eosinophils adhere to laminin, fibronectin, and fibrinogen, and eosinophils cultured on tissue fibronectin have extended survival in vitro (53,54). This occurs as a result of autocrine generation of granulocyte-macrophage colony-stimulating factor (GM-CSF) and interleukin-3 (IL-3) and survival can be blocked with antibodies to the VLA-4 integrins or the cytokines (54). Interactions with matrix proteins, such as fibronectin, increased f-met peptide induced eosinophil respiratory burst but down-regulated the response to tumor necrosis factor–$\alpha$ (TNF-$\alpha$) and platelet-activating factor (PAF) (55–58). These apparent discrepancies emphasize, once again, the complex nature of the dialogue between adhesion molecules and other stimuli and reinforce the need for further studies.

## A.   Effects of Ligation of the VLA-4 Integrins on Human Basophils

With such strong evidence that integrin ligation can modulate many aspects of cell function in two cell types so closely related to the basophil, it is perhaps not surprising that interactions with the matrix affect responses in this cell. We have shown that ligation of the VLA-4 integrins CD49d/CD29 with monoclonal antibodies is sufficient to trigger histamine release in the basophils of selected subjects (59 and see Fig. 1). Our initial studies suggested a relationship between asthma, or another chronic inflammatory condition such as atopic dermatitis, and response to integrin ligation. However, subsequent studies have failed to reveal an unambiguous relationship between asthma severity or plasma IgE levels and the integrin-dependent histamine release (60). In contrast, despite extensive investigations we have no evidence that cross-linking of VLA-4 integrins can initiate degranulation in human lung mast cells (59). This may indicate fundamental differences between tissue mast cells and leukocytes or simply reflect the fact that our lung tissue is obtained from the normal margins of tissue removed for carcinoma and most of these patients are not asthmatic.

We also examined the effects of ligation of VLA-4 integrins on the subsequent response to anti-IgE cross-linking in both basophils and lung mast cells. Some-

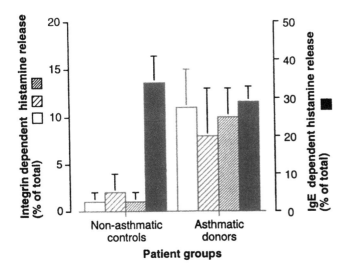

**Figure 1** Effect of integrin clustering on histamine release from asthmatic and non-asthmatic donors. Basophils from eight non-atopic, non-asthmatic controls or 11 asthmatic patients were challenged with an a monoclonal antibody to CD29 (☐), CD49d (▨), or CD9 (☐) followed by an F(ab$_2$)' fragment of rabbit anti–mouse IgG to further cross-link the integrins. Cross-linking of all three of the integrins triggered significant ($P<.05$) histamine release in the basophils of asthmatic patients but failed to affect the non-asthmatic controls. The response to 1 μ g/mL anti-IgE (■) was not different in the two groups. Results are presented as mean± SEM.

what to our surprise, we found that ligation of either CD29 or CD49d reduced the response to an optimal concentration of anti-IgE without affecting the responses to f-met peptide or A23187 (59). This cross talk between IgE and integrin-dependent responses occurred in all basophil donors regardless of allergic and asthmatic status, even when there was no evidence of integrin-dependent degranulation. There was a similar inhibition of IgE-induced histamine release in the lung mast cells. This contrasts with the RBL cell, where adhesion to fibronectin enhanced the subsequent response to IgE cross-linking. The reasons for the discrepancies between the responses of human mast cells and basophils and the RBL cells are not clear. Basophils are isolated directly from the bloodstream and are nonadherent, whereas the RBL cells are an adherent cell line and must be treated with trypsin to remove them from the culture flask; however, human lung mast cells must be treated with collagenase to disrupt the tissue structure. Another possibility is that the intracellular signals generated by integrin clustering may induce some form of IgE desensitization, as there are fundamental differences between the desensitization processes in the RBL cells and human cells (61–63).

## B. Effects of the Ligation of Other Integrins on Basophil Responses

A second basophil surface marker, CD9, is also reported to act as a receptor for fibronectin (19), and we have found similar results when CD9 is ligated. Clustering of CD9 is sufficient to initiate histamine release in the basophils of some asthmatic patients (see Fig. 1). Ligation of CD9 also modulates the response to a subsequent IgE challenge (see Fig. 1). The basophil responses to CD9 ligation closely resembled the effects of clustering the VLA-4 integrins, and it has been reported that CD9 may associate with VLA-4 on the cell surface (29). These similarities may suggest that CD9 and CD29/CD49d utilize similar signal transduction pathways or that ligation of CD9 causes co-clustering of the VLA-4 integrin and triggers histamine release via this pathway.

As mentioned above, in addition to the β1 and β2 integrins, basophils also possess ICAM-1, 2, and 3, members of the immunoglobulin superfamily and ligands for the β2 integrins. Sianni et al. (64) have shown that though ligation of ICAM-3 does not initiate basophil degranulation directly, it increased a subsequent response to anti-IgE. The effects of cross-linking of ICAM-3 were

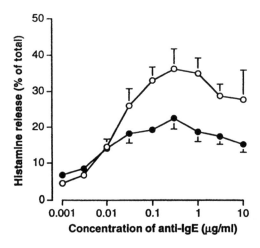

**Figure 2**   Effect of cross-linking CD9 on anti-IgE induced histamine release. Basophils from six non-atopic donors were incubated either with a buffer control (o) or with a monoclonal antibody to CD9 followed by an F(ab$_2$)' fragment of rabbit anti–mouse IgG to cross-link CD9 (●). The cells were incubated for 60 min at 37°C and then challenged with different concentrations of anti-IgE. The supernatant was recovered 45 min later and histamine measured fluorimetrically. Cross-linking of CD9 closely resembled CD29 or CD49d (data not shown) and significantly suppressed histamine release at the optimum concentration of anti-IgE.

detectable within 1 min and gradually declined over the next 30 min. This contrasts with the studies on cross-linking of VLA-4, where the effects on anti-IgE induced release were maximal at 60 min and could not be detected at all at 5 min (59). It is tempting to speculate that the differences in the kinetics reflect the very distinct roles that ICAM-3 and VLA-4 play on the cell surface and in particular the length of time that they are likely to be engaged during the normal process of adhesion and recruitment. Finally, cross-linking of ICAM-3 also increases the binding of basophils to activated and unstimulated endothelial cells (64), supporting the hypothesis that engagement of one adhesion protein may increase the activity of other molecules in the adhesion cascade.

## C. Effects of Interactions with the Extracellular Matrix

Many of the studies of integrin clustering have used cell–matrix interactions as a model system. This has the advantage of clear physiological relevance, but interpreting the responses may be complicated by signals transduced from more than one adhesion molecule. We wished to know if interaction of human mast cells and basophils with extracellular matrix proteins triggered similar responses to the ligation of the integrins using monoclonal antibodies. Our initial experiments examined the effects of fibronectin on basophil histamine release. We found that addition of tissue fibronectin initiated histamine release in a population of basophil donors and that this correlated with the response to CD49d (59,60). Intriguingly, the dose response curve to fibronectin showed marked heterogeneity. In some donors the maximum response came at either 30 or 10 µl/mL fibronectin with clear evidence of superoptimal inhibition; in other donors, the maximum response occurred at 100 µg/mL. Similar observations have been made in human eosinophils (65) where the maximum response to tissue fibronectin occurred at 25 µg/ml.

## IV. MECHANISMS OF INTEGRIN SIGNAL TRANSDUCTION

The mechanisms by which integrin clustering initiate or modify basophil degranulation remain highly speculative. Cross-linking of the integrins has been shown to activate a wide range of second messenger pathways including activation of tyrosine kinases (66–68), activation of protein kinase C (69), intracellular $Ca^{2+}$ fluxes (70,71), alterations in cytoplasmic pH (72), and reorganization of the cytoskeleton (73–75). These studies were carried out in a range of very distinct cells under different conditions and involved a range of different integrins and suggest that there is no single unifying signal transduction pathway that will explain all the current observations. However, the more detailed studies in rodent mast cell lines and human eosinophils provide convenient models on which to begin to dissect the signal transduction pathways. Studies in the RBL cell suggest that p125[FAK] and related proteins (42,76) play important roles early in the signal transduction cascade. We have shown that many of the pharmacologic agents that modulate IgE-dependent histamine release also affect the integrin-dependent responses (59,77). Investigations into the signal transduction mechanisms utilized by integrins suggest that integrin-dependent responses can be blocked by inhibitors of tyrosine kinases. Both genistein, a broad-spectrum tyrosine kinase inhibitor, and piceatannol, a putative inhibitor of the tyrosine kinase p72*syk* (78), were able to completely ablate integrin-dependent responses (see Table 1). In addition, wortmannin, an inhibitor of phospholipase D, and phosphatidylinositol-3-kinase (PI3-K) completely ablated the response to integrin clustering and anti-IgE. The $IC_{50}$ for wortmannin was below 1 nM, suggesting that the drug was acting on PI3-K.

Like IgE-dependent degranulation, integrin-dependent release requires the presence of extracellular $Ca^{2+}$, and the histamine release increases between 0.1 and 5 mM $Ca^{2+}$, reaching a maximum at 5 mM $Ca^{2+}$. Preliminary experiments using digital video microscopy indicate that clustering of the integrins on the basophil surface leads to the mobilization of $Ca^{2+}$, even in basophils when there is no accompanying degranulation (unpublished observations).

**Table 1**  Effect of Different Pharmacologic Agents on Anti-Integrin and Anti-IgE Induced Histamine Release in Human Basophils from Asthmatic Donors

|  | Cross-linking of CD29 | Cross-linking of CD49d | Cross-linking of IgE |
|---|---|---|---|
| Optimum extracellular $Ca^{2+}$ | 5 mM | 5 mM | 2–5 mM |
| Genistein (10 μM) | ↓↓ | ↓↓ | ↓↓ |
| Piceatannol (10 μM) | ↓↓ | ↓↓ | ↓↓ |
| Wortmannin (10 nM) | ↓↓ | ↓↓ | ↓ |

Basofils from 6 to 8 asthmatic donors were treated with the different drugs for 15 min at 37°C and challenged for 45 min with either 5 μg/mL mouse anti–human CD29 or 5 μg/mL mouse anti–human CD49 followed by an F(ab₂)' fragment of rabbit anti–mouse IgG to cross-link the integrins and initiate histamine release. In each experiment we also examined the effect of the drug on anti-IgE induced release for comparison. Control release was 11±4% for CD29, 8±4% for CD49, and 21±6% for anti-IgE (1 μg/mL for 45 min). ↓↓ indicates that inhibition exceeded 80%; ↓ indicates that inhibition exceeded 50%.

## V. ALTERATIONS IN GENE EXPRESSION FOLLOWING INTEGRIN LIGATION

Several studies have indicated that ligation of integrins by matrix proteins can affect gene expression and protein synthesis in both mast cells (79) and eosinophils (53,54). To date, much of the evidence is indirect, relying on increased levels of protein synthesis in adherent cells. Currently most of the research is focused on the effects of cell–matrix interaction on cytokine gene regulation, reflecting the key role that cytokines play in the development of chronic allergic inflammation. It seems likely that within 1 to 2 years we will have a much clearer understanding of how cell-matrix interactions affect cytokine gene transcription in basophils and other cells and the potential role in the development of chronic allergic inflammation.

In addition to the cytokines, there is evidence that contact with the matrix via the integrins induces the synthesis and release of the matrix metalloproteinases. These enzymes are critical for cell migration, facilitating leukocyte movement through the extracellular matrix (80,81). It seems logical that the interactions of cells with the matrix either alone or in combination with the chemotactic agents would initiate the release of enzymes that break down the matrix and facilitate cell migration. The matrix metalloproteinases have also been implicated in the proteolytic cleavage of selectins (82,83) and other molecules from the cell surface (84,85), and again this may facilitate cell migration through the tissues by removing these ligands once they have fulfilled their function. Preliminary data from our laboratory and our collaborators support the hypothesis that interactions of both eosinophils and basophils with extracellular matrix proteins such as laminin and fibronectin lead to an increase in the release of matrix metalloproteinases into the supernatant (J.K. Shute, University of Southampton, personal communication).

## VI. SUMMARY

Though our understanding of how cell–matrix interactions affect cell function has come a long way recently, there can be little doubt that we have only scratched the surface of this important area. The complexity of the adhesion process, with coordinated interactions between different adhesion proteins on distinct cell types over widely different time scales, ensures that we still have some way to go before we can fully understand cell–matrix interactions and their possible role in disease pathology.

## ACKNOWLEDGMENT

This work was supported by the National Asthma Campaign.

## REFERENCES

1. T. Ishizaka, R. De Bernardo, H. Tomioka, I. M. Lichtenstein, and K. Ishizaka, *J. Immunol.*, *8*:1000 (1972).
2. D. W. MacGlashan, Jr., and L. M. Lichtenstein, *J. Immunol.*, *124*:2519. (1980).
3. D. W. MacGlashan, Jr., J. M. White, S. K. Huang, S. J. Ono, J. T. Schroeder, and L. M. Lichtenstein, *J. Immunol.*, *152*:3006 (1994).
4. T. Brunner, A. L. de Weck, and C. A. Dahinden, *J. Immunol.*, *147*:237 (1991).
5. T. C. Sim, H. Li, L. M. Reece, and R. Alam, *J. Allergy Clin. Immunol.*, *97*:358 (1996). Abstract.
6. P. Markvardsen, T. Bjerke, N. Rudiger, P. O. Schiotz, N. Gregeren, J. Justesen, and K. Paludan, *Scand. J. Lab. Invest.*, *55*:252 (1995).
7. A. S. Gounni, B. Lamkhloued, K. Ochiai, Y. Tanaka, E. Delaporte, A. Capron, J. P. Kinet, and M. Capron, *Nature*, *367*:183 (1994).
8. D. Maurer, E. Fiebiger, B. Rieninger, B. Wolff Winiski, M. H. Jouvin, O. Kilgus, J.-P. Kinet, and G. Stingl, *J. Exp. Med.*, *179*:745 (1994).
9. O. M. Kon, B. S. Sihra, J. A. Grant, and A. B. Kay, *J. Allergy Clin. Immunol.*, *97*:358 (1996).
10. E. N. Charlesworth, A. F. Hood, N. A. Soter, A. Kagey-Sobotka, P. S. Norman, and L. M. Lichtestein, *J. Clin. Invest.*, *83*:1519 (1989).
11. C.-B. Guo, M. C. Liu, S. J. Galli, B. S. Bochner, A. Kagey-Sobotka, and I. M. Lichtestein, *Am. J. Respir. Cell. Mol. Biol.*, *10*:384 (1994).
12. H. Agis, W. Füreder, H. C. Bankl, M. Kundi, W. R. Sperr, M. Willheim, G. Boltz-Nitulescu, J. H. Butterfield, K. Kishi, K. Lechner, and P. Valent, *Immunology*, *87*:535 (1996).
13. P. Valent and P. Bettelheim, *Adv. Immunol.*, *52*:333 (1992).
14. C. Stain, H. Stockinger, M. Scharf, U. Jäger H. Gössinger, K. Lechner, and P. Bettelheim, *Blood*, *70*:1872 (1987).
15. M. P. Bodger and L. A. Newton, *Br. J. Haematol.*, *67*:281 (1987).
16. B. S. Bochner, D. W. McGlashan, Jr., G. V. Marcotte, and R. P. Schleimer, *J. Immunol.*, *142*:3180 (1989).
17. B. S. Bochner, F. W. Luscinskas, M. A. Gimbrone, W. Newman, S. A. Sterbinsky, C. P. Derse-Anthony, D. Klunk, and R. P. Schleimer, *J. Exp. Med.*, *173*:1553 (1991).
18. M. E Hemler, *Ann. Rev. Immunol.*, *8*:365 (1990).
19. D. A. Wilkinson, T. J. Fitzgerald, and L. K. Jennings, *FASEB J.*, *9*:A1500 (1995).
20. A. Masellis-Smith and A. R. Shaw, *J. Immunol.*, *152*:2768 (1994).
21. M. Letarte, J. G. Seehafer, A. Greaves, A. Masellis-Smith, and A. R. Shaw, *Leukemia*, *7*:93 (1993).
22. S. K. Lo, S. Lee, R. A. Ramos, R. Lobb, M. Rosa, G. Chi-Rosso, and S. D. Wright, *J. Exp. Med.*, *173*:1493 (1991).
25. T. W. Kuijpers, B. C. Hakkert, M. Hoogerwerf, J. F. M. Leeuwenberg, and D. Ross, *J. Immunol.*, *147*:1369 (1991).
26. T. F. Tedder, Cell-surface-receptor *Am. J. Resp. Cell Mol. Biol.*, *5*:305 (1991).
27. T. K. Kishimoto, M. A. Jutila, E. L. Berg, and E. C. Butcher, *Science*, *245*:1238 (1989).
28. M. L. Dustin and T. A. Springer, *Nature*, *341*:619 (1989).

29. E. Martz, *Human Immunol.*, *18*:3 (1987).
30. D. J. Erle and R. Pytela, *Am. J. Resp. Cell Mol Biol.*, *6*:459 (1992).
31. S. A. Sporn, D. F. Eirman, C. E. Johnson, J. Morris, G. Martin, M. Ladner, and S. Haskill, *J. Immunol.*, *144*:4434 (1990).
32. T. H. Lin, C. Rosales, K. Modale, J. B. Bolen, S. Haskill, and R. L. Juliano, *J. Biol. Chem.*, *270*:16189 (1995).
33. R. Raghow, *FASEB J.*, *8*:823 (1994).
34. M. Stewart, M. Thiel, and N. Hogg, *Curr. Opin. Cell. Biol.*, *7*: 690 (1995).
35. J. Dastych, J. J. Costa, H. L. Thompson, and D. D. Metcalfe, *Immunology*, *73*:478 (1989).
36. H. L. Thompson, P. D. Burbelo, B. Segui-Real, Y. Yamada, and D. D. Metcalfe, *J. Immunol.*, *143*:2323 (1989).
37. D. D. Metcalfe, *Int. Arch. Allergy Immunol.*, *107*:60 (1995).
38. D. Taub, J. Dastych, N. Inamura, J. Upton, D. Kelvin, and D. Metcalfe, *J. Immunol.*, *154*:2393 (1995).
39. M. M. Hamawy, S. E. Mergenhagen, and R. P. Siraganian, *Immunology Today*, *15*:62 (1994).
40. M. M. Hamawy, C. Oliver, S. E. Mergenhagen, and R. P. Siraganian, *J. Immunol.*, *149*:615 (1992).
41. M. M. Hamawy, S. E. Mergenhagen, and R. P. Siraganian, *J. Biol. Chem.*, *268*:5227 (1993).
42. M. M. Hamawy, S. E. Mergenhagen, and R. P. Siraganian, *J. Biol. Chem.*, *268*:6851 (1991).
43. A. S. Koster, M. Enstad, F. A. M. Redegeld, and F. P. Nikjkamp, *Am. Rev. Resp. Dis.*, *153*:A632 (1996).
44. G. Nilsson, J. H. Butterfield, K. Nilsson, and A. Siegbahn, Stem *J. Immunol.*, *153*:3717 (1994).
45. P. J. Bianchine, P. R. Burd, and D. D. Metcalfe, *J. Immunol.*, *11*:3665 (1992).
46. T. Kingshi and T. A. Springer, *Blood*, *83*:1033 (1994).
47. J. E. Meredith, B. Fazeli, and M. A. Schwartz, *Mol. Cell Biol.*, *4*:953 (1993).
48. S. M. Frisch and H. Francis, *J. Cell Biol.*, *124*:619 (1994).
49. S. K. Akiyama, H. Larjava, and K. M. Yamada, *Cancer Res.*, *50*:1601 (1990).
50. S. Dedhar and R. Saulnier, *J. Cell Biol.*, *110*:481 (1990).
51. L. J. Walsh, M. S. Kaminer, G. S. Lazarus, R. M. Lavker, and G. F. Murphy, *Lab. Invest.*, *65*:433 (1991).
52. S. Montefort, C. Graziou, D. Goulding, R. Polosa, D. O. Haskard, P. H. Howarth, S. T. Holgate, and M. P. Carroll, *J. Clin. Invest.*, *93*:1411 (1994).
53. A. R. Anwar, R. Moqbel, G. M. Walsh, A. B. Kay, and A. J. Wardlaw, *J. Exp. Med.*, *177*:839 (1993).
54. G. M. Walsh, F. A. Symon, and A. J. Wardlaw, *Clin. Exp. Allergy*, *25*:1128 (1995>
55. A. Tourkin, T. Anderson, E. C. Lekoy, and S. Hoffman, *Cell. Adhesion Commun.*, *1*:161 (1993).
56. H. Kita, S. Horie, and G. J. Gleich, *J. Allergy Clin. Immunol.*, *93*:212A (1994).
57. S. P. Neeley, K. J. Hamann, and T. Dowlin, *Am. Rev. Resp. Dis.*, *147*:A242 (1993).
58. M. Kato, R. T. Abraham, G. J. Gleich, and H. Kita, *Am. J. Respir. Crit. Care Med.*, *153*:A58 (1996).

59. S. E. Laven, K. Goldring, L. H. Thomas, and J. A. Warner, *Am. J. Respir. Cell Mol. Biol.*, *14*:95 (1996).
60. K. Goldring, R. Djukanovic, K. Hurst, and J. A. Warner, *J. Allergy Clin. Immunol.*, *97*:264 (1996).
61. D. W. MacGlashan, Jr., Personal communication.
62. J. A. Warner and D. W. MacGlashan, Jr., *Immunol. Lett.*, *18*:129 (1988).
63. D. W. MacGlashan, Jr., *Mol. Immunol.*, *28*:585 (1991).
64. S. Saini, J. White, W. M. Gallatin, P. A. Hoffman, L. M. Lichtenstein, and B. S. Bochner, *J. Allergy Clin. Immunol.*, *97*:264 (1996).
65. M. Bosse, C. Ferland, J. Chakir, L. P. Boulet, M. Laviolette, *Am. J. Respir. Crit. Care Med.*, *153*:A58 (1996).
66. L. Kornberg, H. S. Earp. C. Turner, C. Prokop, and R. L. Juliano, *Proc. Natl. Acad. Sci. U.S.A.*, *88*:8392 (1991).
67. S. J. Shattil and J. S. Brugge, *Curr. Opin. Cell Biol.*, *3*:869 (1991).
68. A. Golde, J. S. Brugge, and S. J. Shattil, *J. Cell Biol.*, *111*:3117 (1990).
69. S. Jaken, K. Leach, and T. Klauck, *J. Cell Biol.*, *109*:697 (1989).
70. M. E. Jaconi, J. M. Theler, W. Schlegal, R. D. Appel, S. D. Wright, and P. D. Lewis, *J. Cell Biol.*, *112*:1249 (1991).
71. R. Tucker, K. Meade-Coburn, and D. Ferris, *Cell Calcium*, *11*:201 (1990).
72. M. A. Schwartz, C. Lechene, and D. E. Ingber, *Proc. Natl. Acad. Sci. U.S.A.*, *88*:7849 (1991).
73. M. P. Stewart, C. Cabanas, and N. Hogg, *J. Immunol.*, *156*:1810 (1996).
74. K. M. Yamada and S. Miyamoto, *Curr. Opin. Cell Biol.*, *7*:681 (1995).
75. A. Richardson and J. T. Parsons, *Bioassays*, *17*:229 (1995).
76. M. M. Hamawy, K. Minoguchi, W. G. Swaim, S. E. Mergenhagen, and R. P. Siragainian, *FASEB J.*, *9*:A781 (1995).
77. A. J. Lennan, K. Goldring, S. E. Lavens, and J. A. Warner, *FASEB J.*, *9*:A224 (1995).
78. J. M. Oliver, D. L. Burg, G. G. Deanin, J. L. Mclaughlin, and R. H. Geahlen, *J. Biol. Chem.*, *269*:29697 (1994).
79. H. L. Thompson, P. D. Burbelo, G. Gabriel, Y. Yamada, and D. D. Metcalfe, *J. Clin. Invest.*, *87*:619 (1991).
80. D. Leppert, S. L. Hauser, L. Kishiyama, S. An, L. Zeng, and E. J. Goetzl, *FASEB J.*, *9*:1473 (1995).
81. N. Malik, B. W. Greenfield, A. F. Wahl, and P. A. Kiener, *J. Immunol.*, *156*:3952 (1996).
82. T. A. Bennett, E. B. Lynam, L. A. Sklar, and S. Rogelj, *J. Immunol.*, *156*:3093 (1996).
83. M. De Luca, L. C. Dunlope, R. K. Andrews, J. V. Flannery, R. Ettling, D. A. Cumming, G. M. Veldman, and M. C. Berndt, *J. Biol. Chem.*, *270*:26734 (1995).
84. A. J. Gearing, P. Beckett, M. Christodoulou, M. Churchill, J. Clements, A. H. Davidson, A. H. Drummond, W. A. Galloway, R. Gilbert, J. L. Gordon, et al, *Nature*, *370*:555 (1994).
85. N. Kayagaki, A. Kawaski, T. Ebata, H. Ohmoto, S. Ikeda, S. Inoue, K. Yoshino, K. Okumura, and H. Yagita, *J. Exp. Med.*, *182*:1777 (1995).

# 8

# Cell Adhesion Molecules in Mast Cell Adhesion and Migration

**Harissios Vliagoftis and Dean D. Metcalfe**   *National Institute of Allergy and Infectious Diseases, National Institutes of Health, Bethesda, Maryland*

## I.  INTRODUCTION

Adhesion between cells, in the form of homotypic or heterotypic aggregation, as well as adhesion to extracellular matrix proteins may influence cell proliferation, migration, differentiation, survival (1), cytokine production (2), and gene expression (3). These consequences of adhesive interactions between cells or between cells and matrix have been shown to be of critical importance in the immune system, where they have an impact on the differentiation, recruitment, and activation of immune cells (4).

It has been known for some time that mast cells differentiate upon co-culture with fibroblasts into cells that resemble connective tissue mast cells (CTMC) in that they synthesize heparin and specific proteases (5,6). This differentiation has been shown to be dependent on direct contact of mast cells with the fibroblast monolayer. This adhesive interaction involves the c-kit receptor on mast cells and fibroblast surface-bound c-kit ligand, also known as stem cell factor (SCF) (7,8). Similar adhesive interactions between c-kit and SCF have been shown to exist between hematopoietic stem cells and bone marrow stroma cells (9). The observation of this mast cell–fibroblast interrelationship was the first convincing in vitro evidence that cell-to-cell interactions involving surface molecules could affect both mast cell phenotype and biologic function. Since then, a large number of studies have examined the display of adhesion molecules on mast cells and the biologic consequences of this expression.

Mast cells in humans develop from precursor cells, found in bone marrow (10) and the peripheral blood (11), that are CD34+ and c-kit+. Mast cell precursors migrate into tissues, where they differentiate into mast cells under the influ-

ence of locally produced growth factors. Adhesion molecules relate to the recruitment of these cells into tissues where they exhibit characteristic distribution patterns. Mast cell precursors from the vascular compartment must transmigrate
through the vascular endothelium and then target specific tissue structures. Adhesion to extracellular matrix (ECM) is a part of this process.

The adhesive function of additional molecules such as CD40 ligand and major
histocompatibility complex (MHC) class II molecules have also been investigated. In summarizing the literature, this chapter will start with adhesion molecules that have been described on mast cells and then focus on the biologic
functions of mast cells that adhesive interactions may modulate.

## II.  EXPRESSION OF ADHESION MOLECULES

### A.  Integrins

Integrins are heterodimeric molecules that consist of an $\alpha$ and a $\beta$ chain. More
than 11 $\alpha$ chains and seven $\beta$ chains have been described. There is significant
variability in the expression of integrins in mast cell lines, between mouse mast

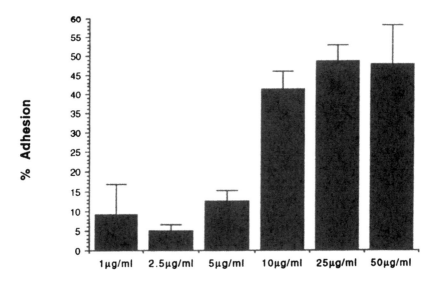

**Laminin   concentration**

**Figure 1**   Dose response of bone marrow cultured mast cell (BMCMC) adhesion to
laminin. Ninety-six-well plates were coated with the indicated concentrations of laminin
in 100 μl of DMEM overnight at 4°C. Four × 10$^4$ BMCMC were added in the wells and
activated with 50 ng/ml of phorbol myristate acetate (PMA). The adhesion assay was
carried out for 1 hour. The results are presented as the mean ± SD for three experiments
done in duplicate.

cells and human mast cells, and in mast cells isolated from different tissues. This variability may reflect the diverse functions of mast cells in different tissues, as well as specific requirements for adhesion receptors during differentiation and during the localization of mast cells in tissues. In this regard, mast cells have been shown to express integrins of the $\beta1$, $\beta2$, $\beta3$, and $\beta7$ families.

Murine bone marrow cultured mast cells (BMCMC) and both factor-dependent or factor-independent mast cell lines have been shown variously to adhere to laminin (12), fibronectin (13) and vitronectin (14). PT-18 murine mast cells and 3- to 4-week-old BMCMC adhere to murine laminin only after phorbol myristate acetate (PMA) activation. This adhesion increases in the presence of interleukin (IL)-3 (12). A dose response of PMA-induced adhesion of BMCMC to increasing amounts of plate-bound laminin is shown in Fig. 1.

FcεRI aggregation-induced adhesion of BMCMC to laminin occurs in a time- and dose-dependent manner. As shown in Fig. 2, the kinetics of BMCMC adhe-

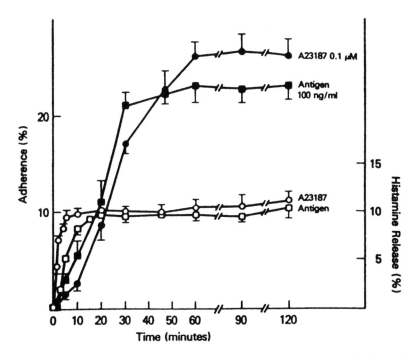

**Figure 2**   Kinetics of bone marrow cultured mast cell (BMCMC) adhesion to laminin and of histamine release after activation with A23187 or FcεRI aggregation. Histamine release (*open symbols*) and adhesion to laminin (*closed symbols*) after activation with 0.1 µM A23187 (*circles*) or 100 ng/ml of antigen (*squares*) was measured for varying time intervals up to 2 hours. Values represent mean plus standard deviation for four experiments performed in triplicate. (From Ref. 15). (Copyright 1994/1990, The American Association of Immunologists).

sion to laminin is slower than the kinetics of histamine secretion, whether mast cells are activated through FcεRI aggregation or the calcium ionophore A23187. Preincubation with transforming growth factor (TGF) $\beta_1$ significantly increases the FcεRI-dependent adhesion to laminin, with maximal effect seen after incubation with 2 ng/mL of TGFβ$_1$ for 24 hours (15). Preincubation of C57 mast cells with 2 ng/mL of TGFβ$_1$ for 24 hours induced these cells to adhere to laminin (Fig. 3). TGFβ$_1$ also increases PMA-induced adhesion and spreading of mast cells on laminin (Fig. 3). Other cytokines and growth factors were tested that did not induce adhesion of mast cells to laminin. These included platelet-derived growth factor (PDGF), fibroblast growth factor, epidermal growth factor, endothelial cell growth factors, tumor necrosis factor (TNF)-α, interferon gamma, granulocyte-monocyte colony-stimulating factor, and IL-1, IL-2, IL-3, and IL-4 (15).

The adhesion of BMCMC to laminin was inhibited with antibodies against laminin or against a 67-kD laminin receptor. The 67-kD laminin receptor was first isolated from BL6 murine melanoma cells (16); it shows high affinity for laminin and appears to play a role in tumor invasion and metastasis (17). BMCMC also express α6 integrin (18), which functions as a major laminin receptor for many cell types. Mouse macrophages after activation adhere to laminin through α6β1 integrin (19); human eosinophils express α6 and spontaneous eosinophil adhesion to laminin is inhibited by the GoH3 monoclonal anti-α6 antibody (20); and human natural killer (NK) cells adhere to laminin after activation with phorbol esters or though cross-linking of surface CD16, and this adhesion is inhibited by an anti-α6 antibody (21).

E8 is a proteolytic fragment of the long arm of laminin released by digestion with elastase. It has been shown that the binding site for α6 integrin is located within E8. BMCMC and C57 cells (Fig. 4) adhere to E8 after PMA activation in a dose-dependent fashion. PT-18 mast cells and BMCMC also adhere to the peptide PA22-2, a 22-amino-acid peptide that belongs to the long arm of the A chain (22). This area of the laminin molecule has been shown to be active in the adhesion of melanocytes and to induce the development of experimental metastasis of the lungs by murine melanoma cells (23).

Many of the integrins that function as receptors for ECM proteins, although constitutively expressed on the surface of mast cells, do not function unless the cells are activated—such as with FcεRI cross-linking (24) or exposure to PMA (13), the ionophore A23187 (15), or stem cell factor (SCF) (25). As noted, cultured mast cells adhere to laminin only after activation. Various stimuli also regulate the adhesion of mast cells to fibronectin (13). The mouse mast cell line MCP5/L adheres spontaneously to fibronectin, but this adhesion is up-regulated after these cells are activated with PMA (13) (Fig. 5A). The kinetics of spontaneous and PMA-induced adhesion are different (Fig. 5B). Adhesion of mast cells to fibronectin is inhibited by RGD-containing peptides, indicating that an RGD-binding integrin mediates this interaction (13). In contrast, murine BMCMC and

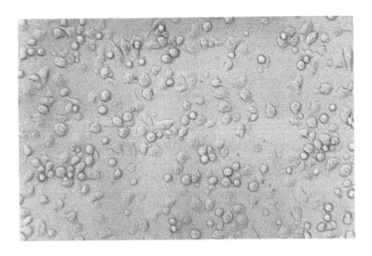

**(a)**

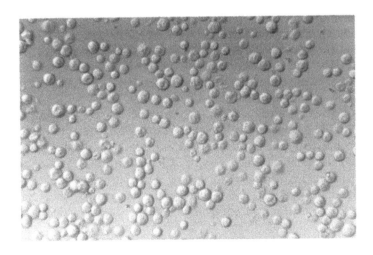

**(b)**

**Figure 3** Adhesion of C57 mast cells to laminin 25 µg/ml after culture with transforming growth factor (TGF)-β 2 ng/ml for 18 hours (a), after phorbol myristate acetate (PMA) activation for 1 hour (b), or after PMA activation of cells cultured in the presence of TGF-β for 18 hours (c).

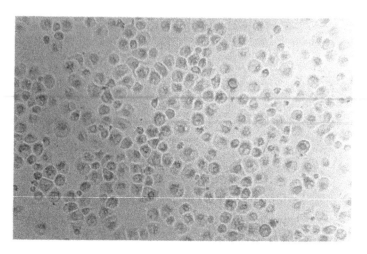

**(c)**

**Figure 3** Continued

C57 mast cells spontaneously adhere to vitronectin (14). This adhesion is inhibited by divalent ion chelation, by the tetrapeptide Arg-Gly-Asp-Ser, and by a polyclonal antibody against the $\alpha v \beta 3$ integrin.

Stem cell factor, apart from being a primary growth factor for mast cells, induces murine BMCMC and MCP5/L mast cells to adhere to both fibronectin (25,26) and laminin. Stem cell factor–induced adhesion of BMCMC to fibronectin is dose-dependent and is inhibited by anti-SCF antibodies (Fig. 6). Stem cell factor–induced mast cell adhesion to matrix occurs at 1–2 ng/mL, a concentration of SCF found in body fluids. For this reason, SCF may be the physiologic stimuli that prevents mast cells from detaching from matrix and circulating as mature cells in the blood. The effect of SCF on BMCMC adhesion to fibronectin is dependent on extracellular calcium, is inhibited by the peptide GRGDSP but not control peptides, and is inhibited by the tyrosine kinase inhibitor genistein (25). Genistein does not inhibit mast cell adhesion induced by PMA.

The effector pathways by which receptor tyrosine kinases regulate cell–matrix adhesion have been studied in more detail in mast cells using wild-type and mutant forms of the PDGF receptor. This receptor is closely related to c-kit, is expressed in mast cells, and is as potent as c-kit in stimulating adhesion to fibronectin (27). Adhesion stimulated through the PDGF receptor is regulated through two independent pathways involving phosphatidyl-inositol 3 kinase (PI3K) and phospholipase C-gamma 1 (PLC gamma)-protein kinase C. Receptors mutated so that they could not associate with PI3K and PLC gamma continued to stimulate mast cell growth, but not adhesion, indicating a crucial role for

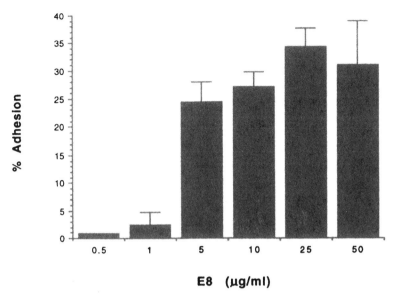

**E8  (µg/ml)**

**Figure 4**  Dose response of C57 mast cell adhesion to E8. Ninety-six-well plates were coated with the indicated concentrations of E8 in 100 µL of DMEM overnight at 4°C. Four × 10⁴ C57 mast cells were added in the wells and activated with 50 ng/ml of phorbol myristate acetate (PMA). The adhesion assay was carried out for 1 hour. The results are presented as the mean ± SD for three experiments done in duplicate.

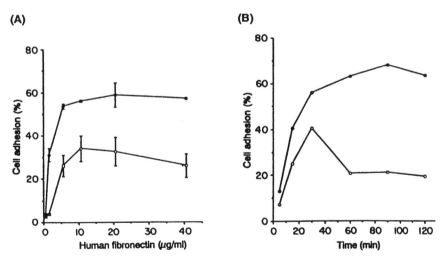

**Figure 5**  (A) Adhesion of MCP-5 mast cells to wells coated with the indicated concentrations of human fibronectin in the presence (*solid circles*) or absence (*open circles*) of 50 ng/ml of phorbol myristate acetate (PMA). (B) Kinetics of MCP-5 mast cell adhesion to fibronectin (20 µg/ml) in the presence (*closed circles*) or absence (*open circles*) of PMA. (From Ref. 13).

**(A)**

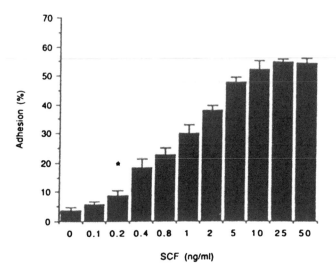

**(B)**

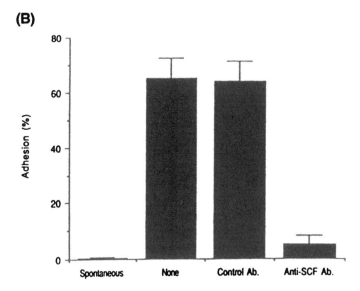

**Figure 6**  Murine bone marrow cultured mast cell (BMCMC) adhesion to plates coated with 10 μg/ml of fibronectin in the presence of increasing concentrations of stem cell factor (SCF) (A). Inhibition of SCF-induced adhesion by anti-SCF antibodies but not control antibodies (B). (From Ref. 25.) (Copyright 1994/1990, The American Association of Immunologists).

these molecules in regulating adhesion rather than cell growth (27). In another approach, transfection of mast cells from Wsh/Wsh mice that do not express endogenous c-kit, with wild-type or mutated murine c-kit, revealed that Y719, necessary for the association of c-kit with PI3-kinase, is also necessary for the induction of adhesion through SCF (28). This mutation, however, impaired only partially the SCF-induced proliferation and survival of mast cells.

RBL-2H3 and phorbol ester–stimulated rat peritoneal mast cells also adhere, as do murine mast cells, to fibronectin, vitronectin, and fibrinogen. RBL-2H3 cells express fibronectin receptor integrins including VLA-4, VLA-5, and the vitronectin receptor ($\alpha v \beta 3$), as estimated by immunofluorescent staining. Antibodies reactive with the rat $\alpha 4$, $\alpha 5$, or $\beta 3$ integrins inhibit RBL cell adhesion to fibronectin (29).

The murine homologue of human $\beta 7$ has been cloned from an RBL cDNA library screened with a probe that recognized genes preferentially expressed by mast cells. The expression of $\beta 7$ increases in IL-3 dependent murine BMCMC as they become more mature in culture. Expression is maximum after 3 weeks (30). Murine BMCMC also express mRNA for $\alpha 4$, $\alpha 5$, and $\beta 1$. The expression of $\alpha 4$ was similar in the time course of expression with that of $\beta 7$, although at 3 weeks the expression of $\alpha 4$ was lower than expression at 2 weeks. The surface expression of $\beta 7$ diminished after 3 weeks, and this decrease did not correlate with levels of $\beta 7$ transcripts made by these cells. It was thus postulated that another $\alpha$ subunit expressed at 3 weeks could associate with the excess $\beta 7$ when $\alpha 4$ is down-regulated. The murine $\alpha M290$ or $\alpha E$ had been shown to associate with $\beta 7$ in intraepithelial lymphocytes and function as a ligand for E-cadherin that is expressed on the surface of mucosal epithelial cells. Bone marrow cultured mast cells cultured in the presence of TGF-$\beta$ expressed $\alpha E$ on their surface (31). Applying these observations to recruitment of mast cells into the gastrointestinal tract led to the speculation that maturing mast cells may be able to express $\alpha 4 \beta 1$ and $\alpha 4 \beta 7$ and thus localize to the lamina propria in the same way that T cells localize to this area. There under the influence of locally produced TGF-$\beta$, mast cells could express $\alpha E \beta 7$ and thus adhere to mucosal epithelial cells through interaction with E-cadherin.

Human skin mast cells have been reported to express VLA-3, VLA-4, and VLA-5, but not VLA-1, VLA-2, or VLA-6 (32). These mast cells adhered to laminin and fibronectin in vitro but did not adhere to collagen type I or type IV, as has been reported for murine mast cells. The adhesion to fibronectin was inhibited by antibodies to all three $\beta 1$ integrins expressed. Adhesion to laminin was inhibited by antibodies against VLA-3 (32).

Human mast cells obtained from fetal liver and cultured with stem cell factor express CD51/CD61 ($\alpha v \beta 3$), an active vitronectin receptor (33). These fetal liver–derived mast cells also express surface $\beta 1$, $\alpha 4$, $\alpha 5$, and very low levels of $\beta 2$ integrins as detected by flow cytometry. Anti-$\alpha v \beta 3$ antibody partially inhibited the adhesion of these cells to laminin.

Human skin mast cells also express α6 integrin, as determined by immunohistochemistry. These dermal mast cells are intimately associated with the laminin in the microvascular and neural basement membranes in vivo and pericellular laminin complexes in vitro (34). Upon degranulation induced by morphine sulfate or ionophore A23187 in vitro, mast cell–laminin complexes dissociated and these mast cells did not adhere to laminin substrates. In cutaneous mastocytosis/urticaria pigmentosa, dermal mast cells reportedly did not express laminin receptors in tissues, as detectable by immunohistochemistry, and were not always found in close association with laminin of vascular basement membranes. Similar observations were made following degranulation of mast cells in normal foreskin organ culture with morphine. The authors concluded that CTMC–laminin interactions may be important determinants of mast cell localization in tissue compartments (34) as has been suggested (12).

Many observations concerning the human mast cell expression of integrins use mast cells dispersed from normal tissues using proteases or examine mast cells that are grown in culture. Other investigators have examined HMC-1 cells, the only human mast cell culture available and which was established from a patient with mast cell leukemia (35). Unfortunately, this cell does not express high-affinity IgE receptors and is not typical of human mast cells isolated from tissues. Table 1 summarizes published data on adhesion molecule expression on human mast cells (36–41).

## B.  Ig-like Molecules, ICAM-1

Human lung and uterine mast cells express ICAM-1 (CD54) (40). Reported expression is low. In fact, some monoclonal antibodies against ICAM-1 fail to yield positive results on flow cytometry (42). HMC-1 mast cells constitutively express mRNA for ICAM-1 and show surface staining with anti–ICAM-1 antibodies (43). ICAM-1 expression is up-regulated after treatment with IL-4. IL-4 has a similar effect on macrophages but not on basophils, fibroblasts, or lymphocytes. IL-13, a cytokine with properties similar to IL-4, also increased the expression of ICAM-1 on HMC-1 cells (38). IL-13 had no effect on the expression of integrins on the surface of HMC-1 cells (38). This is unlike what has been shown for monocytes (44), where IL-13 up-regulates the expression of CD11b, CD11c, CD18, CD29, and CD49e. Treatment of HMC-1 cells with IFN-γ (2000 U/mL) or TNF-α (200 ng/ml) increased the expression of ICAM-1 but had no effect on the expression of CD11a, CD11b, or CD11c. Only IFN-γ increased the expression of CD18 (45). HMC-1 cells also express LFA-1 (37) and LFA-3 (39).

## C.  Other Adhesion Molecules

With a syngeneic dorsal skinfold chamber model of the microcirculation, it has been shown that murine BMCMC roll on venular endothelial cells but not on ar-

**Table 1** Adhesion Molecules on Human Mast Cells

| | HMC-1 (37) | Fetal liver (36) | HMC-1 (38) | Cord blood (38) | HMC-1 (39) | s-mc (32) | l-mc (39) | u-mc (39) | s-mc (39) | l-mc (40) | u-mc (41) |
|---|---|---|---|---|---|---|---|---|---|---|---|
| CD2 | + | nt | nt | nt | nt | nt | – | – | – | nt | nt |
| CD11a | + | nt | nt | nt | – | nt | – | – | – | <20% | 18 ± 11% |
| CD11b | – | – | nt | nt | – | nt | – | – | – | <20% | 14 ± 5% |
| CD11c | (+) | – | nt | nt | – | nt | – | – | – | <20% | 75 ± 12% |
| CD18 | + | – | nt | nt | – | nt | – | – | – | <20% | 72 ± 7% |
| CD29 | + | + | 80% | 50% | + | + | + | + | + | nt | 99 ± 2% |
| CD36 | nt | nt | nt | nt | – | nt | – | nt | – | <20% | nt |
| CD44 | + | nt | nt | nt | + | nt | + | + | + | 80% | nt |
| CD49a | – | nt | 0 | 0 | nt | – | nt | nt | nt | nt | nt |
| CD49b | – | nt | 0 | 0 | – | – | – | – | nt | nt | 7 ± 3% |
| CD49c | – | nt | 0 | 41% | nt | + | nt | nt | nt | nt | nt |
| CD49d | + | nt | 31% | 16% | + | + | + | + | + | nt | 97 ± 2% |
| CD49e | + | nt | 30% | 2% | + | + | + | + | + | nt | 87 ± 6% |
| CD49f | – | nt | 0 | 0 | – | – | – | – | nt | nt | 5% |
| CD51 | nt | nt | nt | nt | + | nt | + | + | + | nt | 86 ± 12% |
| CD54 | + | – | nt | nt | + | nt | + | + | + | 20-40% | 87 ± 10% |
| CD61 | + | nt | nt | nt | + | nt | + | + | + | nt | 75 ± 15% |
| LFA-3 | nt | nt | nt | nt | + | nt | + | + | nt | nt | nt |

l-mc = lung mast cells, u-mc = uterus mast cells, s-mc = skin mast cells, nt = not determined.

terioles or capillaries (46). BMCMC did not express L-selectin or P-selectin, as determined by flow cytometry. The rolling depended on the interaction of mast cells with the P-selectin expressed on the endothelial cells, as neutralizing anti–P-selectin antibodies, but not anti–L-selectin antibodies, were able to inhibit rolling when administered in vivo. It is possible that this interaction could facilitate the transmigration of mast cell precursors into tissues. Murine BMCMC do express and store sergylin, a receptor for CD44 adhesion molecules, in their granules (47). Sergylin secreted from mast cells could regulate lymphoid cell adherence and activation.

## D. Stem Cell Factors in Adhesion

Other surface receptors may function as adhesion molecules when their ligands are expressed on the surface of other cells that are in close proximity with mast cells. As noted, it has been shown that c-kit, a receptor protein tyrosine kinase unique to hematopoietic progenitor cells, melanocytes, and mast cells, acts as an adhesion molecule between mast cells and fibroblasts; the latter express surface-bound SCF (7). As SCF is a major growth and differentiation factor for murine as well as human mast cells, this adhesive interaction may be important for the differentiation and survival of mast cells in tissues.

The $S1^d$ mutation in mice removes the transmembrane and intracellular portions of SCF, together with nine amino acids of the extracellular segment. Mast cells adhere to COS cells transfected with the normal cDNA for SCF. They do not adhere to COS cells transfected with the cDNA for $S1^{d.}$ This effect is evident in $S1/S1^d$ mice. Mice with this mutation lack the membrane-associated form of SCF, but have normal amounts of the soluble form (48). These mice, although they express the soluble form of SCF, have only 1% of the normal number of tissue mast cells.

Dog mastocytoma mast cells adhere to cultured dog tracheal epithelial cells ($35 \pm 13\%$), but adhere poorly to types I and IV collagen, or to fibronectin (less than 7.5% mean adhesion in all cases) (49). In tracheal tissue sections, mast cells adhered preferentially to epithelial cells in surface epithelium or in submucosal glands, but not to basal membrane or connective tissue. Adhesion to cultured epithelial cells was a characteristic of a subpopulation of mast cells; it could persist for more than 48 hours, did not require energy or the presence of divalent cations, and was not mediated by a known family of leukocyte-associated adhesion glycoproteins. Whether SCF was involved is unknown. Mast cell adhesion to airway epithelium could lead to higher numbers of intraluminal mast cells. Thus, inflammatory mediators released from activated mast cells would reach high local concentrations and in turn influence airway function in certain lung diseases (49).

Mast cells also express other molecules that do not belong to any family of adhesion molecules, but may participate in adhesive interactions between different kinds of cells with varied biologic functions. HMC-1 cells and freshly isolated human lung mast cells express CD40L. They may in turn be able to induce IgE production from B cells in the presence of IL-4 through the interaction of CD40L with CD40 on the surface of B cells (50). Mast cells are also able to present antigen to T cell lines through MHC class II and other accessory molecules (51).

## III. BIOLOGIC EFFECTS OF ADHESION

### A. Cell Proliferation and Differentiation

Murine IL-3 dependent BMCMC are safranin-negative. When attached to fibroblasts through SCF for 2 weeks, they differentiate into safranin-positive cells. These cells also now synthesize heparin (6). If IL-3 is still present, these cells proliferate and produce more histamine than cells proliferating in the absence of adhesion to fibroblasts.

IL-3 dependent murine BMCMC adhere to plate-bound vitronectin, through the $\alpha v\beta 3$ integrin. These BMCMC increase proliferation by 40% when adherent. This adhesion does not sustain the survival of BMCMC in the absence of IL-3 (14).

Astrocyte monolayers support rat peritoneal mast cell viability for up to 30 days without any effect on phenotype or the release characteristics of mast cells (52). This effect requires contact between the two cell types. It is possible that a soluble factor is also produced by astrocytes to support the survival of mast cells. Osteoblast-like cells similarly support rat peritoneal mast cell viability and functional activity as a result of heterotypic adhesion (53).

### B. Protein Tyrosine Phosphorylation

There are several studies of the phosphorylation of proteins in mast cells after adhesion to extracellular matrix proteins. Adhesion of RBL cells to fibronectin results in the phosphorylation of a 105–115 kD protein (54) that has been shown to be p125FAK (55). Adhesion also results in the phosphorylation of paxillin, but only after FcεRI aggregation (56). Adhesion of BMCMC to vitronectin also induces the phosphorylation of p125FAK. Phosphorylation increases when mast cells are simultaneously activated through FcεRI aggregation. Adhesion of murine mast cells to laminin, and also to the peptide PA22-2 that belongs to the long arm of the A chain of laminin, induces the phosphorylation of a 90–95 kD protein. This protein has not been identified.

### C. Degranulation

Adhesion of RBL cells to fibronectin increases the degree of secretion following activation through FcεRI aggregation (57). Fibronectin alone, either soluble or plate-bound, has no secretory effect on mast cells.

Co-culture of mast cells with fibroblasts and following adhesion through c-kit induces changes in mast cell responsiveness to various stimuli for degranulation. BMCMC adherent to fibroblasts for 3 weeks have a two to three-fold increase in the release of histamine after immunologic or calcium ionophore–induced activation (58). The production of $LTB_4$ and $PDG_2$ was also increased under the same conditions.

The release of β-hexosaminidase, a granular enzyme, is triggered by FcεRI aggregation. Release is enhanced by adhesion of RBL-2H3 cells to either immobilized fibronectin, or to plate-bound monoclonal antibodies against α4, α5 or β3. Fibronectin enhancement of β-hexosaminidase release is inhibited by antibodies against α4, α5, or β3, by RGD peptide, and by CS-1 peptide. The last two molecules abrogate VLA-5/αvβ3 and VLA-4 adhesion to fibronectin respectively (29). Fibronectin receptor integrins expressed on rat mast cells also regulate mast cell activation in vivo. Passive cutaneous anaphylaxis induced by IgE anti-DNP and DNP-BSA is inhibited by concurrent local injection of the above three antibodies (29).

Adhesion of C57 cells to laminin similarly increases the FcεRI aggregation–induced secretion of β-hexosaminidase. The peptide PA22-2 does not induce secretion from mast cells. PA22-2 also does not influence the secretion induced by FcεRI aggregation.

## D.  Gene Expression and Cytokine Production

It has been reported in other cell systems that adhesion to extracellular matrix proteins through surface integrins induces transcription of genes. There are few data examining gene expression in mast cells following adhesion. Spontaneous adhesion of C57 mast cells to PA22-2 did not increase mRNA for IL-3 and IL-4 as evaluated by reverse transcriptase polymerase chain reaction (RT-PCR). One group has reported that adhesion of mouse mast cells to fibronectin through the α5β1 receptor increases the production of IL-3. This adhesion-related expression of IL-3 could relate to an autocrine activity that would increase mast cell survival after adhesion (59).

## IV.  MIGRATION

Mast cells migrate in response to extracellular matrix proteins. The receptors that participate in migration are not defined. Thus, unactivated PT18 cells have been shown to chemoattract toward laminin but not collagen type IV or fibronectin in a Boyden chamber assay (60). Murine BMCMC show weak chemoattraction toward these proteins. This chemoattraction is enhanced if mast cells are activated via FcεRI aggregation or exposure to PMA or calcium ionophore. This effect is not altered by IL-3 or histamine release from mast cells. The peptide PA22-2, part of the A chain of the laminin molecule, exhibits 40% of the activity of the whole molecule (60).

Mouse mast cells have also been shown to exhibit random migration on laminin, fibronectin, and matrigel-coated surfaces. Adhesion and movement requires activation through FcεRI aggregation (24). A summary of these results is shown in Table 2.

Mast cells also express mRNA for laminin and collagen type IV and produce and release the protein products (61). Production of these extracellular matrix proteins might facilitate the recruitment and retention of mast cells into the areas of inflammation (24).

Stem cell factor induces the migration of cultured human mast cells as well as HMC-1 cells through its interaction with c-kit (62). Migration is dependent on the adhesion of mast cells to extracellular matrix proteins, especially fibronectin. Mast cells do not chemoattract over the filters if these are not coated with matrix.

## V. ADHESION AND MIGRATION OF MAST CELL PRECURSORS

CD34+ cells express adhesion molecules. CD34+ cells isolated from human bone marrow and from cord blood express LFA-1 (CD11a/CD18), α4, α5, β1, and αv

**Table 2** Response of Bone Marrow Cultured Mast Cells (BMCMC) Following FcεRI-mediated Activation on Selected Substrates

| Substrate | Stimulus | Adherence | Cell shape on matrix | Cell migration on matrix |
|-----------|----------|-----------|---------------------|--------------------------|
| 3% BSA | DNP-BSA | No | — | — |
| | None | No | — | — |
| Laminin | DNP-BSA | Yes | Round to flattened | Yes |
| | None | No | — | — |
| IKVAV peptide | DNP-BSA | Yes | Round to flattened | No |
| | None | Yes | Round to flattened | No |
| Fibronectin | DNP-BSA | Yes | Round to flattened | Yes |
| | None | No | — | — |
| Collagen IV | DNP-BSA | No | — | — |
| | None | No | — | — |
| Matrigel | DNP-BSA | Yes | Round to flattened | Yes |
| | None | No | — | — |

Summary of mast cell behavior on the extracellular matrix substrates laminin, collagen IV, fibronectin, matrigel, the synthetic peptide IKVAV, and plastic coated with 3% BSA either in the absence of stimulation or stimulation with 100 ng/ml of DNP-BSA following passive sensitization with 1 µg/ml IgE-DNP (FcεRI-mediated activation). Cells were studied for a total of 1000 min on each substrate. Unsensitized BMCMC exposed to DNP-BSA gave results identical to those observed when sensitized cells were placed on various substrates without DNP-BSA.

integrins, LFA-3, and ICAM-1 (CD54) (63). The expression of adhesion molecules differs according to the stage of maturation of the precursor cells studied. Expression of β1 and β2 integrins by stem cells is important for their adhesion to bone marrow stroma cells (64). During maturation, disappearance of these molecules could affect the release of these cells from the bone marrow and their localization in tissues. The specific progenitor cell for human mast cells has not been studied in terms of adhesion. The steps in migration of this cell into the peripheral tissues is thus not clear. It is known, however, that 1-week-old murine BM-CMC sorted to be FcεRI positive adhere to laminin after PMA activation (15). This observation indicates that the adhesion of early mast cells to laminin may be important in mast cell trafficking and could relate to the differentiation of immature mast cells after they are localized in tissues.

## VI.  MAST CELL EFFECT ON ADHESION AND MIGRATION OF OTHER CELLS

Mast cells are the primary resident effector cells in allergic inflammation. They also are regarded as effector cells in other immunologic and inflammatory responses in which they trigger the migration of other blood cells.

Mast cell activation leads to the release of TNF-α and IL-4. These molecules modulate the expression of adhesion molecules on endothelial cells and on leukocytes. Activation of mast cells in human skin cultures with immunologic and non-immunologic stimuli leads to the induction of ELAM-1 in adjacent endothelial cells (64). This effect is abrogated with cromolyn and with pretreatment with antibodies to TNF-α (65). Human dermal mast cells have been shown to store significant amounts of biologically active TNF-α that they release upon degranulation (66). Release of this stored TNF-α leads to up-regulation of ELAM-1 expression in adjacent endothelial cells (66).

Perfusion of the rat mesentery with compound 48/80, a mast cell degranulating agent, leads to a dose-dependent rise in the number of rolling and adherent cells in the local vasculature (67). These effects are significantly reduced by the mast cell stabilizers ketotifen and cromolyn; and by chronic treatment with compound 48/80, which depletes mast cell mediators. Mast cell degranulation, in this study, induced P-selectin–dependent leukocyte rolling and CD18-dependent leukocyte adhesion to the endothelium via histamine and PAF, respectively (67).

Local challenge of the rat mesentery with compound 48/80 increases the leukocyte rolling fraction, decreases rolling velocity, and induces firm leukocyte adhesion in postcapillary venules, as shown by intravital microscopy (68). In another study evaluating the effect of histamine in a similar system, it was reported that histamine has the same effects; in addition, it increased the clearance of albumin from blood to the superfusate (69). In the first study, the effects of compound 48/80 were inhibited by a monoclonal anti–P-selectin antibody, but not by combined treatment with $H_1$ and $H_2$ histamine receptor antagonists. Moreover, the response to compound 48/80 was not duplicated by exogenous histamine or 5-hydroxytryptamine (serotonin). This was said to indicate that mediator(s) other than histamine and serotonin evoke P-selectin–dependent leukocyte rolling, and thereby promoted firm leukocyte adhesion in mast cell–dependent inflammation (68). However, in the second study the effect that was inhibited by $H_1$ antagonists appeared to be mediated by P-selectin. Antibodies against P-selectin and also soluble sialyl-Lewis$^x$ oligosaccharide inhibited all effects except those from increased extravasation (69). This mast cell–dependent leukocyte recruitment and microvascular permeability was inhibited by the addition of nitric oxide (70).

The inflammatory cell populations and adhesion molecule expression in early phases of delayed-type hypersensitivity (DTH) elicited by 2,4-dinitrochlorobenzene has also been examined (71). The first discernible event, at 1 hour, was mast cell degranulation, followed by induction of endothelial leukocyte adhesion molecule (ELAM-1) expression on dermal postcapillary venules at 2 hours. Endothelial leukocyte adhesion molecule expression peaked at 24 hours and declined by 48 hours. In contrast, endothelial expression of intercellular adhesion molecule-1 (ICAM-1) remained at constitutive levels. Intrafollicular T-cell migration occurred independent of ICAM-1 expression and commenced as early as 4 hours after challenge. Mature, activated CD4-positive lymphocytes that expressed a helper-inducer/memory phenotype predominated in the early lesions. These results demonstrate an association in vivo between mast cell degranulation and ELAM-1 expression in DTH (71).

Mast cells also affect the expression of adhesion molecules on fibroblasts. Fibroblasts exposed to mast cell–conditioned medium expressed increased levels of ICAM-1 and VCAM-1 at 4 hours, with highest levels at 16 hours. These fibroblasts bound 5 times more T cells than nontreated fibroblasts (72). This effect was 80% inhibited by TNF-α–neutralizing antibodies. On the other hand, preincubation of CD4+ lymphocytes with histamine or the supernatant of immunologically activated mast cells resulted in a 40–50% decrease in the PMA-induced adhesion of these cells to both fibronectin and laminin (73). This effect was blocked by $H_2$-antagonists.

The effect of histamine has been examined directly on bronchial epithelial cells obtained by bronchial brushing from 22 nonasthmatic subjects (74). The activation of epithelial cells was assessed by immunocytochemical analysis of the expression of membrane markers ICAM-1 and HLA-DR and the release of fibronectin. There was a highly significant increase in the number of cells expressing ICAM-1 (from 10±11% to 32±20% after treatment with histamine) and HLA-DR (from 8±7% to 23±20%) and in the release of fibronectin. Cycloheximide blocked these effects, suggesting that histamine requires protein synthesis for its action. Pyrilamine ($H_1$ antagonist) and ranitidine ($H_2$ antagonist) at a concentration of 10 μM decreased the effect of histamine. This study suggested that mast cells present in the airways may have a role in the activation of epithelial cells through the release of histamine (74).

## VII.  SUMMARY

Mast cell adhesion and migration are critical events in mast cell biology. Adhesion molecules guide in localizing mast cells to tissues and modifying their biologic responsiveness. Research in this area has focused principally on the expression and biologic significance of integrins and the SCF receptor c-kit on murine and human mast cells. Additional work is clearly needed to understand more clearly how adhesive interactions direct mast cell responsiveness and function.

## REFERENCES

1. Z. Zhang, K. Vuori, J. C. Reed, and E. Ruoslahti, *Proc. Natl. Acad. Sci. U.S.A*, *92*:6161 (1995).
2. T. H. Lin, A. Yurochko, L. Kornberg, J. Morris, J. J. Walker, S. Haskill, and R. L. Juliano, *J. Cell Biol.*, *126*:1585 (1994).
3. P. Huhtala, M. J. Humphries, J. B. McCarthy, P. M. Tremble, Z. Werb, and C. H. Damsky, *J. Cell. Biol.*, *129*:867 (1995).
4. C. R. Mackay and B. A. Imhof, *Immunol. Today*, *14*:99 (1993).
5. C. E. Bland, H. Ginsburg, J. E. Silbert, and D. D. Metcalfe, *J. Biol. Chem.*, *257*:8661 (1982).
6. S. F. Levi, K. F. Austen, P. M. Gravallese, and R. L. Stevens, *Proc. Natl. Acad. Sci. U.S.A*, *83*:6485 (1986).
7. S. Adachi, Y. Ebi, S. Nishikawa, S. Hayashi, M. Yamazaki, T. Kasugai, T. Yamamura, S. Nomura, and Y. Kitamura, *Blood*, *79*:650 (1992).
8. S. Adachi, T. Tsujimura, T. Jippo, M. Morimoto, K. Isozaki, T. Kasugai, S. Nomura, and Y. Kitamura, *Exp. Hematol.*, *23*:58 (1995).
9. H. Kodama, M. Nose, S. Niida, S. Nishikawa, and S. Nishikawa, *Exp. Hematol.*, *22*:979 (1994).
10. A. S. Kirshenbaum, S. W. Kessler, J. P. Goff, and D. D. Metcalfe, *J. Immunol.*, *146*:1410 (1991).
11. M. Rottem, T. Okada, J. P. Goff, and D. D. Metcalfe, *Blood*, *84*:2489 (1994).
12. H. L. Thompson, P. D. Burbelo, R. B. Segui, Y. Yamada, and D. D. Metcalfe, *J. Immunol.*, *143*:2323 (1989).
13. J. Dastych, J. J. Costa, H. L. Thompson, and D. D. Metcalfe, *Immunology*, *73*:478, (1991).
14. P. J. Bianchine, P. R. Burd, and D. D. Metcalfe, *J. Immunol.*, *149*:3665 (1992).
15. H. L. Thompson, P. D. Burbelo, and D. D. Metcalfe, *J. Immunol.*, *145*:3425 (1990).
16. N. C. Rao, S. H. Barsky, V. P. Terranova, and L. A. Liotta, *Biochem. Biophys. Res. Commun.*, *111*:804 (1983).

17. L. A. Liotta, H. P. Horan, C. N. Rao, G. Bryant, S. H. Barsky, and J. Schlom, *Exp. Cell Res.*, *156*:117 (1985).
18. H. Vliagoftis and D. D. Metcalfe, *J. Allergy Clin. Immunol.*, *95*:293 (1995).
19. L. M. Shaw, J. M. Messier, and A. M. Mercurio, *J. Cell. Biol.*, *110*:2167 (1990).
20. S. N. Georas, B. W. McIntyre, M. Ebisawa, J. L. Bednarczyk, S. A. Sternbinsky, R. P. Schleimer, and B. S. Bochner, *Blood*, *82*:2872 (1993).
21. A. Gismondi, F. Mainiero, S. Morrone, G. Palmieri, M. Piccoli, L. Frati, and A. Santoni, *J. Exp. Med.*, *176*:1251 (1992).
22. H. L. Thompson, P. D. Burbelo, Y. Yamada, H. K. Kleinman, and D. D. Metcalfe, *Immunology*, *72*:144 (1991).
23. T. Kanemoto, R. Reich, L. Royce, D. Greatorex, S. H. Adler, N. Shiraishi, G. R. Martin, Y. Yamada, and H. K. Kleinman, *Proc. Natl. Acad. Sci. U.S.A*, *87*:2279 (1990).
24. H. L. Thompson, L. Thomas, and D. D. Metcalfe, *Clin. Exp. Allergy*, *23*:270 (1993).
25. J. Dastych and D. D. Metcalfe, *J. Immunol.*, *152*:213 (1994).
26. T. Kinashi and T. A. Springer, *Blood*, *83*:1033 (1994).
27. T. Kinashi, J. A. Escobedo, L. T. Williams, K. Takatsu, and T. A. Springer, *Blood*, *86*:2086 (1995).
28. H. Serve, N. S. Yee, G. Stella, L. L. Sepp, J. C. Tan, and P. Besmer, *Embo J.*, *14*:473 (1995).
29. M. Yasuda, Y. Hasunuma, H. Adachi, C. Sekine, T. Sakanishi, T. Hashimoto, C. Ra, H. Yagita, and K. Okumura, *Int. Immunol.*, *7*:251 (1995).
30. M. F. Gurish, A. F. Bell, T. J. Smith, L. A. Ducharme, R. K. Wang, and J. H. Weis, *J. Immunol.*, *149*:1964 (1992).
31. T. J. Smith, L. A. Ducharme, S. K. Shaw, C. M. Parker, M. B. Brenner, P. J. Kilshaw, and J. H. Weis, *Immunity*, *1*:393 (1994).
32. M. Columbo, B. S. Bochner, and G. Marone, *J. Immunol.*, *154*:6058 (1995).
33. Y. Shimizu, A. M. Irani, E. J. Brown, L. K. Ashman, and L. B. Schwartz, *Blood*, *86*:930 (1995).
34. L. J. Walsh, M. S. Kaminer, G. S. Lazarus, R. M. Lavker, and G. F. Murphy, *Lab. Invest.*, *65*:433 (1991).
35. J. H. Butterfield, D. Weiler, G. Dewald, and G. J. Gleich, *Leuk. Res.*, *12*:345 (1988).
36. G. Nilsson, K. Forsberg, M. P. Bodger, L. K. Ashman, K. M. Zsebo, T. Ishizaka, A. M. Irani, and L. B. Schwartz, *Immunology*, *79*:325 (1993).
37. G. Nilsson, T. Blom, G. M. Kusche, L. Kjellen, J. H. Butterfield, C. Sundstrom, K. Nilsson, and L. Hellman, *Scand. J. Immunol.*, *39*:489 (1994).
38. G. Nilsson and K. Nilsson, *Eur. J. Immunol.*, *25*:870 (1995).
39. W. R. Sperr, H. Agis, K. Czerwenka, W. Klepetko, E. Kubista, N. G. Boltz, K. Lechner, and P. Valent, *Ann. Hematol.*, *65*:10 (1992).
40. P. Valent, O. Majdic, D. Maurer, M. Bodger, M. Muhm, and P. Bettelheim, *Int. Arch. Allergy Appl. Immunol.*, *91*:198 (1990).
41. C. B. Guo, S. A. Kagey, L. M. Lichtenstein, and B. S. Bochner, *Blood*, *79*:708 (1992).
42. P. Valent and P. Bettelheim, *Adv. Immunol.*, *52*:333 (1992).
43. P. Valent, D. Bevec, D. Maurer, J. Besemer, F. Di Padova, J. H. Butterfield, W. Speiser, O. Majdic, K. Lechner, and P. Bettelheim, *Proc. Natl. Acad. Sci. U.S.A*, *88*:3339 (1991).

44. R. de Waal Malefyt, C. G. Figdor, R. Huijbens, S. Mohan-Peterson, J. Culpepper, W. Dang, G. Zurawski, and J. de Vries, J. *Immunol.*, *151*:6370 (1993).
45. S. Weber, B. Ruh, K. S. Kruger, and B. M. Czarnetzki, *Arch. Dermatol. Res.*, *287*:695 (1995).
46. P. Siramarao, W. Anderson, B. A. Wolitzky, and D. H. Broide, *Lab. Invest.*, *74*:634 (1996).
47. S. N. Toyama, H. Sorimachi, Y. Tobita, F. Kitamura, H. Yagita, K. Suzuki, and M. Miyasaka, *J. Biol. Chem.*, *270*:7437 (1995).
48. J. G. Flanagan, D. C. Chan, and P. Leder, *Cell*, *64*:1025 (1991).
49. S. Varsano, S. C. Lazarus, W. M. Gold, and J. A. Nadel, J. *Immunol.*, *140*:2184 (1988).
50. J. F. Gauchat, S. Henchoz, G. Mazzei, J. P. Aubry, T. Brunner, H. Blasey, P. Life, D. Talabot, R. L. Flores, J. Thompson, et al., *Nature*, *365*:340 (1993).
51. C. C. Fox, S. D. Jewell, and C. C. Whitacre, *Cell. Immunol.*, *158*:253 (1994).
52. P. A. Seeldrayers, L. A. Levin, and D. Johnson, J. *Neuroimmunol.*, *36*:239 (1992).
53. F. Levi Shaffer and S. Z. Bar, *Immunology*, *69*:145 (1990).
54. M. M. Hamawy, S. E. Mergenhagen, and R. P. Siraganian, *J. Biol. Chem.*, *268*:5227 (1993).
55. M. M. Hamawy, S. E. Mergenhagen, and R. P. Siraganian, *J. Biol. Chem.*, *268*:6851 (1993).
56. M. M. Hamawy, W. D. Swaim, K. Minoguchi, A. de Feijter, S. E. Mergenhagen, and R. P. Siraganian, *J. Immunol.*, *153*:4655 (1994).
57. M. M. Hamawy, C. Oliver, S. E. Mergenhagen, and R. P. Siraganian, *J. Immunol.*, *149*:615 (1992).
58. F. Levi Schaffer, E. T. Dayton, K. F. Austen, A. Hein, J. P. Caulfield, P. M. Gravallese, F. T. Liu, and R. L. Stevens, *J. Immunol.*, *139*:3431 (1987).
59. C. Ra, M. Yasuda, H. Yagita, and K. Okumura, *J. Allergy Clin. Immunol.*, *94*: 625 (1994).
60. H. L. Thompson, P. D. Burbelo, Y. Yamada, H. K. Kleinman, and D. D. Metcalfe, *J. Immunol.*, *143*:4188 (1989).
61. H. L. Thompson, P. D. Burbelo, G. Gabriel, Y. Yamada, and D. D. Metcalfe, *J. Clin. Invest.*, *87*:619 (1991).
62. G. Nilsson, J. H. Butterfield, K. Nilsson, and A. Siegbahn, *J. Immunol.*, *153*:3717 (1994).
63. S. Saeland, V. Duvert, C. Caux, D. Pandrau, C. Favre, A. Valle, I. Durald, P. Charbord, and J. de Vries, *Exp. Hematol.*, *20*:24 (1992).
64. J. Teixido, M. E. Hemler, J. S. Greenberger, and P. Anklesaria, *J. Clin. Invest.*, *90*:358 (1992).
65. L. M. Klein, R. M. Lavker, W. L. Matis, and G. F. Murphy, *Proc. Natl. Acad. Sci. U.S.A*, *86*:8972 (1989).
66. L. J. Walsh, G. Trinchieri, H. A. Waldorf, D. Whitaker, and G. F. Murphy, *Proc. Natl. Acad. Sci. U.S.A*, *88*:4220 (1991).
67. J. P. Gaboury, B. Johnston, X. F. Niu, and P. Kubes, *J. Immunol.*, *154*:804 (1995).
68. H. Thorlacius, J. Raud, S. Rosengren-Beezley, M. J. Forrest, P. Hedqvist, and L. Lindbom, *Biochem. Biophys. Res. Commun.*, *203*:1043 (1994).

69. H. Asako, I., Kurose, R. Wolf, S. DeFrees, Z. L. Zheng, M. L. Phillips, J. C. Paulson, and D. N. Granger, *J. Clin. Invest.*, *93*:1508 (1994).

70. J. P. Gaboury, X. F. Niu, and P. Kubes, *Circulation*, *93*:318 (1996).

71. H. A. Waldorf, L. J. Walsh, N. M. Schechter, and G. F. Murphy, *Am. J. Pathol.*, *138*:477 (1991).

72. H. Meng, M. J. Marchese, J. A. Garlick, A. Jelaska, J. H. Korn, J. Gailit, R. A. Clark, and B. L. Gruber, *J. Invest. Dermatol.*, *105*:789 (1995).

73. R. Hershkoviz, O. Lider, D. Baram, T. Reshef, S. Miron, and Y. A. Mekori, *J. Leukoc. Biol.*, *56*:495 (1994).

74. A. M. Vignola, A. M. Campbell, P. Chanez, P. Lacoste, F. B. Michel, P. Godard, and J. Bousquet, *Am. J. Respir. Cell Mol. Biol.*, *9*:411 (1993).

# 9
# Eosinophil–Endothelial Interactions and Transendothelial Migration

**Motohiro Ebisawa**   *National Sagamihara Hospital, Kanagawa, Japan*

**Bruce S. Bochner and Robert P. Schleimer**   *Johns Hopkins Asthma and Allergy Center, Johns Hopkins University School of Medicine, Baltimore, Maryland*

## I.  INTRODUCTION

Eosinophils are one of the major effector cells in allergic reactions and allergic diseases, such as asthma and rhinitis (1). Many studies have been conducted in vivo or in vitro to solve the question of why eosinophils preferentially accumulate at sites of allergic inflammation. Within the past decade, a tremendous growth in our knowledge of leukocyte and endothelial adhesion molecules has occurred. Adhesion molecules are now known to participate in essentially every step in cell recruitment.

Extravasation of leukocytes to sites of inflammation is now thought to consist of at least three sequential steps (2). Circulating leukocytes undergo tethering and rolling on the endothelial surface, a process that involves formation of numerous weak, reversible bonds between selectins and their carbohydrate-containing counter-ligands. This initial binding may be followed by the induction of firm adhesion, which is mediated by leukocyte integrins interacting with endothelial adhesion molecules such as intercellular adhesion molecule-1 (ICAM-1) and vascular cell adhesion molecule-1 (VCAM-1) (3). During the third step, the leukocytes transmigrate between the vascular endothelial cells into the tissue (4).

There have appeared many excellent reviews in which various aspects of the function of adhesion molecules in immune responses and inflammation are discussed (5–7). This chapter covers the known functions of leukocyte and endothelial adhesion molecules in the process of eosinophil recruitment. The studies to be discussed in this chapter have mostly been performed in vitro with isolated human umbilical vein endothelial cells and purified human peripheral blood or lung eosinophils.

## II.   EOSINOPHIL–ENDOTHELIAL INTERACTIONS

## A.   Molecular Mechanisms of Eosinophil-Endothelial Attachment

Several cell surface glycoproteins on endothelial cells are able to mediate adhesion. These molecules are grouped on the basis of shared structural characteristics. On endothelial cells, those belonging to the immunoglobulin gene superfamily, which consists of more than a dozen structures, include ICAM-1, ICAM-2, VCAM-1, and platelet-endothelial cell adhesion molecule-1 (PECAM-1) (Table 1) (8). Another family of adhesion molecules, of which there are three known members, is the selectin gene superfamily (9). Two members of this family, E-selectin (originally endothelial-leukocyte adhesion molecule-1 [ELAM-1]) and P-selectin (formerly GMP-140) can be expressed on endothelial cells upon stimulation (Table 1). Endothelial cells constitutively express ICAM-1, ICAM-2, and PECAM-1 on their surface. In contrast to these molecules, VCAM-1 and E-selectin are not expressed on resting endothelial cells, and various cytokines are known to induce their expression in addition to ICAM-1 (Table 1) (10). P-selectin is not expressed on the resting endothelium, and is rapidly expressed after stimulation with histamine or thrombin (11). E-selectin and P-selectin have lectin domains to bind carbohydrate structures, such as those containing sialylated Lewis X antigen (sLe$^x$) (11).

On eosinophils, additional families of adhesion molecules have been recognized (Table 1). One of these families, the integrins, consists of more than a dozen heterodimers with distinct $\alpha$ and $\beta$ chains, and are responsible for adhesion to endothelial cells, complement fragments, extracellular matrix proteins, and other ligands (12). For example, the very late activation antigen-4 (VLA-4; CD49d/29, $\alpha4\beta1$) and $\alpha4\beta7$ can bind to both VCAM-1 and the CS-1 region of fibronectin (13). The $\alpha4\beta7$ integrin also can bind to another adhesion molecule, mucosal addressin cell adhesion molecule-1 (MAdCAM-1) that has structural homology to both VCAM-1 and ICAM-1 (14). The third member of the selectin family, L-selectin (formerly LECAM-1), is found on eosinophils. L-selectin binds to carbohydrates, although the exact ligand on postcapillary venules remains unknown. Current understanding of the interactions between endothelium and eosinophils is reviewed and discussed in the context of rolling adhesion, firm adhesion, and transendothelial migration (TEM).

## B.   Rolling Adhesion

## 1.   Role of L-selectin

It is not well understood how eosinophils roll on inflamed endothelium under conditions of flow. Knol et al. reported the role of L-selectin in the adhesion of eosinophils to cytokine-activated endothelium under rotational conditions, which

**Table 1** Characteristics of Eosinophil–Endothelial and Eosinophil–Matrix Adhesion Molecule Interactions

| Adhesion molecule | Inducing cytokines[a] | Eosinophil ligands[b] |
|---|---|---|
| ***Endothelial structures*** | | |
| E-selectin | IL-1, TNF | sLe$^x$ and s-di-Le$^x$–containing glycolipids (e.g., myeloglycans) |
| P-selectin | TNF | sLe$^x$-containing glycoproteins (e.g., PSGL-1) |
| Unknown inducible ligand | IL-1, TNF | L-selectin |
| ICAM-1 | IL-1, TNF, IFN-γ | CD11a/CD18 (LFA-1) CD11b/CD18 (Mac-1) |
| ICAM-2 | None | CD11a/CD18 |
| PECAM-1 | None | Unknown (? PECAM-1) |
| VCAM-1 | IL-1, TNF, IL-4, IL-13 | α4β1 (CD49d/CD29, VLA-4), α4β7 |
| MAdCAM-1 | Unknown | α4β7 |
| ***Extracellular matrix proteins*** | | |
| Fibronectin (containing CS-1) | Unknown | α4β1 (CD49d/CD29, VLA-4) |
| Laminin | Unknown | α6β1 (CD49f/CD29, VLA-6) |

[a]Stimuli other than cytokines can induce endothelial adhesion molecule expression, such as bacterial endotoxin for ICAM-1, VCAM-1, and E-selectin, and histamine and thrombin for P-selectin.
[b]Other eosinophil integrins with as yet unknown function include CD11c/CD18 (p150,95) and αdβ2.

mimic conditions of flow (15). Eosinophil adhesion to interleukin (IL)-1β–activated endothelium under nonstatic conditions was examined using a method similar to the one described by Spertini et al. (16), which is essentially a modification of the Stamper-Woodruff assay (17). Endothelial monolayers in 24-well plates were preincubated with or without 1 ng/mL of IL-1β for 4 hours at 37°C. The adhesion assay was performed at 4°C and at 60 rpm on a horizontal rotator platform. Binding of neutrophils and eosinophils under rotational condition was dependent on the activation state of the endothelial cells, in that activation of monolayers resulted in increased adherence of both cell types. Blocking monoclonal antibodies (mAb) recognizing CD18 or L-selectin were examined in the experiment: mAb against CD18 had no effect on adhesion of either cell type, whereas L-selectin mAb (LAM 1-3, LAM 1-6, LAM 1-11) inhibited eosinophil binding. Surprisingly, one of the L-selectin mAb, LAM 1-11, inhibited eosinophil adhesion but not neutrophil adhesion under nonstatic conditions. These results indicated that eosinophils utilize L-selectin for their rolling on inflamed endothelium and raised the possibility that eosinophils use an epitope on L-selectin that is not used by neutrophils. Eosinophils that have shed their L-selectin, either as a result of activation in vivo (18,19) or in vitro (20), adhered poorly in this assay (15).

L-selectin has been shown to mediate eosinophil rolling on endothelium in cytokine-inflamed rabbit mesentery by Broide and coworkers using intravital video microscopy (21). In this study, and in contrast to neutrophil rolling, eosinophils have been shown to utilize VLA-4 for rolling adhesion in addition to L-selectin. This latter observation has been confirmed using lymphocytes and monocytes, which also express VLA-4, and illustrates a heretofore unrecognized role of integrins in leukocyte rolling (22,23). Recent studies suggest that L-selectin can bind to a heavily glycosylated region of murine MAdCAM-1 at a site distinct from the immunoglobulin domains (24). Whether this also occurs with human MAdCAM-1 and L-selectin on eosinophils or other leukocytes is unknown. Finally, recent studies have failed to find any inhibitory effects of L-selectin mAb infusion in a guinea pig model of allergen-induced pulmonary inflammation, even though a mAb to VLA-4 was extremely effective in inhibiting eosinophil accumulation in the airways (25).

## 2. The Role of Sialic Acid–Containing Counter-Ligands for Selectins

Recent data suggest that endothelial P-selectin and E-selectin mediate rolling of neutrophils under nonstatic conditions in vitro and in vivo (11). Unlike L-selectin, both P-selectin and E-selectin will support neutrophil adhesion under static conditions as well (11). Eosinophils have also been shown to bind as well as neutrophils to P-selectin under static conditions; adhesion was inhibited by mAb to the lectin domain of P-selectin (26,27). Based on the effects of treatment with neuraminidase, other glycosidases, or proteases, adhesion to endothelial selectins appears to be mediated by sialylated glycoprotein ligands on the surface of the eosinophils and neutrophils (27,28). Recently, a specific ligand for P-selectin has been identified and termed PSGL-1 (P-selectin glycoprotein ligand-1) (29); whether this is expressed on eosinophils and functions in eosinophil–P-selectin binding has not been reported. The ability of several glycosidases to reduce neutrophil binding was greater than that seen for eosinophils, suggesting some differences in the level of expression or biochemical composition of their P-selectin ligands (27). Using the modified Stamper-Woodruff assay mentioned above, investigators have found that eosinophils adhere to nasal polyp endothelium in tissue sections in a P-selectin–dependent manner (30). Although ICAM-1, E-selectin, and P-selectin were well expressed by nasal polyp endothelium, eosinophil adhesion was almost completely inhibited by mAb against P-selectin. For E-selectin, additional enzyme studies suggest that a glycolipid-anchored dimeric form of sLe$^x$ (s-di-Le$^x$) may be most relevant for adhesion of eosinophils and neutrophils (31,32). Because neutrophils express nearly 10 times as much s-di-Le$^x$ as neutrophils, this may explain why E-selectin is a better ligand for neutrophils than for eosinophils. These data indicate the potential importance of selectins in tissue eosinophilia.

## C.  Firm Adhesion

Exposure of endothelial cells to IL-1, tumor necrosis factor–$\alpha$ (TNF-$\alpha$), phorbol esters, or bacterial lipopolysaccharide (LPS) causes them to express adhesive properties for neutrophils (33–35). Subsequent studies revealed that the same basic results occur with eosinophils as with neutrophils (36). Studies with protein and mRNA synthesis inhibitors revealed that de novo induction of the expression of adhesion molecules on endothelial cells increases the adherence of granulocytes. Activation of endothelium by these stimuli induces expression of endothelial adhesion molecules, including ICAM-1, E-selectin, and VCAM-1. Specific F(ab')$_2$ preparation of antibodies to ICAM-1 and E-selectin inhibited adherence of eosinophils, neutrophils, and basophils to IL-1 stimulated endothelial monolayers by approximately 20% to 30% (37). In contrast, anti–VCAM-1 antibody was extremely effective in inhibiting eosinophil adherence but had no effect on neutrophil adherence, suggesting that eosinophils, but not neutrophils, recognize VCAM-1. It had been established that a counter-ligand for VCAM-1 on lymphocytes was VLA-4 (13); therefore, immunofluorescence and flow cytometry were used to examine VLA-4 expression on eosinophils. As expected, eosinophils, but not neutrophils, were found to express VLA-4 (37). Furthermore, antibodies to VLA-4 inhibited eosinophil adhesion to IL-1–stimulated endothelium (38,39). To independently confirm that eosinophils can adhere to VCAM-1, the ability of these cells to adhere to a truncated recombinant form of VCAM-1, which had been immobilized to plastic plates, was tested. Eosinophils adhered more avidly to this surface compared to control surfaces, and the adhesion was inhibited using specific antibodies against VCAM-1 and VLA-4 (40,41).

These results raised the possibility that specific induction of VCAM-1 on endothelial cells would selectively induce eosinophil (but not neutrophil) adherence. It had been previously suggested that IL-4 might selectively lead to the induction of VCAM-1 expression without any significant effect on the expression of E-selectin and ICAM-1 (42). Incubation of endothelial cells with IL-4 had no effect on neutrophil adhesion but did induce eosinophil adhesion in a dose-dependent manner (41). Eosinophil adhesion to IL-4–activated endothelial cells was inhibited by more than 70% with either VCAM-1 or VLA-4 antibodies. Indeed, eosinophil accumulation is often seen at sites of IL-4 deposition in vivo (43–45). Recently, a cytokine derived from Th2-lymphocytes, termed IL-13, has been identified with homology to IL-4 at the amino acid sequence level and in its spectrum of biologic activities (46). In addition to IL-4, IL-13 has been recently shown to selectively induce VCAM-1 on endothelial cells, resulting in selective induction of eosinophil but not neutrophil adhesion to endothelium (47). Finally, it is clear that synergy can occur among cytokines for induction of VCAM-1 expression, in that the combination of IL-4 with either TNF or IL-1 will result in extremely high levels of surface expression with little or no induction of ICAM-1 or E-selectin (48–50) (see Fig. 1).

## III.  TRANSENDOTHELIAL MIGRATION

Transendothelial migration (TEM) is a process by which cells entering a local inflammatory site traverse the vascular endothelium by migrating between adjacent cells (51). Several in vitro studies including ours have begun to analyze the molecular mechanisms regulating eosinophil TEM.

### A.  Role of Adhesion Molecules

We have employed an in vitro endothelial monolayer system that allows objective evaluation of the effect of cytokines and the role of adhesion molecules on both eosinophils and endothelial cells in the process of eosinophil TEM (52). To evaluate the role of endothelial adhesion molecules in IL-1β–induced eosinophil TEM, IL-1β–activated human endothelial monolayers were pretreated with

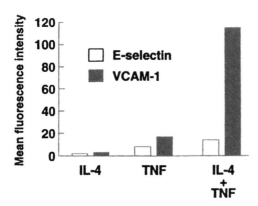

**Figure 1**  Cytokines can be synergistic and selective in their ability to induce vascular cell adhesion molecule-1 (VCAM-1) expression on vascular endothelial cells. Monolayers were cultured for 24 hr in the presence of 0.1 ng/mL of the indicated cytokines, either alone or in combination, and indirect immunofluorescence and flow cytometry was performed to determine expression of E-selectin and VCAM-1. These concentrations of cytokines, by themselves, cause little if any adhesion molecule expression, but their combination is markedly synergistic for VCAM-1 expression. Data are from a single experiment representative of three.

F(ab')$_2$ preparations of mAb against ICAM-1, E-selectin, and VCAM-1, alone and in combination, at 37°C for 45 min prior to the TEM assay. Only anti–ICAM-1 inhibited IL-1β–induced eosinophil TEM as a single mAb treatment. Although neither anti–E-selectin nor anti–VCAM-1 alone inhibited IL-1β–induced eosinophil TEM, the combination of these antibodies with anti–ICAM-1 inhibited the migration significantly better than anti–ICAM-1 alone. Among the combinations of mAb used for pretreatment of human umbilical vein endothelial cell (HUVEC), in all cases where ICAM-1 was blocked, significant inhibition was observed.

In the process of eosinophil TEM, the VCAM-1/VLA-4 interaction seemed to have less role compared with ICAM-1/CD 11/18. However, when the CD18 molecule is dysfunctional or absent, as is the case in leukocyte adhesion deficiency disease type I (53), eosinophils and other VLA-4 positive cells can probably utilize the VCAM-1/VLA-4 pathway in the process of TEM because eosinophils are seen at extravascular tissue sites even though neutrophils are not. The influence of the state of endothelial activation on chemotactic factor regulated on activation, normal T cell expressed and secreted (RANTES)-induced TEM was assessed by using control and IL-1β–activated endothelial monolayers (54). In these studies, RANTES-induced TEM across control endothelial monolayers was inhibited virtually completely by a mAb directed against CD18. When endothelial cells were activated first with IL-1, substantially greater net TEM was observed. The CD18-specific Ab only partially inhibited (approximately 60%–70% inhibition) eosinophil TEM in this case, however. Although a mAb against VLA-4 did not influence eosinophil TEM by itself, in the presence of the CD18 Ab this second mAb completely blocked the residual TEM. This result suggests that the interaction of VLA-4 with VCAM-1 or another counter-ligand on activated endothelial cells may play a role in chemotactic factor–induced TEM.

Other stimuli have been examined for their ability to affect eosinophil TEM. Moser et al. reported a role of both VLA-4 and β2 integrins in IL-4-induced eosinophil TEM (55). In another study, platelet-activating factor (PAF) was used as a chemotactic factor to promote eosinophil TEM across unstimulated endothelium (56). An additional study used a mAb (8A2) that activates, rather than inhibits, β1 integrin function to study its influence on eosinophil migration (57). Treatment with this mAb dramatically inhibited eosinophil chemotaxis and TEM, presumably by enhancing adhesion and thereby preventing migration by tightly immobilizing cells to the endothelial surface.

Experiments were performed to determine whether either β1 or β2 integrins are important for IL-1β–induced eosinophil TEM (52). Eosinophils were pretreated with either anti-CD18 (β2 integrin) or anti-CD29 (β1 integrin) prior to the transendothelial migration assay. Anti-CD18 inhibited IL-1β-induced eosinophil transendothelial migration by more than 80%, whereas anti-CD29 failed to inhibit the migration. The inhibitory effect of anti-CD18 was almost equal to that of the combination of anti-CD11a and anti-CD11b. Anti-CD11b alone inhibited the eosinophil migration by 50%, whereas anti-CD11a alone inhibited by 30%. Most recently, an important role for PECAM-1 (CD31) has been documented during the process of neutrophil and monocyte TEM, apparently at the diapedesis step (58–61). The role of this adhesive pathway in eosinophil TEM has not been examined.

Changes in the expression of cell surface adhesion molecules occur on eosinophils during their movement from the circulation into tissues. A comparison of levels of adhesion molecules on eosinophils recovered from blood and either sputum or bronchoalveolar lavage (BAL) after antigen challenge revealed increased expression of CD11b, CD11c, CD35 and diminished levels of L-selectin, without any change in VLA-4 or CD32 (18,19,62,63). Similar phenotypic changes have been observed after eosinophil TEM in vitro (52,64). However, these events seem to be a common consequence of cell recruitment in general as they are observed on other leukocyte types such as neutrophils (18,65,66).

In summary, stimulation of endothelial cells with IL-1β, TNF-α, and perhaps IL-4 induced significant eosinophil TEM, suggesting that endothelial cells may actively participate in eosinophil recruitment. The initial attachment of eosinophils to IL-1β-activated endothelial cells involves VCAM-1, E-selectin, and ICAM-1 (37); the subsequent TEM process relies heavily on ICAM-1/CD11/18 (52).

## B.  Role of Cytokines and Chemokines

Eosinophil-active cytokines have been previously shown to potentiate eosinophil adhesion to various surfaces in a β2 integrin–dependent manner (67), and have been reported to be both chemotactic/chemokinetic and priming factors for eosinophils (68,69). To investigate the role of eosinophil-active cytokines in the process of eosinophil recruitment, we examined the effects of these cytokines on eosinophil TEM. When freshly isolated eosinophils were cultured in the presence of IL-5 for 48 hours, eosinophil TEM through both nonactivated and IL-1β–activated endothelial monolayers was significantly enhanced (70). Furthermore, co-incubation (2 hr) of eosinophils with IL-5, granulocyte-mocrophage colony-stimulating factor (GM-CSF), and IL-3 enhanced eosinophil TEM through nonactivated endothelial monolayers. The enhancement of eosinophil TEM by IL-5, GM-CSF, or IL-3 was completely CD18 dependent. The observation that primed eosinophils are more efficient in undergoing TEM is consistent with another report (71). Both IL-5 and GM-CSF have been found to be elevated in the serum of patients with angioedema and eosinophilia and BAL fluids from allergic subjects after antigen challenge (72,73). GM-CSF has also been found to be elevated in antigen-challenged skin blister fluids (74). Eosinophils taken from the lungs of allergic subjects challenged 18 to 20 hr previously with antigen showed enhanced TEM, indicating that they may have been exposed to these same cytokines in vivo (70).

Recently, a family of chemoattractant peptides, termed chemokines, has been recognized to play an important role in the migration and transendothelial passage of leukocytes (75). Initially, IL-8 was reported to have some chemotactic effect on primed eosinophils;however, this effect was quite modest (76,77). To determine whether other members of the chemokine family may have preferential eosinophil-activating properties, a panel of chemokines was tested for their ability to promote eosinophil and neutrophil TEM. Among the nine chemokines tested, RANTES was by far the most effective in inducing eosinophil TEM but not neutrophil TEM (54). Independent (and earlier) studies demonstrated the chemotactic actions of RANTES (78–80). Recently, several other chemokines have been shown to be eosinophil-selective chemoattractants, including MCP-3, MCP-4, and eotaxin (81–83). In the TEM assay, the effects of RANTES on eosinophils were synergistic with IL-5 (54). Concerning the effect on adhesion molecules, RANTES is somewhat different from other known chemotactic factors, such as PAF and fMLP, in that it failed to induce up-regulation of CD11b and shedding of L-selectin on eosinophils (54). In addition, RANTES required a concentration gradient—indicating that RANTES has chemotactic activity but not chemokinetic activity.

## IV. SUMMARY AND CONCLUSION

Largely on the basis of in vitro studies, several distinct but interrelated hypothetical mechanisms have been developed that provide a basis for selective eosinophil recruitment in allergic reactions. Such selectivity may be explained in part by: 1.) endothelial expression of VCAM-1 induced by IL-1, TNF-$\alpha$, IL-4, IL-13, or their combinations; 2.) priming of circulating eosinophils by eosinophil-activating cytokines such as IL-3, IL-5, and GM-CSF, which promotes the up-regulation of eosinophil adhesion and TEM; 3.) generation of chemotactic factors such as the chemokines RANTES, MIP-1$\alpha$, MCP-3, MCP-4, or eotaxin, which promote migration of eosinophils but not neutrophils; and 4.) prolongation of eosinophil survival by GM-CSF, IL-3 or IL-5.

In conclusion, the combination of these classes of cytokines and adhesion molecules on eosinophils and endothelial cells may play a crucial role in allergic reactions.

## REFERENCES

1. G. J. Gleich, *J. Allergy Clin. Immunol.*, *85*:422 (1990).
2. E. C. Butcher, *Cell*, *67*:1033 (1991).

3. R. R. Lobb, *Adhesion: its role in inflammatory disease* (J. M. Harlan and D. Y. Liu, eds.), W. H. Freeman and Company, New York, p. 1 (1992).

4. F. W. Luscinskas, M. I. Cybulsky, J.-M. Kiely, C. S. Peckins, V. M. Davis, and M. A. Gimbrone Jr., *J. Immunol.*, *146*:1617 (1991).

5. T. M. Carlos and J. M. Harlan, *Blood*, *84*:2068 (1994).

6. T. A. Springer, *Cell*, *76*:301 (1994).

7. T. A. Springer, *Annu. Rev. Physiol.*, *57*:827 (1995).

8. T. A. Springer, *Nature*, *346*:425 (1990).

9. M. Bevilacqua, E. Butcher, B. Furie, M. Gallatin, M. Gimbrone, J. Harlan, K. Kishimoto, L. Lasky, R. McEver, J. Paulson, S. Rosen, B. Seed, M. Siegelman, T. Springer, L. Stoolman, T. Tedder, A. Varki, D. Wagner, I. Weissman, and G. Zimmerman, *Cell*, *67*:233 (1991).

10. M. P. Bevilacqua, *Annu. Rev. Immunol.*, *11*:767 (1993).

11. M. P. Bevilacqua and R. M. Nelson, *J. Clin. Invest.*, *91*:379 (1993).

12. S. M. Albelda and C. A. Buck, *FASEB J.*, *4*:2868 (1990).

13. M. J. Elices, L. Osborn, Y. Takada, C. Crouse, S. Luhowskyj, M. E. Hemler, and R. R. Lobb, *Cell*, *60*:577 (1990).

14. A. M. Shyjan, M. Bertognolli, C. J. Kenney, and M. J. Briskin, *J. Immunol.* (in press), (1996).

15. E. F. Knol, F. Tackey, T. F. Tedder, D. A. Klunk, C. A. Bickel, and B. S. Bochner, *J. Immunol.*, *153*:2161 (1994).

16. O. Spertini, F. W. Luscinskas, G. S. Kansas, J. M. Munro, J. D. Griffin, M. A. Gimbrone, and T. F. Tedder, *J. Immunol.*, *147*:2565 (1991).

17. H. B. Stamper, Jr., and J. J. Woodruff, *J. Exp. Med.*, *144*:828 (1976).

18. S. N. Georas, M. C. Liu, W. Newman, W. D. Beall, B. A. Stealey, and B. S. Bochner, *Am. J. Respir. Cell Mol. Biol.*, *7*:261 (1992).

19. H. J. J. Mengelers, T. Maikoe, B. Hooibrink, T. W. Kuypers, J. Kreukniet, J.-W. Lammers J., and L. Koenderman, *Clin. Exp. Allergy*, *23*:196 (1993).

20. S. P. Neeley, K. J. Hamann, S. R. White, S. L. Baranowski, R. A. Burch, and A. R. Leff, *Am. J. Respir. Cell Mol. Biol.*, *8*:633 (1993).

21. P. Sriramarao, U. H. von Andrian, E. C. Butcher, M. A. Bourdon, and D. H. Broide, *J. Immunol.*, *153*:4238 (1994).

22. F. W. Luscinskas, G. S. Kansas, H. Ding, P. Pizcueta, B. E. Schleiffenbaum, T. F. Tedder, and M. A. Gimbrone, *J. Cell Biol.*, *125*:1417 (1994).

23. R. Alon, P. D. Kassner, M. W. Carr, E. B. Finger, M. E. Hemler, and T. A. Springer, *J. Cell. Biol.*, *128*:1243 (1995).

24. E. L. Berg, L. M. McEvoy, C. Berlin, R. F. Bargatze, and E. C. Butcher, *Nature*, *366*:695 (1993).

25. A. D. Fryer, R. W. Costello, R. R. Lobb, T. A. Tedder, D. A. Steeber, and B. S. Bochner, (submitted), (1996).

26. M. A. Vadas, C. M. Lucas, J. R. Gamble, A. F. Lopez, M. P. Skinner, and M. C. Berndt, *Eosinophils in Allergy and Inflammation* (G. J. Gleich and A. B. Kay, eds.), Marcel Dekker, Inc., New York, p. 69 (1993).

27. M. Wein, S. A. Sterbinsky, C. A. Bickel, R. P. Schleimer, and B. S. Bochner, *Am. J. Respir. Cell Mol. Biol.*, *12*:315 (1995).

28. G. R. Larsen, D. Sako, T. J. Ahern, M. Shaffer, J. Erban, S. A. Sajer, R. M. Gibson, D. D. Wagner, B. C. Furie, and B. Furie, *J. Biol. Chem.*, *267*:11104 (1992).
29. D. Sako, X.-J. Chang, K. M. Barone, G. Vachino, H. M. White, G. Shaw, G. M. Veldman, K. M. Bean, T. J. Ahern, B. Furie, D. A. Cumming, and G. R. Larsen, *Cell*, *75*:1179 (1993).
30. F. A. Symon, G. M. Walsh, S. R. Watson, and A. J. Wardlaw, *J. Exp. Med.*, *180*:371 (1994).
31. B. S. Bochner, S. A. Sterbinsky, C. A. Bickel, S. Werfel, M. Wein, and W. Newman, *J. Immunol.*, *152*:774 (1994).
32. B. E. Collins, B. S. Bochner, and R. L. Schnaar, *Glycoconjugate J.*, *12*:535 (1995).
33. C. J. Dunn and W. E. Fleming, *Eur. J. Rheumatol. Inflamm.*, *7*:80 (1984).
34. M. P. Bevilacqua, J. S. Pober, M. E. Wheeler, R. S. Cotran, and M. A. Gimbrone Jr., *J. Clin. Invest.*, *76*:2003 (1985).
35. R. P. Schleimer and B. K. Rutledge, *J. Immunol.*, *136*:649 (1986).
36. A. M. Lamas, C. R. Mulroney, and R. P. Schleimer, *J. Immunol.*, *140*:1500 (1988).
37. B. S. Bochner, F. W. Luscinskas, M. A. Gimbrone Jr., W. Newman, S. A. Sterbinsky, C. Derse-Anthony, D. Klunk, and R. P. Schleimer, *J. Exp. Med.*, *173*:1553 (1991).
38. G. M. Walsh, J. Mermod, A. Hartnell, A. B. Kay, and A. J. Wardlaw, *J. Immunol.*, *146*:3419 (1991).
39. A. Dobrina, R. Menegazzi, T. M. Carlos, E. Nardon, R. Cramer, T. Zacchi, J. M. Harlan, and P. Patriarca, *J. Clin. Invest.*, *88*:20 (1991).
40. P. F. Weller, T. H. Rand, S. E. Goelz, G. Chi-Rosso, and R. R. Lobb, *Proc. Natl. Acad. Sci. U.S.A*, *88*:7430 (1991).
41. R. P. Schleimer, S. A. Sterbinsky, J. Kaiser, C. A. Bickel, D. A. Klunk, K. Tomioka, W. Newman, F. W. Luscinskas, M. A. Gimbrone Jr., B. W. McIntyre, and B. S. Bochner, *J. Immunol.*, *148*:1086 (1992).
42. M. H. Thornhill, U. Kyan-Aung, and D. O. Haskard, *J. Immunol.*, *144*:3060 (1990).
43. R. Moser, P. Groscurth, J. M. Carballido, P. L. B. Bruijnzeel, K. Blaser, C. H. Heusser, and J. Fehr, *J. Lab. Clin. Med.*, *122*:567 (1993).
44. R. I. Tepper, P. K. Pattengale, and P. Leder, *Cell, 57*:503 (1989).
45. R. I. Tepper, R. L. Coffman, and P. Leder, *Science, 257*:548 (1992).
46. A. N. J. McKenzie, J. A. Culpepper, R. D. Malefyt, F. Briere, J. Punnonen, G. Aversa, A. Sato, W. Dang, B. G. Cocks, S. Menon, J. E. de Vries, J. Banchereau, and G. Zurawski, *Proc. Natl. Acad. Sci. U.S.A, 90*:3735 (1993).
47. B. S. Bochner, D. A. Klunk, S. A. Sterbinsky, R. L. Coffman, and R. P. Schleimer, *J. Immunol.*, *154*:799 (1995).
48. B. Masinovsky, D. Urdal, and W. M. Gallatin, *J. Immunol.*, *145*:2886 (1990).
49. M. H. Thornhill and D. O. Haskard, *J. Immunol.*, *145*:865 (1990).
50. M. F. Iademarco, J. L. Barks, and D. C. Dean, *J. Clin. Invest.*, *95*:264 (1995).
51. C. W. Smith, *Adhesion: its role in inflammatory disease* (J. M. Harlan and D. Y. Liu, eds.), W. H. Freeman and Company, New York, p. 83 (1992).
52. M. Ebisawa, B. S. Bochner, S. N. Georas, and R. P. Schleimer, *J. Immunol.*, *149*:4021 (1992).
53. D. C. Anderson, F. C. Schmalstieg, M. J. Finegold, B. J. Hughes, R. Rothlein, L. J.

Miller, S. Kohl, M. F. Tosi, R. L. Jacobs, T. C. Waldrop, A. S. Goldman, W. T. Shearer, and T. A. Springer, *J. Infect. Dis.*, *152*:668 (1985).

54. M. Ebisawa, T. Yamada, C. Bickel, D. Klunk, and R. P. Schleimer, *J. Immunol.*, *153*:2153 (1994).

55. R. Moser, J. Fehr, and P. L. B. Bruijnzeel, *J. Immunol.*, *149*:1432 (1992).

56. T. B. Casale, R. A. Erger, and M. M. Little, *Am. J. Respir. Cell Mol. Biol.*, *8*:77 (1993).

57. T. W. Kuijpers, E. P. J. Mul, M. Blom, N. L. Kovach, F. C. A. Gaeta, V. Tollefson, M. J. Elices, and J. M. Harlan, *J. Exp. Med.*, *178*:279 (1993).

58. A. A. Vaporciyan, H. M. DeLisser, H.-C. Yan, I. I. Mendiguren, S. R. Thom, M. L. Jones, P. A. Ward, and S. M. Albelda, *Science.*, *262*:1580 (1993).

59. W. A. Muller, S. A. Weigl, X. Deng, and D. M. Phillips, *J. Exp. Med.*, *178*:449 (1993).

60. S. Bogen, J. Pak, M. Garifallou, X. H. Deng, and W. A. Muller, *J. Exp. Med.*, *179*:1059 (1994).

61. W. A. Muller, *J. Leukoc. Biol.*, *57*:523 (1995).

62. T. T. Hansel and C. Walker, *Clin. Exp. Allergy*, *22*:345 (1992).

63. C. Kroegel, M. C. Liu, W. M. Hubbard, L. M. Lichtenstein, and B. S. Bochner, *J. Allergy Clin. Immunol.*, *93*:725 (1994).

64. C. Walker, S. Rihs, R. K. Braun, S. Betz, and P. L. B. Bruijnzeel, *J. Immunol.*, *150*:4061 (1993).

65. T. K. Kishimoto, M. A. Jutila, E. L. Berg, and E. C. Butcher, *Science*, *245*:1238 (1989).

66. M. A. Jutila, L. Rott, E. L. Berg, and E. C. Butcher, *J. Immunol.*, *143*:3318 (1989).

67. G. M. Walsh, A. Hartnell, A. J. Wardlaw, K. Kurihara, C. J. Sanderson, and A. B. Kay, *Immunology*, *71*:258 (1990).

68. J. M. Wang, A. Rambaldi, A. Biondi, Z. G. Chen, C. J. Sanderson, and A. Mantovani, *Eur. J. Immunol.*, *19*:701 (1989).

69. R. A. J. Warringa, L. Koenderman, P. T. M. Kok, J. Kreukniet, and P. L. B. Bruijnzeel, *Blood*, *77*:2694 (1991).

70. M. Ebisawa, M. C. Liu, T. Yamada, M. Kato, L. M. Lichtenstein, B. S. Bochner, and R. P. Schleimer, *J. Immunol.*, *152*:4590 (1994).

71. R. Moser, J. Fehr, L. Olgiati, and P. L. B. Bruijnzeel, *Blood*, *79*:2937 (1992).

72. J. H. Butterfield, K. M. Leiferman, J. Abrams, J. E. Silver, J. Bower, N. Gonchoroff, and G. J. Gleich, *Blood*, *79*:688 (1992).

73. B. S. Bochner, B. Friedman, G. Krishnaswami, R. P. Schleimer, L. M. Lichtenstein, and C. Kroegel, *J. Allergy Clin. Immunol.*, *88*:629 (1991).

74. W. Massey, B. Friedman, M. Kato, P. Cooper, A. Kagey-Sobotka, L. M. Lichtentein, and R. P. Schleimer, *J. Immunol.*, *150*:1084 (1993).

75. J. J. Oppenheim, C. O. C. Zachariae, N. Mukaida, and K. Matsushima, *Annu. Rev. Immunol.*, *9*:617 (1991).

76. R. A. J. Warringa, H. J. J. Mengelers, J. A. M. Raaijmakers, P. L. B. Bruijnzeel, and L. Koenderman, *J. Allergy Clin. Immunol.*, *91*:1198 (1993).

77. R. Sehmi, A. J. Wardlaw, O. Cromwell, K. Kurihara, P. Waltmann, and A. B. Kay, *Blood*, *79*:2952 (1992).

78. A. Rot, M. Krieger, T. Brunner, S. C. Bischoff, T. J. Schall, and C. A. Dahinden, *J. Exp. Med.*, *176*:1489 (1992).
79. Y. Kameyoshi, A. Dorschner, A. I. Mallet, E. Christophers, and J. M. Schroder, *J. Exp. Med.*, *176*:587 (1992).
80. T. Yamada, M. Ebisawa, J. D. W. MacGlashan, C. Bickel, J. J. Oppenheim, B. S. Bochner, and R. P. Schleimer, *J. Allergy Clin. Immunol.*, *91*:692 (1993).
81. C. A. Dahinden, T. Geiser, T. Brunner, V. von Tscharner, D. Caput, P. Ferrara, A. Minty, and M. Baggiolini, *J. Exp. Med.*, *179*:751 (1994).
82. C. Stellato, L. Beck, L. Schweibert, H. Li, J. White, and R. P. Schleimer, *J. Allergy Clin. Immunol.*, *97*:304 (1996).
83. P. J. Jose, D. A. Griffiths-Johnson, P. D. Collins, D. T. Walsh, R. Moqbel, N. F. Totty, O. Truong, J. J. Hsuan, and T. J. Williams, *J. Exp. Med.*, *179*:881 (1994).

# 10

# Eosinophil Interactions with Extracellular Matrix Proteins

*Effects on Eosinophil Function and Cytokine Production*

**Garry M. Walsh and Andrew J. Wardlaw** *University of Leicester, School of Medicine, Glenfield Hospital, Leicester, England*

## I. INTRODUCTION

Tissue infiltration by often large numbers of eosinophils is one of the consistent features of asthma and related diseases such as allergic rhinitis or atopic dermatitis (1,2). The ability of eosinophils to generate an array of pro-inflammatory mediators, in particular the cytotoxic granule proteins such as eosinophil major basic protein (MBP), has led to the hypothesis that asthma is an ongoing mucosal inflammatory disease, a major component of which is the tissue damage mediated by eosinophil-derived mediators (3,4). The mechanisms responsible for selective eosinophil accumulation in the airways in asthma are complex but probably involve a combination of selective adhesion to bronchial postcapillary venular endothelium, followed by transmigration into the bronchial submucosa under the influence of chemotactic factors such as platelet-activating factor (PAF), leukotriene $B_4$, and C-C chemokines such as regulated on activation, normal T cell expressed and secreted (RANTES) (5,6).

It is now widely accepted that adhesion to the blood vessel by pro-inflammatory cells represents an essential stage in their emigration into tissue, a process controlled by integrin receptors on leukocytes binding to counter-structures on the endothelial cell (7,8). Once eosinophils have left the circulation they come into contact with the proteins of the extracellular matrix (ECM). Although there have been several studies that have identified adhesion molecules that mediate eosinophil interaction with endothelial cells, the mechanisms and functional consequences of adhesion between eosinophils and ECM proteins have received relatively little attention.

## II.  THE EXTRACELLULAR MATRIX

This chapter has focused on eosinophil interaction with the proteins of the ECM because a key feature of many of these molecules is their ability to interact with cell-surface receptors on eosinophils and other pro-inflammatory cells with profound functional consequences. The ECM proteins represent a highly complex and diverse group, and for a more detailed review of these proteins than is possible here, the interested reader is directed to other references (9,10).

The ECM is a relatively stable structure that underlies the epithelia and surrounds connective tissue cells; it is made up of at least four major classes of macromolecules, the collagens, proteoglycans, elastin, and the glycoproteins. The nature and function of the connective tissue determines the relative proportions of each of these constituent molecules. Each of these major classes includes members that, in turn, often represent large families of molecules. Thus, the ECM represents a complex web of large fibrillar proteins that do more than merely provide a support for resident or infiltrating pro-inflammatory cells. Cellular interactions with ECM proteins have been shown, in a variety of systems, to have a profound effect on both leukocyte and general cellular functions in widely varying aspects of biology including embryogenesis, platelet aggregation, lymphocyte homing, wound healing, and metastatic activity by malignant cells (11). It is increasingly clear that interactions between integrins and matrix proteins have a very profound effect on the function of diverse cell types. Adhesion of cells to ECM proteins has been shown to result in tyrosine phosphorylation; cytoplasmic alkalinization in platelets, fibroblasts, lymphocytes, and endothelial cells; activation of lymphocytes and stimulation of leukocyte secretion. For example, adhesion of monocytes to fibronectin resulted in generation of granulocyte-macrophage colony-stimulating factor (GM-CSF) (12), and adhesion to collagen resulted in secretion of tumor necrosis factor-$\alpha$ (TNF-$\alpha$) (13), suggesting the possibility of linkage between different integrin receptors and transcription of specific genes. Furthermore, monocytes adherent to collagen demonstrated enhanced phagocytosis of opsonized bacteria by upregulating the function of complement and Fc receptors (14). Integrins can therefore act as true signaling receptors for a number of cell types, leading to a variety of cellular responses depending on the matix protein ligand, the cell type, and other co-stimulatory signals to which the cell is subjected. However, interaction of many integrins with ECM proteins are not specific, with some integrin receptors binding to several different matrix proteins. For example, gpIIb/IIIa recognizes fibrinogen, fibronectin, von Willebrand factor, and vitronectin; the vitronectin receptor can bind vitronectin, collagen, fibronectin, fibrinogen, and thrombospondin. A more detailed overview of the receptors involved in leukocyte–ECM interactions can be found elsewhere in this book (see Chap. 4).

## A.  Fibronectins

Fibronectins (reviewed in 15) are abundant in the ECM and represent a complex and heterogeneous group of molecules that play an important role in processes as diverse as embryogenesis, thymocyte maturation, and T-cell function. Fibronectin has been shown to be involved in the morphogenesis of lung epithelium (16); its expression is increased in adults during lung injury and repair (17,18); human bronchial epithelial cells are capable of fibronectin production (19); and it may also play an important role in the pathogenesis of lung fibrosis (20,21). Fibronectin is encoded by a single gene, but alternative splicing of the primary RNA transcript results in polypeptide diversity that appears to be regulated in a cell type–specific fashion (22,23). Three separate exons, namely EIIIA, EIIIB, or IIICS, are subject to alternative splicing. The IIICS region of fibronectin contains a number of sites recognized by the integrin $\alpha 4 \beta 1$ (very late activation antigen-4 [VLA-4]), namely CS-1 and CS-5 (24). The 25-amino-acid sequence CS-1 is the most active binding site in the IIICS region (25), and it appears to be the minimum sequence on the fibronectin molecule required for recognition by $\alpha 4 \beta 1$ (26,27). Plasma fibronectin lacks the IIICS cell-binding region in at least half its subunits, whereas tissue fibronectin has it in both subunits (28).

## B.  Laminins

Laminins are a family of large complex glycoproteins that are found primarily in basement membranes. These molecules have many functions, including axonal outgrowth and maintenance of the phenotype and polarization of epithelial cells. Their structure consists of three genetically distinct chains, A, B1, and B2, coded for by different but closely related genes. The chains assemble to form a molecule with a cruciform shape and a molecular weight of 600–800 kD. At least two of these three subunits have alternative forms, and these can assemble in all combinations, Thus, at least four isomeric forms of laminin have been identified that exhibit tissue-specific patterns of expression (29). The mechanisms by which laminin interacts with cells is complex: it is recognized by different integrin receptors including, $\alpha 1 \beta 1$, $\alpha 2 \beta 1$, $\alpha 3 \beta 1$, $\alpha 6 \beta 1$, and $\alpha v \beta 3$. With the exception of $\alpha 6 \beta 1$, which appears to be specific for laminin (30), all these other integrins will bind other ECM proteins. This promiscuity is due in part to a common recognition motif Arg-Gly-Asp (RGD) found in a number of matrix proteins, including laminin (31). However, cellular interaction with laminin is complicated by the fact that it has at least six short peptide integrin recognition sites in the three subunits whose expression varies among the different laminin isoforms. Thus, the laminins represent a complex group of molecules that have a host of mechanisms for cellular interaction (32).

## C.  Other ECM Proteins

Other ECM proteins include the collagens, which are a highly specialized glyco-protein family with a least 16 distinct members that are thought to be encoded by at least 30 genes (9). Collagens are very widespread, being found in the stroma, basement membranes, cartilage, tumors, skin, tendons, and, in the case of asthma, the thickened layer of connective tissue found below the epithelium (primarily collagen type IV). Vitronectin has been identified in both the ECM and plasma, the latter being found as either a single or two-chain form held together by disulfide bridges. The form found in tissues is not yet known. Vitronectin is encoded by a single gene and has a variety of functions, being involved in phago-cytosis, tissue repair, and immune function. Cellular attachment to vitronectin is mediated by $\alpha v \beta 3$, $\alpha IIb \beta 3$, and $\alpha v \beta 5$ via an RGD sequence (33).

Other stromal matrix proteins include hyaluronate and fibrinogen. The latter is not strictly an ECM protein, but fibrinogen is deposited at inflammatory sites where it can influence fibroblast function. Although the integrins are also the principal receptors for ECM proteins, not all matrix protein receptors are inte-grins—e.g., CD44, a lymphocyte homing receptor, also binds to hyaluronate (34). Hyaluronate is a large molecule with a molecular weight approaching 10 million. It is a major component of developing tissue and supports both cell pro-liferation and migration.

## D.  Eosinophils and the Extracellular Matrix

Eosinophils are primarily tissue-dwelling cells, therefore interaction with the ECM proteins are likely to influence their functional ability. Tissue eosinophils can be considered to be activated in that they express mRNA for a number of cy-tokines not seen with resting peripheral blood eosinophils (35). They express re-ceptors such as CD25 (36) and CD69, which are undetectable on unstimulated freshly isolated cells (37), and contain the activated form of ECP detected by mAb EG2 (38,39). The precise mechanisms responsible for eosinophil activation in vivo are not well understood. Although in vitro evidence suggests that the process of transmigration through the endothelium can in itself cause a degree of eosinophil activation (40), there is now good evidence that ECM proteins are also likely to play an important role in eosinophil activation and function. Table 1 il-lustrates the receptors for matrix proteins that have been identified on human eosinophils.

## III.  EOSINOPHIL ADHESION TO ECM PROTEINS

It is only in recent years that eosinophil/matrix interactions have been addressed in any depth. One early study found that eosinophil adhesion to cultured fibrob-lasts was partially inhibited by RGD-containing peptides, suggesting a role for fi-

**Table 1**  Eosinophil Matrix Protein Receptors

| Eosinophil receptor | Matrix protein |
|---|---|
| Mac-1 (CR3) | Fibrinogen |
| VLA-4 ($\alpha4\beta1$) | Fibronectin |
| VLA-6 ($\alpha6\beta1$) | Laminin |
| $\alpha4\beta7$ | Fibronectin |

broblast-derived ECM proteins in the adhesive interaction with eosinophils (41). Resting eosinophils have been shown to adhere to fibronectin in a VLA-4 ($\alpha4\beta1$)–dependent manner and the number of adherent cells was enhanced by incubation with PAF in a dose-dependent fashion (42). Both indirect immunofluoresence and flow cytometry or immunoprecipitation and sodium dodecyl polyacrilamide gel electrophoresis have demonstrated the expression of $\alpha6\beta1$ by human eosinophils that were found to adhere to laminin via $\alpha6\beta1$ without prior stimulation (43). Tourkin et. al extended these observations to show that IL-5-stimulation resulted in greater than fourfold enhancement of the number of eosinophils adherent to laminin compared with resting cell adhesion. Furthermore, culture of eosinophils on laminin for up to 6 days resulted in generation of a hypodense eosinophil phenotype very similar to that achieved by culture with IL-5 (44). In contrast, other reports have shown only negligible adherence by unstimulated eosinophils to fibronectin or laminin; prior eosinophil activation with PAF (45) or a $\beta1$ integrin–activating antibody (46) was required before significant adhesion to these ECM proteins was observed. Recently, we have examined eosinophil adhesion to a number of ECM proteins coated onto plastic microtiter wells and have found that unstimulated eosinophils did not adhere to fibrinogen, collagen, hyaluronate, or vitronectin compared with significant adhesion to tissue fibronectin, laminin, or rhVCAM-1 (unpublished observations). The discrepancies between these findings from a number of groups might be partly explained by the sourc of donors. For example, the study by Kita et al. (45) used eosinophils isolated from nonallergic donors; in the other studies, eosinophils were purified from individuals with a range of allergic symptoms. Thus, in vivo activation of eosinophils may contribute to their ability to adhere to matrix proteins. Another complication is that eosinophils also express Mac-1 (CD11b), which is a promiscuous receptor with a number of ligands. Engagement of this receptor has been shown to affect eosinophil function (47), and this may provide a partial explanation for the conflicting data from different groups. Additionally, as stated above, both fibronectin and laminin exist in a number of different forms depending on their cellular source. Thus, differing results from studies performed by separate groups might be explained by the source of their ECM proteins.

The $\alpha4$ chain of the VLA-4 complex can also be found in association with the $\beta7$ subunit as well as with the $\beta1$ chain. Integrin $\alpha4\beta7$ is thought to be important in lymphocyte homing to Peyer's patches and the intestinal lamina propria, where it interacts with its recently identified ligand—the mucosal addressin MAdCAM-1—on the postcapillary venules (48,49) (see Chap. 5). MadCAM-1 belongs to the Ig-like adhesion receptor superfamily. It has substantial homology with both VCAM-1 and ICAM-1 in its two amino-terminal domains and, in the mouse, the gene for MAdCAM-1 has been located on chromosome 10 (50). In addition, $\alpha 4\beta 7$ will also bind to VCAM-1 and the CS-1 region of fibronectin (51–53). Using immunostaining and flow cytometry, researchers have found that human eosinophils constitutively express $\alpha 4\beta 7$ (54). Eosinophil expression of $\alpha 4\beta 7$ was comparable to that of $\alpha 4\beta 1$ and appeared to be relatively homogenous. This pattern compares closely with that seen in other cells such as natural killer (NK) cells and newborn blood T and B cells. In contrast, human adult blood T and B cells have heterogenous $\alpha 4\beta 7$ expression, which appears related to the degree of cellular activation and "memory" differentiation (52–54). We have recently extended this observation to show that activation of eosinophils with PAF did not increase expression of $\alpha 4\beta 1$ or $\alpha 4\beta 7$. Furthermore, we have also demonstrated important functional roles for $\alpha 4\beta 7$ in that it can contribute to eosinophil adhesion to fibronectin, MAdCAM-1, and VCAM-1. However, $\alpha 4\beta 7$ function on eosiophils appeared to be especially activation-dependent, because stimulation with PAF or mAb TS2/16 [an anti-$\beta 1$ mAb that enhances the ligand-binding ability of both VLA-4 and VLA-5 in a number of cell types (55,56)] was a requirement for $\alpha 4\beta 7$-dependent adhesion to these ligands (57).

## IV. ECM PROTEINS—EFFECT ON EOSINOPHIL DEGRANULATION AND MEDIATOR RELEASE

Eosinophil interaction with the ECM can result in outside-in signaling through integrin receptors leading to eosinophil priming and mediator release (8). Extracellular matrix proteins have been shown to affect the response of eosinophils to physiological soluble stimuli. An early study demonstrated that formyl-Met-Leu-Phe (fMLP)-dependent superoxide production by eosinophils was enhanced when they were adherent to fibrinogen but inhibited in cells adherent to endothelial cells (58). Another study found that fibrinogen co-immobilized on microtiter plates with IgG enhanced eosinophil-derived neurotoxin (EDN) release compared with eosinophils adherent to IgG alone, whereas neither fibronectin nor laminin had any significant effect on IgG-dependent EDN release (59). Neeley et al. have shown that VLA-4–dependent adhesion of eosinophils to fibronectin augmented fMLP-dependent eosinophil peroxidase (EPO) release (60). In contrast, Kita et al. (45) have demonstrated that PAF, C5a, or IL-5 stimulation of eosinophils adherent to human serum albumin (HSA) resulted in EDN release that was inhibited if the cells were adherent to fibronectin or laminin. Modulation of eosinophil EDN release by fibronectin or laminin was found to be secreta-gogue-dependent; i.e., C5a and IL-5–dependent degranulation were inhibited but IgA- or PMA-induced degranulation was not. These studies emphasize the importance of the nature of the secretagogue used when investigating the effect of ECM proteins on eosinophil degranulation.

In addition to having effects on granule-derived mediator release, eosinophil adherence to fibronectin can also influence the generation of membrane-derived mediators such as leukotrienes. For example, we have demonstrated that eosinophils adhere to fibronectin in a α4β1-dependent manner, a process that resulted in short-term priming of eosinophils for enhanced $LTC_4$ release following stimulation with suboptimal concentrations of calcium ionophore. Fibronectin-enhanced $LTC_4$ release was inhibited by anti–VLA-4 mAb and was not seen when eosinophils were adherent to bovine serum albumin (BSA)-coated wells (42). Eosinophil inflammatory mediator production can also be induced by direct ligation of integrins by antibodies. For example, Laudanna et al. demonstrated cross-linking of VLA-4, LFA-1, CR3, or CD18 by immobilized mAb resulted in eosinophil spreading and superoxide production (61). In addition, Nagata and colleagues have shown that eosinophil VLA-4 dependent adhesion to VCAM-1 activates superoxide generation (62). These may represent important observations because eosinophil interaction with integrins or VCAM-1 expressed by resident cells in the tissues—including fibroblasts and epithelial cells—may have important consequences for eosinophil function.

## V. ECM PROTEINS AND EOSINOPHIL CYTOKINE GENERATION

A number of studies have suggested that both peripheral blood and tissue eosinophils might be important sources of pro-inflammatory cytokines, including IL-3, GM-CSF (63,64), and IL-5 (65,66). Furthermore, it has been appreciated for some time that these cytokines enhance eosinophil survival cultured in vitro for up to 2 weeks (67–69). Thus, eosinophil persistence in the tissues may be prolonged in the presence of IL-3, IL-5, and GM-CSF. However, the physiological triggers of autocrine cytokine production by eosinophils are not fully understood. One potential mechanism is integrin-dependent interaction with ECM proteins. For example, Anwar et al. have demonstrated that α4β1-dependent eosinophil adhesion to plasma fibronectin prolonged eosinophil survival in culture for up to 4 days as the result of autocrine production of IL-3 and GM-CSF. Additionally, eosinophils adherent to fibronectin for 24 hours expressed mRNA for GM-CSF as determined by in situ hybridization (70).

As stated above, plasma fibronectin lacks the IIICS cell binding region recognized by α4β1 and α4β7 in at least half of its subunits, whereas tissue fibronectin has it in both subunits (28). In a recent study, we therefore compared the survival of human eosinophils cultured on tissue fibronectin with that on plasma fibronectin (71). Eosinophils cultured on tissue fibronectin–coated wells had significantly greater survival than cells cultured on plasma fibronectin, and the greater eosinophil viability observed on tissue fibronectin was shown to be due to autocrine generation of IL-5, as well as GM-CSF and IL-3. We were surprised to see less impressive eosinophil survival on plasma fibronectin than that reported in our earlier study and excluded the possibility that different batches of plasma fibronectin may have contained variable amounts of the II-ICS alternatively spliced form or that cytokine contamination of either fibronectin preparation might have influenced our results. An alternative explanation was that mononuclear contamination of less than 1% might have contributed to eosinophil survival on plasma fibronectin, because monocytes/macrophages have the capacity to produce GM-CSF when adherent to fibronectin (12), utilizing both $\alpha 4\beta 1$ and $\alpha 5\beta 1$ to bind to fibronectin via the CS-1 and the classic cell-binding RGD sequence, respectively (72). This interpretation was strengthened by our observation that addition of autologous mononuclear cells to the purified eosinophil suspensions to give a final mononuclear concentration of only 1% or 0.5% resulted in eosinophil survival on plasma fibronectin comparable to that seen with tissue fibronectin. These findings emphasize that although the immunomagnetic negative selection method of Hansel et al. (73) has advantages over older eosinophil separation methods (which relied on density gradient centrifugaton and exploited differences in eosinophil density compared to other leukocytes), it does have the disadvantage that it is the mononuclear cell, rather than the neutrophil, which is the potential contaminating cell. This demonstrates the problems mononuclear cell contamination of eosinophil preparations can cause and emphasizes the care needed when purifying eosinophils using CD16-dependent immunomagnetic selection.

The observations from this study provide additional evidence that ECM proteins are important physiological triggers for autocrine cytokine production by eosinophils resident in tissues. Furthermore, data from experiments using the anti–CS-1 mAb 90.45 suggest that eosinophil survival involves interaction with the CS-1 motif in the IIICS region of fibronectin (71). This is consistent with the observation that eosinophil transmigration through endothelium can be inhibited by peptides against the CS-1 region of fibronectin (46). It can be envisaged, therefore, that eosinophil survival and persistence might be enhanced in inflamed tissues if there existed a greater amount of III-CS–containing fibronectin than that seen in normal tissue. CS-1–containing fibronectin variants have been demonstrated to be present on the endothelium of synovium from subjects with rheumatoid arthritis (RA). In contrast, normal synovium expressed little CS-1–containing fibronectin. Furthermore, this study demonstrated that anti–CS-1 and $\alpha 4\beta 1$ mAb inhibited binding of T lymphoblastoid cells (74); another group demonstrated similar inhibition of α4β1-dependent lymphocyte adhesion to RA synovium (75). The possibility exists, therefore, that CS-1–containing fibronectin variants might be important in the accumulation and/or persistence of eosinophils and other pro-inflammatory cells in the inflamed mucosae around the airways in asthma.

As stated above, we have demonstrated that activated eosinophils adhere to fibronectin, VCAM-1, and MAdCAM-1 in a α4β7-dependent manner. We have therefore investigated the contribution of α4β7 to prolonged eosinophil survival on tissue fibronectin. Eosinophils were cultured on optimal concentrations of tissue fibronectin for 3 days. Enhanced eosinophil survival was significantly inhibited by mAb against α4β7, as well as by mAb to VLA-4 and the β1 chain (Fig. 1).

Tourkin et al. [44] have shown that eosinophils cultured on laminin exhibit prolonged survival similar to that reported for fibronectin, and this process was inhibited by mAb to α6β1. This phenomenon appeared to be the result of eosinophil autocrine production of IL-3 and GM-CSF because antibodies to these cytokines inhibited laminin-induced survival. Interestingly, an antibody specific for IL-5 had no effect on laminin-dependent survival. As mentioned above, eosinophils cultured on tissue fibronectin generate IL-5 in addition to IL-3 and

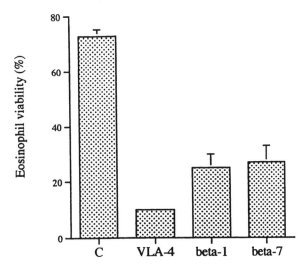

**Figure 1** The effect of monoclonal antibodies (mAb) to very late activation antigen-4 (VLA-4) (HP1/2), β1 (BP6), and β7 (Act-1) on the viability of human eosinophils (>99.9% pure) cultured in 96-well microtiter plates coated with tissue or plasma fibronectin (25 μg/well) for 3 days. Each column represents the mean ± SEM of four experiments.

GM-CSF (71). The data from these two studies suggest that the profile of viability-enhancing autocrine cytokine production by eosinophils may be profoundly influenced by the nature of the ECM substrates to which they are adherent and that their prolonged survival may be caused by triggering through multiple integrin receptors. Figure 2 shows a summary of work performed in our laboratory comparing eosinophil survival on different ECM substrates compared with cells cultured on plates coated with soluble human VCAM-1.

## VI. CONCLUSION

A number of ECM proteins have been found to have a profound effect on eosinophil function. Eosinophils adhere to both fibronectin and laminin in an integrin-specific fashion. Depending on the nature of the secretagogue, ECM proteins can either enhance or modulate eosinophil mediator release. In the longer term, culture of eosinophils on fibronectins or laminin can prolong their survival as a result of autocrine cytokine generation. The latter may represent an important homeostatic mechanism by which eosinophil persistence in tissues is enhanced and may partially explain the selective tissue accumulation of this important pro-inflammatory cell.

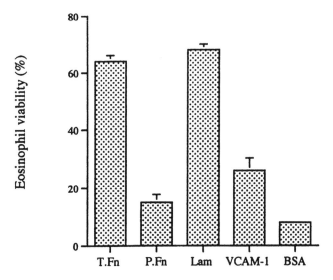

**Figure 2**  A comparison of the survival of human eosinophils (>99.9% pure) cultured in 96-well microtiter plates coated with optimal concentrations of tissue fibronectin (T. Fn), plasma fibronectin (P. Fn), laminin (Lam), soluble vascular cell adhesion molecule-1 (VCAM-1), or bovine serum albumin (BSA). Each column represents the mean ± SEM of four experiments.

# REFERENCES

1. A. J. Wardlaw, S. Dunnette, G. J. Gleich, J. V. Collins, and A. B. Kay, *Am Rev. Respir. Dis.*, *138:62* (1988).
2. J. Bousquet, P. Chanez, V. Lacoste, G. Barneon, N. Gavanian, I. Enander, P. Venge P, Ahlstedt S, Simony-Lafontaine J, Godard P, and F-B Michel F-P, *N. Engl. J. Med.*, *323*:1033 (1990).
3. A. J. Wardlaw and A. B. Kay, *Allergy*, *42*:321 (1987).
4. G. J. Gleich, *J. Allergy Clin. Immunol.*, *85*:422 (1990).
5. B. S. Bochner and R. P. Schleimer, *J. Allergy Clin. Immunol.*, *94*:427, 1995.
6. A. J. Wardlaw, G. M. Walsh, and F. A. Symon, *Allergy*, *49*:797, (1994).
7. T. A. Springer, *Cell*, *76*:301 (1994).
8. E. C. Butcher, *Cell*, *67*:1033 (1991).
9. E. D. Hay (ed.), *Cell biology of the extracellular matrix*. Plenum Press, New York (1991).
10. Y. Shimuzu and S. Shaw, *FASEB J*, *5*:2292 (1991).
11. D. Schubert, *Trends Cell Biology*, *2:63* (1992).
12. B. Thorens, J. J. Mermod, and P. Vassali, *Cell*, *48*:671 (1988).
13. D. F. Eierman, C. E. Johnson, and J. S. Haskill, *J. Immunol.*, *142*:1970 (1989).
14. S. L. Newman and M. A. Tucci, *J. Clin.Invest.*, *86*:703 (1990).
15. R. O. Hynes (ed.), *Fibronectins*. Springer-Verlag Publishing Company, New York, p. 365 (1990).
16. D. C. Dean, *Am. J. Respir. Cell Mol. Biol.*, *1*:5 (1989)
17. S. I. Rennard and R. G. Crystal, *J. Clin. Invest.*, *69*:119 (1981).
18. T. J. Broekelman, A. H. Limper, T. V. Colby, and J. A. Macdonald. *Proc. Natl. Acad. Sci. USA*, *88*:6642 (1991).
19. E. Harkonen, I. Virtanen, A. Linnala, L. L. Laitinen, and V. L. Kinnula, *Am. J. Respir. Cell Mol.*, *13*:109 (1995).
20. R. Raghow, *FASEB I.*, *8*:823 (1994).
21. E. J. Kovacs and L. A. DiPietro, *FASEB J.*, *8*:854 (1994).
22. A. R. Kornblihtt, K. Umezawa, K. Vibe-Pedersen, and F. Baralle, *EMBO J.*, *4*:1755 (1985).
23. R. P. Hershberger and L. A. Culp, *Mol. Cell Biol.*, *10*:662 (1990).
24. A. P. Mould, A. Komioriya, K. M. Yamada, and M. J. Humphries, *J. Biol. Chem.*, *266*:3579 (1991).
25. M. J. Humpries, S. K. Akiyama, A. Komiriya, K. Olden, and K. M. Yamada, *J. Cell Biol.*, *103*:2637 (1986).
26. E. A. Wayner, A. Garcia-Pardo, M. J. Humphries, J. A. McDonald, and W. G. Carter, *J. Cell Biol.*, *109*:1321 (1989).
27. J. L. Guan and R. O. Hynes, *Cell*, *60*:53 (1990).
28. A. P. Mould, A. Wheldon, E. Komoriya, E. A. Wayner, K. M. Yamada, and M. J. Humphries, *J. Biol. Chem.*, *265*:4020 (1990).
29. E. Engvall, D. Earwicker, T. Haaparanta, E. Ruoslahti, and J. R. Sanes, *Cell Regul.*, *1*:731 (1990).
30. I. Sorokin, A. Sonnenberg, M. Aumailley, R. Timpl, and P. Ekblom, *J. Cell Biol.*, *111*:1265 (1990).

31. R. O. Hynes, *Cell*, *69*:11 (1992).
32. K. M. Yamada, *Cell biology of extra-cellular matrix*. (E.D. Hay, ed.), Plenum Press, New York, p. 111 (1991).
33. S. Suzuki, A. Oldberg, E. G. Hayman, M. D. Pierschbacher, and E. Ruoslahti, *EMBO J.*, *4*:2519 (1985).
34. K. Miyake, C. B. Underhill, J. Lesley, and P. W. Kincade, *J. Exp. Med.*, *172*:69 (1990).
35. P. Desremaux, A. Janin, J. F. Colombel, L. Prin, J. Plumas, D. Emilie, D. G. Torpier, A. Capron, and M. Capron, *J. Exp. Med.*, *175*:293 (1992).
36. A. Hartnell, R. Moqbel, G. M. Walsh, B. Bradley, and A. B. Kay *Immunol.*, *69*:264 (1990).
37. A. Hartnell, D. S. Robinson, A. B. Kay, and A. J. Wardlaw. *Immunology*, *80*:281 (1993).
38. M. Azzawi, B. Bradley, P. K. Jeffery, A. J. Frew, A. J. Warldaw, G. Knowles, B. As-soufi, J. V. Collins, S. R., Durham, and A. B. Kay, *Am. Rev. Respir. Dis.*, *142*:1407 (1990).
39. A. M. Bentley, G. Menz, C. Storz, D. S. Robinson, B. Bradley, P. K. Jeffery, S. R. Durham, and A. B. Kay, *Am. Rev. Resp. Dis.*, *146*:500 (1992).
40. C. Walker, S. Rihs, R. K. Braun, S. Betz, and P. J. Bruijnzeel. *J. Immunol.*, *150*:4061 (1993).
41. A. Shock, K. F. Rabe, G. Dent, R. C. Chambers, A. J. Gray, K. F. Chung, P. J. Barnes, and G. J. Laurent. *Clin. Exp. Immunol.*, *86*:185 (1991).
42. A. R. E. Anwar, G. M. Walsh, O. Cromwell, A. B. Kay, and A. J. Wardlaw,. *Immunol.*, *82*:222 (1994).
43. S. N. Georas, B. W. McIntyre, M. Ebisawa, J. L. Bednarczyk, S. A. Sterbinsky, R. P. Schleimer, and B. S. Bochner, *Blood*, *82*:2872 (1993).
44. A. Tourkin, T. Anderson, E. Carwile LeRoy, and S. Hoffman. *Cell Adhes. Commun.*, *1*:161 (1993).
45. H. Kita, S. Horie S, and G. J. Gleich *J. Immunol.*, *156*:1174 (1996).
46. T. W. Kuijpers, E. P. J. Mul, M. Blom, N. L. Kovach, C. A. Gaeta, V. Tollefson, M. J. Elices, and J. M. Harlan, *J. Exp. Med.*, *178*:279 (1993).
47. S. Horie and H. Kita, *J. Immunol.*, *152*:5457 (1994).
48. C. Berlin, E. L. Berg, M. J. Briskin, D. P. Andrew, P. J. Kilshaw, B. Holzmann, I. L. Weissman, A. Hamann, and E. C. Butcher, *Cell*, *74*:185 (1993).
49. M. J. Briskin, L. M. McEvoy, and E. C. Butcher, *Nature*, *331*:461 (1993).
50. S. O. Sampaio, M. Takeuchi, C. Mei, U. Francke, E. C. Butcher, and M. J. Briskin, *J. Immunol.*, *155*:2477 (1995).
51. C. Ruegg, A. A. Postigo, E. E. Sikorski, E. C. Butcher, R. Pytela, and D. J. Erle, *J. Cell Biol.*, *117*:179 (1992).
52. A. A. Postigo, P. Sanchez-Mateos, A. I. Lazarovits, F. Sanchez-Madrid, and M. O. de Landazuri, *J. Immunol.*, *151*:2471 (1993)
53. B. M. C. Chan, M. J. Elices, E. Murphy, and M. E. Hemler, *J. Biol. Chem.*, *267*:8366 (1992).
54. D. J. Erle, M. J. Briskin, E. C. Butcher, A. Garcia-Pardo, A. I. Lazarovits, and M. Tidswell, *J. Immunol.*, *153*:517 (1994).

55. G. C. Arroyo, P. Sanchez-Mateos, M. R. Campanero, I. Martin-Padura, E. Dejana, and F. Sanchez-Madrid, *J. Cell Biol.*, *117*:659 (1992).
56. A. Arroyo, A. Garcia-Pardo, and F. Sanchez-Madrid, *J. Biol. Chem.*, *268*:9863 (1993).
57. G. M. Walsh, F. A. Symon, A. I. Lazarovits, and A. J. Wardlaw, *Immunol.* *89*:112 (1996).
58. P. Dri, R. Cramer, P. Spessotto, M. Romano, and P. Patriarca, *J. Immunol.*, *147*:613 (1991).
59. M. Kaneko, S. Horie, M. Kato, G. J. Gleich, and H. Kita. *J. Immunol.*, *155*:2631 (1995).
60. S. Neeley, K. J. Hamann, T. L. Dowling, K. T. McAllister, S. R. White, and A. R. Leff, *Am. J. Respir. Cell Mol. Biol.*, *11*:206 (1994).
61. C. Laudanna, P. Melotti, C. Bonizzato, G. Piacentini, A. Boner, M. C. Serra, and G. Berton,. *Immunology*, *80*:273 (1993).
62. M. Nagata, J. B. Sedgwick, M. E. Bates, H. Kita, and W. W. Busse, *J. Immunol.*, *155*:2194 (1995).
63. H. Kita, T. Ohnishi, Y. Okubo, D. Weiler, J. S. Abrahms, and G. J. Gleich, *J. Exp. Med.*, *174*:745 (1991).
64. R. Moqbel, Q. Hamid, S. Ying, J. Barkhans, A. Hartnell, A. Tsicopoulos, A. J. Wardlaw, and A. B. Kay, *J. Exp. Med.*, *174*:749 (1991).
65. Q. Hamid., M. Azzawi, S. Ying, R. Moqbel, A. J. Wardlaw, C. J. Corrigan, B. Bradley, S. R. Durham, J. V. Collins, P. K. Jeffery, D. J. Quint, and A. B. Kay, *J. Clin. Invest.*, *87*:1541 (1991).
66. D. H. Broide, M. M. Paine, and G. S. Firestein, *J. Clin. Invest.*, *90*:1414 (1992).
67. W. F. Owen, M. E. Rothenburg, D. S. Silberstein, J. C. Gasson, R. L. Stevens, K. F. Austin, and R. J. Soberman, *J. Exp. Med.*, *166*:129 (1987).
68. M. E. Rothenberg, J. Petersen, R. L. Stevens, D. S. Silberstein, D. T. McKenzie, K. F. Austin, and W. F. Owen, *J. Immunol.*, *143*:2311 (1989).
69. M. E. Rothenberg, W. F. Owen, D. S. Silberstein, J. Woods, R. J. Soberman, K. F. Austen, and R. L. Stevens, *J. Clin. Invest.*, *81*:1986 (1988).
70. A. R. E. Anwar, R. Moqbel, G. M. Walsh, A. B. Kay, and A. J. Wardlaw, *J. Exp. Med.*, *177*:839 (1993).
71. G. M. Walsh, F. A. Symon, and A. J. Wardlaw, *Clin. Exp. Allergy*, *25*:1128 (1995).
72. M. E. Hemler, *Annu. Rev. Immunol.*, *8*:365 (1990).
73. T. T. Hansel, I. J. M. De Vries, T. Iff, S. Rihs, M. Wandzilak, S. Betz, K. Blaser, and C. Walker, *J. Immunol. Methods*, *145*:105 (1991).
74. M. J. Elices, V. Tsai, D. Strahl, et al., *J. Clin. Invest.*, *93*:405 (1994).
75. A. C. H. Dinther-Janssen, S. T., Pals, R. J. Scheper, and C. J. L. M. Meijer, *Ann. Rheum. Dis.*, *52*:672 (1993).

# 11

# Eosinophil–Epithelial Interactions and Transepithelial Migration

**Mary K. Schroth**   *University of Wisconsin Children's Hospital, Madison, Wisconsin*

**James M. Stark**   *Children's Hospital Medical Center, Cincinnati, Ohio*

**Julie B. Sedgwick and William W. Busse**   *University of Wisconsin Medical School, Madison, Wisconsin*

## I.  INTRODUCTION

During allergic and asthmatic inflammation, multiple changes are observed within the airway, including recruitment of circulating cells to the bronchial mucosa and lumen, and destruction of epithelium. The evidence that eosinophils have an active role in these processes is apparent from many observations. For example, eosinophils and eosinophil-derived products are found in increased numbers in the blood, sputum, and airways of symptomatic asthmatic subjects. Furthermore, a correlation exists between the presence of eosinophils (i.e., the number of eosinophils in the circulation, bronchoalveolar lavage [BAL] fluid, and bronchial biopsy), and asthma severity. There is also evidence that cellular products of eosinophils are released during allergic reactions and these products can produce the pathological and physiological changes seen in asthma. Finally, following the late asthmatic response to allergens, a presumed marker of allergic inflammation, increased numbers of eosinophils and their products are found in BAL fluid and mucosal biopsies.

Airway epithelium is also an active participant in allergic inflammation. Epithelial cells are biologically active, express adhesion receptor proteins, and produce cytokines during inflammation. For example, human bronchial epithelial cells produce interleukin-6 (IL-6) (1,2), IL-8 (1–3), granulocyte-macrophage colony-stimulating factor (GM-CSF)(1,2), eotaxin (4,5), and regulated on activation, normal T cell expressed and secreted (RANTES) factor (6–9). These products can influence eosinophil function. Epithelial cells also release chemoattractants for lymphocytes and monocytes during asthma exacerbations (10,11); greater amounts of 15-HETE, prostaglandin $E_2$ ($PGE_2$), and fibronectin are generated in asthmatic patients when compared with normal control subjects (12). Pro-inflammatory cytokines, such as interferon gamma (IFN-γ), IL-1β, and tumor necrosis factor–α (TNF-α), induce airway epithelial cells to express intercellular adhesion molecule-1 (ICAM-1) (13–16), which can then facilitate movement of eosinophils, lymphocytes, monocytes, and neutrophils from the circulation to the airway lumen (13,14,16). Airway epithelial cells also express class I and class II major histocompatibility (MHC) antigens (17). The loss of the epithelial cell layer has been associated with asthma, and these changes are associated with increases in airway hyperresponsiveness (18). Collectively, these observations demonstrate the protective role of airway epithelial cells for the airway smooth muscle, and these effects are noted both mechanically and biochemically.

Eosinophils and epithelial cells interact during airway inflammation at multiple levels and multiple stages of airway injury (Fig. 1). Airway epithelial cells are activated by a variety of stimuli (Table 1). For example, viral infections, antigens, allergens, and environmental pollutants influence epithelial cell function through an induction of epithelial cell cytokine production and increased expression of surface receptors. Alternatively, mast cells, macrophages, and lymphocytes generate mediators that act on the epithelial cells, and under the influence of these mediators, epithelial cell function is also altered. Thus, epithelial cell derived cytokines, in concert with similar products of endothelial cells and intra-airway inflammatory cells, serve to activate and attract eosinophils into the airway lumen.

The process of eosinophil recruitment to the airway is complex and interactive at many tissue levels, and serves to direct, prime, and activate the migrat-

**Table 1**  Airway Epithelial Cell Activators

| Direct activators | |
|---|---|
| Allergen/IgE | |
| Viral infections | |
| Pollutants | |

| Indirect activators | |
|---|---|
| Cell type | Products |
| Mast cells | Histamine, tryptase, TNF-α |
| Macrophages | TNF-α, IL-1β, histamine-releasing factors, $O_2^-$ |
| Lymphocytes | IFN-γ |

Abbreviations: TNF-α = tumor necrosis factor, PAF = platelet-activating factor, IFN-γ = interferon gamma.

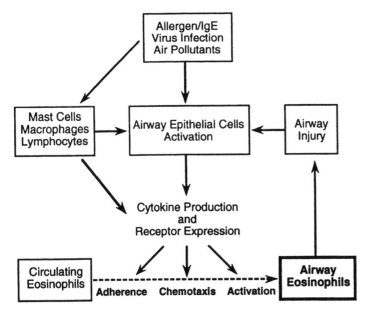

**Figure 1** Activation of airway epithelial cells and recruitment of circulating eosinophils to the airway. Airway epithelial cells are activated by direct and indirect stimuli. This activation results in epithelial cell production of cytokines and expression of surface receptors. The cytokines produced by the migrating cells within the lungs as well as the epithelium and endothelium and cell receptor expression alter the eosinophil through adhesion, chemotaxis, and activation, and the eosinophil arrives in the airway as the phenotypically altered cell.

ing cell. Integral in these processes is activation of adhesion proteins on the migrating cells and the barrier tissues: endothelium, matrix, and epithelium. When these migrating cells arrive in their "new environment," they are activated, primed, and degranulated; with degranulation there is release of highly charged cationic proteins including major basic protein (MBP), eosinophil cationic protein (ECP), and eosinophil peroxidase (EPO). These toxic mediators damage epithelium and can both induce dramatic changes in function of epithelial cells and stimulate them to generate additional pro-inflammatory mediators.

This chapter will review the interactions of eosinophils and airway epithelial cells and examine the unique features of epithelial cell and eosinophil activation during allergic inflammation as well as the sequelae of these events on the pathobiological features of asthma.

## II.  EPITHELIUM ACTIVATION

## A.  Indirect Influences

Mast cells reside within the lung epithelium and interstitial tissues and contain an important profile of preformed mediators. The mast cell is activated by cross-linking of its high-affinity IgE receptors (FcεRI) through allergen-specific binding (19); activation may also occur through non-IgE dependent events, e.g., osmotic factors, opiates, and other stimuli (20). Following activation, mast cells release preformed mediators, including histamine and tryptase, and begin to synthesize cytokines such as TNF-α (20,20a). Histamine is a potent vasodilator and airway smooth muscle contractor. Significantly increased levels of histamine are found in BAL fluid of subjects with allergic asthma and following antigen exposure (21). Histamine can stimulate airway epithelial cells to produce IL-6 (22), IL-8 (22), fibronectin (23), and GM-CSF (22,24), and enhance expression of ICAM-1 and human leukocyte, D related (HLA-DR) on cultured bronchial cells from asthmatic subjects (25). The increased expression of ICAM-1 on bronchial epithelial cells can further facilitate leukocyte recruitment (16). HLA-DR expression likely signifies epithelial cell activation and may serve to participate in antigen presentation (26).

Tryptase, an enzyme specific to mast cells, is stored in mast cell granules and is complexed to heparin proteoglycans (27). Mast cell tryptase (MCT), like histamine, is released following allergen provocation and is increased in BAL fluid of asthmatic subjects (28). MCT is a potent mitogen that can cause airway epithelial cells to proliferate (29). Furthermore, MCT stimulates epithelial cells to generate IL-8 and enhances ICAM-1 expression (29).

Activation of mast cells, as well as macrophages, can result in the release of TNF-α within the asthmatic airway. TNF-α can potentiate the inflammatory process by up-regulating epithelial cell production of IL-6 (1), IL-8 (1,3), RANTES (6,7,9), and endothelin (30–32), and increased ICAM-1 expression (15). In addition, TNF-α enhances mast cell (33) and eosinophil cytotoxicity (34,35), and up-regulates endothelial cell expression of E-selectin, VCAM-1, and ICAM-1 (36–38). The end result is the establishment of conditions favorable for the recruitment of leukocytes to the lung.

Alveolar macrophages are the most prevalent cells in the airways of normal (39) and asthmatic individuals (40). Similar to mast cells, alveolar and pulmonary tissue macrophages can be activated by IgE through cross-linking of IgE receptors; in contrast, these are low-affinity IgE (FcεRII) receptors (20a). Alternative activation may also occur through non-IgE dependent events including viral respiratory infections (41–43), and particulates such as cigarette smoke and pollutants (44). Following activation, macrophages release multiple inflammatory products, including IL-1β (45–48) and TNF-α (20a,46,49,49a). IL-1β and

TNF-$\alpha$ levels are higher in the BAL fluid of subjects with asthma as compared with normal subjects (49,50). In addition, histamine-releasing factors are secreted by alveolar macrophages and monocytes, which induce mast cells to produce histamine (51). As previously discussed, TNF-$\alpha$ promotes airway epithelial cell cytokine production and ICAM-1 expression. Similarly, IL-1$\beta$ induces epithelial cells to produce GM-CSF (1,24,52), IL-6 (1), IL-8 (1,3), IL-11 (53), RANTES (6,7), and endothelin (30–32), and to express ICAM-1(13,15,54). IL-1$\beta$ can induce release of an epithelial-derived relaxing factor that inhibits airway smooth muscle contraction in response to bronchoconstrictors (55); this factor, however, remains controversial.

Lymphocytes also play an important role in asthma airway inflammation, and T lymphocytes are characteristically identified in BAL fluid from subjects with asthma (56). Furthermore, bronchial biopsies demonstrate that the bronchial mucosa is replete with T lymphocytes (57). Although products of a subclass of T helper lymphocytes, type 2 helper (Th2) cells, produce IL-4 and IL-5, which are critical to the development of allergic inflammation, IFN-$\gamma$, a product of Th1-like cells and an inhibitor of Th2 cell proliferation, is a potent activator of epithelial cells and induces production of RANTES (7) and expression of ICAM-1 and HLA-DR receptors (58).

## B.  Direct Influences

The respiratory tract is continuously exposed to allergens, environmental pollutants and irritants, and viral pathogens that have the potential to directly alter airway epithelial cell function. Low-affinity IgE receptors (Fc$\epsilon$RII, CD23) have been identified on the bronchial epithelial cells of some allergic asthmatic subjects (59). Following sensitization, CD23 positive cells also expressed endothelin (59), suggesting that epithelial cells from some allergic asthmatics may be directly activated by an IgE-mediated mechanism. In addition, after antigen challenge in sensitized guinea pigs, eotaxin, a potent eosinophil chemoattractant, is detected in BAL fluid and in airway epithelial cells (4,5,60–63).

Following respiratory syncytial virus (RSV) infection, normal human nasal epithelium produce more IL-8 (3). In an immortalized alveolar type II cell line, A549 cells, RSV infection increase IL-6 (64,65), IL-8 (64–66), IL-11 (53), and GM-CSF (64) production and ICAM-1 expression (67). HLA-DR expression is also identified on airway epithelial cells after RSV infection (68). Other respiratory viruses also alter epithelial cell function; for example, parainfluenza virus type 2 infection of human airway epithelial cells increases ICAM-1 expression (69). Major group rhinoviruses (RV) utilize the ICAM-1 receptor to bind to cells (70,71). Once attached, rhinoviruses enter airway epithelial cells, replicate, and release additional virus (72). In contrast to other respiratory viruses, damage to epithelium is variable following rhinovirus infection. In addition, RV16 infection induces RANTES production in bronchial epithelial cells but does not alter ICAM-1 or HLA-DR expression (58).

Environmental pollutants may directly alter airway epithelial cell function. Ozone treatment of airway epithelial cell cultures increases IL-6, IL-8, and fibronectin production (73) as well as CD18-dependent and ICAM-1–independent neutrophil adherence (74). Exposure of epithelial cells to nitrogen dioxide induces production of both mRNA and protein for IL-8, TNF-$\alpha$, and GM-CSF (75). Moreover, structural damage, such as loss of cilia and secretory granules, necrosis, and sloughing of airway epithelium, follows both acute and chronic exposure to $NO_2$ or ozone (76).

Airway epithelial cells are activated both directly by environmental conditions and indirectly by migratory cells that then produce mediators to alter epithelial cell function. As a result of these processes, respiratory epithelial cells produce cytokines and cellular receptors that have the potential to regulate eosinophil function as the cell undergoes chemotaxis, adherence, and eventual activation.

## III. EOSINOPHIL ACTIVATION AND ADHESION

Airway eosinophils are quantitatively and phenotypically distinct from blood cells. In their migration to the airway, eosinophils must pass through a number of barriers, including the vascular endothelium, the interstitial compartment, and the airway epithelium before arriving in the airway lumen. During these "travels," multiple and critical cell–cell, cell–matrix, and cell–cytokine interactions occur (77–83). It is these interactions that are likely determinants of the cell's eventual function and contribution to airway injury.

### A. Eosinophil–Airway Epithelial Interactions

The respiratory epithelium expresses a limited number of known adhesive ligands. Of these, only intercellular adhesion molecule-1 (ICAM-1, CD54) has been identified on the respiratory epithelium in vitro (84,85) and in vivo in biopsies from asthmatic subjects (86,87). ICAM-1 is a single-chain glycoprotein belonging to the immunoglobulin superfamily; it consists of five immunoglobulin-like extracellular domains, a single membrane-spanning domain, and a short cytoplasmic tail (88,89). Migratory inflammatory cells, including eosinophils (90–93), neutrophils (13,69), T lymphocytes (94), and basophils (95), bind to ICAM-1 via the $\beta2$ integrin family of proteins, which are heterodimers of two subunits containing a common $\beta$ chain (CD18) and one of three alpha chains (CD11a, CD11b, CD11c) (96). Inflammatory cell–ICAM-1 adhesion is mediated by CD11a/CD18 (LFA-1) binding to the amino-terminal domain of ICAM-1 (domain 1, which also serves as a receptor for rhinovirus), and by CD11b/CD18 (Mac-l) binding to ICAM-1 domain 3 (71,89,97–102).

As discussed previously, airway epithelium can be induced to express ICAM-1 through a number of stimuli, including cytokine exposure (IL-1β, TNF-α, IFN-γ (13,14,37,54,84,103)), respiratory virus infection (67,69,85), or oxidant injury (74). ICAM-1 expression by the respiratory epithelium appears to be more sensitive to IFN-α stimulation than to TNF-α or IL-1β (14,104). In addition, eosinophil MBP can induce ICAM-1 expression, and perpetuate a cycle of injury and inflammation (105). In contrast to vascular endothelium, primary airway epithelium has not been demonstrated to express significant levels of members of the selectin family (E-selectin, or P-selectin), or VCAM-1 in vitro (84) or in vivo (78,87,106) (Table 2). However, there is preliminary evidence by flow cytometry that an epithelial cell line, BEAS-2B, can express VCAM-1 in response to combined TNF-α and IL-4 stimulation (107).

There are suggestive data that respiratory epithelium may express another non–ICAM-1 ligand recognized by CD18 integrins, although this receptor has not yet been identified (13,69,74,84). Anti-CD18 monoclonal antibody (mAb) prevents neutrophil (13,69,74) and eosinophil (84) adhesion to respiratory epithelial cell monolayers, whereas blocking antibody to ICAM-1 cannot fully prevent cell adhesion. Furthermore, mice genetically modified to be deficient in ICAM-1 demonstrate normal pulmonary inflammatory responses in a pneumonia model, implying that inflammatory cells utilize a non–ICAM-1 ligand for migration into the airway. The identity of these non–ICAM-1 adhesion molecules expressed by the respiratory epithelium is currently under investigation.

We have investigated eosinophil adhesive interactions with the respiratory

**Table 2** Adhesion Ligands for Eosinophils on Vascular Endothelial Cells and Respiratory Epithelial Cells

| Ligand | CD designation | Vascular endothelium | Respiratory epithelium | Inducings cytokines | Eosinophil counter-ligand |
|---|---|---|---|---|---|
| ICAM-1 | CD54 | + | + | IL-1, TNF, IFN-γ | CD11a/CD18, CD11b/CD18 |
| VCAM-1 | CD106 | + | – | IL-1, TNF, IL-4, IL-13 | VLA-4 |
| E-selectin | CD62E | + | – | IL-1, TNF, IFN-γ, Substance P | sLE$^x$ |
| P-selectin | CD62P | + | – | Thrombin, histamine, substance P | Le$^x$, sLe$^x$ |

Abbreviations: IL = interleukin; TNF-α = tumor necrosis factor–α, IFN-γ = interferon gamma; VLA-4 = very late activation antigen-4; sLe$^x$ = sialylated Lewis$^X$; Le$^x$ = Lewis X.

epithelium (84,85). Isolated blood eosinophils adhere poorly to unstimulated epithelial cell monolayers (primary human bronchial epithelial cells or A549 cells). However, adhesion significantly increases if the eosinophils are first activated with the protein kinase stimulus, phorbol myristate acetate (PMA) (Fig. 2). Treatment of respiratory epithelial cell monolayers with TNF-$\alpha$ or IL-1$\beta$ alone has no effect on adhesion when the eosinophils are unactivated; however, activation with PMA significantly promotes eosinophil adhesion to cytokine-treated epithelial cell cultures (84) (Fig. 2). Eosinophil adhesion to RSV-infected epithelial monolayers demonstrates similar dependence on activation (PMA treatment) (85).

Activation-dependent eosinophil adhesion to the respiratory epithelial cultures is dependent on $\beta2$ integrins. When eosinophils are incubated with blocking antibodies to CD18, PMA-stimulated eosinophil adhesion to RSV (85) or TNF-treated cell monolayers of epithelial cells is significantly decreased (84). In contrast, blocking antibody to CD11a, nonspecific isotype–matched control antibody, or binding-control antibody (W6/32), does not alter eosinophil adhesion to RSV-infected monolayers (85). Data from a number of laboratories indicate that

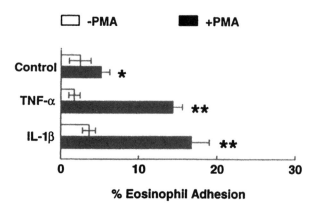

**Figure 2** Eosinophil adhesion to respiratory epithelial cell monolayers: requirement for eosinophil activation. Respiratory epithelial cell cultures (A549 cells) were treated with tumor necrosis factor–$\alpha$ (TNF-$\alpha$)(5 ng/ml) or interleukin (IL) 1$\beta$ (200 pg/mL) for 24 hours. Purified human eosinophils were exposed to control or cytokine-treated monolayers and were allowed to adhere. Data are presented as percent of adherent eosinophils following a 30-min exposure to the A549 monolayers. During the adhesion assay, phorbol myristate acetate (PMA, 1 ng/ml) was added as indicated. There was little eosinophil adhesion to control or treated monolayers in the absence of PMA. In contrast, eosinophil adhesion was significantly increased by activation with PMA. * = $P < .01$ relative to the absence of PMA. ** = $P < .01$ relative to adhesion in the absence of PMA, and also relative to adhesion to control cells in the presence of PMA.

CD11b/CD18 expression on eosinophils is increased by activation with PMA (84) or with cytokines, including GM-CSF and IL-5 (108,109). These in vitro findings parallel our earlier observations of airway eosinophils obtained following segmental bronchoprovocation in which BAL eosinophils were found to have increased CD11b/CD18 expression compared to peripheral blood eosinophils and that the airway cells were functionally up regulated (110–112). These data demonstrate that altered adhesion molecule expression accompanies cell recruitment to the lung during allergic inflammation, and implies that eosinophil CD11b/CD18 expression may play an important role in eosinophil-mediated lung injury.

Although eosinophil–epithelial adhesion may facilitate epithelial cell damage, epithelial cells generate cytokines that can also promote eosinophil–endothelial adhesion. It is this effect that has the potential to greatly enhance eosinophil recruitment to the airways.

## B.   Eosinophil–Vascular Endothelial Interactions

In contrast to airway epithelium, vascular endothelium expresses a number of inducible adhesion ligands that are recognized by eosinophils, neutrophils, lymphocytes, and other migratory inflammatory cells (Table 2). Several of these ligands, including ICAM-1 (37,54,90,113), vascular cell adhesion molecule-1 (VCAM-1) (114–117), and E-selectin (36,118,119), are up-regulated by a similar array of cytokines (e.g., IL-1$\beta$ and TNF-$\alpha$). Subtle differences in the kinetics of expression and adhesion molecules expressed on vascular endothelium likely contribute to the specificity of cellular recruitment to the lung. This topic is discussed in detail in Chapter 9.

## C.   In Vivo Studies

There is a significant body of literature to suggest that eosinophil–epithelial interactions are important in the pathophysiology of asthma, allergic rhinitis, eosinophilic pneumonia, and RSV bronchiolitis. Wegner and his colleagues studied eosinophil adhesive interactions in the lung using cynomolgus monkeys with a naturally occurring hypersensitivity to *Ascaris suum* extract as a model for asthma (16,120). The level of airway hyperresponsiveness following repeated inhalation of *Ascaris* antigen was mediated primarily by eosinophils, ICAM-1, and CD11b/CD18 (16). Monkeys exposed to three alternate-day inhalations of *Ascaris* extract and who received intravenous anti–ICAM-1 mAb, R6.5, throughout the study had significantly attenuated eosinophil infiltration and diminished airway hyperresponsiveness following antigen challenge, compared with monkeys receiving mAb that bound to an ICAM-1 epitope not involved in leukocyte adhesion (105). Similar treatment with antibody to CD11b/CD18 did not prevent eosinophil infiltration but prevented the usual increase in airway responsiveness. Antibody to E-selectin prevented neither eosinophil infiltration nor increased airway responsiveness (105). Immunohistochemical studies of sections of airways from chronically *Ascaris*-treated animals demonstrated increased expression of ICAM-1 on vascular endothelium and airway epithelium, whereas E-selectin was demonstrated only on the vascular endothelium (16). These studies demonstrated that ICAM-1 and CD11b/CD18, but not E-selectin, play key roles in the development of acute eosinophilic airway inflammation.

Because blocking antibodies were administered intravenously or intraperitoneally, the role of vascular endothelial versus respiratory epithelial adhesion cannot be defined. Relatively few studies have addressed the role of airway adhesion molecules using direct administration of the blocking antibodies to the airway through intratracheal administration. Wegner et al. (121) found that intratracheal administration of blocking antibody to ICAM-1 given to sensitized monkeys significantly decreased eosinophil influx, EPO activity, and airway hyperreactivity in some, but not all, treated monkeys. They detected serum IgM antibody against their blocking ICAM-1 monoclonal antibody in three out of four monkeys that did not have decreased airway hyperresponsiveness following anti–ICAM-1 administration, implying that the activity of the blocking antibodies was decreased by host immune responses. This study provides further evidence for the role of airway epithelial adhesion molecules in airway inflammation, and also implies a potential therapeutic intervention for asthma.

Several studies have provided data that demonstrate the need to further define the role(s) of adhesion molecules in lung defense and lung inflammation. In vitro studies using blocking antibodies, studies of patients lacking expression of various adhesion molecules, and studies of transgenic mouse models provide data that other adhesion molecules and other adhesion mechanisms may be important in eosinophil migration to the airway. It is evident that ICAM-1 expression alone does not cause recruitment of eosinophils into the airway. Rodents (mice and rats) express high levels of ICAM-1 on their alveolar type I epithelium at baseline, yet they do not exhibit airway inflammation (122,123). However, bacterial pneumonia up-regulates ICAM-1 expression by the pulmonary alveolar type II cells and the capillary endothelium, and respiratory inflammation develops. Therefore, ICAM-1 expression in conjunction with a second signal, such as cytokine expression, is likely required for recruitment and retention of inflammatory cells within the airway.

## D. Chemotaxis Mediators

Airway epithelial cells actively participate in the recruitment of peripheral eosinophils into the bronchial mucosa and airway lumen. In addition to eosinophil migration through adhesion to cytokine-promoted vascular endothelial adhesion molecules, epithelial cells produce several chemotactic cytokines in response to inflammation and inflammatory mediators (1,2,8,124). Primary cultures of bronchial epithelial cell explants from nonallergic subjects constitutively produce IL-1β, IL-8, TNF-α, GM-CSF, and RANTES (125). Moreover, IL-8 protein can be identified in supernates and cell lysates from 4-day cultures of nasal epithelium from nonallergic patients (126). In patients with asthma, IL-8 and GM-CSF were identified by both mRNA and protein (2) and RANTES by mRNA in bronchial epithelial cells (8).

The constitutive expression of chemotactic cytokines by epithelial cells can be further enhanced by inflammatory cytokines. In addition to their promotion of endothelial and epithelial adhesion, TNF-α and IL-1β amplify the production of IL-8 and GM-CSF by bronchial epithelium (1). IL-8 production by nasal epithelium increased within 6–20 hours of stimulation with respiratory syncytial virus or inclusion of TNF-α or IL-1β in the culture medium (3). Similarly, stimulation of epithelial cell lines, A549 and BEAS-2B, with TNF-α and/or IFN-γ resulted in increased mRNA and protein release for RANTES (6,7) and IL-8 (127). Epithelial generation of all three of these cytokines is inhibited by corticosteroids (6,128,129).

RANTES, IL-8, and GM-CSF all appear to be directly involved in the chemotaxis of eosinophils. RANTES is a member of the C-C family of chemokines and is a potent eosinophil (6,9) as well as T lymphocyte and monocyte (128,130) chemoattractant. RANTES does not affect neutrophils and, unlike most other chemokines, is sufficient to induce CD18-dependent eosinophil transmigration across endothelial cell monolayers (92). Preincubation of eosinophils with IL-5 synergistically augmented RANTES chemotaxis (92), whereas the addition of either anti–IL-5 or anti-RANTES inhibited the chemotactic activity of BAL fluids from atopic patients (131). It has been postulated that IL-5 and RANTES act in conjunction: IL-5 to stimulate chemokinetics and RANTES to stimulate chemotaxis (131).

In addition to chemotaxis, RANTES augments eosinophil production of superoxide anion ($O_2^-$) (132,133), release of ECP (134), and adhesion to ICAM-1 (135). Kapp and associates (133) report that RANTES selectively stimulates the eosinophil respiratory burst to occur more quickly and to a greater degree than either GM-CSF or IL-5. However, the same study found no effect of RANTES on eosinophil degranulation. RANTES effects on CD11b/CD18 expression are not as clear. Alam et al. (136) reported that RANTES enhanced CD11b/CD18 expression, but Kakazu and coworkers (135) found no such effect. CD11b/CD18 integrins are crucial to many eosinophil functions such as $O_2^-$ generation and degranulation (137,138,139); therefore, alterations in the expression of this adhesion marker would greatly alter not only cell adhesion but the cell's function and hence, participation in inflammation.

IL-8 is a member of the C-X-C family of chemokines, which consists of more than 15 structurally related 8- to 10-kD proteins that have secretagogue and chemotactic activities (93). Although initially identified as a neutrophil chemoattractant protein (140,141), IL-8 also has chemoattractant effects on eosinophils (142,143). In a guinea pig model, intradermal injection of IL-8 induced a dose-dependent accumulation of eosinophils over 4 hours (144). In human studies, IL-8 induced eosinophil migration through endothelial and epithelial cell monolayers (143). Interestingly, only eosinophils from atopic patients were responsive to IL-8 chemotaxis (142,143). Therefore, the patient source may determine the observed eosinophil response to IL-8. Sehmi and coworkers (145) report that only eosinophils from eosinophilic patients, and not normal control subjects, were responsive to IL-8 in a chemokinetic mechanism. This differentiation between eosinophils from different sources may reflect in vivo IL-5 priming in eosinophilic patients. Thus, consistent with a report by Schweizer et al. (146), eosinophils from normal, non-atopic volunteers migrated only to RANTES. However, following priming with 10 pM IL-5, the same cells became chemotactically responsive to IL-8 and enhanced their response to RANTES.

Although RANTES and IL-8 are chemotactic for eosinophils, neither cytokine enhances eosinophil adhesion to endothelial cells (92). Expression of GM-CSF in bronchial biopsy specimens from asthma patients was found to correlate with the histamine provocation dose (a marker of bronchial reactivity) and eosinophil presence in the epithelium (129). Although GM-CSF itself is chemotactic for eosinophils only at nanomolar doses (147), at picamolar doses it augments the cells' response to both IL-8 (148) and platelet-activating factor (PAF) (149). GM-CSF is potent in enhancing cell function (150–152) and prolonging survival of eosinophils (153,154). GM-CSF also induces eosinophil expression of CDlla/CD18 and CD11b/CD18 integrins necessary for the cells' adhesion to ICAM-1 on endothelial cells (155,156). This may, in part, be the mechanism of GM-CSF enhancement of eosinophil transmigration through the endothelium (92) and hence, recruitment into the matrix and epithelium. CD11b/CD18 integrin expression is also necessary for eosinophil functions such as degranulation and the respiratory burst (138,139,157).

Eotaxin is a relatively new C-C chemokine identified in guinea pig lungs following an aerosol allergen challenge (62,63). When injected into the lungs or skin, eotaxin induces the selective recruitment of eosinophils (62,63). High levels of constitutive eotaxin mRNA have been identified in guinea pig lungs and these are increased in response to allergen during the late-phase response (5). Human eotaxin has been identified and cloned; it has the same receptor as RANTES and macrophage chemotactic protein-3 (61). Finally, human eotaxin has recently been identified as an early response gene in cytokine-stimulated epithelial and endothelial cells (158). Stimulation of the epithelial cell line, BEAS-2B, with IL-1$\beta$ or TNF-$\alpha$ yields a marked increase in eotaxin mRNA.

## IV. EOSINOPHIL-INDUCED EPITHELIAL CELL DAMAGE

Asthma has been studied with both animal models and human subjects through bronchoscopy and bronchoalveolar lavage. Increased numbers of eosinophils in the airways and epithelial mucosa have been repeatedly associated with bronchial hyperresponsiveness (159) and epithelial damage (160,161). Infiltration of eosinophils into the bronchial mucosa correlate with the opening of epithelial tight junctions and a widening of intercellular spaces; these factors, in turn, correlate with changes in airway responsiveness (162). Instillation of MBP or EPO into the tracheas of *Ascaris suum*–sensitized cynomolgus monkeys results in a significant increase in airway resistance as measured by increased methacholine sensitivity (163). Instillation of eosinophil-derived neurotoxin (EDN) or ECP had no such effect on bronchial hyperreactivity. In patients with symptomatic asthma, bronchoalveolar lavage fluid levels of eosinophils and MBP are elevated and correlate with bronchial hyperreactivity (164).

There are at least three postulated ways epithelium, either altered or shed due to eosinophil interaction, can affect airway hyperresponsiveness (165): 1. enhanced permeability, 2. removal of metabolic barriers, and 3. loss of an inhibitory factor or production of a constriction factor. Similarly, there appear to be multiple mechanisms for eosinophil-dependent epithelial cell damage. It is unlikely that a single mechanism is solely responsible. Eosinophils recovered from atopic patients by BAL following segmental bronchoprovocation with allergen demonstrate increased $O_2^-$ production, adherence to endothelial cells, CD11b/CD18 expression, HLA-DR expression, and intracellular calcium flux (110). Activation of these primed eosinophils in proximity to the airway epithelium results in the release of multiple, highly reactive inflammatory mediators capable of damaging surrounding tissue. These mediators include cationic granule proteins, reactive oxygen metabolites, and proteinases (166).

### A. Cationic Granule Proteins

As noted previously, eosinophils release numerous toxic products, including four highly cationic granule proteins: MBP, EPO, EDN, and ECP (167). These granule proteins alter epithelial cell function in a variety of ways (168). MBP is the most extensively studied of the eosinophil granule proteins for its effect on airway epithelium. In patients with symptomatic asthma, elevated MBP concentrations positively correlate to airway hyperresponsiveness (159); extracellular ECP and EPO produce epithelium shedding (169). In a rat model, instillation of MBP increases airway responsiveness to inhaled methacholine (165). It appears that MBP can affect epithelium in two ways: 1. by damaging airway epithelium (170,171) or 2. by altering epithelial cell function (172,173).

Long-term exposure to high levels of MBP may be necessary to have a cytotoxic effect on epithelial cells by causing desquamation and ciliostasis (169,174–176). MBP lyses pneumocytes, whereas EPO can cause pneumocyte detachment (177). Although MBP can be cytotoxic to epithelium, it does not appear that epithelial cell damage is necessary for increased airway responsiveness (178,179). Although MBP appears to be nontoxic for short-term incubations (e.g., 1 hour), it can alter epithelial cell function under these conditions. MBP inhibition of epithelial cell function included inhibition of ATPase activity (170), increased prostaglandin secretion (180), up-regulated ICAM-1 expression (181), and altered chloride channel function to increase chloride and water secretion into the airways (182). In addition, MBP alters airway smooth muscle contractility indirectly through epithelium and its ability to stimulate production of a contracting factor or to inhibit production of a relaxing factor (153). Either way could be the mechanism of MBP-epithelium dependent contraction of smooth muscle and increased smooth muscle sensitivity to acetylcholine (173,183).

Hulsmann and coworkers (184) determined that the MBP analogue poly-L-arginine could also increase transepithelial fluxes of tracer molecules and focal loss of apical epithelial cells when applied to the luminal aspect of peripheral airways from nonasthmatic subjects. Because other highly cationic proteins were capable of the same effect, these aspects of MBP action may be due to its high cationic charge rather than its enzymatic activity. Consistent with the hypothesis that high cationic density may mediate MBP effects on respiratory epithelium, investigators have demonstrated that negatively charged polyanions such as poly-L-glutamine (185) and heparin are protective (186–188). Hence, MBP may be a major factor in the epithelial cell damage characteristic of asthma, not due to a specific biological mechanism, but rather due to its release by the eosinophils in close proximity to the epithelium.

## B.  Reactive Oxygen Metabolites

Oxygen metabolism has long been accepted as an important mechanism of granulocyte toxicity (189). Like the neutrophil, eosinophils have the potential to undergo a respiratory burst with the resultant release of multiple highly reactive oxygen metabolites (190,191). Although the eosinophil respiratory burst is similar to the neutrophil's in many ways, it is often greater in magnitude, sensitive to different agonists, and yields more reactive physiological metabolites (Fig. 3) (192–197). Blood eosinophils from asthma patients have a stimulus-dependent enhancement of oxygen free radical generation compared with eosinophils from healthy control subjects (198,199). This trend is continued in the airways after allergen challenge. Eosinophils isolated from BAL fluid 48 hours after segmental allergen challenge generated more $O_2^-$ following FMLP activation than the corre-

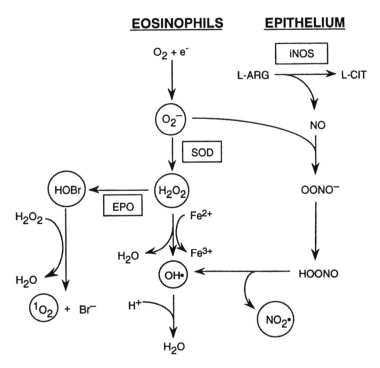

**EOSINOPHILS**          **EPITHELIUM**

**Figure 3** Generation of oxygen metabolites within the airway. The eosinophil respiratory burst is a source of reactive oxygen free radicals. Airway epithelium expresses inducible nitric oxide synthase (iNOS), which catalyzes the five-electron oxidation of L-arginine to L-citrulline and NO. In inflamed tissue, NO quickly reacts with superoxide anion to form the toxic peroxynitrite (OONO⁻), which promotes lipid and sulfhydryl oxidation. OONO⁻ is rapidly decomposed to the free radicals HO• and NO₂•, which are potent biological oxidants.

sponding blood eosinophils (110). In an allergic sheep model, Liberman and associates (200) found similarly increased levels of $O_2^-$ localized in and around the airway epithelium after allergen challenge.

The eosinophil respiratory burst may be one source of reactive oxygen free radicals; nitric oxide (NO) production by epithelial cells may be another. Bronchial epithelial cells from normal, nonallergic subjects express inducible nitric oxide synthase (iNOS) mRNA and produce protein at nanomolar levels (201). However, removal of epithelial cells from the in vivo airways or treatment with inhaled corticosteroids or β-adrenergic agonists results in the loss of iNOS (201–203). This suggests that factors present in vivo are necessary for continued iNOS expression. Cytokines that have been found to augment iNOS expression include TNF-α and IL-1β (203,204), which have been discussed above as produced by epithelial cells and important to eosinophil recruitment. Both constitutive nitric oxidase synthase (cNOS) and iNOS have been identified in human alveolar and bronchial epithelial cells cultured with TNF-$\alpha$, IFN-$\gamma$, or IL-1$\beta$ (203,205). In contrast, other cytokines, such as TGF-$\beta$, IL-4, and IL-10, inhibit the expression of iNOS (206).

iNOS catalyzes the five-electron oxidation of L-arginine to L-citrulline and NO (Fig. 3) (207–211). Nitric oxide can regulate lung function by acting as a cytotoxin, and a mediator of smooth muscle tone, edema, leukocyte accumulation, and bronchodilation and vasodilation to regulate gas exchange (206,212,213). Bronchial biopsies from patients with asthma express more iNOS localized to the airway epithelium than control biopsies (214). Moreover, higher levels of NO are expired by patients with various pulmonary diseases, including asthma and bronchiectasis (206,215–218). In inflamed tissue, NO quickly reacts with $O_2^-$ to form the toxic peroxynitrite (ONOO⁻), which then promotes lipid and sulfhydryl oxidation (206,218,219). ONOO⁻ is rapidly decomposed to the free radicals HO⁻ and $NO_2^-$ which are also very potent biological oxidants. Thus, up-regulated $O_2^-$ generation by infiltrating eosinophils combined with enhanced NO production by cytokine-stimulated epithelial cells may result in the formation of a highly reactive oxidating system leading to tissue damage characteristic of asthma.

## C.  Proteinases

In vitro, stimulated eosinophils augment permeability of bronchial mucosa (151,220). In studies on albumin flux across the epithelium of bovine bronchial mucosa in vitro, the addition of activated eosinophils to either side of the tissue resulted in a rapid (60 min) increase in albumin permeability and extensive disruption of the epithelium, including loss of columnar epithelium from the basal cell layer or underlaying matrix (220). The eosinophil effects in this in vitro model were independent of leukotriene activity, required eosinophil–epithelial cell contact, and were not reproduced by incubation with physiological levels of the MBP cationic analogue poly-L-arginine (221). The results were, however, similar to those resulting from the application of bacterial collagenase to the epithelium, suggesting that metalloproteinases are involved. Bronchial mucosa release gelatinase activity from both apical and basolateral sides of the tissue, whereas eosinophils are much less active (222). However, when eosinophils, stimulated or unstimulated, were added to the basolateral side of the mucosa for 60 min, there was an increased gelatinolytic activity in the conditioned medium. Because it was not necessary for the eosinophils to be activated, this function is distinct from degranulation and mediator release. The increased gelatin-degrading activity did not result in obvious tissue damage or increased permeability. Although gelatinases did not appear to be solely responsible for eosinophil-mediated epithelial injury, they may be important in destabilizing the structural integrity of the airway mucosa and enhancing the effects of other cytotoxic mediators.

## V. EOSINOPHIL CYTOKINES

Recent evidence suggests that eosinophils produce a multitude of pro-inflammatory cytokines. Initial studies have demonstrated that human eosinophils express and release TNF-$\alpha$ (223) and IL-1$\alpha$ (224), which have the capacity to induce adhesion molecule expression and cytokine production by endothelial and epithelial cells. Eosinophils may also synthesize IL-3 (225), GM-CSF (225,226), and IL-5 (227,228). Although the T lymphocyte is the major source of these cytokines, eosinophils may contribute to overall cytokine production in airway inflammation. The potential of eosinophils to produce and respond to these cytokines suggests an autocrine mechanism, whereby eosinophils maintain their own viability and effector function following recruitment and, thus, perpetuate chronic airway inflammation.

## VI. CONCLUSION

Although a great deal of information is known about leukocyte interactions with the vascular endothelium, much less information is known about leukocytes and, in particular, the interactions of eosinophils with the respiratory epithelium (Fig. 1). Available information indicates that there are a number of differences in the adhesive ligands expressed and utilized by the migrating eosinophil in the process of reaching the airway lumen. During the process of migration into the airway lumen, the circulating eosinophil is exposed to a number of cytokines and adhesion molecules produced by the epithelial cells, and becomes altered both phenotypically and physiologically into the activated airway eosinophil. Once in the airway, the eosinophil can further perpetuate the inflammatory state and airway damage, through the release of cationic granule protein, reactive oxygen radicals, and cytokine production, and thus perpetuate a vicious circle of unending airway inflammation. Understanding the mechanisms involved in recruiting and activating leukocytes, particularly eosinophils, in the lung will be important in designing specific therapies for inflammatory diseases such as asthma.

## REFERENCES

1. O. Cromwell, Q. Hamid, C. J. Corrigan, J. Barkans, Q. Meng, and P. D. Collins, *Immunology*, 77:330 (1992).
2. M. Marini, E. Vittori, J. Hollemborg, and S. Mattoli, *J. Allergy Clin. Immunol.*, 89:1001 (1992).
3. S. Becker, H. S. Koren, and D. C. Henke, *Am. J. Respir. Cell Mol. Biol.*, 8:20 (1993).

4. P. D. Collins, S. Marleau, D. A. Griffiths-Johnson, P. J. Jose, and T. J. Williams, *J. Exp. Med.*, *182*:1169 (1995).
5. M. E. Rothenberg, A. D. Luster, C. M. Lilly, J. M. Drazen, and P. Leder, *J. Exp. Med.*, *181*:1211 (1995).
6. O. J. Kwon, P. J. Jose, R. A. Robbins, T. J. Schall, T. J. Williams, and P. J. Barnes, *Am. J. Respir. Cell Mol. Biol.*, *12*:488 (1995).
7. C. Stellato, L. A. Beck, G. A. Gorgone, D. Proud, T. J. Schall, S. J. Ono, L. M. Lichtenstein, and R. P. Schleimer, *J. Immunol.*, *155*:410 (1995).
8. J. L. Devalia, J. H. Wang, R. J. Sapsford, and R. J. Davies, *Eur. Respir. J.*, *7*:98S (1994).
9. J. H. Wang, J. L. Devalia, C. Xia, R. J. Sapsford, and R. J. Davies, *Am. J. Respir. Cell Mol. Biol.*, *14*:27 (1996).
10. S. Koyama, S. I. Rennard, S. Shoji, D. Romberger, J. Linder, R. Ertl, and R. A. Robbins, *Am. J. Physiol.*, *257*:L130 (1989).
11. R. A. Robbins, S. Shoji, J. Linder, G. L. Grossman, L. A. Allington, L. W. Klasen, and S. I. Rennard, *Am. J. Physiol.*, *257*:L109 (1989).
12. A. M. Vignola, P. Chanez, A. M. Campbell, J. Bousquet, F. B. Michel, and P. Godard, *Allergy*, *48*:32 (1993).
13. M. F. Tosi, J. M. Stark, C. W. Smith, A. Hamedani, D. C. Gruenert, and M. D. Infeld, *Am. J. Respir. Cell Mol. Biol.*, *7*:214 (1992).
14. D. C. Look, S. R. Rapp, B. T. Keller, and M. J. Holtzman, *Am. J. Physiol.*, *263*:L79 (1992).
15. P. G. M. Bloemen, M. C. van den Tweel, P. A. J. Henricks, F. Engels, S. S. Wagenaar, A. A. J. J. L. Rutten, and F. P. Nijkamp, *Am. J. Respir. Cell Mol. Biol.*, *9*:586 (1993).
16. C. D. Wegner, R. H. Gundel, P. Reilly, N. Haynes, L. G. Letts, and R. Rothlein, *Science*, *247*:456 (1990).
17. P. M. Taylor, M. L. Rose, and M. H. Yacoub, *Transplantation*, *48*:506 (1989).
18. F. M. Cuss and P. J. Barnes, *Am. Rev. Respir. Dis.*, *136*:S32 (1987).
19. A. M. Dvorak, I. Hammel, E. S. Schulman, S. P. Peters, D. W. MacGlashan, Jr., R. P. Schleimer, H. H. Newball, K. Pyne, H. F. Dvorak, L. M. Lichtenstein, and S. J. Galli, *J. Cell Biol.*, *99*:1678 (1984).
20. E. S. Schulman, *Crit. Rev. Immunol.*, *13*:35 (1993).
20a. Y. Ohkawara, K. Yamauchi, Y. Tanno, G. Tamura, H. Ohtani, H. Nagura, K. Ohkuda, and T. Takishima, *Am. J. Respir. Cell Mol. Biol.*, *7*:385 (1992).
21. N. N. Jarjour, W. J. Calhoun, L. B. Schwartz, and W. W. Busse, *Am. Rev. Respir. Dis.*, *144*:83 (1991).
22. H. Takizawa, T. Ohtoshi, T. Kikutani, H. Okazaki, N. Akiyama, M. Sato, S. Shoji, and K. Ito, *Int. Arch. Allergy Immunol.*, *108*:260 (1995).
23. T. L. Noah, A. M. Paradiso, M. C. Madden, K. P. McKinnon, and R. B. Devlin, *Am. J. Respir. Cell Mol. Biol.*, *5*:484 (1991).
24. L. Churchill, B. Friedman, R. P. Schleimer, and D. Proud, *Immunology*, *75*:189 (1992).
25. A. M. Vignola, A. M. Campbell, P. Chanez, P. Lacoste, F. B. Michel, P. Godard, and J. Bousquet, *Am. J. Respir. Cell Mol. Biol.*, *9*:411 (1993).
26. A. M. Vignola, A. M. Campbell, P. Chanez, J. Bousquet, P. Paul-Lacoste, F. B. Michel, and P. Godard, *Am. Rev. Respir. Dis.*, *148*:689 (1993).

27. S. C. Alter, D. D. Metcalfe, T. R. Bradford, and L. B. Schwartz, *Biochem. J.*, *248*:821 (1987).

28. D. H. Broide, G. J. Gleich, A. J. Cuomo, D. A. Coburn, E. C. Federman, L. B Schwartz, and S. I. Wasserman, *J. Allergy Clin. Immunol.*, *88*:637 (1991).

29. J. A. Cairns and A. F. Walls, *J. Immunol.*, *156*:275 (1996).

30. M. Endo, Y. Uchida, H. Marsumoto, N. Suzuki, A. Nomura, F. Hirata, and S. Hasegawa, *Biochem. Biophys. Res. Commun.*, *186*:1594 (1992).

31. V. Ackerman, S. Carpi, A. Bellini, G. Vassalli, M. Marini, and S. Mattoli, *J. Allergy Clin. Immunol.*, *96*:618 (1995).

32. J. Nakano, H. Takizawa, T. Ohtoshi, S. Shoji, M. Yamaguchi, A. Ishii, M. Yanagi-sawa, and K. Ito, *Clin. Exp. Allergy*, *24*:330 (1994).

33. R. C. Benyon, E. Y. Bissonette, and A. D. Befus, *J. Immunol.*, *147:* 2253 (1991).

34. A. Slungaard, G. M. Vercellotte, G. Walker, R. D. Nelson, and H. E. Jacob, *J. Exp. Med.*, *171*:2025 (1990).

35. D. A. Silberstein and J. R. Davis, *Proc. Natl. Acad. Sci. USA*, *83*:1055 (1986).

36. M. P. Bevilacqua, S. Stengelin, M. A. Gimbrone, Jr., and B. Seed, *Science*, *243*:1160 (1989).

37. J. S. Pober, M. A. Gimbrone Jr., L. A. Lapierre, D. L. Mendrick, W. Fiers, R. Roth-lein, and T. A. Springer, *J. Immunol.*, *137*:1893 (1986).

38. L. Osborn, C. Hession, R. Tizard, C. Vassallo, S. Luhowskyj, G. Chi-Rosso, and R. Lobb, *Cell*, *59*:1203 (1989).

39. The BAL Cooperative Study Group., *Am. Rev. Respir. Dis.*, *141*:S169 (1990).

40. P. Chanez, J. Bousquet, and I. Couret, *Am. Rev. Respir. Dis.*, *144*:923 (1991).

41. J. R. Paunska, F. Midulla, and N. M. Cirino, *Am. J. Physiol.*, *259*:L396 (1990).

42. W. J. Calhoun, K. Murphy, N. N. Jarjour, E. C. Dick, C. A. Stevens, and W. W. Busse, *Am. Rev. Respir. Dis.*, *145*:A35 (1992).

43. J. R. Panuska, M. I. Hertz, H. Taraf, A. Villani, and N. M. Cirino, *Am. Rev. Respir. Dis.*, *145*:934 (1992).

44. P. M. Henson and R. B. Johnston, Jr., *J. Clin. Invest.*, *79*:669 (1987).

45. C. Mazingue, V. Carriere, J. Dessaint, F. Detoueuf, T. Turz, C. Auriault, and A. Capron, *Clin. Exp. Immunol.*, *67*:587 (1987).

46. L. Borish, J. J. Mascali, and L. J. Rosenwasser, *J. Immunol.*, *146*:63 (1991).

47. L. Borish, J. J Mascali, J. Dishuck, W. R. Beam, R. J. Martin, and L. J. Rosen-wasser, *J. Immunol.*, *149*:3078 (1992).

48. J. L. Pujol, B. Cosso, J. P. Daures, J. Clot, F. B. Michel, and P. Goddard, *Int. Arch. Allergy Appl. Immunol.*, *91*:207 (1990).

49. K. Schollmeier, *Am. J. Respir. Cell Mol. Biol.*, *3*: 11. (1990).

49a. J. Vilcek and T. H. Lee, *J. Biol. Chem.*, *266*: 7313. (1991).

50. S. Mattoli, V. L. Mattoso, M. Soloperto, L. Allegra, and A. Fasoli, *J. Allergy Clin. Immunol.*, *87*:794 (1991).

51. M. C. Liu, D. Proud, L. M. Lichtenstein, D. W. MacGlashan, Jr., R. P. Schleimer, N. F. Adkinson, Jr., A. Kagey-Sobotka, E. S. Schulman, and M. Plaut, *J. Immunol.*, *136*:2588 (1986).

52. M. Marini, M. Soloperto, M. Mezzetti, A. Fasoli, and S. Mattoli, *Am. J. Respir. Cell Mol. Biol.*, *4*:519 (1991).

53. J. A. Elias, T. Zheng, O. Einarsson, M. Landry, T. Trow, N. Rebert, and J. Panuska, *J. Biol. Chem.*, *35*:22261 (1994).

54. M. L. Dustin, R. Rothlein, A. K. Bhan, C. A. Dinarello, and T. A. Springer, *J. Immunol.*, *137*:245 (1986).
55. J. Tamaoki, I. Yamawaki, K. Takeyama, A. Chiyotani, F. Yamauchi, and K. Konno, *Am. J. Respir. Crit. Care Med.*, *149*:134 (1994).
56. L. Borish and B. Z. Joseph, *Med. Clin. North Am.*, *76*:765 (1992).
57. A. J. Frew and A. B. Kay, *J. Allergy Clin. Immunol.*, *85*:533 (1990).
58. M. K. Schroth, P. A. Frindt, and J. E. Gern, *Am. J. Respir. Cell Mol. Biol.*, *153*:A728 (1996).
59. A. M. Campbell, A. M. Vignola, P. Chanez, P. Godard, and J. Bousquet, *Immunology*, *82*:506 (1994).
60. P. J. Jose, I. M. Adcock, D. A. Griffiths-Johnson, N. Berkman, T. N. C. Wells, T. J. Williams, and C. A. Power, *Biochem. Biophys. Res. Commun.*, *205*:788 (1994).
61. P. D. Ponath, S. Qin, D. J. Ringler, I. Clark-Lewis, J. Wang, N. Kassam, H. Smith, X. Shi, J. A. Gonzalo, W. Newman, J. C. Gutierrez-Ramos, and C. R. Mackay, *J. Clin. Invest.*, *97*:604 (1996).
62. P. J. Jose, D. A. Griffiths-Johnson, P. D. Collins, D. T. Walsh, R. Moqbel, N. F. Totty, O. Truong, J. J. Hsuan, and T. J. Williams, *J. Exp. Med.*, *179*:881 (1994).
63. D. A. Griffiths-Johnson, P. D. Collins, A. G. Rossi, P. J. Jose, and T. J. Williams, *Biochem. Biophys. Res. Commun.*, *197*:1167 (1993).
64. T. L. Noah and S. Becker, *Am. J. Physiol.*, *265*:L472 (1993).
65. R. Arnold, B. Humbert, H. Werchau, H. Gallati, and W. König, *Immunology*, *82*:126 (1994).
66. M. A. Fiedler, K. Wernke-Dollries, and J. M. Stark, *Am. J. Physiol.*, *269*:L865 (1995).
67. R. Arnold, H. Werchau, and W. König, *Int. Arch. Allergy Immunol.*, *107*:392 (1995).
68. S. Becker, J. Soukup, and J. R. Yankaskas, *Am. J. Respir. Cell Mol. Biol.*, *6*:369 (1992).
69. M. F. Tosi, J. M. Stark, A. Hamedani, C. W. Smith, D. C. Gruenert, and Y. T. Huang, *J. Immunol.*, *149*:3345 (1992).
70. D. E. Staunton, V. J. Merluzzi, R. Rothlein, R. Barton, S. D. Marlin, and T. A. Springer, *Cell*, *56*:849 (1989).
71. D. E. Staunton, M. L. Dustin, H. P. Erickson, and T. A. Springer, *Cell*, *61*:243 (1990).
72. M. C. Subauste, D. B. Jacoby, S. M. Richards, and D. Proud, *J. Clin. Invest.*, *96*:549 (1995).
73. R. B. Devlin, K. P. McKinnon, T. Noah, S. Becker, and H. S. Koren, *Am. J. Physiol.*, *266*:L612 (1994).
74. M. F. Tosi, A. Hamedani, J. Brosovich, and S. E. Alpert, *J. Immunol.*, *152*:1935 (1994).
75. J. L. Devalia, A. M. Campbell, and R. J. Sapsford, *Am. J. Respir. Cell Mol. Biol.*, *9*:271 (1993).
76. P. Chitano, J. J. Hosselet, C. E. Mapp, and L. M. Fabbri, *Eur. Respir. J.*, *8*:1357 (1995).
77. S. M. Albelda, *Am. J. Respir. Cell Mol. Biol.*, *4*:195 (1991).
78. A. R. Leff, K. J. Hamann, and C. D. Wegner, *Am. J. Physiol.*, *260*:L189 (1991).
79. J. A. Elias and R. J. Zitnik, *Am. J. Respir. Cell Mol. Biol.*, *7*:365 (1992).

80. C. H. Smith, J. N. W. N. Barker, and T. H. Lee, *Am. Rev. Respir. Dis.*, *148*:S75 (1993).
81. J. M. Dayer, P. Isler, and L. P. Nicod, *Am. Rev. Respir. Dis.*, *148*:S70 (1993).
82. G. A. Zimmerman, D. E. Lorant, T. M. McIntyre, and S. M. Prescott, *Am. J. Respir. Cell Mol. Biol.*, *9*:573 (1993).
83. T. A. Springer, *Cell*, *76*:301 (1994).
84. V. Godding, J. M. Stark, J. B. Sedgwick, and W. W. Busse, *Am. J. Respir. Cell Mol. Biol.*, *13*:555 (1995).
85. J. M. Stark, V. Godding, J. B. Sedgwick, and W. W. Busse, *J. Immunol.*, *156*:4774 (1996).
86. A. M. Bentley, S. R. Durham, D. S. Robinson, G. Menz, C. Storz, O. Cromwell, A. B. Kay, and A. J. Wardlaw, *J. Allergy Clin. Immunol.*, *92*:857 (1993).
87. Y. Ohkawara, K. Yamauchi, N. Maruyama, H. Hoshi, I. Ohno, M. Honma, Y. Tanno, G. Tamura, K. Shirato, and H. Ohtani, *Am. J. Respir. Cell Mol. Biol.*, *12*:4 (1995).
88. D. E. Staunton, S. D. Marlin, C. Stratowa, M. L. Dustin, and T. A. Springer, *Cell*, *52*:925 (1988).
89. D. Simmons, M. W. Makgoba, and B. Seed, *Nature*, *331*:624 (1988).
90. U. Kyan-Aung, D. O. Haskard, R. N. Poston, M. H. Thornhill, and T. H. Lee, *J. Immunol.*, *146*:521 (1991).
91. W. S. Simonet, T. M. Hughes, H. Q. Nguyen, L. D. Trebasky, D. M. Danilenko, and E. S. Medlock, *J. Clin. Invest.*, *94*:1310 (1994).
92. M. Ebisawa, B. S. Bochner, S. N. Georas, and R. P. Schleimer, *J. Immunol.*, *149*:4021 (1992).
93. B. S. Bochner and R. P. Schleimer, *J. Allergy Clin. Immunol.*, *94*:427 (1994).
94. S. Nakajima, D. C. Look, W. T. Roswit, M. J. Bragdon, and M. J. Holtzman, *Am. J. Physiol.*, *267*:L422 (1994).
95. B. S. Bochner, S. A. Sterbinsky, E. F. Knol, B. J. Katz, L. M. Lichtenstein, D. W. MacGlashan, Jr., and R. P. Schleimer, *J. Allergy Clin. Immunol.*, *94*:1157 (1994).
96. T. Kishimoto, K. Larson, A. L. Corbi, M. L. Dustin, D. E. Staunton, and T. A. Springer, *Adv. Immunol.*, *114*:149 (1990).
97. M. L. Dustin and T. A. Springer, *J. Cell Biol.*, *107*:321 (1988).
98. S. D. Marlin and T. A. Springer, *Cell*, *51*:813 (1987).
99. C. W. Smith, S. D. Marlin, R. Rothlein, C. Toman, and D. C. Anderson, *J. Clin. Invest.*, *83*:2008 (1989).
100. M. S. Diamond, D. E. Staunton, A. R. de Fougerolles, S. A. Stacker, J. Garcia-Aguilar, M. L. Hibbs, and T. A. Springer, *J. Cell Biol.*, *111*:3129 (1990).
101. M. S. Diamond, D. E. Staunton, S. D. Marlin, and T. A. Springer, *Cell*, *65*:961 (1991).
102. M. S. Diamond and T. A. Springer, *J. Cell Biol.*, *120*:545 (1993).
103. J. A. Patel, M. Kunimoto, T. C. Sim, R. Garofalo, T. Elliot, R. Baron, T. Chonmaitree, P. L. Ogra, and F. Schmalstieg, *Am. J. Respir. Cell Mol. Biol.*, *13*:602 (1995).
104. D. C. Look, M. R. Pelletier, and M. J. Holtzman, *J. Biol. Chem.*, *269*:8952 (1994).
105. C. D. Wegner, R. Rothlein, and R. H. Gundel, *Agents Actions*, *34*:529 (1991).
106. F. L. Jahnsen, G. Haraldsen, J. P. Aanesen, R. Haye, and P. Brandtzaeg, *Am. J. Respir. Cell Mol. Biol.*, *12*:624 (1995).

107. J. Atsuta, B. S. Bochner, S. A. Sterbinsky, and R. P. Schleimer, *J. Allergy Clin. Immunol.*, 97:293 (1996).

108. S. P. Neeley, K. J. Hamann, S. R. White, S. L. Baranowski, R. A. Burch, and A. R. Leff, *Am. J. Respir. Cell Mol. Biol.*, 8:633 (1993).

109. J. B. Sedgwick, S. F. Quan, W. J. Calhoun, and W. W. Busse, *J. Allergy Clin. Immunol.*, 96:375 (1995).

110. J. B. Sedgwick, W. J. Calhoun, R. F. Vrtis, M. E. Bates, P. K. McAllister, and W. W. Busse, *J. Immunol.*, 149:3710 (1992).

111. C. Kroegel, M. C. Liu, W. C. Hubbard, L. M. Lichtenstein, and B. S. Bochner, *J. Allergy Clin. Immunol.*, 93:725 (1994).

112. S. N. Georas, M. C. Liu, W. Newman, L. D. Beall, B. A. Stealey, and B. S. Bochner, *Am. J. Respir. Cell Mol. Biol.*, 7:261 (1992).

113. A. M. Lamas, C. M. Mulroney, and R. P. Schleimer, *J. Immunol.*, 140:1500 (1988).

114. B. S Bochner, F. W. Luscinskas, M. A. Gimbrone, Jr., W. Newman, S. A. Sterbinsky, C. P. Derse-Anthony, D. Klunk, and R. P. Schleimer, *J. Exp. Med.*, 173:1553 (1991).

115. A. Dobrina, R. Menegazzi, T. M. Carlos, E. Nardon, R. Cramer, T. Zacchi, J. M. Harlan, and P. Patriarca, *J. Clin. Invest.*, 88:20 (1991).

116. G. M. Walsh, J. J. Mermod, A. Hartnell, A. B. Kay, and A. J. Wardlaw, *J. Immunol.*, 146:3419 (1991).

117. P. F. Weller, T. H. Rand, S. E. Goelz, G. Chi-Rosso, and R. R Lobb, *Proc. Natl. Acad. Sci. USA*, 88:7430 (1991).

118. M. P. Bevilacqua, J. S. Pober, D. L. Mendrick, R. S. Cotran, and M. A. Gimbrone, Jr, *Proc. Natl. Acad. Sci. USA*, 84:9238 (1987).

119. T. F. Tedder, D. A. Steeber, A. Chen, and P. Engel, *FASEB J.*, 9:866, (1995).

120. M. Marini, E. Avoni, J. Hollemborg, and S. Mattoli, *Chest*, 102:661 (1992).

121. C. D. Wegner, R. Rothlein, C. C. Clarke, N. Haynes, C. A. Torcellini, A. M. LaPlante, D. R. Averill, L. G. Letts, and R. H. Gundel, *Am. Rev. Respir. Dis.*, 143:A418 (1991).

122. P. J. Christensen, S. Kim, R. H. Simon, G. B. Toews, and R. Paine, III, *Am. J. Respir. Cell Mol. Biol.*, 8:9 (1993).

123. A. R. Burns, F. Takei, and C. M. Doerschuk, *J. Immunol.*, 153:3189 (1994).

124. J. A. Gonzalo, G. Q. Jia, V. Aguirre, D. Friend, A. J. Coyle, N. A. Jenkins, G. S. Lin, H. Katz, A. Lichtman, N. Copeland, M. Kopf, and J. C. Gutierrez-Ramos, *Immunity*, 4:1 (1996).

125. A. W. Stadnyk, *FASEB J.*, 8:1041 (1994).

126. J. S. Kenney, C. Baker, M. R. Welch, and L. C. Altman, *J. Allergy Clin. Immunol.*, 93:1060 (1994).

127. H. Nakamura, K. Yoshimura, H. A. Jaffe, and R. C. Crystal, *J. Biol. Chem.*, 266:19611 (1991).

128. T. J. Schall, K. Bacon, K. J. Toy, and D. V. Geoddel, *Nature*, 347:669 (1990).

129. J. H. Wang, C. J. Trigg, J. L. Devalia, S. Jordan, and R. J. Davies, *J. Allergy Clin. Immunol.*, 94:1025 (1994).

130. T. J. Schall, *Cytokine*, 3:165 (1991).

131. J. Venge, M. Lampinen, L. Håkansson, S. Rak, and P. Venge, *J. Allergy Clin. Immunol.*, 97:1110 (1996).

132. J. Chihara, N. Hayashi, T. Kakazu, T. Yamamoto, D. Kurachi, and S. Nakajima, *Int. Arch. Allergy Immunol.*, 104:52 (1994).

133. A. Kapp, G. Zeck-Kapp, W. Czech, and E. Schopf, *J. Invest. Dermatol.*, *102*:906 (1994).
134. A. J. M. Van Oosterhout, A. Van Der Poel, L. Koenderman, D. Roos, and F. P. Nijkamp, *Mediators of Inflammation*, *3*:53 (1994).
135. T. Kakazu, J. Chihara, A. Saito, and S. Nakajima, *Int. Arch. Allergy Appl. Immunol.*, *108*:9 (1995).
136. R. Alam, S. Stafford, P. Forsythe, R. Harrison, D. Faubion, M. A. Lett-Brown, and J. A. Grant, *J. Immunol.*, *150*:3442 (1993).
137. M. M. Kaneko, S. Morie, M. Kato, G. J. Gleich, and H. Kita, *J. Immunol.*, *155*: 2631 (1995).
138. C. Laudanna, P. Melotti, C. Bonizzato, G. Piacenttini, A. Boner, M. C. Serra, and G. Berton, *Immunology*, *80*:273 (1993).
139. S. Horie and H. Kita, *J. Immunol.*, *152*:5457 (1994).
140. C. A. Hébert and J. B. Baker, *Cancer Invest.*, *11*:743 (1993).
141. M. Baggiolini, A. Walz, and S. L. Kunkel, *J. Clin. Invest.*, *84*:1045 (1989).
142. R. A. J. Warringa, H. J. J. Mengelers, J. A. M. Raaijmakers, P. L. B. Bruijnzeel, and L. Koenderman, *J. Allergy Clin. Immunol.*, *91*:1198 (1993).
143. R. A. Erger and T. B. Casale, *Am. J. Physiol.*, *268*:L117 (1995).
144. P. D. Collins, V. B. Weg, L. H. Faccioli, M. L. Watson, R. Moqbel, and T. J. Williams, *Immunology*, *79*:312 (1993).
145. T. Sehmi, O. Cromwell, A. J. Wardlaw, R. Moqbel, and A. B. Kay, *Clin. Exp. Immunol.*, *23*:1027 (1993).
146. R. C. Schweizer, B. A. C. Welmers, J. A. M. Raaijmakers, P. Zanen, J. W. J. Lammers, and L. Koenderman, *Blood*, *83*:3697 (1994).
147. R. A. J. Warringa, L. Koenderman, P. T. Kok, J. Kreukniet, and P. L. B. Bruijnzeel, *Blood*, *77*:2694 (1991).
148. J. L. Devalia, R. J. Sapsford, C. Rusznak, and R. J. Davies, *Am. J. Respir. Cell Mol. Biol.*, *7*:270 (1992).
149. M. B. Resnick, S. P. Colgan, C. A. Parkos, C. Delp-Archer, D. McGuirk, P. F. Weller, and J. L. Madara, *Gastroenterology*, *108*:409 (1995).
150. D. S. Silberstein, W. F. Owen, J. C. Gasson, J. F. DiPersio, D. W. Golde, J. C. Bina, R. Soberman, K. F. Austen, and J. R. David, *J. Immunol.*, *137*:3290 (1986).
151. I. Fabian, Y. Kletter, S. Mor, C. Geller-Bernstein, M. Ben-Yaakov, B. Volovitz, and D. W. Golde, *Br. J. Haematol.*, *80*:137 (1992).
152. C. F. Howell, J. Pujol, A. E. G. Crea, R. Davidson, A. J. H. Gearing, P. Godard, and T. H. Lee, *Am. Rev. Respir. Dis.*, *140*:1340 (1989).
153. G. J. Gleich, *J. Allergy Clin. Immunol.*, *85*:422 (1990).
154. L. A. Burke, M. P. Hallsworth, T. M. Litchfield, R. Davidson, and T. H. Lee, *J. Allergy Clin. Immunol.*, *88*:226 (1991).
155. A. F. Lopez, C. G. Begley, D. J. Williamson, D. J. Warren, M. A. Vadas, and C. J. Sanderson, *J. Exp. Med.*, *163*:1085 (1986).
156. V. Graves, T. Gabig, L. McCarthy, E. F. Strour, T. Leemhuis, and D. English, *Blood*, *80*:776 (1992).
157. A. Egesten, U. Gullberg, I. Olsson, and J. Richter, *J. Leukoc. Biol.*, *53*:287 (1993).
158. E. A. Garcia-Zepeda, M. E. Rothenberg, R. T. Ownbey, J. Celestin, P. Leder, and A. D. Luster, *Nature Med.*, *2*:449 (1996).

159. G. J. Gleich and C. Adolphson, *Agents Actions Suppl.*, *43*:223 (1993).
160. B. R. Horn, E. D. Robin, J. Theodore, and A. Van Kessel, *N. Engl. J. Med.*, *292*:1152 (1975).
161. E. Frigas and G. J. Gleich, *J. Allergy Clin. Immunol.*, *77*:527 (1986).
162. Y. Ohaski, S. Motojima, T. Fukuda, and S. Makino, *Am. Rev. Respir. Dis.*, *145*:1469 (1992).
163. R. H. Gundel, L. G. Letts, and G. J. Gleich, *J. Clin. Invest.*, *87*:1470 (1991).
164. A. J. Wardlaw, S. Dunnette, G. J. Gleich, J. V. Collins, and A. B. Kay, *Am. Rev. Respir. Dis.*, *137*:62 (1988).
165. A. J. Coyle, D. Uchida, S. J. Ackerman, W. Mitzner, and C. G. Irvin, *Am. J. Respir. Crit. Care Med.*, *150*:S63 (1994).
166. P. M. Henson and R. B. J. Johnston, *J. Clin. Invest.*, *79*:669 (1987).
167. G. J. Gleich and C. F. Adolphson, *Adv. Immunol.*, *39*:177 (1995).
168. N. N. Jarjour and W. J. Calhoun, *J. Allergy Clin. Immunol.*, *89*:60 (1992).
169. S. Motojima, E. Frigas, D. A. Loegering, and G. J. Gleich, *Am. Rev. Respir. Dis.*, *139*:801 (1989).
170. A. T. Hastie, D. A. Loegering, G. J. Gleich, and F. Kueppers, *Am. Rev. Respir. Dis.*, *135*:848 (1987).
171. K. Hisamatsu, T. Ganbo, T. Nakazawa, Y. Murakami, G. J. Gleich, K. Makiyama, and H. Koyama, *J. Allergy Clin. Immunol.*, *86*:52 (1990).
172. N. A. Flavahan, N. R. Slifman, G. J. Gleich, and P. M. Vanhoutte, *Am. Rev. Respir. Dis.*, *138*:685 (1988).
173. S. R. White, S. Ohno, N. M. Munoz, G. J. Gleich, C. Abrahams, J. Solway, and A. R. Leff, *Am. J. Physiol.*, *259*:L294 (1990).
174. E. Frigas, D. A. Loegering, and G. J. Gleich, *Lab. Invest.*, *42*:35 (1980).
175. W. V. Filley, K. E. Holley, G. M. Kephart, and G. J. Gleich, *Lancet*, *2*:11 (1982).
176. K. Pazdrak, D. Schreiber, P. Forsythe, L. Justement, and R. Alam, *J. Exp. Med.*, *181*:1827 (1995).
177. G. H. Ayars, L. C. Altman, G. J. Gleich, D. A. Loegering, and E. B. Baker, *J. Allergy Clin. Immunol.*, *76*:595 (1985).
178. A. J. Coyle, S. J. Ackerman, and C. G. Irvin, *Am. Rev. Respir. Dis.*, *147*:896 (1993).
179. A. J. Coyle, W. Mitzner, and C. G. Irvin, *J. Appl. Physiol.*, *74*:1761 (1993).
180. S. R. White, K. S. Sigriss, and S. M. Spaethe, *Am. J. Physiol.*, *265*:L234 (1993).
181. L. Altman, G. Ayars, C. Baker, and D. Luchtel, *J. Allergy Clin. Immunol.*, *92*:527 (1994).
182. D. B. Jacoby, I. F. Ueki, J. J. Widdicombe, D. A. Loegering, G. J. Gleich, and J. A. Nadel, *Am. Rev. Respir. Dis.*, *137*:13 (1988).
183. J. D. Brofman, S. R. White, J. S. Blake, N. M. Munoz, G. J. Gleich, and A. R. Leff, *J. Appl. Physiol.*, *66*:1867 (1989).
184. A. R. Hulsman, H. R. Raatgeep, J. C. Denhollander, W. H. Bakker, P. R. Saxena, and J. C. Dejongste, *Am. J. Respir. Crit. Care Med.*, *153*:841 (1996).
185. R. L. Barker, R. H. Gundel, G. J. Gleich, J. L. Checkel, D. A. Loegering, L. R. Pease, and K. J. Hamann, *J. Clin. Invest.*, *88*:798 (1991).
186. T. Raochat, J. Casale, G. W. Hunninghake, and M. W. Peterson, *Am. J. Physiol.*, *255*:603 (1988).

187. S.-W. Change, J. Y. Westcott, J. E. Henson, and N. F. Voelkel, *J. Appl. Physiol.*, *62*:1932 (1987).
188. M. W. Peterson and D. Gruenhaupt, *J. Appl. Physiol.*, *68*:220 (1990).
189. S. J. Klebanoff, *Ann. Intern. Med.*, *93*:480 (1980).
190. S. H. Pincus, *J. Invest. Dermatol.*, *80*:278 (1983).
191. J. Elsner, M. Oppermann, W. Czech, G. Dobos, E. Schopf, J. Norgauer, and A. Kapp, *Eur. J. Immunol.*, *24*:518 (1994).
192. J. B. Sedgwick, R. F. Vrtis, M. F. Gourley, and W. W. Busse, *J. Allergy Clin. Immunol.*, *81*:876 (1988).
193. D. B. Learn and E. P. Brestel, *Agents Actions*, *12*:485 (1982).
194. M. Yazdanbakhsh, C. M. Eckmann, and D. Roos, *J. Immunol.*, *135*:1378 (1985).
195. C. Kroegel, J. C. Virchow, W. Luttmann, C. Walker, and J. A. Warner, *Eur. Respir. J.*, *7*:519 (1994).
196. Sedgwick,J. B.,*Asthma and Rhinitis* (W. W. Busse and S. T. Holgate, eds.),Blackwell Scientific, Boston, p. 285 (1996).
197. S. J. Weiss, S. T. Test, C. M. Eckmann, D. Roos, and S. Regaini, *Science*, *234*:200 (1986).
198. P. Chanez, G. Dent, T. Yukawa, P. J. Barnes, and K. F. Chung, *Eur. Respir. J.*, *3*:1002 (1990).
199. J. B. Sedgwick, K. M. Geiger, and W. W. Busse, *Am. Rev. Respir. Dis.*, *142*:120 (1996).
200. H. Liberman, A. T. Mariasy, D. Sorace, S. Suster, and W. M. Abraham, *Lab. Invest.*, *72*:348 (1995).
201. F. H. Guo, H. R. De Raeve, T. W. Rice, D. J. Stuehr, F. B. Thunnissen, and S. C. Erzurum, *Proc. Natl. Acad. Sci. U.S.A.* *92*:7809 (1995).
202. R. A. Robbins, D. R. Springall, J. B. Warren, O. J. Kwon, L. D. K. Buttery, A. J. Wilson, I. M. Adcock, V. Riveros-Moreno, S. Moncada, J. M. Polak, and P. J. Barnes, *Biochem. Biophys. Res. Commun.*, *198*:835 (1994).
203. R. A. Robbins, P. J. Barnes, D. R. Springall, J. B. Warren, O. J. Kwon, L. D. K. Buttery, A. J. Wilson, D. A. Geller, and J. M. Polak, *Biochem. Biophys. Res. Commun.*, *203*:209 (1994).
204. B. Jain, I. Rubinstein, R. A. Robbins, and J. H. Sisson, *Am. J. Physiol.*, *268*:L911 (1995).
205. K. Asano, C. B. E. Chee, B. Gaston, C. M. Lilly, C. Gerard, J. M. Drazen, and J. S. Stamler, *Proc. Natl. Acad. Sci. USA*, *91*:10089 (1994).
206. F. P. Nijkamp and G. Folkerts, *Arch. Int. Pharmacodyn. Ther.*, *329*:81 (1995).
207. M. A. Tayeh and M. A. Marlette, *J. Biol. Chem.*, *264*:19654 (1989).
208. D. J. Stuehr, N. S. Kwon, C. F. Nathan, O. W. Griffith, P. L. Feldman, and J. Wiseman, *J. Biol. Chem.*, *266*:6259 (1991).
209. P. A. Bush, N. E. Gonzalez, J. M. Griscavage, and L. J. Ignarro, *Biochem. Biophys. Res. Comm.*, *185*:960 (1992).
210. U. Forstermann, L. D. Gorsky, J. S. Pollock, K. Ishii, H. H. Schmidt, M. Heller, and F. Murad, *Mol. Pharmacol.*, *38*:7 (1990).
211. N. S. Kwon, C. F. Nathan, and D. J. Stuehr, *J. Biol. Chem.*, *264*:20496 (1989).
212. C. F. Nathan, *FASEB J.*, *6*:3051 (1992).

213. M. M. Teixeira, T. J. Williams, and P. G. Hellewell, *Br. J. Pharmacol.*, *110*:1515 (1993).
214. Q. Hamid, D. R. Springall, V. Riveros-Moreno, P. Chanez, P. Howarth, A. Redington, J. Bousquet, P. Godard, S. Holgate, and J. M. Polak, *Lancet*, *342*:1510 (1993).
215. S. A. Kharitonov, A. U. Wells, B. J. O'Connor, P. J. Cole, D. M. Hansell, R. B. Logan-Sinclair, and P. J. Barnes, *Am. J. Respir. Crit. Care Med.*, *151*:1889 (1995).
216. S. A. Kharitonov, B. J. O'Connor, D. J. Evans, and P. J. Barnes, *Am. J. Respir. Crit. Care Med.*, *151*:1894 (1995).
217. S. A. Kharitonov, D. Yates, R. A. Robbins, R. Logan-Sinclair, E. A. Shinebourne, and P. J. Barnes, *Lancet*, *343*:133 (1994).
218. P. H. Howarth, A. E. Redington, D. R. Springall, U. Martin, S. R. Bloom, J. M. Polak, and S. T. Holgate, *Int. Arch. Allergy Immunol.*, *107*:228 (1995).
219. T. M. Wizemann, C. R. Gardner, J. D. Laskin, S. Quinones, S. K. Durham, N. L. Goller, S. T. Ohnishi, and D. L. Laskin, *J. Leukoc. Biol.*, *56*:759 (1994).
220. C. A. Herbert, D. Edwards, J. R. Boot, and C. Robinson, *Br. J. Pharmacol.*, *110*:840 (1993).
221. C. A. Herbert, D. Edwards, J. R. Boot, and C. Robinson, *Br. J. Pharmacol.*, *104*:391 (1991).
222. C. A. Herbert, M. J. P. Arthur, and C. Robinson, *Br. J. Pharmacol.*, *117*:667 (1996).
223. J. J. Costa, K. Matossian, and M. B. Resnick, *J. Clin. Invest.*, *91*:2673 (1993).
224. V. Del Pozo, B. DeAndres, and E. Martin, *J. Immunol.*, *144*:3117 (1990).
225. H. Kita, T. Ohnishi, Y. Okubo, D. Weiler, J. S. Abrams, and G. J. Gleich, *J. Exp. Med.*, *174*:743 (1991).
226. R. Moqbel, Q. Hamid, and S. Ying, *J. Exp. Med.*, *174*:749 (1991).
227. H. Broide, N. Paine, and G. Firestein, *J. Clin. Invest.*, *90*:1414 (1992).
228. S. P. Dubucquoi, P. Desreumaux, and A. Janin, *J. Exp. Med.*, *179*:703 (1993).

# 12

# Regulation of Eosinophil Mediator Release by Adhesion Molecules

**Hirohito Kita**    *Mayo Clinic and Mayo Foundation, Rochester, Minnesota*

## I.  INTRODUCTION

At present, eosinophils are recognized as pro-inflammatory cells and likely play a major role in allergic diseases, such as bronchial asthma and atopic dermatitis. The eosinophil is an important source of cytotoxic proteins, lipid mediators, and oxygen metabolites, which have the potential to induce pathology in disease. Recent studies on cell adhesion molecules suggest that in addition to adhesive interactions, adhesion molecules modulate intracellular signaling pathways and regulate effector functions of the cells. The purpose of this chapter is to discuss the potential roles of adhesion molecules as regulators of mediator release from eosinophils.

## II.  EOSINOPHIL MEDIATORS

Eosinophils generate an array of inflammatory mediators, including granule proteins, proteolytic enzymes, lipid mediators, and oxygen metabolites. These are summarized in Table 1. Eosinophils are also potential sources of cytokines and growth factors, as described in detail in Chapter 10.

### A.  Granule Proteins

Eosinophils contain two main types of secretory granules: the specific and the small granule. The specific granules contain major basic protein (MBP), which is localized to the crystalloid core of granules. The matrix of specific granules contains eosinophil cationic protein (ECP), eosinophil peroxidase (EPO), eosinophil-

**Table 1**   Secretory Products of Eosinophils

| Granule proteins | Lipid mediators |
|---|---|
| Major basic protein | Leukotriene $B_4$ (small amount) |
| Eosinophil-derived neurotoxin | Leukotriene $C_4$ |
| Eosinophil cationic protein | Leukotriene $C_5$ |
| Eosinophil peroxidase | 5-HETE |
| β-glucuronidase | 5,15- and 8,15-diHETE |
| Acid phosphatase | 5-oxo-15-hydroxy 6,8,11,13 ETE |
| Arylsulfatase B | Prostaglandin E1 and E2 |
| **Reactive oxygen intermediates** | 6-keto-prostaglandin F1 |
| $O_2^-$ | Thromboxane B2 |
| $H_2O_2$ | PAF |
| Hydroxy radicals | |
| Singlet oxygen | |
| **Enzymes** | |
| Elastase (questionable) | |
| Charcot-Leyden crystal protein | |
| Collagenase | |
| 92 kD gelatinase | |

HETE, hydroxy eicosatetraenoic acid; ETE, eicosatetraenoic acid; diHETE, dihydroxy eicosate-traenoic acid; $O_2^-$, superoxide anion; PAF, platelet-activating factor.

derived neurotoxin (EDN), and β-glucuronidase. The small granules, in contrast, contain several enzymes, including acid phosphatase and arylsulfatase B.

Major basic protein consists of a single polypeptide chain of 117 amino acids; it is rich in arginine, has a molecular mass of 14 kD, and has a calculated isoelec-tric point of 10.9 (1). Major basic protein shows a variety of activities on non-mammalian and mammalian cells. For example, purified MBP damages schistosomula of *S. mansoni* (2). When eosinophils were incubated with schisto-somula in the presence of antibody against *S. mansoni*, MBP was released and deposited on the surface of the parasites (3). Major basic protein also damages other parasites, such as *Trichinella spiralis* and *Brugia pahangi* (4,5). Major ba-sic protein is toxic to human epithelial cells (6) and causes ciliostasis and exfoli-ation of respiratory epithelial cells (7); this latter effect mimics the pathology of asthma (8). Major basic protein augments contraction of the tracheal smooth muscle induced by acetylcholine in vitro (9), and inhalation of MBP causes bronchial hyperresponsiveness in primates (10). Interestingly, incubation of hu-man peripheral blood leukocytes with MBP causes a dose-dependent release of histamine that is calcium-, temperature-, and energy-dependent, suggesting non-cytotoxic activation of basophils by MBP (11). Similarly, MBP activates media-tor release from mast cells (11), neutrophils (12), alveolar macrophages (13), platelets (14), and eosinophils themselves (15). Thus MBP is a cytostimulant as well as being cytotoxic to cells.

Eosinophil cationic protein consists of a single polypeptide chain of 160 amino acids; it has a molecular mass of 15 kD, an isoelectric point of approximately 10.8, and RNase activity (16). Eosinophil cationic protein is also a potent toxin for schistosomula of *S. mansoni* (17); ECP is approximately 10 times more active than MBP (18). Eosinophil cationic protein produces complete fragmentation and disruption of the schistosomula, whereas MBP produces distinctive ballooning and detachment of tegumental membrane. In contrast, ECP apparently has much less activity on mammalian cells, such as epithelial cells, basophils, platelets, and eosinophils when compared to MBP. Both ECP and EDN produce a neurotoxic reaction when injected into the cerebrospinal fluid of rabbits (19).

Eosinophil-derived neurotoxin is a 134-amino-acid polypeptide with an molecular mass of 15 kD (isoelectric point 8.9) (20), and it is identical to human urinary RNase and human liver RNase (21,22). The amino acid sequence of EDN reveals marked homology to ECP with identity at 37 of 54 N-terminal residues (23). The most striking activity of EDN is its neurotoxicity. When EDN is injected intrathecally into rabbits or guinea pigs, it causes the Gordon phenomenon, a syndrome that begins with stiffness and mild ataxia, followed by incoordination and marked ataxia (19). Eosinophil-derived neurotoxin is considerably less toxic for parasites and mammalian cells compared to the other eosinophil granule proteins.

Eosinophil peroxidase differs from neutrophil or monocyte myeloperoxidase (MPO) in its absorption spectrum and its heme prosthetic groups (24). Analysis of EPO reveals the molecular masses of 12 and 53 kD for the heavy and light subunits, respectively; their calculated isoelectric points are 10.8 and 10.7, respectively (25). In the presence of $H_2O_2$, EPO is able to oxidize halides to form reactive hypohalous acids (26). Analyses of EPO's halide preference shows that eosinophils preferentially utilize bromide over chloride (27). This EPO/$H_2O_2$/halide system kills a variety of microorganisms, such as schistosomula of *S. mansoni* and *Escherichia coli* (28,29). Studies of cultured human pneumocytes (30), nasal epithelial cells (31), and tumor cells (32) indicate that the EPO/$H_2O_2$/halide system causes toxicity to mammalian cells. Eosinophils by themselves can generate $H_2O_2$ (33) suggesting that this EPO/$H_2O_2$/halide system is an effective system to mediate toxicity toward numerous targets. In the absence of $H_2O_2$ and halide, EPO is also toxic to some targets although the activity seems somewhat limited. Eosinophil peroxidase, as well as MBP, is a potent stimulus for human platelets and eosinophils (14,15)

## B.  Proteases

Eosinophil collagenase degrades type I and type III collagen, the two major connective tissue components of human lung parenchyma (34,35). Using in situ hybridization and immunohistochemistry, it was demonstrated that eosinophils are a major source of a 92-kD matrix metalloproteinase (MMP)-9 (gelatinase B) in basal cell carcinoma lesions (36,37). Recently, this gelatinase was localized to eosinophils infiltrating into the lesions of patients with bullous pemphigoid (38), and it cleaved the type XVII collagen, a transmembrane molecule of the epidermal hemidesmosome, suggesting that production and release of gelatinase by eosinophils contributes to tissue damage of bullous pemphigoid. Because the MMP-9 also degrades tissue matrix proteins, such as type IV collagen, proteoglycans, and possibly laminin (39), this protease may play a role in the pathophysiology of asthma.

## C.   Lipid Mediators

In eosinophils, the predominant metabolite via the 5-lipoxygenase pathway is leukotriene (LT)$C_4$ (40). Eosinophils can generate relatively large amounts of $LTC_4$, but only negligible amounts of $LTB_4$ (40). In contrast, neutrophils can produce large amounts of $LTB_4$, but little if any $LTC_4$. The other major 5-lipoxygenase metabolite of eosinophils is 5-hydroxy-6,8,11,14-eicosatetraenoic acid (HETE) (41), which can be metabolized to $LTC_4$. These mediators contract airway smooth muscle, promote the secretion of mucus, alter vascular permeability, and elicit eosinophil and neutrophil infiltration. In eosinophils stimulated by the addition of exogenous arachidonic acid, 15-lipoxygenase metabolites are detectable, including 5,15- and 8,15-dihydroxyeicosatetraenoic acid (diHETE) and a novel chemotactic lipid, 5-oxo-15-hydroxy 6,8,11,13 eicosatetraenoic acid [5-oxo-15(OH)-ETE] (42–44). Analysis shows that eosinophils contain high levels of ether phospholipids [the stored precursor to platelet-activating factor (PAF)], about fourfold more than neutrophils, suggesting that the eosinophil is a good PAF producer (45). Much of this material is in the form of 1-alkyl-2-lyso-*sn*-glycero-3-phosphocholine. Eosinophils produce PAF by an acetylation reaction using the enzyme 1-alkyl-2-lyso-*sn*-glycero-3-phosphocholine:acetyl-CoA acetyl transferase (46). Eosinophils produce at least three molecular species of PAF, the predominant species being 1-hexadecyl-2-acetyl-glycero-3-phosphocholine (16:0). Platelet-activating factor has a number of important pharmacologic activities, including the activation of platelets and neutrophils, and the induction of bronchoconstriction. Thus, eosinophil-derived PAF has possible roles in the pathophysiology of bronchial asthma.

## D.   Oxidative Products

When challenged with an appropriate stimulus, eosinophils, like neutrophils and monocytes, produce oxygen-derived free radicals. In fact, the oxidative products of eosinophils include superoxide anion, hydroxyl radicals, $H_2O_2$, and singlet oxygen (47,48). When membrane-associated oxidase is activated, it catalyzes the single electron reduction of oxygen using NADPH as an electron donor. The product of the NADPH:$O_2$ oxidoreductase–catalyzed reaction is the superoxide anion, a powerful oxidizing and reducing agent with bactericidal properties (49). Superoxide anions can undergo spontaneous oxidation to $0_2$, or reduction to $H_2O_2$ in the presence of superoxide dismutase. The $H_2O_2$ produced by eosinophils shows potent biological activity in conjunction with EPO, as described above. Furthermore, in the presence of ferrous ions, the superoxide anion and $H_2O_2$ interact to form the membrane-perturbating hydroxy radical, a reactive and unstable oxidizing species. It is noteworthy that the activity of the NADPH:$O_2$ oxidoreductase is significantly higher in eosinophils than it is in neutrophils (50,51), which may be relevant to the pathophysiology of eosinophil-mediated diseases such as bronchial asthma (52).

## III. ROLES OF EOSINOPHIL MEDIATORS IN ALLERGIC DISEASES

As the numerous activities and biological effects of eosinophils and their granule proteins have been discovered, a role for the eosinophil in asthma has become increasingly well defined (8). Asthma is associated with blood, tissue, and sputum eosinophilia, and the degree of eosinophilia is correlated with bronchial hyperresponsiveness (53). The eosinophil granule proteins, MBP, EPO, and ECP (8,30,54), all damage respiratory epithelium and pneumocytes (30). In particular, the effects of MBP on respiratory epithelium mimic the pathology of asthma (7). In fatal asthma, immunostaining bronchial mucosa shows large numbers of eosinophils (55) and striking MBP deposition (56). Pulmonary segmental allergen challenge in allergic individuals causes eosinophil recruitment and release of eosinophil granule proteins; a dramatic increase in vascular permeability is also found (57). Patients with asthma have increased MBP concentrations in their bronchoalveolar lavage (BAL) fluids when compared with control subjects (58). Furthermore, the level of MBP correlates with the number of desquamated epithelial cells and the degree of bronchial hyperresponsiveness. Airway hyperresponsiveness in asthma may be the result of the ability of MBP to selectively block M2 cholinergic receptors (59). Major basic protein and EPO may directly stimulate the respiratory epithelium on contact, which, in turn, causes smooth muscle contraction and increased sensitivity of the muscle to methacholine (10); recent studies suggest this may be the result of bradykinin generation (60). Finally, airway hyperresponsiveness in guinea pigs sensitized and challenged with ovalbumin was prevented by neutralization of MBP by a specific antiserum, suggesting that eosinophil degranulation plays an important role in the alterations of bronchopulmonary function in the guinea pig (61).

Atopic dermatitis is a common, chronic skin disease that is often accompanied by peripheral blood eosinophilia (62). Despite a paucity of intact eosinophils infiltrating the skin, eosinophil granule protein deposition is prominent in lesions of atopic dermatitis (63,64). Furthermore, both eosinophils and eosinophil granule proteins may be elevated in the peripheral blood of patients with atopic dermatitis; both factors correlate with disease severity and decrease with clinical improvement (65–67). Eosinophil infiltration and deposition of eosinophil granule proteins also appear to be importantly associated with the pathology of allergic rhinitis. When compared with nonallergic, normal individuals, allergic patients had more eosinophils in both the nasal mucosa and in nasal lavage fluid (68, 69). Eosinophils entered the nasal mucosa within hours of allergen provocation and degranulated, releasing MBP and EDN that could be recovered in nasal lavage fluids (70). Marked eosinophil infiltration and MBP deposition have also been found in areas of epithelial desquamation in paranasal sinus tissue (71). Thus, the presence of eosinophil products in the inflammatory lesions and the blood of patients with allergy, and the emerging knowledge of the functions and mediators of activated eosinophils, support a role for the eosinophil in amplifying inflammation and in altering the physiology of allergic diseases.

## IV. STIMULI FOR EOSINOPHIL MEDIATOR RELEASE

The striking feature of inflammatory reactions involving eosinophils is the marked deposition of granule proteins in the tissues, often in the presence of relatively small numbers of intact eosinophils. However, the mechanism of eosinophil degranulation in the tissues is still poorly understood. In vitro, immunoglobulins immobilized onto a large surface are one of the most potent agonists for eosinophil degranulation. Initial studies of eosinophil degranulation used a model of parasites to mimic host defense against parasite infection. When eosinophils were incubated with antiserum-coated schistosomula of *S. mansoni*, degranulation occurred, as evidenced by release of MBP (2,72). However, investigation of eosinophil degranulation in this system was complicated by the presence of the viable worms and various substances released by them. Therefore, large surfaces (e.g., Sepharose 4B beads) coated with immunoglobulins have been subsequently used as an in vitro model of eosinophil degranulation. Current evidence indicates that Sepharose beads coated with IgG, IgA, and sIgA stimulate eosinophil degranulation (73). Among these immunoglobulins, sIgA is the most effective inducer of degranulation. Recent studies of antibody-dependent eosinophil killing of *S. mansoni*, making use of chimeric mouse/human antibodies, demonstrate that $IgA_2$ is a highly potent stimulus for eosinophils (74). IgE also seems important for eosinophil activation. Eosinophils isolated from patients with eosinophilia degranulated in response to anti-IgE antibody or IgE-coated parasites (75). However, it is still uncertain whether a similar response to IgE is observed with eosinophils from normal subjects or patients with allergy. By using sera from patients with ragweed-sensitive hay fever, we found that ragweed-specific IgG, but not IgE, is important for allergen-dependent eosinophil degranulation in vitro (76). Among the IgG subclasses tested, IgG1 and IgG3 induced eosinophil degranulation, whereas IgG4 did not. The ability of IgG2 to induce degranulation was unique to certain donors of eosinophils (76). Furthermore, the ability of eosinophils to secrete granule proteins in response to immobilized immunoglobulins is markedly enhanced by priming with relatively low concentrations of soluble mediators such as chemotactic factors and cytokines. For example, IL-5, granulocyte-macrophage colony-stimulating factor (GM-CSF), and interleukin (IL)-3, but not IL-1, IL-2, IL-4, IL-6 and tumor necrosis factor (TNF), increased sIgA- or IgG-induced eosinophil degranulation (77).

Eosinophil degranulation can also be induced by soluble stimuli. Cytokines, especially those with eosinophilopoietic activity, cause eosinophil degranulation. For example, eosinophils incubated for four days with IL-5 release 30% to 60% of their granule proteins (78). GM-CSF also potently induces eosinophil degranulation after several hours of incubation when CD11b/CD18-dependent cellular adhesion is available, suggesting costimulatory effects of cytokines and adhesion molecules on eosinophil degranulation (see below) (79). The release of granule proteins during culture with IL-5 may be an important mechanism for deposition of these cationic toxins in various diseases where IL-5 plays a role. Eosinophil-chemotactic cytokines, such as regulated on activation, normal T cell expressed and secreted (RANTES) and macrophage inflammatory protein (MIP)-1$\alpha$, also induce eosinophil degranulation, although the effects are less pronounced than those of IL-5 or GM-CSF (80, 81). Interestingly, eosinophil granule proteins themselves, including MBP and EPO, stimulate eosinophils and cause degranulation in a noncytotoxic manner, suggesting an autocrine mechanism of eosinophil degranulation (15). The other soluble stimuli for eosinophil degranulation include: serum-opsonized zymosan (82); formyl-Met-Leu-Phe (fMLP) (83); the lipid mediator, PAF (84) and 5-oxoETE (85); the complement fragments C5a and C3a (86, 87); the naturally occurring peptides substance P and mellitin (88); calcium ionophore A23187 (89); and phorbol myristate acetate (PMA) (89).

Eosinophils produce the sulfidopeptides LTC$_4$ and LTD$_4$ in response to a variety of stimuli, including calcium ionophore A23187 (90), fMLP (90), PAF (91), and IgG-dependent stimuli (92). Furthermore, eosinophils release LTC$_4$ on exposure to *S. mansoni* coated with parasite-specific IgE or IgG antibodies (93). The secretion of PAF from eosinophils is caused by chemotactic stimuli, such as C5a, as well as zymosan and A23187 (46). Platelet-activating factor biosynthesis by eosinophils is increased three- to fourfold by preincubation of eosinophils with GM-CSF for 72 hours. Eosinophil oxidative metabolism is stimulated by fMLP (47), PAF (50), LTB$_4$ (94), C3a (95), zymosan (96), IgG-coated surfaces (97), calcium ionophore A23187 (50), and PMA (50). Thus, as summarized in Table 2, eosinophil mediator release can be provoked by a variety of stimuli, including

**Table 2**  Stimuli for Eosinophil Mediator Release

| Stimuli | Degranulation | Lipid mediator production | Superoxide production | Cytotoxicity against Ab-coated targets | Priming[a] |
|---|---|---|---|---|---|
| **Immunoglobulins** | | | | | |
| IgG | ↑ | ↑ | ↑ | ↑ | |
| IgE | ↑ | ↑ | | ↑ | |
| IgA | ↑ | | | | |
| Secretory IgA | ↑ | | | | |
| **Complement** | | | | | |
| C3a | ↑ | | ↑ | | |
| C5a | ↑ | ↑ | | | |
| Zymosan | ↑ | ↑ | ↑ | | |
| **Lipid mediators** | | | | | |
| PAF | ↑ | ↑ | ↑ | ↑ | ↑ |
| LTB4 | | | ↑ | | |
| **Cytokines** | | | | | |
| IL-3 | ↑ | | ↑ | ↑ | ↑ |
| IL-5 | ↑ | ↑ | ↑ | ↑ | ↑ |
| GM-CSF | ↑ | ↑ | ↑ | ↑ | ↑ |
| TNF-α | | ↑ | | ↑ | |
| **Chemokines** | | | | | |
| RANTES | ↑[b] | | | | |
| MCP3 | ↑[b] | | | | |
| MIP-1α | ↑[b] | | | | |
| **Other molecules** | | | | | |
| fMLP | ↑ | ↑ | ↑ | | ↑ |
| MBP | ↑ | ↑ | | | |
| EPO | ↑ | | | | |
| Substance P | ↑ | | | | |
| Mellitin | ↑ | | | | |
| A23187 | ↑ | ↑ | ↑ | | |
| PMA | ↑ | | ↑ | | |

Induction and/or upregulation of eosinophil function is indicated by "↑".
[a]"Priming" means enhancement of eosinophil function induced by other agonists.
[b]At high concentrations.
Ab, antibody; PAF, platelet-activating factor; LTB4, leukotriene B4; GM-CSF, granulocyte-macrophage colony-stimulating factor; TNF-α, tumor necrosis factor–α; MCP-3, monocyte chemotactic protein-3; MIP-1α, macrophage inflammatory protein-1α; fMLP, formyl-Met-Leu-Phe; MBP, major basic protein; EPO, eosinophil peroxidase; PMA, phorbol myristate acetate.

immunoglobulins immobilized onto a large surface, cytokines, lipid mediators, and complement fragments, and several pharmacologic agonists. Therefore, the question arises as to whether these stimuli use completely separate pathways for signal transduction to stimulate mediator release or whether there is a common pathway(s). Our own studies and those of other investigators suggest that the latter may be the case and that cellular adhesion may be a prerequisite signal for effector functions of eosinophils.

## V.  EOSINOPHIL ADHESION RECEPTORS AND LIGANDS

Human eosinophils express a number of adhesion receptors. As shown in Table 3, eosinophils constitutively express L-selectin and carbohydrate ligands for endothelial selectins (98). Ligands for L-selectin consist of heavily O-glycosylated, mucin-like glycoprotein receptors, in which large amounts of O-linked sugars are

**Table 3**  Eosinophil Adhesion Molecules and Their Ligands

| Eosinophil structures | Surface molecules (CD designation) | Ligands |
|---|---|---|
| Integrins | $\alpha L\beta 2$, LFA-1 (CD11a/CD18) | ICAM-1, ICAM-2, ICAM-3 |
| | $\alpha M\beta 2$, Mac-1 CR3 (CD11b/CD18) | C3bi, ICAM-1 |
| | $\alpha X\beta 2$, p150, 95 (CD11c/CD18) | C3bi |
| | $\alpha d\beta 2$ | ICAM-3 |
| | $\alpha 4\beta 1$, VLA-4 (CD49d/CD29) | VCAM-1, fibronectin |
| | $\alpha 6\beta 1$, VLA-6 (CD49f/CD29) | Laminin |
| | $\alpha 4\beta 7$ | MAdCAM-1, VCAM-1, fibronectin |
| Selectins | L-selectin | GlyCAM-1, CD34 |
| Carbohydrates | PSGL-1 | P-selectin |
| | Sialyl-Lewis$^X$ (CD15s)[a] | E-selectin, P-selectin |
| | Sialyl-dimeric Lewis$^X$ | E-selectin |
| | Leukosialin (CD43) | |
| | Pgp-1 (CD44) | Hyaluronic acid |
| Immunoglobulin- like | PECAM-1 (CD31) | PECAM, heparan sulfate |
| | ICAM-1 (CD54)[b] | LFA-1, Mac-1 |
| | ICAM-3 (CD50) | LFA-1, $\alpha d\beta 2$ |
| | LFA-3 (CD58) | CD2 |

LFA, lymphocyte function-associated antigen; VLA, very late antigen; PSGL-1, P-selectin glycoprotein ligand-1; ICAM, intercellular cell adhesion molecule; VCAM-1, vascular cell adhesion molecule-1; MadCAM-1, mucosal adressin cell adhesion molecule-1; GlyCAM-1, glycosylated cell adhesion molecule-1; PECAM-1, platelet-endothelial cell adhesion molecule-1.
[a]Levels of expression are very low compared to neutrophils.
[b]Expressed only on activated eosinophils.

presented on a rigid protein backbone (99). The inhibition assay using a panel of monoclonal antibodies suggests that eosinophils utilize a different functional epitope of L-selectin for this ligand binding than neutrophils (100). The counter-ligands for endothelial selectins are also different for eosinophils and neutrophils. Neutrophils bind more avidly than eosinophils to purified E-selectin (101). Neutrophils express large amounts of sialyl-Lewis$^x$, whereas the expression of this epitope on eosinophils is extremely low or undetectable. A major proportion of the E-selectin ligand on the surface of eosinophils appears to be sialyl-dimeric Lewis$^x$ (101). The counter-ligands for P-selectin on eosinophils and neutrophils are similarly sialylated, protease-sensitive, endo-beta-galactosidase-resistant structures, presumably P-selectin glycoprotein ligand (PSGL)-1 (102). Eosinophil binding to E- and P-selectin does not appear to be affected by the state of cell activation. Furthermore, the levels of expression of L-selectin were decreased in eosinophils activated in vivo or in vitro by lipid mediators or cytokines (100,103).

Integrins are composed of $\alpha$ and $\beta$ transmembrane subunits selected from among 16 $\alpha$ and 8 $\beta$ subunits that heterodimerize to produce more than 20 different receptors (104). Unlike selectins, ligand recognition by integrins requires cellular activation. Cells are able to regulate the function of integrins by signaling in the opposite direction ("inside-out signaling") and the receptors are converted from a form with low affinity for their ligands to one with high affinity. This so-called "activation" of the integrin is required for firm interaction between integrins and their ligands (105), and this is also the case with eosinophils. The levels of expression of integrins are also dependent on the activation state of the cells. Eosinophils express several integrins that can bind to immunoglobulin family receptors expressed on the endothelium and to components of extracellular matrix proteins (Table 2). The resultant bond is much firmer than the selectin carbohydrate bond and results in the eosinophils' flattening and transmigration between endothelial cells (106). Eosinophils express $\beta1$ (CD29) (107, 108), $\beta2$ (CD18) (109), and $\beta7$ integrins (110). Among the $\beta1$ integrins, $\alpha4\beta1$ [very late antigen (VLA)-4, CD49d/CD29], which binds the vascular cell adhesion molecule (VCAM)-1, and fibronectin play important roles in eosinophil migration from bloodstream into the tissues and in modulation of eosinophil function. Among the $\beta2$ integrins, $\alpha L\beta2$ (lymphocyte function-associated antigen [LFA]-1, CD11a/CD18) and $\alpha M\beta2$ (Mac-1, CD11b/CD18) integrins both bind the intercellular adhesion molecule (ICAM)-1 expressed on various types of cells and also play important roles for eosinophil migration as well as effector functions of the cells. Eosinophils also express another $\beta2$ integrin, $\alpha X\beta2$ (p150, 95), and a recently identified novel $\beta2$ integrin, $\alpha d\beta2$ (111); however, the roles of these integrins in eosinophil functions are still unknown.

## VI. REGULATION OF EOSINOPHIL MEDIATOR RELEASE BY β2 INTEGRINS

### A. Eosinophil Interaction with Plastic Surfaces

Recent studies have suggested that adhesion molecules function not only to anchor cells in specific locations within tissues and regulate their movement, but also to regulate effector functions of leukocytes. Initial studies on involvement of cellular adhesion in eosinophil mediator release were described by Dri et al. (112). They found that eosinophils adherent to plastic surfaces of tissue culture plates undergo respiratory burst in the absence of any exogenous stimuli. After coating the polystyrene plates with serum, the extracellular matrix proteins fibronectin, laminin, type I collagen, and type IV collagen, or human umbilical vein endothelial cells (HUVEC) prevented plastic-induced superoxide production by eosinophils (112). Coating the plastic surface with matrix proteins or HUVEC also inhibited superoxide production from activated eosinophils stimulated by fMLP, TNF-α, or PAF. Importantly, under these same conditions, superoxide production induced by PMA was not affected. We subsequently found that the polystyrene plates, especially those modified for high protein binding, caused eosinophil degranulation, and this plastic-induced degranulation was inhibited by coating the plate surface with albumin (113). Therefore, it appears that eosinophils adhere to uncoated plastic surfaces and this plastic adherence results in functional activation of the cells. Caution is therefore needed for the potential of the plate or tube materials to confound in vitro studies.

### B. Eosinophils Stimulated by Soluble Stimuli

Although adhesion to plastic is not physiological, the results clearly indicate that cellular adhesion plays important roles in eosinophil activation. Therefore, we investigated the roles of adhesion molecules in mediating effector functions of eosinophils in more physiologic conditions. Similarly to the extracellular matrix proteins described above, coating the polystyrene wells with human serum albumin (HSA) prevented plastic-induced adhesion and degranulation of resting eosinophils (113) (Fig. 1, panel A). After stimulation by GM-CSF or PAF, eosinophils incubated in HSA-coated wells released large amounts of granule proteins (79) (Table 3). Morphologically, they vigorously adhered to the HSA-coated wells (Fig. 1, panel B), and kinetic studies showed eosinophil degranulation was always preceded by cellular adhesion with a lag period of 15 to 45 min. In the absence of PAF or GM-CSF, the plates alone did not induce eosinophil adherence or subsequent degranulation, suggesting that activation of eosinophils is required for them to adhere to albumin-coated wells. Eosinophil adhesion stimulated by PAF or GM-CSF was inhibited essentially completely by anti-CD18

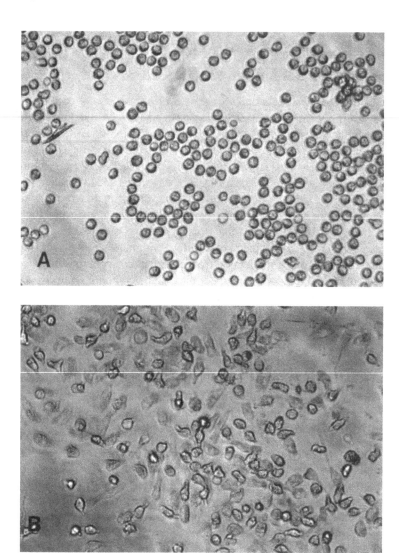

**Figure 1** Morphology of eosinophils incubated in wells coated with albumin. Eosinophils were incubated for 2 hours with medium alone (*panel A*) or granulocyte-macrophage colony-stimulating factor (GM-CSF) (10 ng/mL) (*panel B*) in wells coated with human serum albumin. Eosinophils incubated with medium alone are round-shaped and refractile (*panel A*). When eosinophils are stimulated with GM-CSF, they adhered to the bottom of the wells, spread, and assumed a spindle shape (*panel B*). (Original magnification ×400).

mAb and significantly by anti-αM (CD11b) mAb, indicating that adhesion of activated eosinophils to HSA was αMβ2-dependent (Fig. 2, panel A). Antibody against αL (CD11a) had no effect on eosinophil adhesion to the HSA-coated wells. Similarly, eosinophil degranulation induced by PAF or GM-CSF was abolished by anti-CD18 mAb or anti-CD11b mAb, but not by anti-CD11a mAb (Fig. 2, panel B). Eosinophil superoxide production induced by PAF or GM-CSF was also inhibited by anti-CD18 mAb. In contrast, eosinophil degranulation or superoxide production induced by PMA was not inhibited by anti-CD18 mAb treatment although this antibody inhibited adhesion of PMA-stimulated eosinophils to HSA-coated wells, suggesting differences in the integrin dependency among various agonists (Fig. 2, panels A and B). Furthermore, when the cell suspensions were gently stirred, degranulation of eosinophils stimulated by PAF or GM-CSF was inhibited. Collectively, these observations indicate that eosinophil adhesion mediated by αMβ2 plays a critical role for degranulation and superoxide production of eosinophils stimulated by PAF and GM-CSF.

We also tested the abilities of other soluble stimuli to induce eosinophil degranulation and superoxide production under similar conditions (113) (Table 4). In general, the agonists' abilities to induce degranulation correspond to their capacities to promote eosinophil adhesion. For example, PAF and eosinophil growth factors, including IL-3, IL-5, and GM-CSF, promoted strong adhesion of eosinophils to HSA-coated wells in a β2 integrin-dependent manner. These agents can also stimulate eosinophil degranulation and superoxide production. TNF-α and a chemokine, RANTES, cause weak adhesion of eosinophils to HSA-coated wells; these agents also cause degranulation, but less potently than PAF or eosinophil growth factors. In contrast, IL-1β or interferon gamma (IFN-γ) cause neither adhesion nor degranulation of eosinophils.

Important roles for β2 integrins in eosinophil activation were further supported by experiments using anti-integrin mAb. Laudanna et al. incubated eosinophils with anti–integrin mAb, such as anti-αMβ2 and anti-αLβ2, and reacted cell-bound mAb with protein A immobilized onto tissue culture plates (114). Eosinophils with their integrins ligated in such a manner spread on the surface of tissue culture wells and produced superoxide. We also found that anti–CD11b mAb coupled to polystyrene microbeads triggered degranulation response of eosinophils (115). These findings suggest that engagement of αMβ2, without any other exogenous stimuli, is sufficient to trigger mediator release from eosinophils. Therefore, it is possible that the role of soluble stimuli in eosinophil mediator release may be merely to cause firm integrin-dependent adhesion of eosinophils to the solid surface; this adhesion is both required and sufficient for cellular activation. It is also possible that integrin molecules and soluble stimuli may act synergistically to expand intracellular signaling events required for cellular functions.

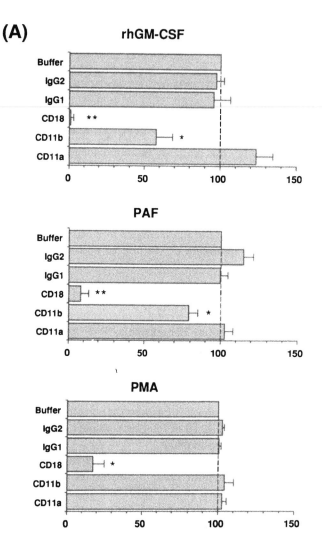

**Figure 2** Effects of monoclonal antibody (mAb) to β2 integrins on eosinophil adhesion (*panel A*) and degranulation (*panel B*). Eosinophils were preincubated with mAb to CD11a, CD11b, or CD18 (10 μg/mL) and stimulated with granulocyte-macrophage colony-stimulating factor (GM-CSF) (10 ng/mL), platelet-activating factor (PAF) (1 μM), or phorbol myristate acetate (PMA) (1 ng/ml) for 2 hours (adhesion assay, *panel A*) or for 4 hours (degranulation assay, *panel B*). Data were normalized to the mean values in the absence of mAb (buffer) for each stimulus group. Significant differences are **(*P* < .01) and *(*P* < .05) from values obtained with eosinophils preincubated with irrelevant isotype-matched Abs, IgG2, and IgG1. (From Ref. 79.)

**(B)**

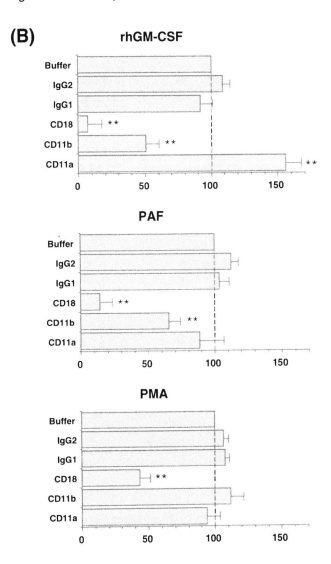

## C. Modulatory Effects of Extracellular Matrix Proteins (ECM)

The importance of β2 integrins for eosinophil functional responses to soluble stimuli was further supported by experiments using extracellular matrix (ECM) proteins, such as fibronectin or laminin (116). Eosinophils were incubated in wells coated with HSA or fibrinogen (ligands for β2 integrin) or fibronectin or laminin (ligands for β1 integrins) and stimulated by physiologic agonists, including PAF, C5a, or IL-5. Eosinophils isolated from normal donors did not adhere to any of these ECM proteins. When eosinophils were stimulated with PAF

**Table 4**   Eosinophil Degranulation Induced by Various Secretagogues

| Stimuli | EDN release | | |
| --- | --- | --- | --- |
| | ng/$10^6$ cells | % of total | (n)[a] |
| Medium alone | 105.1 ± 10.8[b] | 2.3 ± 0.2[b] | (38) |
| *Eosinophil growth factors* | | | |
| IL-3 (10 ng/mL) | 325.3 ± 52.1[c] | 6.5 ± 1.1 | (9) |
| IL-5 (10 ng/mL) | 428.3 ± 70.0[c] | 8.9 ± 1.6 | (9) |
| GM-CSF (10 ng/mL) | 697.3 ± 68.8[c] | 13.2 ± 1.4 | (9) |
| *Pro-inflammatory cytokines* | | | |
| IL-1 (10 ng/mL) | 124.1 ± 31.0 | 3.0 ± 0.6 | (5) |
| TNF-α (100 ng/mL) | 173.2 ± 21.4[c] | 4.0 ± 0.5 | (9) |
| INF-γ (1,000 U/mL) | 107.8 ± 26.1 | 2.7 ± 0.5 | (5) |
| *Chemokines* | | | |
| IL-8 (100 ng/ml) | 116.1 ± 27.5 | 2.4 ± 0.6 | (4) |
| RANTES (100 ng/mL) | 219.4 ± 23.1[c] | 5.6 ± 1.1 | (4) |
| MIP-1α (100 ng/mL) | 131.2 ± 12.1 | 2.8 ± 0.9 | (3) |
| *Others* | | | |
| sIgA (beads) | 701.6 ± 251.4[c] | 14.5 ± 2.2 | (3) |
| PAF (1 μM) | 550.4 ± 49.4[c] | 16.4 ± 3.4 | (4) |
| PMA (1 ng/mL) | 780.4 ± 50.0[c] | 21.5 ± 1.8 | (9) |
| Total EDN | 4404.8 ± 276.3 | (100) | (38) |

Eosinophils were incubated with various stimuli in the wells coated with human serum albumin for 4 hours. The concentrations of eosinophil-derived neurotoxin (EDN) in the supernatants were measured as a marker of eosinophil degranulation.
[a](n) indicates the number of experiments.
[b]Values were presented as mean ± SEM from indicated numbers of experiments (n).
[c]Significantly different ($P < .01$) from values obtained with eosinophils incubated with medium alone.

and incubated in wells coated with laminin or fibronectin, the eosinophils adhered to these ECM using both β1- and β2-integrins. The extent of degranulation of eosinophils adherent to laminin- or fibronectin-coated wells was reduced about 50% compared to cells adherent to HSA-coated wells. These inhibitory effects of laminin and fibronectin on eosinophil degranulation were concentration-dependent. As described earlier, when PAF-stimulated eosinophils are incubated in HSA-coated wells, the eosinophils adhered using only β2 integrins, a crucial molecule for eosinophil degranulation. Therefore, decreased eosinophil degranulation on fibronectin- or laminin-coated wells is likely due to the decrease in β2 integrin-dependent cellular adhesion. Perhaps signaling through adhesion that is partially dependent on β2 integrins (wells coated with laminin or fibronectin) may not be as effective as that through adhesion totally dependent on β2 integrins (wells coated with HSA). Consistent with this notion, eosinophils adherent to immobilized fibrinogen, a known αMβ2 ligand (117), released similar amounts of granule protein as those adherent to the HSA-coated wells (116).

## D. Eosinophils Stimulated by Immobilized Immunoglobulins

β2 integrins, especially αMβ2, are also likely important for eosinophil responses to immunoglobulins immobilized onto a "non-phagocytizable" surface. Receptor ligands immobilized to relatively large surfaces, such as IgG-coated Sepharose beads and parasites, but not particulate ligands, such as aggregated IgG and bacteria, have been known to be effective stimuli for eosinophil degranulation (72,118). The interaction of ligands on the targets with αMβ2 on the eosinophils is important in the eosinophil-mediated damage of opsonized parasites (119). We found that cross-linking of eosinophils' cytophilic IgG by immobilized anti–human IgG induced degranulation, whereas soluble anti–human IgG did not (120). Similarly, eosinophils exposed to immobilized human IgG adhered and degranulated; both adherence and degranulation were inhibited by mAb to CD11b or CD18, but not by mAb to CD29. Eosinophil degranulation induced by Sepharose 4B beads coated with IgG or sIgA was also inhibited by anti–CD18 mAb. Morphologically, eosinophils stimulated by immobilized IgG protruded numerous pseudopods to which αMβ2 was localized; this morphological change was inhibited by mAb to CD18, suggesting a critical role for αMβ2 in eosinophil adhesion and degranulation responses to solid-phase IgG. Interestingly, although mAb to CD18 inhibited superoxide production by eosinophils stimulated by IgG immobilized onto the wells, superoxide production by neutrophils stimulated by the same stimulus was not inhibited by anti–CD18 mAb. This relative independence of neutrophil function on adhesion compared to the eosinophils is consistent with other reports. For example, several FcγR-mediated functions of neutrophils, such as binding to IgG and superoxide production, are only partially dependent on or independent of αMβ2 (121,122). Furthermore, the neutrophil respiratory burst can be induced by soluble immune complexes (123). Therefore, β2 integrin-dependent adhesion is not always necessary for the IgG-induced functions of neutrophils, whereas the IgG-mediated functions of eosinophils are totally dependent on adhesion molecules. Although the reasons for these differences between eosinophils and neutrophils are still unknown, we can speculate that αMβ2 may directly interact with the FcγRII on eosinophils, which is the only IgG receptor expressed on normal eosinophils. In fact, the ability of freshly isolated eosinophils to form rosettes with IgG-sensitized erythrocytes (EA-IgG) is negligible (124), as expected from the low affinity of FcγRII. In contrast, eosinophils ingest opsonized yeast organisms as vigorously as neutrophils after stimulation by GM-CSF (125). Addition of GM-CSF, IL-3, or IL-5 causes an increase in EA-IgG rosettes without significant change in FcγRII expression (124). Furthermore, the kinetics of activation of FcγRII binding to EA-IgG are paralleled by the activation of the binding of αMβ2 to EA-C3bi. In addition, zymosan particles coated with either IgG or iC3b poorly induce eosinophil PAF production, whereas those coated with both IgG and iC3b strongly activate eosinophils (126). These observations suggest that there may be intracellular "cross talk" between the FcγRII and αMβ2 on eosinophils, and synergistic interaction between these receptors may influence effector functions of eosinophils.

## VII.  REGULATION OF EOSINOPHIL MEDIATOR RELEASE BY β1 INTEGRINS

### A.  Interaction with ECM

After migration through the endothelium, eosinophils come into contact with many extracellular matrix proteins. As shown in Table 2, eosinophils bind to matrix proteins mainly through β1 integrins. Eosinophils bind to fibronectin through α4β1 (VLA-4) and α4β7 (110,127), and laminin through α6β1 (VLA-6) (128). This adhesion of eosinophils to matrix proteins modifies the effector functions of the cells. For example, as a result of autocrine stimulation of IL-3 and GM-CSF production, eosinophils survived for prolonged periods when cultured on fibronectin as described in detail in Chapter 10 (127). However, the effects of ECM on eosinophil mediator release are contradictory. Dri et al. (112) demonstrated that eosinophil superoxide production induced by fMLP or substance P is decreased when tissue culture wells are coated with fibronectin or laminin. We also found that eosinophil degranulation stimulated by PAF, C5a, or IL-5 was decreased when eosinophils are incubated in wells coated with fibronectin or albumin compared with wells coated with albumin (116). In contrast, Anwar et al. (129) demonstrated enhancement of calcium-ionophore–stimulated LTC$_4$ generation by eosinophils adherent to fibronectin. Neeley et al. (130) reported that VLA-4–mediated interaction with fibronectin resulted in increased degranulation of eosinophils stimulated by fMLP in the presence of cytochalasin B. Furthermore, Munoz et al. (131) reported that when eosinophils are preincubated in wells coated with fibronectin and subsequently exposed to PAF, they showed increased LTC$_4$ production, resulting in augmented narrowing of explanted human bronchi in vitro. Although the reasons for these discrepancies are unknown, perhaps the most likely explanation is the varied sources of eosinophils. Eosinophils purified from allergic donors showed more adhesion to laminin than those from normal donors in the absence of exogenous stimulus (128). The binding of cellular integrins to their ligands generally requires prior stimulation of the cells and activation of the integrin molecules (132). Therefore, it is possible that eosinophils used for some studies are more activated in vivo than others, resulting in the differences in the interactions with matrix proteins. Another possible explanation for these discrepancies is the secretagogues. For example, cytochalasin B is known to cause a transition of $\alpha M\beta 2$ to the activated conformer (133), and to sustain the increases of intracellular calcium caused by fMLP (134). Therefore, cytochalasin B used in some experiments may modify the physiologic behavior and even the adhesive responses to the cells. Thus, although more studies are needed to clarify these discrepancies, it may be safe to conclude that the interaction of $\beta 1$ integrins with extracellular matrix proteins prepares eosinophils for mediator release when they are subsequently stimulated by secretagogues. In contrast, adhesion through $\beta 1$ integrins per se may not provide such a strong signal for eosinophil function as adhesion through $\beta 2$ integrins.

## B.  Interaction with VCAM-1

VCAM-1 is expressed on endothelial cells, and it is selectively up-regulated by IL-4 and IL-13, important cytokines in allergic inflammation (135). Eosinophils bind to VCAM-1 through $\alpha 4\beta 1$, and this interaction also likely affects the effector functions of eosinophils. Nagata et al. reported that eosinophils incubated in wells coated with VCAM-1 generated modest but significant amounts of superoxide, suggesting that the $\alpha 4\beta 1$/VCAM-1 interaction without exogenous agonists is sufficient to provoke a function of eosinophils (136). Anti–$\alpha 4$ mAb almost completely inhibited eosinophil adhesion to VCAM-1 and subsequent superoxide production, suggesting that $\alpha 4$ is required for eosinophil interaction with VCAM-1. Interestingly, mAb against $\beta 2$ integrin also inhibited VCAM-1–induced superoxide production, but not adhesion, of eosinophils. Therefore, it is possible that initial adhesion to VCAM-1 via $\alpha 4\beta 1$ resulted in modulation of the distribution and function of $\beta 2$ integrins, and hence $\beta 2$ integrins promoted the cell's respiratory burst. Furthermore, in these experiments, it was noted that eosinophil adhesion to VCAM-1 did not stimulate $LTC_4$ generation nor degranulation, consistent with the strong dependency of degranulation response on $\beta 2$ integrins. These observations suggest that enhancement of a specific eosinophil function by $\beta 1$ integrin-mediated adherence is determined not only by the particular adhesion molecules and their ligands, but also by the specific cellular responses.

## VIII.  REGULATION OF EOSINOPHIL MEDIATOR RELEASE BY SELECTIN LIGANDS

Few studies have been performed to investigate the roles of selectins in mediator release from eosinophils. P-selectin immobilized onto a tissue culture plate is a powerful adhesion molecule for eosinophils (137). Indeed, eosinophils adherent to plate-bound P-selectin failed to spread or produce superoxide anion. Furthermore, P-selectin prevented eosinophils from degranulation in response to PMA, suggesting that soluble- or surface-bound P-selectin is an inhibitor of eosinophil function. In contrast, Satoh et al. demonstrated that ligation of Lewis[x] or sialyl-Lewis[x] on eosinophils by mAbs against these molecules stimulated degranulation and the respiratory burst in eosinophils isolated from patients with the idiopathic hypereosinophilic syndrome (138). There are several differences in the experimental models used in these two studies, including the sources of eosinophils and the methods for cellular stimulation. Thus, the roles of selectins and selectin ligands on effector functions of eosinophils are still poorly defined.

## IX.  INTRACELLULAR MECHANISMS OF THE REGULATION OF EOSINOPHIL MEDIATOR RELEASE BY ADHESION MOLECULES

The receptor ligation of eosinophils activates several second messengers, including cAMP, inositol 1,4,5-triphosphate ($IP_3$), intracellular calcium, diacylglycerol (DAG), phosphatidic acid, and arachidonic acid. These second messengers activate target enzymes, which, in turn, lead to the effector functions of eosinophils. In addition to these second messengers, eosinophils also possess a variety of serine/threonine kinases [e.g., protein kinase C (PKC)] and tyrosine kinases (139). Many of these kinases are targets for specific second messengers; conversely, these kinases can regulate production of second messengers. Although the details of eosinophil signal transduction can only be summarized in this limited chapter (140 for detail), the activation of kinases and production of various second messengers are important events for transducing signals and effector functions of eosinophils. Recent studies on hematopoietic cells and interstitial cells suggest that adhesion molecules play regulatory roles in cell function through modulating the intracellular signaling of the cells. Several cellular activation events in eosinophils, known to be crucial for the effector function of the cells, are also regulated by adhesion molecules.

## A.  Activation of Phospholipase C (PLC)

In numerous receptor systems, ligand binding triggers a rapid increase in phosphoinositide hydrolysis induced by phospholipase C (PLC) activation (141). PLC-mediated hydrolysis of phosphatidylinositol 4,5-bisphosphate ($PIP_2$) generates at least two second messengers, $IP_3$ and DAG, which, in turn, trigger intracellular calcium mobilization and PKC activation, respectively (142). Several reports found that phosphoinositide-specific PLC was activated by incubation of eosinophils with PAF (143) or immobilized Ig (144,145), and that activation of this enzyme is strongly associated with the degranulation response of the cells (145). The activation of this PLC is likely modulated by adhesion molecules. We showed that when eosinophils are incubated in plates coated with HSA and stimulated by PAF, they adhered to HSA-coated wells via β2 integrin and showed a sixfold increase in PLC activity (116). In contrast, when plates were coated with laminin, the PAF-stimulated increase in PLC activity in eosinophils was markedly diminished. This decrease in PLC activity was also associated with decreased degranulation response by the cells. Similarly, activation of PLC in eosinophils incubated in HSA-coated plates and stimulated by IL-5 was inhibited when cellular adhesion was inhibited by pretreatment of the cells with anti-CD18 mAb (115). Furthermore, engagement of the αMβ2 on eosinophils by anti-CD11b mAb coupled to polystyrene beads resulted in increased PLC activity in eosinophils. Thus, cellular adhesion through β2 integrins is likely playing an important role in the activation of PLC in eosinophils.

## B. Activation of Protein Tyrosine Kinase

Among the various signaling events involved in the activation of cells, tyrosine phosphorylation initiated by protein tyrosine kinase (PTK) is recognized as a critical event in a variety of cellular signaling pathways (146). Protein tyrosine phosphorylation also seems to be an early physiologic change in activated eosinophils. Stimulation of eosinophils with sIgA or IgG immobilized to Sepharose beads rapidly induces the tyrosine phosphorylation of a number of eosinophil proteins within several minutes (144). A number of intracellular proteins, including MAP kinase, Jak-2 (a tyrosine kinase), and STAT-1 (a nuclear factor) were tyrosine phosphorylated during activation of eosinophils by IL-5 (147–149). Tyrosine phosphorylation of proteins is important for the effector functions of eosinophils. For example, tyrosine phosphorylation was blocked by PTK inhibitors, such as genistein and herbimycin A, and the ability of eosinophils to degranulate in response to immobilized sIgA and IgG was abolished by treatment with these compounds (144). Conversely, when the tyrosine phosphorylation of eosinophil proteins was enhanced by the tyrosine phosphatase inhibitor pervanadate, eosinophil degranulation was enhanced (144). Thus, activation of PTK is provoked by stimulation of eosinophils by various agonists, and this process is likely critical for effector functions of eosinophils. PTK is also involved in eosinophil activation mediated by adhesion molecules. Nagata et al. showed that genistein, a tyrosine kinase inhibitor, did not affect eosinophil adhesion to VCAM-1 immobilized onto tissue culture plates (136). In contrast, genistein did inhibit superoxide production by eosinophils adherent to VCAM-1, suggesting that PTK is involved in intracellular signaling generated by eosinophils adherent to VCAM-1, but that PTK is not involved in adhesion of eosinophils to VCAM-1. In fact, the investigators could identify increased tyrosine phosphorylation of several 30- and 40-kD proteins in eosinophils adherent to VCAM-1. We also found that eosinophils stimulated by anti-CD11b mAb-coupled to polystyrene microbeads showed tyrosine phosphorylation of several intracellular proteins, including paxillin and a 115-kD protein (150). Eosinophils stimulated by IL-5 and adherent to protein-coated surfaces in an αMβ2-dependent manner showed tyrosine phosphorylation of a number of intracellular proteins; tyrosine phosphorylation of this 115-kD protein was specifically inhibited when anti-CD18 mAb prevented eosinophil adhesion. Thus, activation of tyrosine kinases and tyrosine phosphorylation of cellular proteins is likely involved in transmembrane signaling mediated by integrin adhesion molecules in eosinophils. Further studies are required to identify individual proteins involved in this reaction.

## X.  SUMMARY

Recent evidence suggests that eosinophils and their toxic cationic granule proteins play important roles in the pathophysiology of allergic diseases. As shown in Fig. 3, the release of granule proteins and other inflammatory mediators from eosinophils is a multistep event, which is initiated by the ligation of cell surface receptors by various stimuli, including immobilized immunoglobulins, cytokines, and lipid mediators, and is greatly amplified by cellular adhesion. Various secretagogues require cellular adhesion as costimulatory molecules; activation of integrins is a crucial step for eosinophil mediator release shared by various classes of secretagogues. Among various adhesion molecules, β2 inte-

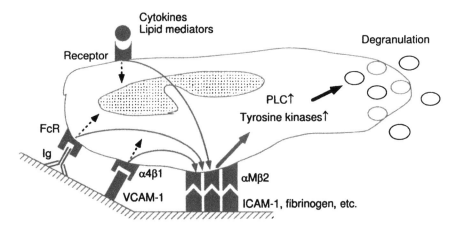

**Figure 3**   Schematic view of the regulatory mechanisms of eosinophil degranulation by adhesion molecules. See text, Summary section, for detail. ICAM-1, intercellular cell adhesion molecule-1; VCAM-1, vascular cell adhesion molecule-1; Ig, immunoglobulin; FcR, Fc receptor for Ig; PLC, phospholipase C.

grins, especially αMβ2, are strongly associated with effector functions of activated eosinophils. Although the roles of β1 integrins for cellular functions are somewhat controversial, they appear to prime and prepare eosinophils for subsequent stimulation by secretagogues, and by themselves, provoke some functions of eosinophils. It is likely that integrins regulate functions of eosinophils by modulating intracellular signaling events, such as activation of tyrosine kinases and PLC. In the future, it will be necessary to identify which adhesion molecules play pivotal roles in eosinophil function in vivo. Furthermore, it will be important to identify the intracellular event(s) which couple adhesion molecules and the secretory machinery of eosinophils. These studies may lead to new approaches for the treatment of patients with various allergic diseases.

# REFERENCES

1. R. L. Barker, G. J. Gleich, and L. R. Pease, *J. Exp. Med.*, *168*:1493 (1988).
2. A. E. Butterworth, D. L. Wassom, G. J. Gleich, D. A. Loegering, and J. R. David, *J. Immunol.*, *122*:221 (1979).
3. A. M. Glauert, A. E. Butterworth, R. F. Sturrock, and V. Houba, *J. Cell Sci.*, *34*:173 (1978).
4. K. J. Hamann, R. L. Baker, D. A. Loegering, and G. J. Gleich, *J. Parasitol.*, *73*:523 (1987).
5. K. J. Hamann, G. J. Gleich, J. L. Checkel, D. A. Loegering, J. W. McCall, and R. L. Baker, *J. Immunol.*, *144*:3166 (1990).
6. G. J. Gleich, E. Frigas, D. A. Loegering, D. L. Wassom, and D. Steinmuller, *J. Immunol.*, *123*:2925 (1979).
7. E. Frigas, D. A. Loegering, and G. J. Gleich, *Lab. Invest.*, *42*:35 (1980).
8. E. Frigas and G. J. Gleich, *J. Allergy Clin. Immunol.*, *77*:527 (1986).
9. J. D. Brofman, S. R. White, J. S. Blake, N. M. Munoz, G. J. Gleich, and A. Leff, *J. Appl. Physiol.*, *66*:1867 (1989).
10. R. H. Gundel, L. G. Letts, and G. J. Gleich, *J. Clin. Invest.*, *87*:1470 (1991).
11. M. C. O'Donnell, S. J. Ackerman, G. J. Gleich, and L. L. Thomas, *J. Exp. Med.*, *157*:1981 (1983).
12. J. N. Moy, G. J. Gleich, and L. L. Thomas, *J. Immunol.*, *145*:2626 (1990).
13. J. A. Rankin, P. Harris, and S. J. Ackerman, *J. Allergy Clin. Immunol.*, *89*:746 (1992).
14. M. S. Rohrbach, C. L. Wheatley, N. R. Slifman, and G. J. Gleich, *J. Exp. Med.*, *172*:1271 (1990).
15. H. Kita, R. I. Abu-Ghazaleh, S. Sur, and G. J. Gleich, *J. Immunol.*, *154*:4749 (1995).
16. R. L. Barker, D. A. Loegering, R. M. Ten, K. J. Hamann, L. R. Pease, and G. J. Gleich, *J. Immunol.*, *143*:952 (1989).
17. D. J. McLaren, J. R. McKean, I. Olsson, P. Venges, and A. B. Kay, *Parasite Immunol.*, *3*:359 (1981).

18. S. J. Ackerman, G. J. Gleich, D. A. Loegering, B. A. Richardson, and A. E. Butterworth, *Am. J. Trop. Med. Hyg.*, *34*:735 (1985).

19. D. T. Durack, S. M. Sumi, and S. J. Klebanoff, *Proc. Natl. Acad. Sci. USA.*, *76*:1979 (1979).

20. K. J. Hamann, R. L. Baker, D. A. Loegering, R. L. Pease, and G. J. Gleich, *Gene*, *83*:161 (1989).

21. J. J. Beintema, J. Hofsteenge, M. Iwama, T. Morita, K. Ohgi, M. Irie, R. H. Sugiyama, G. L. Schieven, C. A. Dekker, and D. G. Glitz, *Biochemistry*, *27*:4530 (1988).

22. S. Sorrentino, G. K. Tucker, and D. G. Glitz, *J. Biol. Chem.*, *263*:16125 (1988).

23. G. J. Gleich, D. A. Loegering, M. P. Bell, J. L. Checkel, S. J. Ackerman, and D. J. McKean, *Proc. Natl. Acad. Sci. USA*, *83*:3146 (1986).

24. B. G. Bolscher, H. Plat, and R. Wever, *Biochim. Biophys. Acta*, *784*:177 (1984).

25. H. Saito, K. Hatake, A. M. Dvorak, K. M. Leiferman, A. D. Donnengerg, N. Arai, K. Ishizaka, and T. Ishizaka, *Proc. Natl. Acad. Sci. USA*, *85*:2288 (1988).

26. A. J. Bos, R. Wever, M. N. Hamers, and D. Roos, *Infect. Immun.*, *32*:427 (1981).

27. A. N. Mayeno, A. J. Curran, R. L. Roberts, and C. S. Foote, *J. Biol. Chem.*, *264*:5660 (1989).

28. E. C. Jong, A. A. F. Mahmound, and S. J. Klebanoff, *J. Immunol.*, *126*:468 (1981).

29. E. C. Jong, W. R. Henderson, and S. J. Klebanoff, *J. Immunol.*, *124*:1378 (1980).

30. G. H. Ayars, L. C. Altman, G. J. Gleich, D. A. Loegering, and C. B. Baker, *J. Allergy Clin. Immunol.*, *76*:595 (1985).

31. G. H. Ayars, L. C. Altman, M. M. McManus, J. M. Agosti, C. Baker, D. L. Luchtel, D. A. Loegering, and G. J. Gleich, *Am. Rev. Respir. Dis.*, *140*:125 (1989).

32. E. C. Jong and S. J. Klebanoff, *J. Immunol.*, *124*:1949 (1980).

33. R. L. Baehner and R. B. Johnston, Jr., *Br. J. Haematol.*, *20*:277 (1971).

34. W. B. Davis, G. A. Fells, X. H. Sun, J. E. Gadek, A. Venet, and R. G. Crystal, *J. Clin. Invest.*, *74*:269 (1984).

35. M. S. Hibbs, C. L. Mainardi, and A. H. Kang, *Biochem. J.*, *207*:621 (1982)

36. M. Stahle-Backdahl, B. D. Sudbeck, A. Z. Eisen, H. G. Welgus, and W. C. Parks, *J. Invest. Dermatol.*, *99*:497 (1992).

37. M. Stahle-Backdahl and W. C. Parks, *Am. J. Pathol.*, *142*:995 (1993)

38. M. Stahle-Backdahl, M. Inoue, G. L. Guidice, and W. C. Parks, *J. Clin. Invest.*, *93*:2022 (1994)

39. C. M. Alexander and Z. Werb, *Cell Biology of Extracellular Matrix*, 2nd ed., (E.D. Hay, ed.), Plenum Press, New York, p. 255 (1991)

40. P. F. Weller, C. W. Lee, D. W. Foster, E. J. Corey, K. F. Austen, and R. A. Lewis, *Proc. Natl. Acad. Sci. USA*, *80*:7626 (1983)

41. A. J. Dessein, T. H. Lee, P. Elsas, J. Ravalese 3rd, D. Silberstein, J. R. David, K. F. Austen, and R. A. Lewis, *J. Immunol.*, *136*:3829 (1986)

42. E. Morita, J.-M. Schroder, and E. Christophers, *J. Immunol.*, *144*:1893 (1990)

43. U. Schwenk, E. Morita, R. Engel, and J. M. Schroder, *J. Biol. Chem.*, *267*:12482 (1992)

44. J. Turk, R. L. Maas, A. R. Brash, L. J. Roberts 2nd, and J. A. Oates. *J. Biol. Chem.*, *257*:7068 (1982).

45. T. Sugiura, K. Mabuchi, A. Ojima-Uchiyama, Y. Masuzawa, N. N. Cheng, T. Fukuda, S. Makino, and K. Waku, *J. Lipid Mediat.*, *5*:151 (1992)

46. A. R. E. Anwar and A. B. Kay, *J. Immunol.*, *119*:976 (1977)

47. J. R. Kanofsky, H. Hoogland, R. Wever, and S. J. Weiss, *J. Biol. Chem.*, *263*:9692 (1988)

48. D. C. Petreccia, W. M. Nauseef, and R. A. Clark, *J. Leukoc. Biol.*, *41*:283 (1987)

49. B. M. Babior, R. S. Kipnes, and J. T. Curnette, *J. Clin. Invest.*, *52*:741 (1973)

50. J. B. Sedgwick, R. F. Vrtis, M. F. Gourley, and W. W. Busse, *J. Allergy Clin. Immunol.*, *81*:876 (1988).

51. T. Yamashita, A. Someya, and E. Hara, *Arch. Biochem. Biophys.*, *241*:447 (1985)

52. D. S. Postma, T. E. Renkema, J. A. Noordhoek, H. Faber, H. J. Sluiter, and H. Kauffman, *Am. Rev. Respir. Dis.*, *137*:57 (1988).

53. J. H. Butterfield, K. M. Leiferman, and G. J. Gleich, *Samter's Immunologic Diseases*, (M. M. Frank, K. F. Austen, H. N. Claman, and E. R. Unanue, eds.), Little, Brown and Company, Boston, p. 501 (1994).

54. S. Motojima, E. Frigas, D. A. Loegering, and G. J. Gleich, *Am. Rev. Respir. Dis.*, *139*:801 (1989)

55. M. Azzawi, P. W. Johnston, S. Majumdar, A. B. Kay, and P. K. Jeffery, *Am. Rev. Respir. Dis.*, *145*:1477 (1992)

56. W. V. Filley, K. E. Holley, G. M. Kephart, and G. J. Gleich, *Lancet*, *2*:11 (1982).

57. D. S. Collins, R. Dupuis, G. J. Gleich, K. R. Bartemes, Y. Y. Koh, M. Pollice, K. H. Albertine, J. E. Fish, and S. P. Peters, *Am. Rev. Respir. Dis.*, *147*:677 (1993).

58. A. J. Wardlaw, S. Dunnette, G. J. Gleich, J. V. Collins, and A. B. Kay, *Am. Rev. Respir. Dis.*, *137*:62 (1988).

59. D. B. Jacoby, G. J. Gleich, and A. D. Fryer, *J. Clin. Invest.*, *91*:1314 (1993).

60. A. J. Coyle, S. J. Ackerman, R. Burch, D. Proud, and C. G. Irvin, *J. Clin. Invest.*, *95*:1735 (1995).

61. J. Lefort, M.-A. Nahori, C. Ruffie, B. B. Vargagtig, and M. Pretolani, *J. Clin. Invest*, *97*:1117 (1996).

62. G. Rajka, *Major Probl. Dermatol.*, *3*:42 (1975).

63. K. M. Leiferman, S. J. Ackerman, H. A. Sampson, H. S. Haugen, P. Y. Venencie, and G. J. Gleich, *N. Engl. J. Med.*, *313*:282 (1985).

64. K. M. Leiferman, T. Fujisawa, B. H. Gray, and G. J. Gleich, *Lab. Invest.*, *62*:579 (1990).

65. N. L. Ott, G. J. Gleich, E. A. Peterson, T. Fujisawa, S. Sur, and K. M. Leiferman, *J. Allergy Clin. Immunol.*, *94*:120 (1994).

66. D. L. Wassom, D. A. Loegering, G. O. Solley, S. B. Moore, R. T. Schooley, A. S. Fauci, and G. J. Gleich, *J. Clin. Invest.*, *67*:651 (1981).

67. A. Kapp, *Allergy*, *48*:1 (1993).

68. M. C. Lim, R. M. Taylor, and R. M. Naclerio, *Am. J. Respir. Crit. Care Med.*, *151*:136 (1995).

69. R. M. Naclerio, F. M. Baroody, A. Kagey-Sobotka, and L. M. Lichtenstein, *J. Allergy Clin. Immunol.*, *94*:1303 (1994).

70. R. Bascom, U. Pipkorn, D. Proud, S. Dunnette, G. J. Gleich, L. M. Lichtenstein, and R. M. Naclerio, *J. Allergy Clin. Immunol.*, *84*:338 (1989).

71. S. L. Harlin, D. G. Ansel, S. R. Lane, J. Myers, G. M. Kephart, and G. J. Gleich, *J. Allergy Clin. Immunol.*, *81*:867 (1988).

72. A. E. Butterworth, M. A. Vadas, D. L. Wassom, A. Dessein, M. Hogan, B. Sherry, G. J. Gleich, and J. R. David, *J. Exp. Med.*, *150*:1456 (1979).

73. R. I. Abu-Ghazaleh, T. Fujisawa, J. Mestecky, R. A. Kyle, and G. J. Gleich, *J. Immunol.*, *142*:2393 (1989).

74. D. W. Dunne, B. A. Richardson, F. M. Jones, M. Clark, K. J. Thorne, and A. E. Butterworth, *Parasite Immunol.*, *15*:181 (1993).

75. M. Capron, H. L. Spiegelberg, L. Prin, H. Bennich, A. E. Butterworth, R. J. Pierce, M. A. Ouaissi, and A. Capron, *J. Immunol.*, *132*:462 (1984).

76. M. Kaneko, M. C. Swanson, G. J. Gleich, and H. Kita, *J. Clin. Invest.*, *95*:2813 (1995).

77. T. Fujisawa, R. Abu-Ghazaleh, H. Kita, C. Sanderson, and G. J. Gleich, *J. Immunol.*, *144*:642 (1990).

78. H. Kita, D. A. Weiler, R. I. Abu-Ghazaleh, C. J. Sanderson, and G. J. Gleich, *J. Immunol.*, *149*:629 (1992).

79. S. Horie and H. Kita, *J. Immunol.*, *152*:5457 (1994).

80. R. Alam, S. Stafford, P. Forsythe, R. Harrison, D. Faubion, M. A. Lett-Brown, and J. A. Grant, *J. Immunol.*, *150*:3442 (1993).

81. A. Rot, M. Krieger, T. Brenner, S. C. Bischoff, T. J. Schall, and C. A. Dahinden, *J. Exp. Med.*, *176*:1489 (1992).

82. I. Winqvist, T. Olofsson, and I. Olsson, *Immunology*, *51*:1 (1984).

83. M. Yazdanbakhsh, C. M. Eckmann, L. Koenderman, A. J. Verhoeven, and D. Roos, *Blood*, *70*:379 (1987).

84. C. Kroegel, T. Yukawa, G. Dent, P. Venge, K. F. Chung, and P. J. Barnes, *J. Immunol.*, *142*:3518 (1989).

85. J. T. O'Flaherty, M. Kuroki, A. B. Nixon, J. Wijkander, E. Yee, S. L. Lee, P. K. Smitherman, R. L. Wykle, and L. W. Daniel, *J. Immunol.*, *157*:336 (1996).

86. P. J. Daffern, P. H. Pfeifer, J. A. Ember, and T. E. Hugli, *J. Exp. Med.*, *181*:2119 (1995).

87. P. Kernen, M. P. Wymann, V. von Tscharner, D. A. Deranleau, P. C. Tai, C. J. Spry, C. A. Dahinden, and M. Baggiolini, *J. Clin. Invest.*, *87*:2012 (1991).

88. C. Kroegel, M. A. Giembycz, and P. J. Barnes, *J. Immunol.*, *145*:2581 (1990).

89. H. Kita, R. I. Abu-Ghazaleh, G. J. Gleich, and R. T. Abraham, *J. Immunol.*, *147*:3466 (1991).

90. W. F. Owen Jr., R. J. Soberman, T. Yoshimoto, A. L. Sheffer, R. A. Lewis, and K. F. Austen, *J. Immunol.*, *138*:532 (1987).

91. P. L. Bruijnzeel, P. T. Kok, M. L. Hamelink, A. M. Kijne, and J. Verhagen, *Prostaglandins*, *34*:205 (1987).

92. R. J. Shaw, S. M. Walsh, O. Cromwell, R. Moqbel, C. J. Spry, and A. B. Kay, *Nature*, *316*:150 (1985).

93. R. Moqbel, A. J. Macdonald, O. Cromwell, and A. B. Kay, *Immunology*, *69*:435 (1990).

94. J. Palmblad, H. Gyllenhammar, J. A. Lindgren, and C. L. Malmsten, *J. Immunol.*, *132*:3041 (1984).

95. J. Elsner, M. Oppermann, W. Czech, G. Dobos, E. Schopf, J. Norgauer, and A. Kapp, *Eur. J. Immunol.*, *24*:518 (1994).

96. L. Koenderman, A. T. Tool, D. Roos, and A. J. Verhoeven, *J. Immunol.*, *145*:3883 (1990).

97. M. K. Bach, J. R. Brashler, E. N. Petzold, and M. E. Sanders, *Agents Actions Suppl.*, *35*:1 (1992).

98. H. J. Mengelers, T. Maikoe, L. Brinkman, B. Hooibrink, J. W. Lammers, and L. Koenderman, *Am. J. Respir. Crit. Care Med.*, *149*:345 (1994).

99. Y. Shimizu and S. Shaw, *Nature*, *366*:630 (1993).

100. E. F. Knol, F. Tackey, T. F. Tedder, D. A. Klunk, C. A. Bickel, S. A. Sterbinsky, and B. S. Bochner, *J. Immunol.*, *153*:2161 (1994).

101. B. S. Bochner, S. A. Sterbinsky, C. A. Bickel, S. Werfel, M. Wein, and W. Newman, *J. Immunol.*, *152*:774 (1994).

102. M. Wein, S. A. Sterbinsky, C. A. Bickel, R. P. Schleimer, and B. S. Bochner, *Am. J. Respir. Cell Mol. Biol.*, *12*:315 (1995).

103. S. P. Neeley, K. J. Hamann, T. L. Dowling, K. T. McAllister, S. R. White, and A. R. Leff, *Am. J. Respir. Cell Mol. Biol.*, *8*:633 (1993).

104. E. A. Clark and J. S. Brugge, *Science*, *268*:233 (1995).

105. M. S. Diamond and T. A. Springer, *Current Biol.*, *4*:506 (1994).

106. G. M. Walsh, A. Hartnell, A. J. Wardlaw, K. Kurihara, C. J. Sanderson, and A. B. Kay, *Immunology*, *71*:258 (1990).

107. B. S. Bochner, F. W. Luscinskas, M. A. Gimbrone Jr., W. Newman, S. A. Sterbinsky, C. P. Derse-Anthony, D. Klunk, and R. P. Schleimer, *J. Exp. Med.*, *173*:1553 (1991).

108. G. M. Walsh, J. J. Mermod, A. Hartnell, A. B. Kay, and A. Wardlaw, *J. Immunol.*, *146*:3419 (1991).

109. A. Hartnell, R. Moqbel, G. M. Walsh, B. Bradley, and A. B. Kay, *Immunology*, *69*:264 (1990).

110. H. C. Wan, A. I. Lazarovits, W. W. Cruikshank, H. Kornfeld, D. M. Center, and P. F. Weller, *Int. Arch. Allergy Immunol.*, *107*:343 (1995).

111. M. Van der Vieren, H. L. Trong, C. L. Wood, P. F. Moore, T. St. John, D. E. Staunton, and W. M. Gallatin, *Immunity*, *3*:683 (1995).

112. P. Dri, R. Cramer, P. Spessotto, M. Romano, and P. Patriarca, *J. Immunol.*, *147*:613 (1991).

113. S. Horie, G. J. Gleich, and H. Kita, *J. Allergy Clin. Immunol.*, *98*:371 (1996).

114. C. Laudanna, P. Melotti, C. Bonizzato, G. Piacentini, A. Boner, M. D. Serra, and G. Berton, *Immunology*, *80*:273 (1993).

115. M. Kato, R. T. Abraham, G. J. Gleich, and H. Kita, *J. Allergy Clin. Immunol.*, *95*:340 (1995).

116. H. Kita, S. Horie, and G. J. Gleich, *J. Immunol.*, *156*:1174 (1996).

117. E. Ruoslahti and M. D. Pierschbacher, *Science*, *238*:491 (1987).

118. T. Ishikawa, K. Wicher, and C. E. Arbesman, *Int. Arch. Allergy Appl. Immunol.*, *46*:230 (1974).

119. M. Capron, M. D. Kazatchkine, E. Fischer, M. Joseph, A. E. Butterworth, J. P. Kusnierz, L. Prin, J. P. Papin, and A. Capron, *J. Immunol.*, *139*:2059 (1987).

120. M. Kaneko, S. Horie, M. Kato, G. J. Gleich, and H. Kita, *J. Immunol.*, *155*:2631 (1995).
121. I. L. Graham, H. D. Gresham, and E. J. Brown, *J. Clin. Invest.*, *81*:365 (1988).
122. G. Sehgal, K. Zhang, R. F. Todd III, L. A. Boxer, and H. R. Petty, *J. Immunol.*, *150*:4571 (1993).
123. E. Crockett-Torabi and J. C. Fantone, *J. Immunol.*, *145*:3026 (1990).
124. L. Koenderman, S. W. Hermans, P. J. Capel, and J. G. van de Winkel, *Blood*, *81*:2413 (1993).
125. A. F. Lopez, D. J. Williamson, J. R. Gamble, C. G. Begley, J. M. Harlan, S. J. Klebanoff, A. Waltersdorph, G. Wong, S. C. Clark, and M. A. Vadas, *J. Clin. Invest.*, *78*:1220 (1986).
126. T. van der Bruggen, P. T. Kok, J. A. Raaijmakers, J. W. Lammers, and L. Koenderman, *J. Immunol.*, *153*:2729 (1994).
127. A. R. Anwar, R. Moqbel, G. M. Walsh, A. B. Kay, and A. J. Wardlaw, *J. Exp. Med.*, *177*:839 (1993).
128. S. N. Georas, B. W. McIntyre, M. Ebisawa, J. L. Bednarczyk, S. A. Sterbinsky, R. P. Schleimer, and B. S. Bochner, *Blood*, *82*:2872 (1993).
129. A. R. Anwar, G. M. Walsh, O. Cromwell, A. B. Kay, and A. J. Wardlaw, *Immunology*, *82*:222 (1994).
130. S. P. Neeley, K. J. Hamann, T. L. Dowling, K. T. McAllister, S. R. White, and A. R. Leff, *Am. J. Respir. Cell Mol. Biol.*, *11*:206 (1994).
131. N. M. Munoz, K. F. Rabe, S. P. Neeley, A. Herrnreiter, X. Zhu, K. McAllister, D. Mayer, H. Magnussen, S. Galens, and A. R. Leff, *Am. J. Physiol.*, *270*:L587 (1996).
132. T. W. Kuijpers, E. P. Mul, M. Blom, N. L. Kovach, F. C. Gaeta, V. Tollefson, M. J. Elices, and J. M. Harlan, *J. Exp. Med.*, *178*:279 (1993).
133. G. S. Elemer and T. S. Edgington, *J. Biol. Chem.*, *269*:3159 (1994).
134. C. Garcia, M. Montero, J. Alvarez, and M. S. Crespo, *J. Biol. Chem.*, *268*:4001 (1993).
135. R. P. Schleimer, S. A. Sterbinsky, J. Kaiser, C. A. Bickel, D. A. Klunk, K. Tomioka, W. Newman, F. W. Luscinskas, M. A. Gimbrone, B. W. McIntyre, and B. S. Bochner, *J. Immunol.*, *148*:1086 (1992).
136. M. Nagata, J. B. Sedgwick, M. E. Bates, H. Kita, and W. W. Busse, *J. Immunol.*, *155*:2194 (1995).
137. M. A. Vadas, C. M. Lucas, J. R. Gamble, A. F. Lopez, M. P. Skinner, and M. C. Berndt, *Eosinophils in allergy and inflammation* (G. J. Gleich and A. B. Kay, eds.), Marcel Dekker, Inc., New York, p. 69 (1994).
138. T. Satoh, A. Knowles, M.-S. Li, L. Sun, J. A. Tooze, G. Zabucchi, and C. J. G. Spry, *Immunology*, *83*:313 (1994).
139. C.-K. Huang, *Membr. Biochem.*, *8*:61 (1989).
140. H. Kita, *Signal transduction in leukocytes: role of G protein-related and other pathways* (P. M. Lad, J. S. Kaptein, and C.-K. E. Lin, eds.), CRC Press, Boca Raton, p. 285 (1995).
141. D. A. Bass, J. C. Lewis, P. Szejda, L. Cowley, and C. E. McCall, *Lab. Invest.*, *44*:403 (1981).
142. A. A. Abdel-Latif, *Pharmacol. Rev.*, *38*:227 (1986).

143. C. Kroegel, E. R. Chilvers, M. A. Giembycz, R. A. J. Challiss, and P. J. Barnes, *J. Allergy Clin. Immunol.*, *88*:114 (1991).

144. M. Kato, R. T. Abraham, and H. Kita, *J. Immunol.*, *155*:357 (1995).

145. H. Kita, R. I. Abu-Ghazaleh, G. J. Gleich, and R. T. Abraham, *J. Immunol.*, *147*:3466 (1991).

146. R. M. Perlmutter, S. D. Levin, M. W. Appleby, S. J. Anderson, and J. Alberola-Ila, *Ann. Rev. Immunol.*, *11*:451 (1993).

147. M. E. Bates, P. J. Bertics, and W. W. Busse, *J. Immunol.*, *156*:711 (1996).

148. K. Pazdrak, S. Stafford, and R. Alam, *J. Immunol.*, *155*:397 (1995).

149. T. van der Bruggen, E. Caldenhoven, D. Kanters, P. Coffer, J. A. Raaijmakers, J. W. Lammers, and L. Koenderman, *Blood*, *85*:1442 (1995).

150. M. Kato, R. T. Abraham, G. J. Gleich, and H. Kita, *Am. J. Respir. Crit. Care Med.*, *153*:A58 (1996).

# 13

# Mast Cell Activation and Leukocyte Recruitment Responses Into Skin Sites

## Role of Cell Adhesion Molecules

**Michael D. Ioffreda and George F. Murphy**    *University of Pennsylvania School of Medicine, Philadelphia, Pennsylvania*

## I. INTRODUCTION

Inflammatory disorders that affect the skin are distinguished histologically by the nature, degree, and microscopic localization of leukocytes, often resulting in a distinctive clinical picture. Cutaneous neoplasms may also have leukocytic infiltrates that help to identify them histologically, such as the plasma cells that aid in the identification of syringocystadenoma papilliferum (1). The exact mechanisms responsible for the influx of characteristic leukocytes in many inflammatory disorders of the skin are poorly understood. Inherent to mechanisms of cutaneous inflammation are factors responsible for leukocyte homing to sites of skin inflammation, including expression of specific cell adhesion molecules (CAMs) on skin microvessels and chemoattractant factors important to transvascular diapedesis. In recent years, research advances in cutaneous immunology have begun to elucidate the mechanisms responsible for recruitment and selection of specific subsets of leukocytes to microvascular beds.

The idea that a special relationship exists between skin and the immune system was first advanced by Streilein (2). The concept of skin-associated lymphoid tissue (SALT) encompasses skin-seeking lymphocytes that function in immune surveillance and may traffic to both normal and inflamed integument. That unique subsets of previously stimulated lymphocytes "home" to particular tissues, including skin (3), was reviewed by Picker and Butcher (4). Fundamental to the principles governing cutaneous leukocyte trafficking is that naive T cells migrate into lymph nodes, whereas memory T lymphocytes migrate primarily into nonlymphoid tissue (5).

How sensitized T cells selectively home to cutaneous microvascular beds represents a fundamental mystery in skin biology. Presumably, local factors influence microvessels in such a way that promotes interaction between the plasma membranes of circulating leukocytes and endothelial cells, culminating in transmural diapedesis of the former into the perivascular space. The molecular events responsible for this phenomenon are undoubtedly complex and coordinated. Candidate cells in the local microenvironment that might influence microvascular endothelium include keratinocytes of the epidermis as well as cells that reside in the immediate perivascular space, including mast cells.

It is now known that many important effects of mast cells on the recruitment of leukocytes in cutaneous inflammation relate to their direct effects on endothelial cells (6,7). In general, mast cells congregate perivascularly about microvascular plexi just below environmental–epithelial interfaces, where they are ideally situated for impacting on leukocyte influx into tissues (8). Cutaneous mast cells have been called "gatekeepers" of inflammation because of their localization in the skin and their effect on recruitment of leukocytes (9). Skin mast cells are located preferentially about postcapillary venules of the superficial vascular plexus, which forms an anastomosing network parallel to and directly beneath the epidermal layer. The same plexus from which capillaries sprout to supply the avascular epidermis with nutrients also serves as a conduit for the different leukocytes that invade the epidermis in various types of dermatitis. For example, T lymphocytes are the primary responders in allergic contact dermatitis, polymorphonuclear leukocytes are characteristic of psoriasis, and eosinophils may invade the epidermis in bullous pemphigoid. In non-epidermotropic disorders, rarely are infiltrates confined to the deep dermis; almost always they exclusively or dually involve the superficial perivascular spaces. Thus, the presence of mast cells about the superficial vascular plexus is an important structural clue to potential functional interaction with endothelial cells that might serve to attract leukocytes from the bloodstream to the most superficial cutaneous layers.

The factors secreted by mast cells are varied. Many of these factors have been identified in murine mast cells, but human mast cells have been found to contain some of them also. Structural and biochemical heterogeneity exists between mast cells at different body sites (reviewed in refs. 10–14). Connective tissue type mast cells (MCTC), present in skin and gut submucosa, contain the serine proteinases chymase and tryptase, whereas mucosal mast cells (MCT) contain only tryptase (15). In addition to such proteinases, mast cells also contain proteoglycans (heparin), vasoactive peptides (histamine), and cytokines (tumor necrosis factor–$\alpha$ (TNF-$\alpha$), transforming growth factor–$\beta$ (TGF-$\beta$), interleukins (IL-3, IL-4, IL-5, IL-6, IL-8), leukotriene C4, platelet-activating factor (PAF), and granulocyte-macrophage colony-stimulating factor (GM-CSF) (16–18). Some mediators are preformed and released on activation, such as

TNF-α (9). The production of others begins only when induced by the proper conditions.

## II. MAST CELL ACTIVATION

Secretagogues are substances that cause mast cells to release their mediators by a process termed degranulation, whereby membrane-bound cytoplasmic granules undergo rapid exocytosis (Fig. 1A–D). Mast cell secretagogues include the neuropeptide substance P, morphine sulfate, compound 48/80, calcium ionophore, the complement components C3a and C5a, major basic protein, and histamine-releasing factors (HRF), including macrophage inflammatory protein-1α (MIP-1α) and monocyte chemoattractant peptide–1 (MCP-1) (16,19,20). Mast cell mediator release can also be initiated by proteins or factors that cross-link IgE bound to high-affinity Fcε receptors on the mast cell surface (21,22). The IgE response is characteristic of mast cell degranulation in type I reactions and is prototypical of the ability of many secretagogues to induce granule discharge within minutes of exposure ("anaphylactic degranulation"). In addition, unmyelinated type C sensory nerve fibers originating in the dorsal root ganglion release neurotransmitters that impact on tissue mast cells as a result of retrograde impulses (reviewed in ref. 20). The neuropeptides substance P, vasoactive intestinal peptide (VIP), neurokinin A and B, somatostatin, calcitonin gene related peptide, and adenosine nucleotides have been recognized as mast cell degranulating agents (20,23), and the potential importance of neurogenic and psychogenic modulation of the cutaneous immune response via mast cells has been emphasized (24,25). There is long-standing evidence that substance P is capable of causing histamine release from skin mast cells (26,27), although heterogeneity appears to exist regarding secretagogue responsiveness of mast cells from different anatomical sites.

In addition to "anaphylactic," or immediate, degranulation, connective tissue mast cells may demonstrate time-dependent, partial, degranulation of the "piecemeal" type, with loss of granules and portions of granules over time, a finding that is potentially relevant to more chronic inflammatory disorders of skin (28,29). The functional significance of this finding with respect to possible selective mediator release by connective tissue mast cells has yet to be determined.

Mast cells from different body sites also respond differently to stimuli. Cutaneous mast cells release histamine after exposure to degranulating doses of morphine sulfate, substance P, and compound 48/80; mucosal mast cells do not respond by degranulating (reviewed in ref. 20). Other factors that are important to mast cell degranulation in vivo include physical stimuli such as cold, heat, and mechanical trauma (30). The array of mast cell activators, the substances they release, and the timing of release all may impact on the inflammatory phenotype of the leukocyte response.

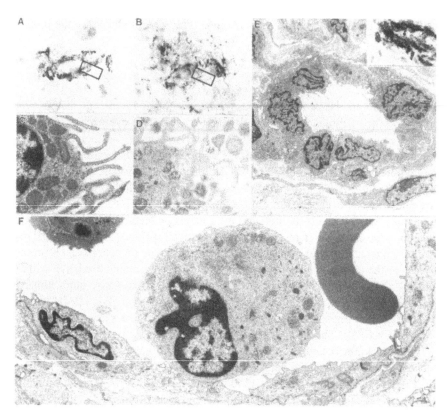

**Figure 1** Morphologic expression of mast cell–endothelial interactions in human skin. Normal perivascular mast cell (A), stained immunohistochemically for the granule proteinase chymase, differs from degranulating mast cell exposed to secretagogue (B); note apparent externalization of granule contents into the perivascular space. Ultrastructural analysis of secretagogue effect on mast cells confirms rapid transformation of normally electron-dense cytoplasmic granules (C) to variable electron-lucent, coalescent granules that undergo exocytosis by merging with the plasma membrane (D). Adjacent dermal postcapillary venule (E) shows "rounding up" of endothelial cells after mast cell degranulation as a result of centripetal endothelial contraction; the vessel now resembles the high endothelial venule of lymph nodes. Molecular alterations, such as induction of E-selectin on the endothelial surface (E) may be demonstrated immuno-histochemically after mast cell degranulation and, in this case, are primarily the consequence of release of tumor necrosis factor–$\alpha$ (TNF-$\alpha$) by the degranulating mast cell. The result of mast cell endothelial interaction may be adhesion of leukocytes to the luminal surface of the activated endothelial cell (F, here demonstrated after experimental mast cell degranulation in human skin xenografted to a SCID mouse).

## III. ENDOTHELIAL CELL ACTIVATION BY MAST CELL PRODUCTS

Mast cell degranulation causes venular endothelial cells to undergo structural changes. These reproducible alterations include consolidation of cytoplasmic actin filaments and cellular contraction or "rounding up," producing intercellular "gaps" that allow transudation of plasma into the interstitium, resulting in edema and fibrin deposition. These morphologic changes were originally termed the "histamine effect" (13,31), though they are now recognized in many immune reactions and correlate with metabolic changes typical of an "activated" endothelial cell. The activated endothelial cells of postcapillary dermal venules bear structural resemblance to high-endothelial venules (HEV) of lymph nodes, which normally serve to sequester blood leukocytes within lymphoid parenchyma. The functional analogy in skin is the role that activated postcapillary venules may assume in cutaneous leukocyte influx. In particular, experimental immune complex–mediated vasculitis characterized by neutrophil recruitment and cutaneous delayed-type hypersensitivity (DTH) reactions mediated by lymphocytes both principally target the postcapillary venule (32). Both responses also share endothelial activation, mast cell degranulation, and early leukocyte influx that are temporally and spatially linked.

In the late 1980s, mast cell degranulation was elucidated as an early event in DTH reactions corresponding to both structural evidence of endothelial activation within postcapillary venules and their expression of cell adhesion molecules (CAMs) (33) prior to leukocyte infiltration. These findings suggested that mast cell degranulation might in some way be linked to cascades of CAM expression on endothelial cells, and that these events could represent early triggers for initiation of cellular tissue inflammation (34).

## IV. THE ENDOTHELIAL CELL–LEUKOCYTE ADHESION CASCADE

### A. Recruitment of Neutrophils

The steps involved in neutrophil accumulation at sites of inflammation take place in a temporally and spatially coordinated fashion called the endothelial cell–leukocyte adhesion cascade, which was reviewed in depth by Albelda et al. (35), and will be presented here to put into perspective the mast cell–mediated effects on this process. It is the process by which leukocytes adhere to activated postcapillary venules expressing sequential endothelial CAMs.

Neutrophil slowing within venules, also known as margination, involves neutrophil rolling along the lumenal endothelial surface at velocities of less than 50 μm/sec (36), and is moderated by CAMs belonging to the selectin family. Members of the selectin family have in common an N-terminal, lectin-like domain capable of binding specific carbohydrate ligands on leukocytes (37). L-selectin (LECAM-1) is constitutively expressed on the surface of neutrophils and loosely binds them to endothelial cells via an as yet unidentified ligand. P-selectin (CD62, GMP140, PADGEM) mediates relatively weak, rolling adhesion between leukocytes and endothelial cells under flow conditions. It is constitutively stored in cytoplasmic, lysosome-like Weibel-Palade bodies and is translocated to the lumenal endothelial surface within minutes of exposure to histamine, thrombin, bradykinin, leukotriene C4, or free radicals (38). Mast cells may be the direct source of histamine and leukotriene C4, which contribute to P-selectin–mediated rolling of neutrophils on postcapillary venules (39,40). The ligand for P-selectin on neutrophils may be P-selectin glycoprotein ligand (PSGL-1) (41). The binding interaction between neutrophils and E-selectin is capable of supporting neutrophil rolling under conditions of flow in vitro, suggesting that E-selectin expressed by cutaneous vessels could potentially contribute to neutrophil rolling in vivo (42). It is expressed by postcapillary venules (43) in response to TNF-$\alpha$, which is produced and secreted by a variety of skin cell types, including mast cells (Fig. 1E). Indeed, mast cell degranulation is known to directly up-regulate expression of E-selectin in a TNF-$\alpha$–dependent manner (9,44).

Interestingly, E-selectin is also induced by IL-1 derived from epidermal keratinocytes (45). In addition to the primary keratinocyte IL-1, IL-1$\alpha$, these cells also make IL-1$\beta$. Both induce E-selectin, although IL-1$\beta$ requires mast cell assistance for activation. A precursor of IL-1$\beta$ is cleaved to an active form by the serine proteinase chymase, a constitutive component of mast cell secretory granules (46). E-selectin is synthesized and expressed on the lumenal endothelial membrane within 6 hours after exposure to a provocative cytokine. The ligand for E-selectin on myeloid cells is the carbohydrate sialyl-Lewis$^x$ (47,48).

At baseline, neutrophils are nonactivated and relatively nonadhesive. Firm adhesion of neutrophils to venules after initial selectin-mediated slowing depends on induction of members of the integrin family of CAMs, and neutrophil rolling appears to be a prerequisite for this to occur. This occurs via activation on neutrophils of the $\beta$2 (CD 18) integrins, leukocyte function–associated antigen (LFA)-1 (CD11a/CD18), and Mac-1 (CD11b/CD18), which mediate binding to up-regulated intercellular adhesion molecule-1 (ICAM-1), normally constitutively expressed on endothelial cells (35). ICAM-1 may be upregulated by TNF-$\alpha$, a mast cell secretory product, as well as by interferon gamma (IFN-$\gamma$) produced by other skin cell types (49); this increased expression is important for neutrophil–endothelial interactions. Diapedesis of neutrophils seems to be strongly dependent on leukocyte integrins, especially CD11b/CD18 (Mac-1), but both CD11a and CD11b may be involved, acting in synergy (50). Neutrophils do not appear to use the VLA-4/VCAM pathway to bind to stimulated endothelial cells (50).

Activation allows neutrophils to shed L-selectin, flatten, and up-regulate

CD11b in preparation for firm adherence to the endothelium. Platelet-activating factor (PAF) may be one mechanism by which neutrophils become activated. Platelet-activating factor is released by human lung mast cells (51), and is also rapidly produced by endothelial cells that may be stimulated by histamine, thrombin, or leukotriene C4, the same factors that may also influence P-selectin transference to the lumenal surface of the endothelial cell. Although human skin mast cells are unable to produce PAF, PAF injected into the skin causes an immediate wheal-and-flare response followed later by neutrophilic inflammation (reviewed in ref. 20). Platelet-activating factor may be involved in the activation of neutrophils bound to endothelium by P-selectin; this activation may include the up-regulation of β2 integrins (52). As neutrophils roll along the vessel wall, they can encounter endothelial cell–derived PAF and become activated, thus shedding L-selectin and intensifying their surface expression of β2 integrins (53). In addition, interleukin-8 (IL-8) may be manufactured by endothelial cells in response to exposure to TNF-α and binds to the wall of stimulated vessels, where it is able to activate neutrophils (reviewed in ref. 35).

Finally, transmigration of neutrophils through the vessel wall into the interstitium is accomplished in part by a chemotactic gradient. One potential chemoattractant is the cytokine IL-8, which is induced by TNF-α release, as stated above (35,54). In addition, Moller and colleagues demonstrated the production of IL-8 by the human mast cell line HMC-1 and by cutaneous mast cells after anti-IgE activation (55).

Platelet-endothelial cell adhesion molecule-1 (PECAM-1, CD31), a member of the immunoglobulin gene superfamily of CAMs, is expressed on the surface of leukocytes, platelets, and endothelial cells. In cultured endothelial cells that recapitulate a partially activated phenotype, PECAM-1 localizes at inter-endothelial cell junctions (56). Possibly through homophilic adhesion, PECAM-1 may play a role in the passage of neutrophils through the gaps between endothelial cells (57,58). In our laboratory, we have observed resting endothelium in situ to express PECAM-1 circumferentially along the entirety of the endothelial plasma membrane. Endothelium stimulated by TNF-α or mast cell degranulation, however, is characterized by aggregation of PECAM-1 to inter-endothelial junctions where diapedesis is expected to occur (unpublished observations). Such redistribution could establish an "adhesion gradient" that contributes to preferential congregation of loosely bound leukocytes to endothelial cell–cell junctions in preparation for their transmural migration. Muller and colleagues showed that anti–PECAM-1 antibodies blocked the transmigration of leukocytes through TNF-α–activated endothelial monolayers in vitro, and Vaporciyan et al. supported this finding in animal models of neutrophil transmigration (58,59). In vivo experiments using mast cell–deficient mice showed that TNF of mast cell origin was responsible for neutrophil emigration (60).

There is evidence to suggest that redundancy exists such that both E- and P-selectin may contribute to leukocyte rolling (61). Studies of E-selectin–deficient mice, which possessed the ability to support neutrophilic inflammation, showed that the anti-murine P-selectin antibody 5H1 could impair neutrophil accumulation and edema in skin in a delayed-type hypersensitivity reaction (61). However, Lawrence noted that neutrophils in vitro roll on substrates bearing purified selectins, and that neutrophil adhesion to E-selectin is much stronger than P-selectin, with neutrophil rolling velocities that are slower and have less variance (62,63). Although both E- and P-selectin have been shown to be important in leukocyte rolling, the expression of *both* selectins in a given inflammatory event may not be essential for leukocyte rolling and transmural diapedesis to occur. With this in mind, E- and P-selectin–dependent leukocyte rolling may represent redundancy of a general mechanism for leukocyte rolling. This fact may be significant in that different types of inflammation may require different CAM combinations and preferentially employ one mechanism or the other. For instance, in cobra venom–induced inflammation, anti–P-selectin antibodies have a blocking effect, whereas anti–E-selectin antibodies do not (64). In vitro studies of rat mesentery treated with the mast cell secretagogue compound 48/80 increased the number of rolling leukocytes, an effect that was impeded by an anti–P-selectin antibody, but not by histamine-receptor antagonists (65). It has been confirmed by several sources that TNF-α could be important in the expression of P-selectin on endothelial cells in vivo (66,67).

Inflammatory cells home to baseline skin as a part of normal cutaneous immune surveillance. Using intravital microscopy in BALB/c and homozygous hairless mice, investigators demonstrated that physiologic leukocyte rolling in the absence of an inflammatory challenge could be blocked by the administration of anti–P-selectin antibody, but not anti–L- or anti–E-selectin antibody, implicating P-selectin in spontaneous leukocyte rolling in skin (68), a finding supported by other groups (69,70). Unstimulated neutrophils at their baseline spontaneously adhere weakly to endothelial cell monolayers (50).

In vivo evidence for the importance of mast cells to neutrophilic recruitment was shown by Gaboury et al., albeit in studies of mucosal mast cells (71). In rat mesentery, perfusion with the secretagogue 48/80 caused a dose-dependent increase in the number of rolling and adherent neutrophils. This effect was reduced by the mast cell stabilizers ketotifen or cromolyn, by chronic depletion of mast cell mediators, by the H1 histamine receptor antagonist diphenhydramine, and by an anti–P-selectin antibody. These results support an important role for histamine and histamine-induced P-selectin from mucosal mast cells in neutrophil rolling in vivo. Furthermore, a PAF-receptor antagonist and an anti-CD18 antibody squelched the mast cell secretagogue–induced rise in neutrophil adhesion. These data suggest that mast cell–derived histamine induces P-selectin, which promotes leukocyte rolling, as well as PAF, which contributes to CD18-dependent leukocyte adhesion.

## B.  Recruitment of Lymphocytes: General Considerations

The adhesion pathway responsible for the recruitment of lymphocytes to the skin differs from that pertaining to neutrophils, although some overlap exists (72). (The steps involved in the recruitment of neutrophils and lymphocytes into skin sites as a consequence of mast cell degranulation are diagrammed in Fig. 2.) Moreover, the activation status of T cells greatly influences the nature of the adhesion cascade responsible for their accumulation in skin. For example, binding of naive human CD4+ T cells to human umbilical vein endothelial cells (HUVEC) occurs primarily via LFA-1/ICAM-1 interaction, whereas memory CD4+ T cells adhere via more complex interactions involving LFA-1/ICAM-1, sialyl-Lewis$^x$/E-selectin, and VLA-4/VCAM-1 (73).

VCAM-1 (vascular cell adhesion molecule-1) is an adhesion molecule relevant to binding of lymphocytes and certain tumor cells (melanoma) to vascular endothelium. Like ICAM-1 and PECAM-1, it is a member of the immunoglobulin gene superfamily (35). VCAM-1 is not expressed constitutively by endothelium, but is induced within 24 hours after exposure to TNF-$\alpha$ and IL-4, both cytokines produced by perivascular mast cells (74). The receptor on leukocytes for VCAM-1 is VLA-4, the $\alpha4\beta1$ integrin that also mediates leukocyte interaction with fibronectin. The functional importance of VLA-4/VCAM-1 binding in cutaneous inflammation is underscored by recent studies showing inhibition of DTH reactions in mice treated with antibodies to VLA-4 (75) or VCAM-1 (76,77).

Lymphocyte–endothelial adhesion not only differs from neutrophil–endothelial adhesion with respect to endothelial ligands (e.g., VCAM-1), but also as a result of expression of specific skin-homing molecules by T cells. Cutaneous lymphocyte–associated antigen (CLA) is a molecule closely related to sialyl-Lewis$^x$ (78,79), and is purported to be the primary T cell ligand for E-selectin (80). It is of interest that IL-6, a cytokine present in mast cells, may up-regulate CLA expression on the lymphocyte surface (81,82). Although only 10% to 25% of circulating CD3+ T cells are CLA-positive, 90% of T cells that infiltrate the skin express this molecule (83). Via interaction with E-selectin, CLA is believed to be in part responsible for accumulation of memory/effector T cells in sites of antigen-specific challenge (84). VLA-4/VCAM-1 interaction also appears to be important in cutaneous homing of CLA-positive T cells, because antibodies to VLA-4 and VCAM-1 interrupt migration of CLA-enriched but not CLA-depleted T cells across cytokine-stimulated HUVEC (84). Both CLA-positive and CLA-negative T cells express VLA-4; therefore, it has been postulated that binding of CLA-positive T cells to activated endothelium may cause changes in VLA-4 that enhance its functional interaction with VCAM-1. In this system, anti-CD18/CD11a and anti–ICAM-1 antibodies block adhesion of both CLA-

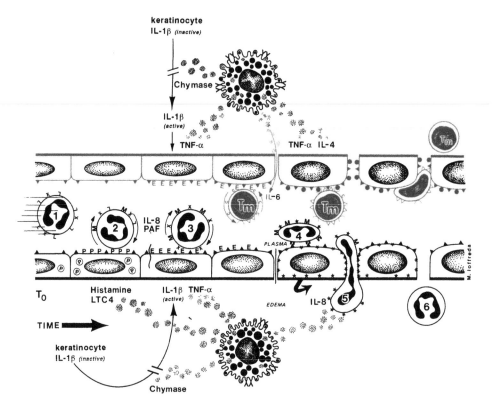

**Figure 2** Mast cell effects on leukocyte recruitment into skin sites, dermal post-capillary venule sectioned longitudinally.

Neutrophil recruitment *(bottom)*. Mast cell activation at $T_0$ releases a variety of mediators that impact on endothelial and leukocyte expression of cell adhesion molecules (CAMs) and release of endothelial-derived cytokines. These mediators include histamine, leukotriene C4 (LTC4), chymase, tumor necrosis factor–$\alpha$ (TNF-$\alpha$), interleukin (IL)-8, IL-4, and IL-6. Circulating neutrophils [1] contain L-selectin (L) and sialyl-Lewis$^x$ (x) on their cell membrane. Neutrophil slowing [2] on the endothelial surface (margination) begins when histamine and LTC4 cause endothelial cells to rapidly translocate P-selectin (P) from cytoplasmic Weibel-Palade bodies to the lumenal membrane. Histamine activation of endothelial cells begins to induce "rounding up," producing intercellular gaps through which plasma escapes into the interstitium, causing dermal edema. Keratinocyte IL-1$\beta$ may be cleaved by chymase to an active form which stimulates, in conjunction with TNF-$\alpha$, E-selectin (E) expression on endothelial cells. E-selectin interaction with sialyl-Lewis$^x$ on neutrophils [3] causes them to roll at a slower velocity along the vessel wall. Histamine may incite endothelial cells to produce PAF (platelet-activating factor), and TNF-$\alpha$ may cause them to produce IL-8, two factors that contribute to neutrophil activation, heralded by up-regulation of Mac-1/LFA-1 (M) and flattening of the cell [4]. Up-regulation of Mac-1/LFA-1 leukocyte associated antigen (LFA)1 is associated with loss of L-selectin from the positive and CLA-negative T cells (84). In summary, CLA expression by lymphocytes fosters migration across stimulated HUVEC via CLA/E-selectin, VLA-4/VCAM-1, and LFA-1/ICAM-1 interactions, whereas CLA-independent T cell migration primarily involves the LFA-1/ICAM-1 pathway (84). In vitro experiments by Meng et al. have shown that mast cells affect VCAM-1 and ICAM-1 expression by vascular endothelial cells, possibly through secretion of preformed TNF-α (85).

Once lymphocytes have diapedesed into the perivascular space, adhesion pathways potentially regulated in part by mast cells continue to contribute to their migratory fate. VCAM-1 and ICAM-1 may affect lymphocyte adhesion to interstitial cells such as fibroblasts (86) and perivascular dermal dendrocytes (87). Fibroblast cultures exposed to mast cell–conditioned media show up-regulation of these molecules, a phenomenon attributed to release of TNF-α upon degranulation (86). TNF-α may also affect CD4+ T lymphocyte localization to the extracellular matrix by binding to fibronectin and up-regulating β1 integrin-mediated T cell adhesion to fibronectin (88). Mast cells themselves appear to be anchored to the perivascular space rich in laminin in part because of their

---

neutrophil surface. Mac-1/LFA-1 interacts with ICAM-1 (▲) to form a stronger bond with the endothelial cell, slowing the neutrophil even further in preparation for diapedesis. ICAM-1 is constitutively expressed by these vessels, but is up-regulated by TNF-α. Neutrophil transmigration [5] follows a chemotactic gradient of IL-8. Transmigration may be facilitated by platelet-endothelial cell adhesion molecule-1 (PECAM-1) (★) redistribution from diffuse endothelial cell membrane expression (constitutive) to concentration at inter-endothelial cell junctions with activation by TNF-α. PECAM-1 binds to leukocyte PECAM-1 via homophilic interaction. Upon completion of transmural diapedesis, a neutrophil [6], having successfully journeyed from the bloodstream, resides at least temporarily in the perivascular dermis.

Memory T lymphocyte (Tm) recruitment (*top*). Some of the same influences and endothelial CAMs responsible for neutrophil influx contribute to skin recruitment of T lymphocytes of the memory subset. E-selectin endothelial expression occurs as the result of activated IL-1β and mast cell TNF-α. Lymphocytes interact with endothelial E-selectin via cutaneous lymphocyte antigen, or CLA (c), an analogue of sialyl-Lewis$^x$, on the lymphocyte surface. Cutaneous lymphocyte associated antigen (CLA) expression may be up-regulated by IL-6. More stable contacts between the lymphocyte and endothelial cell come about as a result of LFA-1 on the lymphocyte cell membrane (ΔΔ) binding with up-regulated ICAM-1 (▲) on the endothelial cell surface, and endothelial induction of vascular cell adhesion molecule-1 (VCAM-1) (•) caused by TNF-α and IL-4; VCAM-1 binds to VLA-4 (∪), an integrin that is constitutively expressed on the lymphocyte cell surface. These latter interactions may be important in lymphocyte transmigration along with less well understood mechanisms. The time frame for lymphocyte influx into skin is often later than that for neutrophil influx.

expression of α6β1 integrin (89). Indeed, manipulation of mast cell localization with respect to extracellular matrix has been suggested as a strategy to modulate allergic inflammatory responses (90).

## C.  Type I Reactions

IgE-mediated reactions have traditionally been synonymous with mast cell/basophil-induced reactions. Indeed, the importance of the mast cell to IgE-mediated inflammatory responses in vivo was emphasized by Wershil and Galli in studies of mast cell–deficient mice (91). However, the presence of high-affinity IgE receptors on other cell types, including Langerhans cells, monocytes, and eosinophils, expands the potential importance of IgE-mediated immune responses (18).

It has been shown in IgE-mediated skin reactions that mast cells exhibit ultrastructural changes characteristic of both anaphylactic and piecemeal degranulation (29). In vitro studies examining the adhesion of leukocytes to cultured HUVEC monolayers stimulated with TNF showed that both ICAM-1 and E-selectin are involved in the binding of eosinophils and neutrophils (92). The fact that similar mechanisms and dose responses are necessary for the adhesion of neutrophils and eosinophils to cytokine-stimulated endothelial cells may explain the recruitment of eosinophils into skin sites of cutaneous type I reactions.

Immediate-type hypersensitivity reactions (type I) are mediated in part by histamine, which is mast cell derived (93). It has been shown that histamine can result in the rapid expression of P-selectin on postcapillary venules. Endothelial expression of P-selectin results in early neutrophil adhesion associated with neutrophil rolling (94). As stated earlier, P-selectin is constitutively stored in Weibel-Palade bodies (38) and is translocated to the lumenal endothelial surface within minutes of exposure to histamine, thrombin, bradykinin, leukotriene C4, or free radicals. A study of the type I hypersensitivity reaction in skin demonstrated that mast cell degranulation had occurred 20 min after intradermally injected pollen antigen in sensitized individuals (95). This coincided with neutrophil accumulation about some dermal vessels that expressed P-selectin along their lumenal surface. Translocation of P-selectin from Weibel-Palade bodies to the lumenal membrane was confirmed by immunoelectron microscopy. Subsequent display of E-selectin on these same vessels was noted by 6 hours. Asako et al. showed that histamine-dependent leukocyte recruitment could be blocked by histamine (H1) receptor antagonists such as hydroxyzine and diphenhydramine as well as blocking antibodies to P-selectin (96). In IgE-mediated type I reactions in murine skin in vivo, mast cell degranulation is associated with tissue edema and subsequent infiltration of neutrophils; in many cases this is followed by the arrival of lymphocytes and eosinophils (97).

In IgE-mediated type I hypersensitivity reactions in the skin, the immediate reaction to allergen is followed by local inflammation that peaks 6 to 12 hours later and is known as the late-phase reaction (LPR) (98). Because of accessibility of the integument, the LPR in the skin has provided a model for study of a similar tissue response to allergen in the lung (99). The inflammatory cells in the LPR include neutrophils and CD4+ T lymphocytes of the memory subset (99). As mentioned above, in IgE-mediated cutaneous type I reactions induced by intradermal pollen injection in allergic individuals, microvenules at antigen injection sites at 20 min initially showed perivascular mast cell degranulation, focal margination of neutrophils, neo-expression of P-selectin, and absence of E-selectin (95). At 6 hours these sites showed increased margination and diapedesis of neutrophils, as well as eosinophils, endothelial expression of E-selectin, and diminished P-selectin expression. Thus, both the early and late components of this response may be at least in part explained by mast cell mediators (histamine induced translocation of P-selectin and TNF-α–induced expression of E-selectin, respectively). There is in vivo evidence in a mouse model using mast cell–deficient *W/W*ᵛ mice demonstrating that neutrophil recruitment in IgE-dependent cutaneous LPR is mast cell dependent, with TNF-α primarily responsible (100).

Leung et al. studied the human LPR in vivo and in vitro, comparing skin of atopic patients with respiratory allergy to non-atopic skin (101). Sequential biopsies performed between 20 min and 24 hours after intradermal injection of allergen showed E-selectin expression beginning at 2–4 hours, followed by leukocyte influx at 3–4 hours. Skin organ cultures established from atopic individuals injected with allergen or saline prior to harvest demonstrated E-selectin only in the allergen-treated sites, albeit without an inflammatory infiltrate. This implicated resident skin cells, as opposed to infiltrating cells, as the source of inciting cytokines. Furthermore, E-selectin expression could be blocked in vitro by polyclonal antisera to IL-1 (α and β) and TNF-α together, but not singly, emphasizing the importance of both cytokines to E-selectin induction.

Werfel and colleagues showed that in the cutaneous LPR in human skin in vivo, the majority of T lymphocytes in the fluid of skin blister chambers were of the memory phenotype and expressed increased levels of VLA-4, IL-2 receptor (CD25), and HLA-DR antigens, indicating that these cells are preferentially recruited to skin sites of allergen challenge in sensitized individuals (102). Similar in vivo studies have also documented up-regulation of E-selectin and ICAM-1 in human LPRs (92). Studies of cutaneous type I reactions also have demonstrated PECAM-1 redistribution from a cell membrane pattern to inter-endothelial junctions (103). In light of experimental evidence suggesting a role for PECAM-1 at endothelial cell–cell junctions in leukocyte diapedesis, the redistribution phenomenon of PECAM-1 at these sites potentially supports this function in cutaneous type I reactions.

## D.  Type IV Reactions

The prototype of the type IV hypersensitivity reaction or the delayed-type hypersensitivity (DTH) reaction in human skin is allergic contact dermatitis to poison ivy (rhus dermatitis). The human cutaneous DTH has an afferent limb whereby epidermal Langerhans cells take up topically applied antigen, process it, and present it to naive T lymphocytes after migrating to draining lymph nodes (initiation). During subsequent exposure to that antigen (challenge), memory/effector T lymphocytes respond by trafficking to the epidermis at the site of topically applied antigen (104). Lewis and colleagues showed that DTH reactions are characterized by mast cell degranulation as early as 4 hours after antigenic challenge, preceding leukocyte influx by at least 12 hours (105). Waldorf et al. demonstrated that mast cell degranulation is temporally coupled to the expression of E-selectin by dermal postcapillary venules, and that this, too, precedes leukocyte influx in human cutaneous DTH (33). In addition, immunohistochemical characterization of the initial leukocytes recruited to E-selectin–positive venules were of the helper-inducer/memory phenotype. Work by Berg et al. indicates that E-selectin may be a specific ligand for T-lymphocyte binding of the memory subclass (78).

Galli and coworkers have demonstrated that murine mast cells are capable of synthesizing and secreting TNF-$\alpha$ (106). In human skin, Walsh et al. have demonstrated that dermal mast cells are a source of preformed TNF-$\alpha$ and contain more stores of this cytokine per cell than monocyte/macrophages or resting keratinocytes (9). This implicated mast cell–derived preformed TNF-$\alpha$ in the induction of endothelial selectins in early phases of cutaneous inflammation. Resting mast cells express relatively little TNF-$\alpha$ mRNA, but message increases after experimental degranulation (107) or during skin inflammation (9).

An organ culture model was developed in which neonatal foreskin samples containing epidermis and dermis were incubated partially submerged in culture medium to study the early phases ($\leq$ 48 hours) of cutaneous inflammation in vitro. Using this model, Klein and coworkers noted E-selectin expression on postcapillary venules after exposure to mast cell secretagogues. This effect could be blocked by anti–TNF-$\alpha$ antibodies, implicating TNF-$\alpha$ as the crucial mediator responsible for this induction (44). Mast cell degranulation was confirmed by loss or relative decreases compared to controls in immunohistochemical reactivity for the mast cell protease, chymase, 1 hour after initial exposure to mast cell secretagogues. The peak of E-selectin expression was at 4–6 hours and waned by 24 hours.

One of the mediators that may result in mast cell degranulation in vivo is the neuropeptide substance P. Matis et al. showed that substance P could degranulate mast cells in skin organ culture in vitro and produce TNF-$\alpha$-like effects (24). Other studies in vivo in the mouse implicated an indirect role for substance P and

CGRP (calcitonin gene-related peptide) in leukocyte influx during allergic contact dermatitis (23).

In order to determine whether mast cell degranulation and E-selectin induction are temporally linked to lymphocyte recruitment in vivo, skin inflammation corresponding to the DTH was studied in healthy male volunteers (33). Subjects were sensitized with topically applied 2, 4-dinitrochlorobenzene and subsequently challenged 2 weeks later. Serial biopsies were performed after challenge at preclinical (1, 2, 4, 6 hours) and clinical (24 and 48 hours) stages of inflammation. Mast cell degranulation was confirmed at 1 hour by immunohistochemical and ultrastructural evidence of chymase depletion. E-selectin was induced at 2–6 hours, followed at 6–24 hours by perivascular recruitment of lymphocytes, which were CD45RO+, confirming their identity as helper-inducer/memory T cells. Epidermotropic migration of T lymphocytes was first observed in follicular epithelium and was ICAM-1 independent. Interfollicular epithelium expressed ICAM-1 between 24 and 96 hours, presumably related to IFN-$\gamma$ from lymphocytes (108). This was associated with epidermotropism of LFA-1-positive lymphocytes. Thus, mast cell degranulation was the earliest event documented in the human DTH reaction, preceding display of endothelial and keratinocyte adhesion molecules and related inflammation.

## E. The SCID Mouse/Human Skin Xenograft (SCID/hsx) Bioassay

A major advance in the understanding of cutaneous inflammation and the role of mast cells in this process came in 1983 with the advent of a genetically mutated strain of mouse, the SCID (severe combined immunodeficient) mouse (109). Due to a homozygous mutation in the *scid* locus, B- and T-lymphocyte development is retarded. Thus, these mice lack mature B and T cells and endogenous immunoglobulin production. These animals permit the xenografting of human skin. Full-thickness human foreskin has been successfully xenografted to SCID mice with graft viability exceeding 4 months (110). Human skin architecture and cytology is structurally and functionally intact, as confirmed by histology, immunohistochemical reactivity, and by electron microscopy (110,111).

The microvasculature of the human xenografts form anastomoses with the murine microvasculature and hence is perfused by murine blood. This was verified by anti–human PECAM-1 (CD31) and anti–murine PECAM antibodies. Perivascular mast cells are found in normal numbers, retain human ultrastructural characteristics, and show reactivity for human chymase and TNF-$\alpha$ for 2 months (112), after which they become hyperplastic. Murine neutrophils are capable of adhering to selectins expressed by the human endothelium because of homology between human and murine leukocyte–endothelial adhesive ligands (113). The

SCID/hsx mouse model permits evaluation of mast cell degranulation on the accumulation of leukocytes in a system where mast cell-activated human vessels are perfused in a living animal.

The intradermal injection of recombinant TNF-α into human skin xenografts has been shown to effect murine neutrophil influx via induction of endothelial E-selectin, and intravenous antibodies to human E-selectin can block neutrophil immigration (114). Recent experiments in our laboratory have recapitulated in the SCID/hsx model many of the findings of prior in vitro experiments that used the organ culture system and cultured human dermal microvascular endothelial cells (HDMEC) (115). Specifically, experimentally induced mast cell degranulation in xenografts was linked to tissue edema, E-selectin expression by microvessels, and accumulation of leukocytes in affected vessels (Fig. 1F). However, lumenal expression of P-selectin was not detected after intradermal injection of the mast cell secretagogue 48/80 using a technique whereby anti-endothelial cell monoclonal antibodies directed against P-selectin were simultaneously injected intravenously. The functional significance of a lack of detectable up-regulation of P-selectin expression in this experimental model has yet to be explored. Intradermal leukocyte accumulation peaked between 2 and 4 hours after secretagogue injection. Furthermore, intravenous injection of a monoclonal antibody to human E-selectin reduced secretagogue-induced dermal accumulation of leukocytes following mast cell degranulation by 78%. Thus, mast cell–induced expression of endothelial E-selectin appears to be a sufficient stimulus for leukocyte recruitment in vivo.

With regard to chronic skin inflammation, there is evidence to suggest that although E-selectin expression in vitro may be transient even in the setting of continuous cytokine stimulation (116), perhaps as a result of endothelial tachyphylaxis, persistent in vivo expression of E-selectin in chronic inflammation may be possible as a result of enhanced stability of certain E-selectin mRNA transcripts secondary to gene regulatory mechanisms (117).

# V.  CONCLUSION AND FUTURE DIRECTIONS

The role of endothelial adhesion molecules has been shown to be important in many forms of cutaneous inflammation (118,119). It is not a prerequisite that mast cells degranulate for inflammation to occur in vivo (120). Although mast cells may play a role in many forms of cutaneous inflammation, certainly other mechanisms exist for mast cell–independent initiation of the inflammatory responses. For instance, lipopolysaccharide, as may be produced by infiltrating bacteria, may directly induce the expression of adhesion molecules on endothelium, including E-selectin and P-selectin (66,76,121). Cytokines produced by keratinocytes (122), monocyte/macrophages, and even endothelial cells themselves (123) probably play a role in endothelial CAM regulation and expression.

Mast cells may not be a part of all inflammation occurring in the skin, but there is strong evidence that mast cells participate in many types of cutaneous inflammation, including the active early recruitment of leukocytes to sites of skin inflammation. There is evidence for common features between mast cell–dependent mechanisms that occur early in different inflammatory processes. For instance, despite differences in kinetics and composition of the inflammatory infiltrate in LPR and the classic DTH, there may be some overlap between the two (99). Mast cell degranulation may serve as an "early common pathway" in the induction of many types of cutaneous inflammation.

Understanding the possible contributory role of mast cells in the regulation of pro-inflammatory endothelial CAMs may also provide insight into potential new therapeutic strategies to blunt skin inflammation at its earliest stages. Although cromolyn is currently used to inhibit lung mast cell secretion in allergic asthma, it is not used in the therapy of mast cell–related skin disease, because of its limited effects on cutaneous mast cells in vivo. Because mast cell degranulation is an active process involving signal transduction and cytoskeletal alterations, it is possible that novel strategies may be developed to abrogate secretion. Substance P is an endogenous secretagogue; therefore, inhibition of its activity could potentially contribute to amelioration of certain "neurogenic" influences in skin inflammation. Blockades of mast cell TNF-α, histamine, or their receptors, and of specific endothelial CAMs by antibodies, have been effective anti-inflammatory strategies in in vivo models (114,115). Recently, antisense oligonucleotides to ICAM-1 have been utilized effectively to down-regulate human skin inflammation in vivo in the SCID/hsx model (124). Such approaches are desirable because they could target pro-inflammatory signals inherent in the mast cell–endothelial axis before the influx of and injury attendant to inflammatory cells homing to the skin.

## REFERENCES

1. G. F. Murphy, *Dermatopathology*, W. B. Saunders Company, Philadelphia, p. 213 (1995)
2. J. W. Streilein, *Immunol. Ser.*, *46*:73 (1989).
3. L. J. Picker, S. A. Michie, L. S. Rott, and E. C. Butcher, *Am. J. Pathol.*, *136*:1053 (1990).
4. L. J. Picker and E. C. Butcher, *Ann. Rev. Immunol.*, *10*:561 (1992).
5. Y. Shimizu, W. Newman, Y. Tanaka, and S. Shaw, *Immunol. Today*, *13*:106 (1992).
6. R. M. Lavker and G. F. Murphy, *Pharmacol. Skin*, *4*:51 (1991).
7. M. D. Tharp, *J. Invest. Dermatol.*, *93*:107S (1989).

8. D. D. Metcalfe and M. Kaliner, *CRC Crit. Rev. Immunol.*, *3*:23 (1981).
9. L. J. Walsh, G. Trinchieri, H. A. Waldorf, D. Whitaker, and G. F. Murphy, *Proc. Natl. Acad. Sci. U.S.A.*, *88*:4220 (1991).
10. M. K. Church, Y. Okayama, and P. Bradding, *Ann. N.Y. Acad. Sci.*, *725*:13 (1994).
11. L. B. Schwartz and K. F. Austen, *Prog. Allergy*, *34*:271 (1984).
12. S. J. Galli, A. M. Dvorak, and H. F. Dvorak, *Prog. Allergy*, *34*:1 (1984).
13. S. J. Galli, *Lab. Invest.*, *62*:5 (1990).
14. N. Weidner and K. F. Austen, *Lab. Invest.*, *63*:63(1990).
15. N. M. Schechter, *Monogr. Allergy* (L. B. Schwartz, ed.), Karger, Basel, p. 114 (1990).
16. S. J. Galli, *N. Engl. J. Med.*, *328*:257 (1993).
17. M. R. Parwaresch, H. P. Horny, and K. Lennert, *Pathol. Res. Pract.*, *179*:439 (1985).
18. J. S. Marshall and J. Bienenstock, *Curr. Opin. Immunol.*, *6*:853 (1994).
19. R. Alam, K. Dhruv, D. Anderson-Walters, and P. A. Forsythe, *J. Immunol.*, *152*:1298 (1994).
20. M. D. Tharp, *Dermatol. Clin.*, *8*:619 (1990).
21. B. Baird, R. J. Shopes, V. T. Oi, J. Erickson, P. Kane, and D. Holowka, *Int. Arch. Allergy Appl. Immunol.*, *88*:23 (1989).
22. M. Plaut, J. H. Pierce, C. J. Watson, J. Hanley-Hydes, R. P. Nordan, and W. E. Paul, *Nature*, *339*:64 (1989).
23. M. Goebeler, U. Henseleit, J. Roth, C. Sorg, *Arch. Dermatol. Res.*, *286*:341 (1994).
24. W. L. Matis, R. M. Lavker, and G. F. Murphy, *J. Invest. Dermatol.*, *94*:492 (1990).
25. G. F. Murphy, *Dermal Immune System* (B. J. Nickoloff, ed.), CRC Press, Boca Raton, p. 227 (1993).
26. O. Hagermark, T. Hokfelt, and B. Pernow, *J. Invest. Dermatol.*, *71*:233 (1978).
27. J. M. Ebertz, C. A. Hirshman, N. S. Kettelkamp, H. Uno, and J. M. Hanifin, *J. Invest. Dermatol.*, *88*:682 (1987).
28. A. M. Dvorak, M. C. Mihm Jr, J. E. Osage, T. H. Kwan, K. F. Austen, and B. U. Wintroub, *J. Invest. Dermatol.*, *78*:91 (1982).
29. M. S. Kaminer, G. F. Murphy, B. Zweiman, and R. M. Lavker, *Clin. Diagn. Lab. Immunol.*, *2*:297 (1995).
30. G. F. Murphy, K. F. Austen, E. Fonferko, and A. L. Sheffer, *J. Allergy Clin. Immunol.*, *80*:603 (1987).
31. K. Williams-Kretschmer, M. H. Flax, and R. S. Cotran, *Lab. Invest.*, *1*:334 (1967).
32. R. S. Cotran, *Current Topics in Inflammation* (G. Majno, R. S. Cotran, M. H. Flax, and N. Kaufman, eds.), Williams & Wilkins, Baltimore, p. 18 (1982).
33. H. A. Waldorf, L. J. Walsh, N. M. Schechter, and G. F. Murphy, *Am. J. Pathol.*, *138*:477 (1991).
34. L. J. Walsh, R. M. Lavker, and G. F. Murphy, *Lab. Invest.*, *63*:592 (1990).
35. S. M. Albelda, C. W. Smith, and P. A. Ward, *FASEB J.*, *8*:504 (1994).
36. M. B. Lawrence and T. A. Springer, *Cell*, *65*:859 (1991).
37. M. P. Bevilacqua and R. M. Nelson, *J. Clin. Invest.*, *91*:379 (1993).
38. E. Larsen, A. Celi, G. E. Gilbert, B. C. Furie, J. K. Erban, R. Bonfanti, and B. Furie, *Cell*, *59*:305 (1989).

39. P. Kubes and S. Kanwar, *J. Immunol.*, *152*:3570 (1994).
40. D. A. Jones, O. Abbassi, L. V. McIntire, R. P. McEver, and C. W. Smith, *Biophys. J.*, *65*:1560 (1993).
41. K. L. Moore, K. D. Patel, R. E. Bruehl, F. Li, D. A. Johnson, H. S. Lichenstein, R. D. Cummings, D. F. Bainton, and R. P. McEver, *J. Cell. Biol.*, *128*:661 (1995).
42. O. Abbassi, T. K. Kishimoto, L. V. McIntire, D. C. Anderson, and C. W. Smith, *J. Clin. Invest.*, *92*:2719 (1993).
43. D. V. Messadi, J. S. Pober, W. Fiers, M. A. Gimbrone Jr., and G. F. Murphy, *J. Immunol.*, *139*:1557 (1987).
44. L. M. Klein, R. M. Lavker, W. L. Matis, and G. F. Murphy, *Proc. Natl. Acad. Sci. U.S.A.*, *86*:8972 (1989).
45. H. Mizutani, R. Black, and T. S. Kupper, *J. Clin. Invest.*, *87*:1066 (1991).
46. H. Mizutani, N. M. Schechter, G. Lazarus, R. A. Black, and T. S. Kupper, *J. Exp. Med.*, *174*:821 (1991).
47. M. L. Phillips, E. Nudelman, F. C. Gaeta, M. Perez, A. K. Singhal, S. Hakomori, and J. C. Paulson, *Science*, *250*:1130 (1990).
48. G. Walz, A. Aruffo, W. Kolanus, M. Bevilacqua, and B. Seed, *Science*, *250*:1132 (1990).
49. M. L. Dustin, R. Rothlein, A. K. Bhan, C. A. Dinarello, and T. A. Springer, *J. Immunol.*, *137*:245 (1986).
50. K. Yong and A. Khwaja, *Blood Rev.*, *4*:211 (1990).
51. R. A. Lewis and K. F. Austen, *Nature*, *293*:103 (1981).
52. D. E. Lorant, M. K. Topham, R. E. Whatley, R. P. McEver, T. M. McIntyre, S. M. Prescott, and G. A. Zimmerman, *J. Clin. Invest.*, *92*:559 (1993).
53. G. A. Zimmerman, T. M. McIntyre, M. Mehra, and S. M. Prescott, *J. Cell Biol.*, *110*:529 (1990).
54. R. A. Swerlick and T. J. Lawley, *J. Invest. Dermatol.*, *100*:111S (1993).
55. A. Moller, U. Lippert, D. Lessman, G. Kolde, K. Hamann, P. Welker, D. Schadendorf, T. Rosenbach, T. Luger, and B. M. Czarnezki, *J. Immunol.*, *151*:3261 (1993).
56. W. A. Muller, C. M. Ratti, S. L. McDonell, and Z. A. Cohn, *J. Exp. Med.*, *170*:399 (1989).
57. S. A. Bogen, H. S. Baldwin, S. C. Watkins, S. M. Albelda, and A. K. Abbas, *Am. J. Pathol.*, *141*:843 (1992).
58. W. A. Muller, S. A. Weigl, X. Deng, and D. M. Phillips, *J. Exp. Med.*, *178*:449 (1993).
59. A. A. Vaporciyan, H. M. DeLisser, H-C. Yan, I. I. Mendiguren, S. R. Thom, M. L. Jones, P. A. Ward, and S. M. Albelda, *Science*, *262*:1580 (1993).
60. Y. Zhang, B. F. Ramos, and B. A. Jakschik, *Science*, *258*:1957 (1992).
61. M. A. Labow, C. R. Norton, J. M. Rumberger, K. M. Lombard-Gillooly, D. J. Shuster, J. Hubbard, R. Bertko, P. A. Knaack, R. W. Terry, M. L. Harbison, F. Kontgen, C. L. Stewart, K. W. McIntyre, P. C. Will, D. K. Burns, and B. A. Wolitzky, *Immunity*, *1*:709 (1994).
62. M. B. Lawrence and T. A. Springer, *J. Immunol.*, *151*:6338 (1993).
63. M. B. Lawrence and T. A. Springer, *Cell*, *65*:859 (1991).

64. M. S. Mulligan, M. J. Polley, R. J. Bayer, M. F. Nunn, J. C. Paulson, and P. A. Ward, *J. Clin. Invest.*, *90*:1600 (1992).
65. H. Thorlacius, J. Raud, S. Rosengren-Beezley, M. J. Forrest, P. Hedqvist, and L. Lindbom, *Biochem. Biophys. Res. Comm.*, *203*:1043 (1994).
66. U. Gotsch, U. Jager, M. Dominis, and D. Vestwever, *Cell Adhes. Commun.*, *2*:7 (1994).
67. A. Weller, S. Isenmann, and D. Vestweber, *J. Biol. Chem.*, *267*:15176 (1992).
68. D. Nolte, P. Schmid, U. Jager, A. Botzlar, F. Roesken, R. Hecht, E. Uhl, K. Messmer, and D. Vestweber, *Am. J. Physiol.*, *267*:H1637 (1994).
69. M. Dore, R. J. Korthuis, D. N. Granger, M. L. Entman, and C. W. Smith, *Blood*, *82*:1308 (1993).
70. T. N. Mayadas, R. C. Johnson, H. Rayburn, R. O. Hynes, and D. D. Wagner, *Cell*, *74*:541 (1993).
71. J. P. Gaboury, B. Johnston, X-F. Niu, and P. Kubes, *J. Immunol.*, *154*:804 (1995).
72. D. Rohde, W. Schluter-Wigger, V. Mielke, P. von den Driesch, B. von Gaudecker, and W. Sterry, *J. Invest. Dermatol.*, *98*:794 (1992).
73. Y. Shimizu, W. Newman, T. V. Gopal, K. J. Horgan, N. Graber, L. D. Beall, G. A. van Seventer, and S. Shaw, *J. Cell Biol.*, *113*:1203 (1991).
74. P. Bradding, I. H. Feather, P. H. Howarth, R. Mueller, J. A. Roberts, K. Britten, J. P. Bews, T. C. Hunt, Y. Okayama, C. H. Heusser, et al., *J. Exp. Med.*, *176*:1381 (1992).
75. M. J. Elices, S. Tamraz, V. Tollefson, and L. W. Vollger, *Clin. Exp. Rheumatol.*, *11*:S77 (1993).
76. D. M. Briscoe, R. S. Cotran, and J. S. Pober, *J. Immunol.*, *149*:2954 (1992).
77. T. B. Issekutz, *Am. J. Pathol.*, *143*:1286 (1993).
78. E. L. Berg, T. Yoshino, L. S. Rott, M. K. Robinson, R. A. Warnock, T. K. Kishimoto, L. J. Picker, and E. C. Butcher, *J. Exp. Med.*, *174*:1461 (1991).
79. H. Rossiter, F. van Reijsen, G. C. Mudde, F. Kalthoff, C. A. Bruijnzeel-Koomen, L. J. Picker, and T. S. Kupper, *Eur. J. Immunol.*, *24*:205 (1994).
80. L. J. Picker, T. K. Kishimoto, C. W. Smith, R. A. Warnock, and E. C. Butcher, *Nature*, *349*:796 (1991).
81. S. Kruger-Krasagakes, A. Moller, G. Kolde, U. Lippert, M. Weber, and B. M. Henz, *J. Invest. Dermatol.*, *106*:75 (1996).
82. L. J. Picker, J. R. Treer, B. Ferguson-Darnell, P. A. Collins, P. R. Bergstresser, and L. W. Terstappen, *J. Immunol.*, *150*:1122 (1993).
83. L. J. Picker, R. J. Martin, A. Trumble, L. S. Newman, P. A. Collins, P. R. Bergstresser, and D. Y. M. Leung, *Eur. J. Immunol.*, *24*:1269 (1994).
84. L. F. Santamaria Babi, R. Moser, M. T. Perez Soler, L. J. Picker, K. Blaser, and C. Hauser, *J. Immunol.*, *154*:1543 (1995).
85. H. Meng, M. G. Tonnesen, M. J. Marchese, W. F. Bahou, R. A. Clark, and B. L. Gruber, *J. Cell Physiol.*, *165*:40 (1995).
86. H. Meng, M. J. Marchese, J. A. Garlick, A. Jelaska, J. H. Korn, J. Gailit, R. A. F. Clark, and B. L. Gruber, *J. Invest. Dermatol.*, *105*:789 (1995).
87. H. Sueki, D. Whitaker, M. Buchsbaum, and G. F. Murphy, *Lab. Invest.*, *69*:160 (1993).

88. R. Alon, L. Cahalon, R. Hershkoviz, D. Elbaz, B. Reizis, D. Wallach, S. K. Akiyama, K. M. Yamada, and O. Lider, *J. Immunol.*, *152*:1304 (1994).

89. L. J. Walsh, M. S. Kaminer, G. S. Lazarus, R. M. Lavker, and G. F. Murphy, *Lab. Invest.*, *65*:433 (1991).

90. M. M. Hamawy, S. E. Mergenhagen, and R. P. Siraganian, *Immunol. Today*, *15*:62 (1994).

91. B. K. Wershil and S. J. Galli, *Adv. Exp. Med. Biol.*, *347*:39 (1994).

92. U. Kyan-Aung, D. O. Haskard, R. N. Poston, M. H. Thornhill, and T. H. Lee, *J. Immunol.*, *146*:521 (1991).

93. M. D. Tharp, R. T. Suvunrungsi, and T. J. Sullivan, *J. Immunol.*, *130*:1896 (1983).

94. J-G. Geng, M. P. Bevilacqua, K. L. Moore, T. M. McIntyre, S. M. Prescott, J. M. Kim, G. A. Bliss, G. A. Zimmerman, and R. P. McEver, *Nature*, *343*:757 (1990).

95. G. Murphy, L. Leventhal, and B. Zweiman, *J. Allergy Clin. Immunol.* [Abstr.], *93*:183 (1994).

96. H. Asako, I. Kurose, S. DeFrees, Z-L. Zheng, M. L. Phillips, J. C. Paulson, and D. N. Granger, *J. Clin. Invest.*, *93*:1508 (1994).

97. B. K. Wershil, Z-S. Wang, J. Gordon, and S. J. Galli, *J. Clin. Invest.*, *87*:446 (1991).

98. E. N. Charlesworth, A. F. Hood, N. A. Soter, A. Kagey-Sobotka, P. S. Norman, and L. M. Lichtenstein, *J. Clin. Invest.*, *83*:1519 (1989).

99. A. J. Frew, V. A. Varney, M. Gaga, and A. B. Kay, *Skin Pharmacol.*, *4*:71 (1991).

100. B. K. Wershil and S. J. Galli, *Immunobiology of Proteins and Peptides VII*, (M. Z. Atassi, ed.), Plenum Press, New York, p. 39 (1994).

101. D. Y. M. Leung, J. S. Pober, and R. S. Cotran, *J. Clin. Invest.*, *87*:1805 (1991).

102. S. Werfel, W. Massey, L. M. Lichtenstein, and B. S. Bochner, *J. Allergy Clin. Immunol.*, *96*:57 (1995).

103. M. D. Ioffreda, S. M. Albelda, D. E. Elder, A. Radu, L. C. Leventhal, B. Zweiman, and G. F. Murphy, *Endothelium*, *1*:47 (1993).

104. J. W. Streilein, S. F. Grammer, T. Yoshikawa, A. Demidem, and M Vermeer, *Immunol. Rev.*, *117*:159 (1990).

105. R. E. Lewis, M. Buchsbaum, D. Whitaker, and G. F. Murphy, *J. Invest. Dermatol.*, *93*:672 (1989).

106. J. D. Young, C. C. Liu, G. Butler, Z. A. Cohn, and S. Galli, *Proc. Natl. Acad. Sci. U.S.A.*, *84*:9175 (1987).

107. J. R. Gordon and S. J. Galli, *Nature*, *346*:274 (1990).

108. M. L. Dustin, K. H. Singer, D. T. Tuck, and T. A. Springer, *J. Exp. Med.*, *167*:1323 (1988).

109. G. C. Bosma, R. P. Custer, and M. J. Bosma, *Nature*, *301*:527 (1983).

110. H-C. Yan, I. Juhasz, J. Pilewski, G. F. Murphy, M. Herlyn, and S. M. Albelda, *J. Clin. Invest.*, *91*:986 (1993).

111. V. Mielke, W. Sterry, and R. Kaufmann, *J. Invest. Dermatol.* [Abstr.], *100*:445 (1993).

112. M. Christofidou-Solomidou, S. M. Albelda, and G. F. Murphy, *J. Invest. Dermatol.* [Abstr.], *102*:611 (1994).

113. M. Becker-Andre, R. H. VanHuijsduijnen, C. Losberger, J. Whelan, and J. F. Delamarter, *Eur. J. Biochem.*, *206*:401 (1992).

114. H-C. Yan, H. M. DeLisser, J. M. Pilewski, K. M. Barone, P. J. Szklut, Chang X-J, Ahern T. J. Langer-Safer P, S. M. Albelda. *J. Immunol., 152*:3052 (1994).
115. M. Christofidou-Solomidou, G. F. Murphy, and S. M. Albelda, *Am. J. Pathol., 148*:177 (1996).
116. M. P. Bevilacqua, J. S. Pober, D. L. Mendrick, R. S. Cotran, and M. A. Gimbrone, *Proc. Natl. Acad. Sci. USA, 84*:9238 (1987).
117. W. Chu, D. H. Presky, R. A. Swerlick, and D. K. Burns, *J. Immunol., 153*:4179 (1994).
118. R. W. Groves, M. H. Allen, J. N. W. N. Barker, D. O. Haskard, and D. M. MacDonald, *Br. J. Dermatol., 124*:117 (1991).
119. A. A. Irani, H. A. Sampson, and L. B. Schwartz, *Allergy, 44*:31 (1989).
120. S. J. Galli and I. Hammel, *Science, 226*:710 (1984).
121. M. P. Bevilacqua, *Annu. Rev. Immunol., 11*:767 (1993).
122. J-M. Schroder, *J. Invest. Dermatol., 105*:20S (1995).
123. R. A. Swerlick, E. Garcia-Gonzalez, Y. Kubota, Y. Xu, and T. J. Lawley, *J. Invest. Dermatol., 97*:190 (1991).
124. M. Christofidou-Solomidou, C. F. Bennett, S. M. Albelda, and G. F. Murphy, *J. Invest., Dermatol.* [Abstr.], *106*:830 (1996).

# 14

# Mast Cell Activation and Leukocyte Rolling Responses

**Paul Kubes**   *Immunology Research Group, University of Calgary Medical Center, Calgary, Alberta, Canada*

## I.  INTRODUCTION

The human body has developed a tremendously effective immune system to identify and destroy invading foreign particles. This requires a detection system that must be present in all tissues and a cytotoxic system that can be summoned and rapidly gain access to the afflicted tissue, thereby minimizing infection. Deficiencies in the development of the inflammatory response lead to a severely compromised host. However, an inappropriate, excessive, or unnecessarily prolonged inflammatory response leads to debilitating or even fatal diseases. Clearly, understanding the mechanisms that underlie the induction and termination of the inflammatory response will be critical for the development of new therapeutics that will modulate numerous pathologies.

An important component of the complex series of events that constitute the inflammatory process is the recruitment of leukocytes. It is now well appreciated that leukocytes migrate into tissues via a cascade of events that can be divided into three components: 1.) tethering to and rolling along the length of endothelium, 2.) firm adhesion to the endothelium, and 3.) transendothelial migration into the interstitial space. The division is based, at least in part, on the fact that each of these events is mediated by a different group of adhesion molecules, some of which are constitutively expressed whereas others require activation [see (1–4)]. It is our view (Fig. 1) that mast cells in the extravascular space of all tissues, when activated, release all of the stimuli to initiate leukocyte rolling, adhesion, and emigration. It is the aim of this chapter to summarize some of the data available to support a role for mast cells as initiators of leukocyte recruitment with a particular focus on leukocyte rolling and the selectins. I also high-

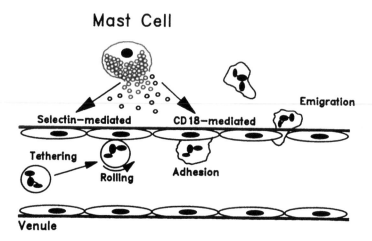

**Figure 1** Leukocytes enter sites of inflammation and injury via a multistep recruitment paradigm. The first step is initial contact with the endothelium, termed tethering, that is followed by a rolling action along the length of the venule. These events are mediated by a group of endothelial adhesion molecules termed E-selectin and P-selectin. This event is a prerequisite to firm adhesion and emigration via the integrins. The expression and/or activation of these adhesion molecules occurs via pro-inflammatory mediators released from perivascular cells such as the mast cells found in the local environment (see text for details).

light some of the areas of controversy and emphasize areas that require further investigation.

## II.  INTRAVITAL MICROSCOPY

The most common method used to study leukocyte behavior in the microcirculation is an approach termed intravital microscopy, which permits direct visualization of the lumen of microvessels (2,5). White cells that have moved from the mainstream of blood and have made direct contact with the lining of the vessel move with sufficiently reduced velocity that their behavior can be observed using a microscope. A standard (nonfluorescent) microscope can be used; however, the tissue must be translucent, thus the use of atypical tissues such as the mesentery (6), the bat wing (7), and the cremaster muscle (8). One can use other tissues such as the liver (9), hamster cheekpouch (10), and skin (11), but this requires the use of fluorescence microscopy where fluorescent probes taken up selectively by leukocytes are used. This approach does not require trans-illumination, but criticism has been levied against this technique because these tracer dyes may inappropriately activate leukocytes (12). Nevertheless, we are gaining an appreciation that various tissues may differ in the mechanisms of leukocyte recruitment, making fluorescence microscopy an important tool.

In addition to being able to visualize the multistep recruitment of leukocytes, one can use intravital microscopy to study numerous other parameters. For example, this system can be used in association with very small concentrations of ruthenium red to detect when mast cells become activated. This stain is taken up to a much greater degree by activated mast cells than all other cells and therefore serves as an on-line detector of the degree of mast cell activation (13,14). Additionally, investigators have used fluorescence microscopy to measure oxidative stress (15) and microvascular permeability of single venules (16).

## III. THE MULTISTEP RECRUITMENT CASCADE

It has been well established that leukocyte infiltration is a multistep mechanism (see summary in Fig. 1) that requires that leukocytes moving at very high speeds in the mainstream of blood make initial transient contact with endothelial cells lining the vessel wall and roll along at a greatly reduced velocity relative to red blood cells. Once cells begin to roll, they can then firmly adhere and finally emigrate out of the vasculature. It should be noted that this is an interdependent series of events, inasmuch as inhibiting leukocyte rolling prevents subsequent leukocyte adhesion and ultimately leukocyte emigration out of the vasculature.

Leukocyte rolling is dependent on the selectin family of adhesion molecules. P-selectin (induced in minutes) and E-selectin (4–6 hr for maximal induction) expressed on activated endothelium each contribute significantly to the rolling event (17–20). Neither adhesion molecule is thought to be expressed constitutively, and therefore a stimulus is required for the expression of either adhesion molecule. This, however, means that it is possible to interfere with the expression of either of these adhesion molecules by inhibiting the stimulus responsible for their expression.

## A. P-selectin

P-selectin is a very likely mediator of the early phase of leukocyte rolling. This selectin is stored in Weibel-Palade bodies of endothelial cells and is rapidly expressed on the cell surface by histamine, cysteinyl leukotrienes, $H_2O_2$, thrombin and numerous other mediators (17). Histamine-induced P-selectin expression supports transient leukocyte adhesion to endothelial cells in static assay systems (17). Moreover, Lawrence and Springer (21) demonstrated that leukocytes, exposed to shear forces, rolled on artificial lipid bilayers containing purified P-selectin but not other adhesion molecules (ICAM-1). A recent publication has shown that exposure of human umbilical vein endothelial cells

(HUVECs) to histamine causes P-selectin–dependent leukocyte rolling under shear conditions in vitro (18). These in vitro studies support the hypothesis that histamine can induce P-selectin expression on endothelium and thereby promote leukocyte rolling. Using intravital microscopy, it has been demonstrated that the leukocyte rolling observed following exteriorization of tissue was dependent on P-selectin (22). It is now well appreciated that addition of mediators that cause P-selectin expression in vitro also cause P-selectin–dependent leukocyte rolling in vivo (22,23). This work is described in the next section of the chapter.

## B.   E-Selectin

Much of the evidence that implicates E-selectin as a mediator of leukocyte rolling in inflamed microvessels is inferred from in vitro studies. E-selectin is significantly expressed after 4 hours on the surface of endothelium exposed to LPS, interleukin (IL)-1, tumor necrosis factor–$\alpha$ (TNF-$\alpha$) and perhaps other cytokines, suggesting that this selectin contributes to the recruitment of leukocytes at a later stage (19). E-selectin expressed on transfected L-cells (fibroblasts), endothelial cell monolayers treated with IL-1, TNF-$\alpha$ or LPS, or substrates bearing purified E-selectin all supported neutrophil rolling in flow chambers (20,24). The dearth of information regarding the contribution of E-selectin to leukocyte–endothelial cell interactions in vivo results from the short-term experimental protocols (less than 2 hr) that have dominated published work employing intravital video microscopy. Olofsson et al. (25) have recently reported that IL-1 treatment of rabbit mesenteries supported leukocyte rolling via E-selectin at 4 hours of cytokine administration. However, E-selectin–dependent leukocyte rolling in postcapillary venules in various inflammatory models has not been documented to date.

## C.   Other Rolling Pathways

L-selectin constitutively expressed on leukocytes has also been shown to support leukocyte rolling (26). This event is not dependent on the release of mediators, and its importance under baseline conditions has been challenged (8,27,28). Recently, it has been suggested that an integrin, the $\alpha_4$ subunit of $\beta_1$-integrin may support non-neutrophil leukocyte rolling in vitro. T cells were shown to roll on IL-1–activated human umbilical vein endothelium and vascular cell adhesion molecule-1 (VCAM-1) transfected L-cells in a planar flow chamber, and this event was inhibitable by antibodies against VCAM-1 and $\alpha_4\beta_1$ (29). More recently, purified VCAM-1 adsorbed to coverslips supported rolling of both T cells and $\alpha_4\beta_1$ transfected cells (30). Finally, $\alpha_4\beta_1$ was shown to support human eosinophil but not neutrophil rolling in rabbit mesenteric venules (31).

## IV. MAST CELLS RECRUIT LEUKOCYTES

## A. Compound 48/80

Initial work (32) to demonstrate that mast cells can indeed have an impact on the vasculature to recruit circulating leukocytes made use of a mast cell degranulating agent, compound 48/80 (CMP 48/80). This agent, when injected into the skin of mice, induced a rapid (i.e., within 1–2 hr) neutrophilic influx into skin. This was followed at 24 hours by a mononuclear infiltrate, suggesting that mast cell degranulation had the capacity to recruit at least two types of leukocytes. Injection of mast cell granules into mouse skin elicited a similar cellular infiltrate as CMP 48/80, suggesting (albeit indirectly) that CMP 48/80 was functioning to recruit leukocytes via degranulation of mast cells. If indeed leukocyte rolling is an essential component of leukocyte recruitment, then, by inference, the CMP 48/80-induced leukocyte infiltration would imply increased leukocyte rolling. Applying intravital microscopy to the rat mesentery, we (14) and others (10,33,34) assessed whether mast cells could contribute to leukocyte rolling. Superfusion of the rat mesentery with CMP 48/80 revealed a rapid increase in the number of rolling leukocytes (relative to untreated preparations) that persisted for the next 60 min (Fig. 2). Stabilization of mast cells with sodium cromoglycate or ketotifen reduced both mast cell activation and leukocyte recruitment (14). Finally, chronically depleting mast cells of their granules in vivo prevented the subsequent recruitment of leukocytes with CMP 48/80 (14). These data as a whole strongly support the view that CMP 48/80-induced leukocyte recruitment is mast cell dependent.

A P-selectin monoclonal antibody, PB1.3, as well as a selectin-binding carbohydrate, fucoidan, completely inhibited this event (Fig. 2), supporting the view that substances released from mast cells were capable of inducing endothelial P-selectin–associated leukocyte rolling. Consistent with the view that leukocyte rolling was essential for leukocyte adhesion was the observation that both PB1.3 and fucoidan also inhibited leukocyte adhesion. In our laboratory, the CMP 48/80-induced leukocyte rolling appeared to be histamine-dependent inasmuch as diphenhydramine (an $H_1$-receptor antagonist) inhibited the mast cell–associated leukocyte rolling (14). However, Thorlacius et al. (33) reported that the CMP 48/80–induced, selectin-dependent leukocyte rolling was not inhibitable by a different $H_1$-receptor antagonist (mepyramine) in combination with an $H_2$ receptor antagonist (cimetidine). The discrepancy between these two studies must be related to the histamine receptor antagonists inasmuch as the preparations were otherwise identical. Inhibitors of other mediators were unremarkable in their ability to prevent the rise in mast cell–induced leukocyte rolling; inhibitors of leukotrienes, platelet-activating factor (PAF), or serotonin did not reduce the increased leukocyte rolling (14,35).

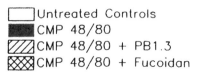

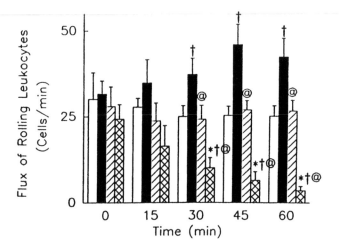

**Figure 2** In single postcapillary venules, compound 48/80 (CMP 48/80; 1 μg/mL) induces leukocyte rolling that is inhibitable by an anti–P-selectin antibody (PB1.3) or fucoidan (25 mg/kg), the selectin-binding carbohydrate. The antihistamine diphenhydramine also reduces leukocyte rolling in this preparation (not shown). Please see reference for details (14). (*P < .05 relative to control.)

## B. Baseline Leukocyte Rolling

It is likely that in nontraumatized tissue there is little, if any rolling and that exteriorization of tissues induces leukocyte rolling. This concept of surgically induced rolling is consistent with the observations that leukocyte rolling is very low in normal tissues that require no manipulation, including intact mouse skin and bat wing (7,36). If leukocyte rolling is not a normal event, then the surgical manipulation must somehow induce this response. The proximity of mast cells to the vasculature (10,37), in conjunction with their intense sensitivity to physical stress, temperature changes, and other physical disturbances that are likely to occur during surgical preparation, make mast cells good candidates to mediate baseline alterations in leukocyte–endothelial cell interactions (38). Therefore, we hypothesized that the surgery associated with intravital microscopy caused mast cell degranulation, which induced baseline leukocyte rolling. To test this hypothesis we pretreated animals (rats) with sodium cromoglycate or ketotifen, two mast cell stabilizers that prevent mast cell degranulation. Sodium cromoglycate

(39) given prior to surgical manipulation (but not after), attenuated by 80% baseline leukocyte rolling (22). Moreover, diphenhydramine, the $H_1$-receptor antagonist, also significantly (>50%) reduced baseline leukocyte rolling. These data collectively suggest that mast cell–derived histamine (and other unidentified agents) promote leukocyte rolling during surgical manipulation. The baseline rolling is also likely related to P-selectin expression inasmuch as therapeutically targeting P-selectin (40) with selective antibodies or inhibiting both L-selectin and P-selectin with fucose polymers such as fucoidin (41) inhibited baseline rolling. Moreover, P-selectin knockout mice do not have baseline rolling. It should be noted that baseline rolling could not be reduced in the rat or cat mesentery by the P-selectin antibody PB1.3 (42,43). However, based on the other antibodies and the P-selectin knockout mouse data, this may be an anomaly of this particular antibody. Overall, these data suggest that the preparation-induced mast cell activation potentially leads to P-selectin expression and increased baseline rolling.

## C. Mediator-induced Leukocyte Rolling

Mast cells produce many mediators thought to increase P-selectin expression. Therefore, these mediators should also increase leukocyte rolling in vivo. We used mast cell–stabilized preparations described above to minimize baseline rolling and maximize responses to different stimuli. In these preparations, histamine caused prolonged leukocyte rolling that was entirely inhibitable by either diphenhydramine, $H_1$-receptor antagonist, or a P-selectin antibody (22). In animals that had not been mast cell stabilized, there was an unremarkable response to histamine, possibly reflecting the fact that histamine released from mast cells during surgical manipulation caused the vasculature to be tachyphylactic to subsequent histamine treatment. Very similar data were simultaneously published by Asako et al. (44), who circumvented the problem of surgically induced mast cell degranulation by choosing those vessels wherein initial baseline leukocyte rolling was extremely low. Animals with a higher level of leukocyte rolling again failed to respond to histamine. Finally, Ley (45) also reported that histamine evoked leukocyte rolling. In this study, the histamine was added prior to exteriorization of the mesentery in an attempt to again dissociate effects induced by histamine from those induced by surgical manipulation. Collectively, these data are consistent with the view that histamine is capable of inducing leukocyte rolling.

Another mediator that is released from mast cells, $LTC_4$ (23), also increased leukocyte rolling. Interestingly, this mediator provided an entirely different pattern of leukocyte rolling than histamine. The leukocytes rolled with a greatly reduced velocity so that they appeared to be crawling rather than rolling per se. Interestingly, the P-selectin antibody again reduced the number of rolling leukocytes; however, the few remaining rolling cells still displayed the profound reduction in rolling velocity. A sialyl-Lewis$^x$ (sLe$^x$) analogue (a selectin ligand) blocked both the increased rolling and decreased rolling velocity induced by LTC$_4$. Whether a second sLe$^x$ ligand contributes to the reduced rolling velocity remains to be determined.

The physiologic relevance of the reduced rolling velocity becomes apparent when a chemotactic factor such as platelet-activating factor (PAF) is co-administered with either histamine or LTC$_4$. In the presence of LTC$_4$, 1 µM PAF is sufficient to induce firm adhesion whereas in the presence of histamine at least 10 µM PAF is required to induce similar amounts of adhesion (23). Both LTC$_4$ and histamine cause similar amounts of rolling, but only LTC$_4$ reduces the rolling velocity. These data have been interpreted to suggest that the LTC$_4$-induced reduction in rolling velocity increases the propensity of rolling cells to adhere. It should be noted that without histamine or LTC$_4$, PAF was not sufficient to induce adhesion in low-rolling preparations, again illustrating the importance of rolling as a prerequisite for subsequent adhesion. Other potential rat mast cell mediators such as serotonin did not increase leukocyte rolling despite the fact that at the concentration tested, a very profound increase in microvascular permeability was noted (35).

Finally, there is strong evidence to suggest that substance P induces leukocyte infiltration (46) and by inference leukocyte rolling. Smith et al. (47) illustrated that an intradermal injection of substance P induced a significant increase in neutrophil accumulation into normal human skin. This was associated with enhanced P-selectin (early) and E-selectin (4–8 hr after injection) expression on dermal vessels. There is no evidence, to my knowledge, that mast cells can produce substance P; however, substance P did not induce leukocyte recruitment in mast cell–deficient mice, suggesting that mast cells are involved in the substance P–induced leukocyte recruitment. Although a role for selectins in this recruitment is likely, a direct test of this hypothesis is lacking.

## V. INFLAMMATORY MODELS OF MAST CELL–INDUCED LEUKOCYTE RECRUITMENT

### A. Oxidant-Induced Leukocyte Rolling

Oxidative stress is a common feature of many different inflammatory diseases. Both leukocytes and the endothelium are capable of producing oxidants, and many of these substances are known to induce P-selectin expression on the surface of endothelium. When superoxide was generated locally in the rat mesentery by infusing xanthine oxidase and hypoxanthine directly into the superior mesenteric artery or by superfusing xanthine oxidase and hypoxanthine directly onto the mesentery (42,48), a significant increase in leukocyte rolling was noted (in the absence of mast cell stabilization). Because hypoxanthine and xanthine oxidase will react to produce both superoxide and $H_2O_2$, we examined whether the increased leukocyte rolling was inhibitable by catalase, the agent that detoxifies $H_2O_2$, or by superoxide dismutase (SOD), the anti-oxidant that detoxifies superoxide. In this series of experiments, SOD but not catalase entirely inhibited the increased leukocyte rolling associated with the infusion of xanthine oxidase and hypoxanthine, suggesting that sufficient amounts of superoxide, but not $H_2O_2$, were generated to induce the rise in rolling. Interestingly, this increase in rolling was entirely inhibited by ketotifen, a mast cell stabilizer (37). These data suggest that superoxide induces leukocyte rolling via mast cells. These data are also consistent with in vitro studies that reveal that $H_2O_2$ is a very weak stimulus of mast cell degranulation (histamine release) whereas the hypoxanthine and xanthine oxidase system is a potent activator of histamine release.

In contrast, Zimmerman et al. (49) demonstrated that $H_2O_2$ as well as tert-butyrl hydroxide (but not superoxide) were capable of directly inducing P-selectin expression on the surface of endothelium (in the absence of mast cells). The expression of P-selectin persisted for prolonged periods and maintained a level of P-selectin–dependent adhesion in static assays for a similar length of time. In vivo, in rat mesenteric experiments, $H_2O_2$ caused a very profound increase in leukocyte rolling (50). The increase was entirely P-selectin dependent and transient when $H_2O_2$ was continuously superfused onto the mesentery. When $H_2O_2$ was superfused for a brief 5-min pulse, the leukocyte rolling persisted for prolonged periods of time, perhaps suggesting that prolonged exposure to $H_2O_2$ impairs or "turns off" P-selectin–dependent rolling. It should be noted that the $H_2O_2$-induced rolling occurred in the presence of mast cell stabilizers, suggesting that mast cells were unlikely to be involved in the $H_2O_2$-induced increase in leukocyte rolling. Collectively, these data suggest that superoxide but not $H_2O_2$ activates mast cells to cause P-selectin–dependent leukocyte rolling.

## B. Ischemia/Reperfusion and Leukocyte Rolling

If oxidants such as superoxide can activate mast cells to induce leukocyte rolling, then inflammatory conditions with an underlying superoxide component may also involve mast cells. Indeed, there are numerous reports on the role of mast cells in ischemia/reperfusion-induced leukocyte recruitment. First, there is evidence to suggest that mast cells become activated during ischemia/reperfusion. Mast cells (no other cells) release a protease termed rat mast cell protease II (RMCP II); using this index of mast cell degranulation, investigators documented that mast cells do indeed degranulate during intestinal ischemia/reperfusion (51). Recently, Kurose et al. (52) visually assessed mast cell integrity 30 min after reperfusion of postischemic mesentery and noted a significant increase in degranulated mast cells that coincided closely with leukocyte recruitment. However, whether the mast cell degranulation was a cause or effect of I/R-induced leuko-cyte recruitment was unclear.

To directly address the issue of whether the degranulated mast cells con-tributed to leukocyte recruitment, mast cell stabilizers that prevented the rise in RMCP II levels were used. In animals pretreated with these reagents, a reduction in myeloperoxidase activity (index of neutrophil infiltration) was noted in the postischemic intestine (51). To determine which phase of the leukocyte recruit-ment was affected by mast cells in I/R, intravital microscopy was performed and leukocyte rolling, adhesion, and emigration were assessed in the cat postischemic mesenteric microvasculature (53). There is a tremendous increase in leukocyte rolling, which reaches 200 cells/min in postischemic venules within 2–3 min of reperfusion of ischemic vessels (Fig. 3). This increased flux of rolling cells is then maintained for the next 60 min. In cromolyn-pretreated animals, the flux of rolling leukocytes increased in the very early (5 min) reperfusion period but by 10 min the number of rolling leukocytes was reduced to baseline levels. Prelimi-nary data from our laboratory have revealed that mast cell–deficient mice also have an impairment in the recruitment of rolling cells in postischemic venules of the cremaster muscle. In normal mice, leukocyte rolling increases to approxi-mately 300 cells/min immediately following reperfusion and then wanes with time. In mast cell–deficient mice, leukocyte rolling increased only to one-third the value noted in normal animals and rapidly returned toward control levels. These data suggest an important contribution of interstitial mast cells as media-tors of the leukocyte rolling in postischemic vasculature.

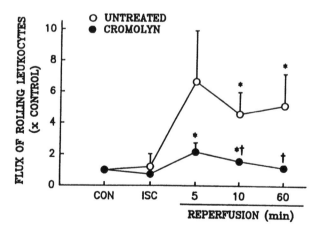

**Figure 3**  Reperfusion of the cat mesentery induces leukocyte rolling that is at least in part attenuated by the mast cell stabilizer sodium cromoglycate. See reference for details (53). (*P < .05 relative to control; +P < .05 relative to untreated group.)

There has been a tremendous amount of work on the involvement of adhesion molecules in the recruitment of leukocytes into postischemic vessels. Much of the evidence implicates P- and L-selectin as important in the initial recruitment of rolling leukocytes in postischemic vessels. Using intravital microscopy, it was observed that the L- and P-selectin–binding carbohydrate, fucoidan, essentially abolished (>90%) leukocyte rolling in postischemic vessels (43). Antibodies against either P-selectin (PB1.3) or L-selectin (DREG 200) reduced the number of rolling leukocytes by approximately 60% in the cat mesentery at both 10 and 60 min of reperfusion (43). Leukocyte rolling was not decreased further in animals given both anti–L-selectin and anti–P-selectin antibody; a 60% reduction in rolling was still observed. In another rat model of splanchnic artery occlusion and reperfusion, an anti–P-selectin antibody (PB1.3) significantly attenuated leukocyte rolling in postischemic venules (54). It is noteworthy that in the mouse cremaster muscle, ischemia/reperfusion induced absolutely no rolling in P-selectin knockout mice or in animals treated with an anti–P-selectin antibody (P. Kubes and S. Kanwar, unpublished observations) suggesting either a more important role for P-selectin in the mouse or perhaps a limitation in cross-reactivity with the human antibodies used in the cat.

At present there is good evidence that there is mast cell–dependent, P-selectin–mediated leukocyte rolling in ischemia/reperfusion. It is tempting to conclude that a mast cell–derived mediator(s) is essential to induce P-selectin–dependent recruitment of rolling cells in postischemic vessels; however, the identity of this mediator(s) remains unknown. Based on our knowledge of mediators of P-selectin expression, it is tempting to postulate that histamine, cysteinyl leukotrienes, and other mediators are involved. Boros et al. (55) have reported increased histamine release from the postischemic intestine. Although the cellular source of the histamine was not identified, mast cells are a primary source of this pro-inflammatory mediator in the small bowel (56), making these cells a likely source. However, to date the role of antihistamines on I/R-induced leukocyte recruitment has not been documented. Lehr et al. (11) have also attempted to reveal the identity of the chemotactic factors that contributed to leukocyte recruitment and microvascular permeability in postischemic skin. Although inhibitors of $LTB_4$ and $LTC_4/LTD_4$ both inhibited microvascular dysfunction, leukocyte recruitment was not affected by an $LTC_4/LTD_4$ receptor antagonist. Although leukocyte rolling was not documented, these data suggest that cysteinyl leukotrienes are unlikely to contribute to leukocyte–endothelium interactions in this tissue during ischemia/reperfusion.

If mast cell degranulation causes leukocyte infiltration into postischemic tissue, then some factor(s) must be activating the mast cells. The increased flux of oxidants previously described at the onset of reperfusion (57) may be responsible for mast cell activation based on the fact that superoxide is known to activate these immunocytes (58). In order to test the hypothesis that oxidants are involved in ischemia/reperfusion-induced mast cell degranulation, animals were pretreated with superoxide dismutase and catalase (51). This protocol prevented the release of RMCP II from mast cells and the subsequent leukocyte infiltration, suggesting that indeed superoxide and/or hydrogen peroxide were instrumental in mast cell degranulation. Although the source of the oxidants remains unknown, the fact that Boros et al. (55) demonstrated that allopurinol blocked by 87% the histamine release from postischemic gut suggests an important role for the oxidant-generating enzyme xanthine oxidase in I/R-induced mast cell activation. Another family of mediators known to activate mast cells are the anaphylatoxins (C3a and C5a). In fact, it has been shown that anaphylatoxins cause histamine release from cardiac mast cells (59,60). C3a and C5a have been shown to play an important role in ischemia/reperfusion of the heart (61); by inference, then, it is tempting to speculate that these anaphylatoxins, produced at the time of reperfusion, can in addition to their direct effects on the vasculature also stimulate mast cells to release histamine and other pro-inflammatory mediators, thereby contributing to leukocyte recruitment and reperfusion injury. However, data to implicate the mast cell–derived mediators responsible for the recruitment of leukocytes following ischemia/reperfusion is currently not available, and this is clearly an area that requires investigation.

## C.  Delayed-Type Hypersensitivity Reactions

Mast cells are important cellular mediators of immediate hypersensitivity reactions. They are also thought to contribute significantly to the delayed recruitment of leukocytes during late phase reactions. This view is primarily based on the fact that challenge of sensitized mice or humans with anti-IgE caused a profound leukocytic influx into skin that peaked at 6–12 hours after antigen exposure (62–64). This leukocyte infiltration, however, was virtually undetectable in genetically, mast cell–deficient *W/W*ᵛ mice (62). The leukocytic infiltrate observed at 6 hours in normal mice challenged with anti-IgE could be reduced by approximately 50% with anti-TNF antiserum, suggesting that this cytokine contributes significantly to the delayed leukocyte infiltration in this allergen-induced, mast cell degranulation system. The time course of this leukocyte recruitment raised the possibility that E-selectin may also be involved in the induction of leukocyte rolling in this condition. Indeed, IgE-dependent activation of human foreskin mast cells has been shown to induce the expression (at 6 hr) of E-selectin (65). Based on the fact that this endothelial adhesion molecule induces leukocyte rolling (20), the authors (65) proposed the scenario that IgE-dependent mast cell degranulation stimulates leukocyte–endothelial cell interactions via TNF-α and E-selectin.

In a human/SCID mouse chimeric model, Christofidou-Solomidou et al. (66)

illustrated that a monoclonal antibody against human E-selectin significantly reduced mast cell degranulation–induced leukocyte recruitment at 4 hours. Although the contribution of P-selectin in this model was not determined, an important role for E-selectin in leukocyte recruitment is proposed. On the other hand, mutant mice lacking E-selectin had no leukocytic impairment in delayed-type hypersensitivity responses; however addition of a P-selectin antibody significantly reduced neutrophil infiltrate in this model (67). It should be noted that leukocyte recruitment and the role of TNF-$\alpha$ and E-selectin or any of the other selectins has not been examined in late phase reactions using intravital video microscopy, in part because of the difficulties with maintaining viable preparations for prolonged periods of time. However, we have recently developed a model of late phase reaction that has permitted us to visualize leukocyte recruitment in the cremaster muscle, 8 hours after the intrascrotal administration of ovalbumin in sensitized animals (68). This experiment revealed that a very significant proportion of the leukocyte rolling could be inhibited by a P-selectin antibody. Moreover, no cells could be seen rolling 8 hours after challenge in P-selectin knockout mice, suggesting an unlikely role for E-selectin in the cremaster muscle microcirculation (unpublished observations). Therefore, the importance of E-selectin in the recruitment of rolling leukocytes remains unclear.

In addition to P-selectin there may be some contribution to leukocyte rolling from other adhesion molecules. At 8 hours after antigen challenge, when many cells are observed rolling in the cremaster microcirculation, administration of a very late activation antigen (VLA)-4 antibody reduced leukocyte rolling by 50% (68). However, the fact that P-selectin knockout mice exhibited no rolling suggests that the VLA-4 rolling was entirely dependent on initial leukocyte-endothelial cell interactions via P-selectin. This is not entirely surprising based on in vitro experiments using flow chambers (30). VLA-4 could not support tethering to its ligand VCAM-1 at shear rates where tethering occurred via the selectins. Once tethered, VLA-4 could support rolling at higher shears. Therefore, there may be an initial selectin requirement for tethering even for VLA-4–dependent leukocyte rolling.

L-selectin may also contribute significantly to the pathology of delayed-type hypersensitivity reactions; however, this event may be important at more delayed time points. The use of mutant mice lacking L-selectin has revealed reduced edema formation in delayed-type hypersensitivity reactions at 24 and 48 hours (69). Cellular infiltrates were not determined in these experiments. Although these data imply a role for L-selectin as a mediator of tissue dysfunction in mast cell–dependent models of inflammation, documentation of impaired leukocyte influx still awaits confirmation. Evidence showing that mast cell-derived mediators contribute directly to the expression of endothelial selectins in these models is an important, as yet undetermined issue.

## D. Bacterial Toxins Mediate Leukocyte Recruitment Through Mast Cell Activation

The use of antibiotic therapy can be associated with a chronic inflammatory condition affecting the colon, termed pseudomembranous colitis. The principal etiologic agent responsible for this iatrogenic form of enterocolitis is *Clostridium difficile*, a gram-negative anaerobic bacillus that produces two protein exotoxins: toxin A and toxin B. Based on animal studies, it is thought that toxin A (Tx-A) mediates the excess fluid secretion and inflammation associated with experimental *C. difficile* enterocolitis. Mast cells have been implicated in the pathobiology of Tx-A–induced mucosal dysfunction, with mast cell degranulation observed in gut mucosa within 15 min after Tx-A exposure. Pretreatment of animals with ketotifen, the mast cell stabilizer, attenuated leukocyte infiltration and tissue necrosis, suggesting a potentially important role for mast cells (70,71).

Using intravital video microscopy, investigators have demonstrated that exposure of rat mesentery to Tx-A results in an increased adherence and emigration of leukocytes, enhanced albumin leakage in postcapillary venules, and the degranulation of perivenular mast cells (72). Although leukocyte rolling per se was not documented in this study, the recruitment of leukocytes elicited by Tx-A was dependent on P-selectin and sialyl-Lewis$^x$ (sLe$^x$), a selectin counter-receptor. Both of the reagents (mAb to P-selectin and soluble sLe$^x$) were also effective in blunting the enhanced albumin leakage normally elicited by the toxin, suggesting that the recruitment of leukocytes was an essential prerequisite for the microvascular alteration. Lodoxamide, a mast cell stabilizer, effectively prevented the mast cell degranulation in rat mesentery that was elicited by Tx-A exposure, and this particular therapy also attenuated leukocyte recruitment and the corresponding reduction in albumin leakage. Similar protective effects were noted following treatment with either diamine oxidase (histaminase) or an H$_1$-receptor antagonist. These observations suggested that histamine represents an important mast cell–derived mediator of Tx-A–induced leukocyte recruitment. Based on these data, it is thought that Tx-A rapidly induces mast cell degranulation, leading to release and accumulation of histamine in the perivenular compartment. The mast cell–derived histamine appears to mediate at least part of the leukocyte–endothelial cell interactions by engaging H$_1$-receptors on endothelial cells to perhaps increase the expression of P-selectin. The expression of this adhesion molecule would recruit rolling leukocytes that would directly contribute to the leukocyte recruitment elicited by *C. difficile* toxin A.

A second toxin that may recruit white cells via mast cells may be *Helicobacter pylori*, a bacterium known to be responsible for the pathogenesis of chronic gastritis and gastric ulceration. This contention is based on the fact that ingestion of this microorganism by human volunteers produces gastritis, and eradication of *H. pylori* in ulcer patients leads to resolution of pathology. The inflammatory potential of *H. pylori* has also been demonstrated using intravital microscopic techniques (73). Exposure of rat mesentery to a water extract of *H. pylori* (HPE) leads to the recruitment of leukocytes, which subsequently emigrate into the adjacent interstitial compartment. The HPE-induced leukocyte recruitment is accompanied by mast cell degranulation and enhanced albumin leakage in mesenteric venules. The mast cell degranulation induced by HPE occurred as early as 10 min after exposure to the extract and was largely prevented by prior treatment with ketotifen. The mast cell stabilizer had no effect on HPE-induced recruitment of leukocytes (described primarily as adherent cells); however, it significantly attenuated the normally associated leukocyte emigration and albumin leakage responses. Although leukocyte rolling was not documented, if rolling had been reduced by ketotifen, then adhesion would likely have also been reduced. Clearly, the use of intravital microscopy in this model hint at numerous important concepts: 1.) mast cell degranulation could contribute selectively to some (emigration) but not all (rolling and adhesion) of the events underlying leukocyte influx into tissues and 2.) mast cell degranulation does not automatically translate into increased leukocyte rolling. The lack of effect of ketotifen on HPE-induced adhesion in vivo is consistent with the view that HPE promotes the adhesion of isolated human neutrophils to monolayers of human umbilical endothelial cells in the absence of mast cells in these cell cultures (74). Interestingly, comparable inflammatory responses were not elicited by exposure of the rat mesentery to a water extract of *E. coli*, suggesting that not all toxins cause mast cell degranulation.

## VI.  CONCLUDING REMARKS

Much attention has been given to understanding the role of selectins in the recruitment of leukocytes to sites of inflammation. Additionally, a significant amount of work has been published on the role of mast cells as important contributors to the inflammatory response. Moreover, a link between activation of mast cells and the functional expression of selectins manifested as leukocyte rolling has been documented. However, critical experiments to demonstrate a link between selectin expression and mast cell activation in inflammatory conditions have not been completed. This is in part the result of the difficulty in measuring mediator release directly from mast cells. Often these cells are situated close to the vasculature so that mediator release will be seen by the vessel but may never reach detectable levels in the systemic circulation. Moreover, the mast cell–derived mediators that induce selectins in many inflammatory conditions remain, for the most part, an unexplored area. Finally, little attention has been given to the importance of mast cells in different organ systems. Certainly organs in direct contact with the external environment, such as the lung and gastrointestinal tract, have a much higher density of mast cells than organs such as the heart; therefore, mast cells in the gut and lung may have a more critical role to play in the recruitment of leukocytes. Improvements in intravital microscopy to visualize leukocytes in tissues such as the heart and lung, stains to distinguish the type of leukocytes rolling in the field of view, and on-line techniques to observe and quantify selectin expression are needed to address some of the causal links between mast cell activation and leukocyte recruitment.

## ACKNOWLEDGEMENTS

Supported by grants from the Canadian Heart and Stroke Foundation of Canada.

## REFERENCES

1. T. A. Springer, *Cell*, *76*, 301 (1994).
2. D. N. Granger and P. Kubes, *J. Leukocyte Biol.*, *55*:662 (1994).
3. S. M. Albelda, C. W. Smith, and P. A. Ward, *FASEB J.*, *8*:504 (1994).
4. G. A. Zimmerman, S. M. Prescott, T. M. McIntyre, *Immunology Today*, *13*:93 (1992).
5. M. D. Menger and HA. Lehr, *Immunology Today*, *14*:519 (1993).
6. S. D. House and H. H. Lipowsky, *Microvasc. Res.*, *34*:363 (1987).
7. H. N. Mayrovitz, R. F. Tuma, and M. P. Wiedeman, *Microvasc. Res.*, *20*:264 (1980).
8. K. Ley, D. C. Bullard, M. L. Arbones, R. Bosse, D. Vestweber, et al., *J. Exp. Med.*, *181*:669 (1995).
9. S. Post, M. D. Menger, M. Rentsch, A. P. Gonzalez, C. Herfarth, et al., *Transplantation*, *54*:789 (1992).
10. J. Raud, S.-E. Dahlen, G. Smedegard, and P. Hedqvist, *Acta. Physiol. Scand.*, *135*:95 (1989).
11. H. A. Lehr, A. Guhlmann, D. Nolte, D. Keppler, and K. Messmer, *J. Clin. Invest.*, *87*:2036 (1991).
12. P. Hansell, E. Berger, J. D. Chambers, and K. E. Arfors, *J. Leukoc. Biol.*, *56*:464 (1994).
13. D. Lagunoff, *J. Histochem. Cytochem.*, *20*:938 (1972).
14. J. P. Gaboury, B. Johnston, X.-F. Niu, and P. Kubes, *J. Immunol.*, *154*:804 (1995).
15. M. Suematsu, T. Tamatani, F. A. Delano, M. Miyasaka, M. Forrest, et al., *Am. J. Physiol.*, *266*:H2410 (1994).
16. I. Kurose, D. C. Anderson, M. Miyasaka, T. Tamatani, J. C. Paulson, et al., *Circ. Res.*, *74*:336 (1994).
17. J.-G. Geng, M. P. Bevilacqua, K. L. Moore, T. M. McIntyre, S. M. Prescott, et al., *Nature*, *343*:757 (1990).
18. D. A. Jones, O. Abbassi, L. V. McIntire, R. P. McEver, and C. W. Smith, *Biophys. J.*, *65*:1560 (1993).
19. M. P. Bevilacqua, J. S. Pober, D. L. Mendrick, R. S. Cotran, and M. A. Gimbrone, *Proc. Natl. Acad. Sci. U.S.A*, *84*:9238 (1987).
20. O. Abbassi, T. K. Kishimoto, L. V. McIntire, D. C. Anderson, and C. W. Smith, *J. Clin. Invest.*, *92*:2719 (1993).

21. M. B. Lawrence and T. A. Springer, *Cell*, *65*:859 (1991).
22. P. Kubes and S. Kanwar, *J. Immunol.*, *152*:3570 (1994).
23. S. Kanwar, B. Johnston, and P. Kubes, *Circ. Res.*, *77*:879 (1995).
24. M. B. Lawrence and T. A. Springer, *J. Immunol.*, *151*:6338 (1993).
25. A. M. Olofsson, K. E. Arfors, L. Ranezani, B. Wolitzky, E. C. Butcher, et al. *Blood*, *84*:2749 (1994).
26. U. H. Von Andrian, J. D. Chambers, L. M. McEvoy, R. F. Bargatze, K. E. Arfors, et al., *Proc. Natl. Acad. Sci. USA*, *88*:7538 (1991).
27. M. B. Lawrence, D. F. Bainton, and T. A. Springer, *Immunity*, *1*:137 (1994).
28. D. Nolte, P. Schmid, U. Jager, A. Botzlar, F. Roesken, et al., *Am. J. Physiol.*, 267:H1637 (1994).
29. D. A. Jones, L. V. McIntire, C. W. Smith, and L. J. Picker, *J. Clin. Invest.*, *94*:2443 (1994).
30. R. Alon, P. D. Kassner, M. W. Carr, E. B. Finger, M. E. Hemler, et al., *J. Cell Biol.*, *128*:1243 (1995).
31. P. Sriramarao, U. H. Von Andrian, E. C. Butcher, M. A. Bourdon, and D. H. Broide, *J. Immunol.*, *153*:4238 (1994).
32. S. Tannenbaum, H. Oertel, W. Henderson, and M. Kaliner, *J. Immunol.*, *125*:325 (1980).
33. H. Thorlacius, J. Raud, S. Rosengren-Beezley, M. J. Forrest, P. Hedqvist, et al., *Biochem. Biophys. Res. Commun.*, *203*:1043 (1994).
34. J. Raud, L. Lindbom, S.-E. Dahlen, and P. Hedqvist, *Am. J. Pathol.*, *134*:161 (1989).
35. J. P. Gaboury, X.-F. Niu, and P. Kubes, *Circulation*, *93*:318 (1996).
36. H. N. Mayrovitz, *Am. J. Physiol.*, *262*:H157 (1992).
37. P. Kubes, S. Kanwar, X.-F. Niu, and J. Gaboury, *FASEB J.*, *7*:1293 (1993).
38. D. D. Metcalfe, J. J. Costa, and P. R. Burd, *Inflammation basic principles and clinical correlates* (J. I. Gallin, I. M. Goldstein, and R. Snyderman, eds.), Raven Press, New York, p. 709 (1992).
39. K. B. P. Leung, K. E. Barrett, and F. L. Pearce, *Agents Actions Suppl.*, *14*:461 (1984).
40. M. Dore, R. J. Korthuis, D. N. Granger, M. L. Entman, and C. W. Smith, *Blood*, *82*:1308 (1993).
41. L. Lindbom, X. Xie, J. Raud, and P. Hedqvist, *Acta Physiol. Scand.*, *146*:415 (1992).
42. J. Gaboury, D. C. Anderson, and P. Kubes, *Am. J. Physiol.*, *266*:H637 (1994).
43. P. Kubes, M. A. Jutila, and D. Payne, *J. Clin. Invest.*, *95*:2510 (1995).
44. H. Asako, I. Kurose, R. Wolf, S. DeFrees, Z-L. Zheng, et al. *J. Clin. Invest.*, *93*:1508 (1994).
45. K. Ley, *Am. J. Physiol.*, *267*:H1017 (1994).
46. H. Yano, B. K. Wershil, N. Arizono, and S. J. Galli, *J. Clin. Invest.*, *84*:1276 (1989).
47. C. H. Smith, J. N. W. N. Barker, R. W. Morris, D. M. MacDonald, and T. H. Lee, *J. Immunol.*, *151(6)*:3274 (1993).
48. R. F. Del Maestro, M. Planker, and K. E. Arfors, *Int. J. Microcirc. Clin. Exp.*, *1*:105 (1982).
49. K. D. Patel, G. A. Zimmerman, S. M. Prescott, R. P. McEver, and T. M. McIntyre, *J. Cell Biol.*, *112*:749 (1991).
50. B. Johnston, S. Kanwar, and P. Kubes, *FASEB J.*, *9(3)*:A34 (1995). [abstract]
51. S. Kanwar and P. Kubes, *Am. J. Physiol.*, *267*:G316 (1994).

52. I. Kurose, R. Wolf, M. B. Grisham, and D. N. Granger, *Circ. Res.*, *74*:376 (1994).
53. S. Kanwar and P. Kubes, *Microcirculation*, *1(3)*:175 (1994).
54. K. L. Davenpeck, T. W. Gauthier, K. H. Albertine, and A. M. Lefer, *Am. J. Physiol.*, *267*:H622 (1994).
55. M. Boros, J. Kaszaki, and S. Nagy, *Eur. Surg. Res.*, *21*:297 (1989).
56. S. E. Crowe and M. H. Perdue, *Gastroenterology*, *103*:1075 (1992).
57. H. Blum, J. J. Summers, M. D. Schnall, C. Barlow, J. S. Leigh, et al., *Ann. Surg.*, *204*:83 (1986).
58. D. Salvemini, E. Masini, A. Pistelli, P. F. Mannaioni, and J. Vane, *J. Cardiovasc. Pharmacol.*, *17 (Suppl 3)*:S258 (1991).
59. A. R. Johnson, T. E. Hugli, and H. J. Muller-Eberhard, *J. Immunol.*, *28*:1067 (1974).
60. U. Hachfeld del Balzo, R. Levi, and M. J. Polley, *Proc. Natl. Acad. Sci. USA*, *82*:886 (1985).
61. H. F. Weisman, T. Bartow, M. K. Leppo, H. C. Marsh, G. R. Carson, et al., *Science*, *249*:146 (1993).
62. B. K. Wershil, Z. S. Wang, J. R. Gordon, and S. J. Galli, *J. Clin. Invest.*, *87*:446 (1991).
63. J. Dolovich, F. E. Hargreave, R. Chalmers, K. J. Shier, J. Gauldie, et al., *J. Allergy Clin. Immunol.*, *52(1)*:38 (1973).
64. G. J. Gleich, *J. Allergy Clin. Immunol.*, *70*:160 (1982).
65. L. M. Klein, R. M. Lavker, W. L. Matis, G. F. Murphy, *Proc. Natl. Acad. Sci. USA*, *86*:8972 (1989).
66. M. Christofidou-Solomidou, G. F. Murphy, S. M. Albelda, *Am. J. Pathol.*, *148(1)*:177 (1996).
67. M. A. Labow, C. R. Norton, J. M. Rumberger, K. M. Lombard-Gillooly, D. J. Shuster, et al., *Immunity*, *1*:709 (1995).
68. S. Kanwar and P. Kubes, *Microcirculation, 43rd Ann. Conf.*:55 [abstr.] (1996).
69. T. F. Tedder, D. A. Steeber, P. Pizcuetta, *J. Exp. Med.*, *181*:2259 (1995).
70. G. C. Triadafilpoulos, C. Pothoulakis, M. O'Brian, and J. T. Lamont, *Gastroenterology*, *92*:273 (1987).
71. C. Pothoulakis, F. Karmeli, C. P. Kelly, R. Eliakim, M. A. Joshi, et al., *Gastroenterology*, *105*:701 (1993).
72. I. Kurose, C. Pothoulakis, J. T. Lamont, D. C. Anderson, J. C. Paulson, et al., *J. Clin. Invest.*, *94*:1919 (1994).
73. I. Kurose, D. N. Granger, D. J. Evans, D. G. Evans, D. Y. Graham, et al., *Gastroenterology*, *107*:70 (1994).
74. N. Yoshida, D. N. Granger, D. J. Evans, D. G. Evans, D. Y. Graham, et al., *Gastroenterology*, *107*:70 (1994).

# 15

# Adhesion Pathways Controlling Recruitment Responses of Lymphocytes During Allergic Inflammatory Reactions In Vivo

**Donald Y. M. Leung**   *The National Jewish Medical and Research Center, and University of Colorado Health Sciences Center, Denver, Colorado*

**Louis J. Picker**   *The University of Texas Southwestern Medical Center, Dallas, Texas*

## I. INTRODUCTION

Chronic allergic diseases such as atopic dermatitis (AD), allergic rhinitis, and asthma are associated with organ-specific infiltration of T helper type 2 (Th2)-like T cells expressing interleukin (IL)-4, IL-5, and IL-13 genes (1,2). These memory T cells play an important role in the induction of local allergen-specific IgE responses and the recruitment of eosinophils into the bronchial airways of asthmatics, the nasal mucosa of patients with allergic rhinitis, or the skin lesions of patients with AD. The potential mechanisms that control recruitment of T lymphocytes to different tissue sites in particular allergic diseases are therefore of considerable interest.

A large body of evidence has demonstrated that memory, but not naive, T cells have the ability to localize efficiently in extralymphoid immune "effector" sites such as the skin or the lung. However, these cells do not migrate to these sites homogeneously. Instead, memory T cells, as a group, display a heterogeneous, complex homing potential, with some subsets showing striking tissue selectivity in their homing behavior. Studies of lymphocyte homing in animal models have repeatedly demonstrated the tissue-selective homing of mucosa-derived memory lymphocytes or blasts to mucosal tissues such as the intestinal lamina propria versus the homing of peripheral-derived memory cells to nonmucosal "peripheral" sites such as the skin (3).

This tissue-selective homing is regulated in large part at the level of lymphocyte interaction with the vascular endothelial cells of postcapillary venules, which is mediated by the interaction of differentially expressed lymphocyte homing receptors and their respective endothelial cell ligands. Adhesion molecules mediating specific retention of lymphocytes in particular tissue microenvironments may also play an important role in this process. This chapter reviews the role of specific cell adhesion molecules likely to be involved in the "tissue-selective" homing of T lymphocytes in inflammatory responses in specific allergic diseases.

## II. CONCEPT OF LYMPHOCYTE HOMING

The capacity of the host to recognize and respond to the many foreign antigens that are encountered at different tissue sites depends on the efficient recruitment of functionally competent T lymphocytes to various lymphoid tissues and extralymphoid sites such as the skin or lung. To accomplish this task, an elaborate system of lymphocyte homing and recirculation has evolved to effectively distribute various T cells and to orchestrate their function in different lymphoid compartments. Most memory T cells recirculate continuously between the blood and tissue compartments. Data gathered over the past 20 years have shown conclusively that this recirculation is not random but instead highly regulated so as to distribute specific T lymphocyte subsets to particular tissues (4–9). This regulation occurs at the level of lymphocyte–endothelial cell interaction, including both the process of recognition/adhesion and the migratory mechanisms mediating diapedesis of the lymphocytes across the vascular wall (see below).

An important concept in T lymphocyte homing/recirculation physiology relates to the differential distribution of virgin as opposed to memory/effector T cells to secondary lymphoid tissues such as Peyer's patches, tonsil, spleen versus extralymphoid sites (e.g., the skin, intestinal lamina propria, or pulmonary interstitium) (4–8). As a rule, virgin T cells are programmed to home selectively to secondary lymphoid tissues that are involved in the antigen-induced activation and differentiation of this lymphocyte subset. Virgin T cells do not show significant heterogeneity in their capacity to migrate to the various types of secondary lymphoid tissue, nor do they migrate with any degree of efficiency to extralymphoid sites. Memory T cells, however, have markedly different homing capabilities. First, memory T cells have the ability to access extralymphoid tissues, and indeed, these cells constitute the large majority of T lymphocytes within these tissue sites. Second, in contrast to virgin T cells, memory T lymphocytes, as indicated above, are heterogeneous in their ability to localize, and thus form subsets with tissue-selective, homing behavior (4–7). The best-defined homing specificities are to mucosal versus peripheral sites (e.g., the intestinal tract versus peripheral lymph nodes and skin) (4), but indirect evidence suggests other homing specificities to tissues such as the lung and joints (9,10).

## III. REGULATION OF LYMPHOCYTE HOMING

It was first demonstrated over 15 years ago that the ability of lymphocyte popula-tions to bind postcapillary venular endothelial cells within Peyer's patches versus peripheral lymph nodes correlated well with their ability to localize in these tis-sues in vivo (4). Since then, a number of important advances have been made in our understanding of the molecular basis of lymphocyte homing. Recent studies indicate that in most instances the extravasation of lymphocytes into specific tis-sues requires a multistep process that is mediated by independent adhesive inter-actions between lymphocyte homing receptors and endothelial cell counter-receptors (reviewed in refs. 6 and 11).

Four successive steps have been identified in the recruitment of lymphocytes into sites of inflammation (Fig. 1). The first step involves so-called *primary adhe-sion*, in which free-flowing lymphocytes interact with endothelial cells via con-stitutively expressed homing receptors and endothelial cell ligands. This primary adhesion is transient and reversible under shear stress. In the case of lympho-cytes, it is manifested by either a slow rolling, or a transient, immediate arrest (12,13). The second step involves the rapid activation of lymphocytes adhering to endothelial cell ligands. These transiently interacting cells are released back into

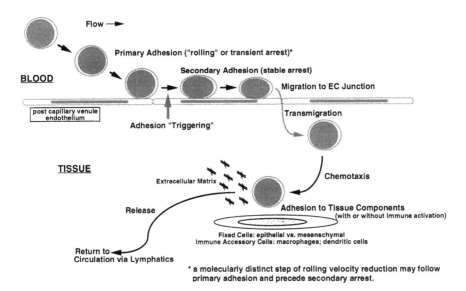

**Figure 1** Multistep model of lymphocyte-endothelial cell recognition and recruitment of lymphocytes from the blood. The requirement for sequential independent interactions provides the specificity for lymphocyte homing to particular tissues. (From Ref. 5.)

the circulation unless *secondary adhesion* (step 3) occurs. This latter event is "triggered," during step 2, by specific activating stimuli, which are thought to be chemoattractant substances operating through G protein–linked receptors (12). Secondary adhesion is thus activation-dependent and mediates a firm attachment that is stable under shear stress. In some instances an intermediate step, mediated by distinct adhesion molecules, may "bridge" primary and secondary adhesion by reducing the initially "too high" rolling velocity of the primary interaction to levels more commensurate with effective secondary adhesion (6). Once activation-dependent secondary adhesion mechanisms are brought fully into play, the cell arrests stably on the endothelial surface, and these secondary adhesion events serve as a prelude to the final step of extravasation—diapedesis into the tissue.

A number of lymphocyte and endothelial cell adhesion molecules have been shown to participate in primary and secondary lymphocyte interactions with the high endothelial venules in secondary lymphoid tissues and/or with activated (cytokine-stimulated) vascular endothelium in extralymphoid sites (5–7). "Triggering" factors are not well characterized, but the chemokine group of chemoattractants has been implicated as possible mediators of this step (11,14). In the context of this multistep model, the recruitment of lymphocytes to specific tissues may be controlled at each of the four major steps of this process (primary adhesion, triggering, secondary adhesion or diapedesis). It is also becoming apparent that most, if not all, of the adhesion molecule pairs participating in homing can be differentially used by lymphocyte subsets to localize to different tissue sites.

Several homing receptor/endothelial cell ligand pairs appear to dominate lymphocyte homing to certain tissues to such an extent that, by themselves, they play a major role in determining a particular homing specificity. These include 1.) L-selectin and its ligand(s) [termed the peripheral lymph node addressin (PNAd)], which—although involved in several homing pathways—appears to play a dominating role in lymphocyte homing to peripheral lymph node, 2.) the cutaneous lymphocyte–associated antigen (CLA) and its counter-receptor E-selectin, which direct lymphocyte homing to skin (see below), and 3.) the $\alpha 4 \beta 7$integrin and its ligand MAdCAM-I (mucosal addressin), which direct lymphocyte homing to Peyer's patch and intestinal lamina propria (5,6). Table 1 lists these and other adhesion molecule interactions that are thought to participate in the homing process.

## IV. THE CUTANEOUS LYMPHOCYTE–ASSOCIATED ANTIGEN—A PARADIGM FOR THE BIOLOGY OF A TISSUE-SELECTIVE LYMPHOCYTE HOMING RECEPTOR

In humans, the CLA antigen has been the best studied of the putative tissue-selective homing receptors. Therefore, we will expand our discussion on the characterization and control of CLA as a prelude to our later review of its role in allergic

**Table 1** Adhesion Molecules Involved in Lymphocyte–Endothelial Recognition

| Lymphocyte homing receptor | Predominant endothelial cell ligands | Role in multistep extravasation | Primary homing pathways |
|---|---|---|---|
| **Selectin-CHO** | | | |
| L-selectin (CD62L) | PNAd (includes CD34, other protein cores) | 1° adhesion | Naive lymphocyte homing to lymph nodes; memory lymphocyte homing to peripheral > mucosal sites of chronic inflammation |
| PSGL-1 | P-selectin | 1° adhesion | ? ? ? |
| CLA | E-selectin | 1° adhesion | Memory T cell homing to skin |
| **Integrin-Ig family** | | | |
| $\alpha_4\beta_7$ | MAdCAM-1 (mucosal addressin) | 1° and 2° adhesion; and "bridging" (rolling velocity reduction) | Naive lymphocyte homing to Peyer's patch and appendix; memory lymphocyte homing to nonpulmonary mucosal sites |
| $\alpha_4\beta_1$ | VCAM-1 | 1° and 2° adhesion | Memory lymphocyte homing to many extra-intestinal inflammatory sites |
| $\alpha_L\beta_2$ (LFA-1) | ICAM-1, –2, ?others | 2° adhesion | Widespread involvement |

Abbreviations: CHO, carbohydrate; CLA, cutaneous lymphocyte–associated antigen; HEV, high endothelial venule; ICAM, intercellular adhesion molecule; LFA, leukocyte function antigen; MAdCAM, mucosal addressin cell adhesion moleule; PNAd, peripheral lymph node addressin; PSGL, P-selectin glycoprotein ligand; VCAM, vascular cell adhesion molecule.

skin responses. In the past 5 years, it has become firmly established that CLA is a lymphocyte homing receptor selectively involved in T-cell trafficking to skin (6,13,15–24). Cutaneous lymphocyte–associated antigen was first identified as an antigenic determinant selectively expressed by a subset of memory T cells in the blood and essentially all T cells in the skin, involving a variety of inflammatory conditions. In contrast, very few T cells expressed CLA in chronically inflamed extracutaneous sites including the heart, lung, synovium, thyroid, or kidney (15).

Examination of T cells from suction blisters placed over cutaneous delayed-type hypersensitivity reactions revealed that essentially the entire skin T-cell population is shifted toward CLA– expression, with mean CLA cell surface densities an average of 23 times higher than on corresponding peripheral blood T cells from the same individual (17,18). Moreover, this striking enrichment of CLA–expressing T cells in skin was observed throughout the development of the delayed-type hypersensitivity reactions, including the earliest T cells emigrating into the blister fluid (17). Thus, the profound enrichment of CLA positive(+) T cells in skin could not be attributed to a specific up-regulation of CLA expression after cutaneous extravasation, but rather is most likely due to the skin-selective homing of this CLA+ T cell subset. The specificity of CLA enrichment in cutaneous inflammatory sites was further demonstrated in separate studies in which it was found that in contrast to the situation in skin, both lung and synovial T cells predominantly lacked CLA (18,19).

The association of T-cell CLA expression and skin localization has been further supported by the observation that the skin-localizing malignant T cells of patients with cutaneous T cell lymphomas selectively express CLA (15). It was further demonstrated that patients with mycosis fungoides had excess numbers of circulating CLA+ cells. Importantly, the number of CLA+ T cells correlated strongly with the extent of skin disease (21), and overt cases of leukemic cutaneous T-cell lymphomas (Sézary syndrome) were characterized by high levels of CLA expression (22).

These data strongly suggested a role for CLA in mediating skin-selective T cell homing, most likely as a specific receptor for an endothelial adhesion molecule. This possibility was confirmed by the discovery that CLA glycoproteins represent the major, if not exclusive, T cell ligands for E-selectin, an IL-1 or tumor necrosis factor (TNF)-inducible endothelial cell adhesion molecule. In this regard, highly purified CLA+ T cells avidly bind E-selectin transfectants, whereas CLA negative(–) T cells do not bind these transfectants (23). In a separate study, it was proved that CLA and E-selectin represented a receptor–ligand pair by demonstrating that purified CLA glycoproteins specifically bound E-selectin transfectants (24). More recently, Rossiter et al. (20) have shown that CLA expression predicts E-selectin binding by examination of a series of T-cell clones, and Jones et al. demonstrated that CLA:E-selectin mediate a typical "rolling" primary interaction of T cells under physiologic conditions of flow (13). It has also been determined that the CLA determinant itself is an oligosaccharide closely related to, but serologically distinct from, sialyl-Lewis[x] (24), and that CLA is distinct from T-cell ligands for the vascular adhesion molecule P-selectin (20, and L. Picker, unpublished data). Cutaneous lymphocyte–associated antigen oligosaccharides appear to "decorate" a restricted set of core proteins, of which low molecular size species of CD45 glycoproteins may constitute a major component (L. Picker, unpublished data).

The clinical significance of CLA:E-selectin interactions is supported by the observation of venular E-selectin expression in a wide variety of chronic inflammatory skin lesions associated with the infiltration of T lymphocytes. In this setting, E-selectin was found to be expressed preferentially at high levels on superficial dermal venules (in association with CLA+ T cell infiltrates), relative to most other tissue sites (23). E-selectin expression is induced in the first few hours after a pro-inflammatory stimulus in the skin (25,26), and is especially persistent on dermal microvascular endothelium (27), likely due to alternative RNA processing (28). Taken together, these data suggest strongly that the CLA:E-selectin interaction is required for efficient T cell homing to skin but not to the vast majority of extracutaneous sites.

The CLA-defined memory T cell subset displays an overall pattern of homing-associated adhesion molecules consistent with skin-selective homing behavior (16,18). Among peripheral blood T cells, CLA is almost exclusively expressed by a 25% to 30% subset of $CD45RA^{low}/RO^{high}$ memory T cells. Almost all CLA+ memory T cells also express the peripheral lymph node homing receptor, L-selectin, in keeping with the intimate functional association between skin and its draining peripheral lymph nodes. This observation is also in keeping with the demonstration that L-selectin ligands are commonly expressed on endothelium in chronic dermatitis, a finding suggesting that L-selectin could play a supplementary role in T cell homing to skin (29). The majority of CLA+ T cells, including all the CLA (bright) cells, also lack the mucosal selective homing receptor $\alpha_4\beta_7$-integrin. Only a minor subset of CLA (dim to intermediate cells) express this gut-selective homing receptor. Interestingly, the other $\alpha_4$-integrin, $\alpha_4\beta_1$ (the primary lymphocyte receptor for VCAM-1 ligand), is strongly expressed by CLA (dim to intermediate+) T cells, but the CLA (bright+) T cell subset shows diminished expression of all $\alpha_4$-integrins, including $\alpha_4\beta_1$.

These data indicate the intensity of CLA expression is related to the expression of other homing receptors. In this regard, it is important to note that whereas CLA (dim to intermediate+) cells can usually be found as a minor T-cell subset in extracutaneous sites of inflammation, few CLA (bright+) T cells are found in such sites. In addition, CLA (bright+) are the usually the first T cells entering skin blisters overlying delayed type hypersensitivity skin reactions, and persist the longest (i.e., present even as the delayed type hypersensitivity response is decreasing). Moreover, in some delayed-type hypersensitivity reactions, usually those of lesser clinical intensity, the CLA (bright+) subset is the predominant T cell to extravasate during the entire course of the reaction. These observations suggest two classes of skin homing T cells that are defined by their relative surface density of CLA: a CLA (bright+) T cell subset that displays strikingly skin-specific homing, and a CLA (dim to intermediate+) subset that are skin-selective in their homing behavior, but retain the ability to extravasate at other tissue sites.

Cutaneous lymphocyte–associated antigen expression is induced during the T cell virgin-to-memory transition in human secondary lymphoid tissues. Moreover, there is a clear tissue-specificity in this regulation: memory T cells generated in the appendix are predominantly L-selectin–/CLA–, whereas those generated in peripheral lymph nodes are predominantly L-selectin+, and 50% to 60% express CLA (17,30). These data suggest that at least one component of the differential regulation of these two homing receptors is related to the anatomical location of the differentiation process.

CLA expression can be differentially regulated on virgin and memory T cells in vitro by microenvironmental signals. When purified CLA-negative naive T cells are stimulated with PHA + IL-2 alone, there is little or no up-regulation of CLA expression (17). Other selected cytokines (IL-1, IL-3, IL-4, IL-5, IL-7, IL-8, IL-10, TNF-$\alpha$, granulocyte-macrophage colony-stimulating factor [GM-CSF], interferon gamma [IFN-$\gamma$]) also lacked CLA up-regulatory activity when added to these cultures (17, and L. Picker, unpublished data). However, two cytokines—transforming growth factor (TGF)-$\beta$I and, to a lesser extent, IL-6—were able to up-regulate CLA expression. These cytokines also increased CLA expression on established memory T cells, both converting CLA– memory T cells to CLA+ and increasing CLA expression on established CLA+ memory cells. Interestingly, L-selectin expression by activated virgin T cells was also increased by these same two cytokines in keeping with the L-selectin+ phenotype of most CLA+ positive memory T cells (30). Taken together, these data were the first to indicate that both activated virgin T cells and reactivated memory T cells are subject to HR expression regulation by specific microenvironmental factors.

In more recent studies, we have observed that when bacterial superantigens (including staphylococcal enterotoxins A and B, toxic shock syndrome toxin-1 and streptococcal pyrogenic exotoxins A and C) are used to stimulate T cells (with accessory cells present), there is a profound CLA up-regulation, without exogenous cytokine addition (31). We further demonstrated that 1.) these toxins had the ability to stimulate accessory cell production of IL-12, 2.) anti–IL-12 (but not anti–TGF-$\beta$1) blocked toxin-induced CLA up-regulation (see Fig. 2), and 3.) adding exogenous IL-12 to PHA-stimulated (accessory cell free) cultures also resulted in strong CLA up-regulation (on both naive and memory T cells). These data not only identified IL-12 as a potent, specific up-regulator of CLA, they also demonstrated that certain T cell stimuli can, by virtue of their effects on accessory cells, simultaneously activate T cells and regulate their functional differentiation. Thus, the nature of the antigen, the cytokine milieu in which the immune responses occurs, and the location of the response may all contribute to the regulation of homing receptors such as CLA.

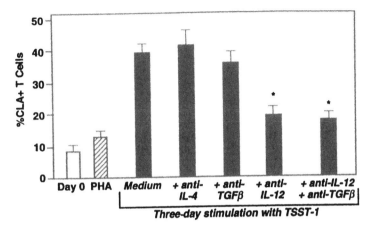

**Figure 2** Cutaneous lymphocyte–associated antigen positive (CLA+) T cell expression induced by toxic shock syndrome toxin-1 (TSST-1) is dependent on interleukin (IL)-12, but not transforming growth factor (TGF)-β. Neutralizing antibody to IL-12 and antibody to TGF-β was added to TSST-1 stimulated (PBMC). The addition of anti–IL-12 caused a significant inhibition of TSST-1 induced CLA+ T cell expansion (*p < 0.01). (From Ref. 31.)

## V. FUNCTIONAL HETEROGENEITY OF T CELLS

Homing behavior is not the only heterogeneous functional characteristic of memory T lymphocytes. The ability to express a certain pattern of cytokines also shows considerable heterogeneity (32–39) This has been best demonstrated in the mouse, where CD4+ memory T cell clones demonstrate distinct patterns of cytokine secretion, forming the TH1 (IL-2, IFN-γ, and TNF-α producing) and TH2 (IL-4, IL-5, IL-6, IL-10, and IL-13 producing) lymphocyte subsets (32,33). The physiologic relevance of these TH1 and TH2 phenotypes has been supported by the in vivo demonstration of polarized "TH1-" and "TH2-like" cytokine responses, which have a profound impact on the outcome of certain experimental infections in mouse models. For example, the TH1 subset is critical for protective immunity against intracellular pathogens such as *Leishmania major*, whereas the TH2 subset appears to be more important in infections with nematodes such as *Trichuris muris*, and in down-regulating TH1 responses (32–34). Although the exact patterns of cytokine secretion by human memory CD4+ T cells differ from that found in mice, the existence of human memory T cell subsets with distinct TH1- versus TH2-like cytokine secretion patterns have also been demonstrated (35–39).

The development of cytokine-defined memory T cell subsets does not appear to occur during thymic development, but rather is thought to occur in secondary lymphoid tissues during the antigen-induced virgin to memory T cell transition. The regulation of TH1- versus TH2 cell development has been an area of intense investigation. It has been shown that microenvironmental factors, both cytokines and "costimulatory" adhesion molecules, influence the differential generation of these subsets (32,40–48). For example, IL-12 promotes—whereas IL-4 suppresses—the development of IFN-γ–producing (TH1-like) T cells. This regulation is also clearly under genetic control, as the tendency to manifest predominant TH1 versus TH2 responses varies in mouse strains with different genetic backgrounds (32). As described above, our own work suggests homing receptor expression is differentially regulated at the same time as the development of T cells with different cytokine potential, in some instances by the same factors. Signals delivered to T cells undergoing virgin to memory cell transition via cytokines and/or cell adhesion events with accessory cells appear to coordinately determine the pattern of cytokine gene expression and homing characteristics of antigen-responsive T cells.

Chronic allergic diseases such as atopic dermatitis, asthma, and allergic rhinitis have in common the generation of allergen-specific IgE responses, and tissue-specific inflammation characterized by the infiltration of memory T cells, eosinophils, and monocyte/macrophages (49). Recent studies have demonstrated that T cells infiltrating into the inflamed tissues of patients with atopic dermatitis, asthma, and allergic rhinitis primarily express IL4, IL-5, and IL-13 (1,2), but not IFN-γ. The ability of such TH2-like T cells to mediate IL-4-dependent cutaneous inflammation has been demonstrated in a murine model by Hauser and coworkers (50). IgE also plays a role in allergic diseases, particularly in triggering allergic responses, and possibly in facilitating allergen capture for allergen processing and presentation to T cells. In this regard, IL-4 and IL-13 also play a key role in immunoglobulin isotype switching to IgE production (51).

Current models of allergic disease (49) suggest that immunologic (e.g., cytokines) and genetic factors likely contribute to the differentiation of Th2-like cells following the stimulation of virgin T cells with allergens. In secondary lymphoid tissue, allergen-specific Th2 cells likely play a key role in inducing IgE responses by allergen-specific B cells, which in turn, serve to "arm" mast cells throughout the body, sensitizing them to the allergen in question. Allergen deposition in skin (either making its way through an excessively permeable epidermis or deposited there from the bloodstream after inhalation or ingestion) would trigger these mast cells or other resident IgE-bearing effector cells (e.g., macrophages) to secrete mediators, producing not only an immediate hypersensitivity reaction, but also TNF-α and other cytokines that can induce the expression of adhesion molecules such as E-selectin, intercellular adhesion molecules (ICAMS), and vascular cell adhesion molecule (VCAM)-1 on the vascular endothelium (25,52). This cytokine-activated microvasculature would then support the extravasation of those memory T cells that express appropriate homing receptors (e.g., in the case of skin, CLA), including (but not limited to) the allergen-specific T cells.

## VI. PATHWAYS CONTROLLING RECRUITMENT OF LYMPHOCYTES DURING ALLERGIC REACTIONS

With regard to allergic diseases, it has been found that T cells migrating into inflamed skin lesions are highly enriched for the CLA– expressing memory T cell subset, whereas memory T cells isolated from the airways of asthmatics are predominantly CLA negative (Fig. 3 and Ref. 18). Thus, the propensity of a given individual to develop atopic dermatitis as opposed to asthma may depend on differences in the skin- versus the lung-seeking behavior of their memory/effector T cells, which, at least in part, is determined by regulation of CLA expression.

In support of a role for CLA in the homing of T cells to allergic skin reactions, the atopic dermatitis skin lesion as well as the allergen-induced cutaneous late phase reaction is associated with the increased expression of leukocyte adhesion molecules such as E-selectin (25,53) Importantly, the induction of adhesion molecules like E-selectin in allergen-stimulated atopic skin can be blocked by neutralizing antibodies to IL-1 and TNF-$\alpha$ (25). Thus, the local release of cytokines represents an important initiating event in the local accumulation of inflammatory cells at the site of allergic reactions. In vitro studies have also demonstrated that following the primary adhesion of CLA+ T cells to E-selectin on cytokine-stimulated endothelial cells, transendothelial migration of such T cells are dependent on very late activation antigen (VLA)-4, vascular cell adhesion molecule (VCAM)-1 and lymphocyte function–associated antigen (LFA)-1/ICAM-1 pathways (54). Of note, VCAM-1 and ICAM-1 expression is also increased in the AD skin lesion (54).

Children with food-induced atopic dermatitis provide an opportunity for determining whether there is a relationship between the tissue specificity of a clinical reaction to an allergen and the expression of homing receptors on T cells activated in vitro by the relevant allergen. In this regard, we have assessed the expression of CLA and L-selectin on peripheral blood T cells from patients with atopic dermatitis and milk-induced eczema, and compared their homing receptor expression—following stimulation with casein, the major milk allergen—to T cells collected from patients with allergic eosinophilic gastroenteritis, milk-induced enterocolitis, or non-atopic healthy controls (55). Following in vitro stimulation with casein, but not *C. albicans*, patients with milk allergy and atopic dermatitis had a significantly greater percentage of CLA+ T cells than controls with milk-induced enterocolitis, allergic eosinophilic gastroenteritis, or non-atopic healthy control subjects (Fig. 4). In contrast, the percentage of L-selectin–expressing T cells did not differ significantly between these groups. These

**(A)**

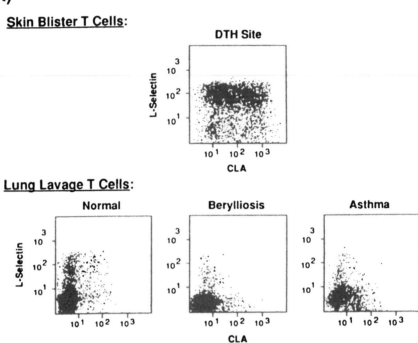

**Figure 3** Characterization of cutaneous lymphocyte–associated antigen (CLA) and L-selectin expression on T cells in pulmonary versus cutaneous inflammatory tissues sites. Representative flow cytometric profiles of the correlated expression of CLA and L-selectin on skin blister T cells associated with the peak of a delayed-type hypersensitivity reaction, lung lavage T cells associated with normal lung, berylliosis, and asthma. Compared to peripheral blood, skin blister T cells are markedly enriched for CLA+ T cells, and about 80–85% are L-selectin+. In contrast, lung lavage T cells are predominantly L-selectin−, and CLA[low to dim]. This phenotype was similar for pulmonary T cells from normal lungs, asthmatics or berylliosis patients. (From Ref. 18.)

data suggest that following casein stimulation, allergic patients with milk-induced skin disease have an expanded population of CLA+ T cells, as compared to non-atopic control subjects or allergic patients without skin involvement. Thus, heterogeneity in the regulation of homing receptor expression on antigen-specific T cells may play a role in determining sites of involvement in tissue-directed allergic responses.

Further evidence for the relationship between CLA and cutaneous T cell responses in atopic dermatitis has been provided by Santamaria Babi et al. (56). These investigators studied several groups of patients with cutaneous hypersensi-

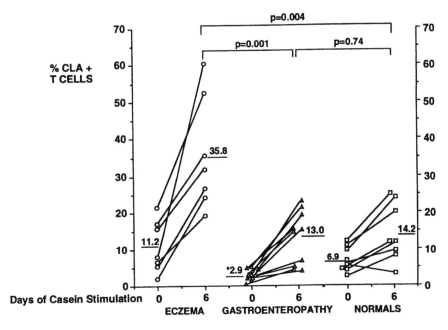

**Figure 4** Casein induces cutaneous lymphocyte–associated antigen positive (CLA+) T cell expression in patients with atopic dermatitis and milk-induced eczema. CLA+ T cell expression was analyzed prior to (day 0) and after 6 days of in vitro stimulation with casein in 7 patients with AD and milk-induced eczema, 10 patients with allergic gastroenteropathy, and 8 normal donors. At baseline, atopic dermatitis patients with milk-induced eczema had significantly greater (*P = .005) CLA+ T cell expression than children with gastroenteropathy. After 6 days of casein stimulation, children with milk-induced eczema had significantly greater CLA+ T cell expression than the other two control groups. (From Ref. 55.)

tivity to defined allergens. These groups included 1.) patients with nickel-induced allergic contact dermatitis, 2.) house-dust-mite-sensitive patients who manifested atopic dermatitis alone, and 3.) house-dust-mite-sensitive patients who manifested asthma with or without an atopic dermatitis component. In these studies, CLA+ versus CLA– memory T cells (CD45RO+) were physically separated and compared for their proliferative responses to these allergens. The results demonstrated that in all individuals with nickel-induced allergic contact dermatitis, the proliferative response to nickel was almost exclusively confined to the CLA+ subset.

In patients with house-dust-mite sensitivity and atopic dermatitis alone, seven of eight patients with measurable proliferative responses to house-dust-mite antigens showed complete or nearly complete restriction of this response to the CLA+ subset. In contrast, patients with house-dust-mite sensitivity and asthma (either alone or with an atopic dermatitis component) showed predominant house-dust-mite antigen–specific proliferative responses in the CLA– subset. The link between CLA expression and skin disease–associated T-cell effector function in atopic dermatitis was further demonstrated by the observation that freshly isolated circulating CLA+ T cells in atopic dermatitis patients, but not in normal control subjects, selectively expressed the HLA-DR activation antigen and demonstrated spontaneous production of IL-4 but not IFN-γ. More recently, this group has also demonstrated that the number of activated (HLA-DR or IL-2 receptor positive) CLA+ T cells is a sensitive marker of disease activity in atopic dermatitis and can be used to monitor response to treatment (57).

Additional data specifically linking CLA expression to T-cell localization in allergic skin reactions came from the observation that CLA is highly expressed on allergen-specific T cell clones obtained from the skin of patients with AD, but is not expressed by non–allergen-specific T cell clones obtained from the blood of the same individuals grown under similar conditions (20). Taken together, all these data provide support for the association of T-cell CLA expression and cutaneous immunity and strongly suggest the existence of regulatory mechanisms capable of segregating individual T-cell antigen specificities into memory subsets defined by their homing receptors.

To date, a primary homing receptor for T cells targeting the lung has not been identified. To further investigate the localization characteristics of memory T cells in asthmatics, Picker et al. (18) quantitated the expression of homing receptor–defined memory T cell subsets in peripheral blood and either skin from subjects with delayed hypersensitivity reactions or bronchoalveolar lavage fluid from asthmatics. Marked differences were observed in the memory T cell subsets localizing to the skin versus the lung. In contrast to cutaneous memory T cells, which were dominated by the CLA+/L-selectin+ subset, lung memory/effector T cells were almost all CLA$^{-to\ low}$/L-selectin–. α4β7-integrin–expressing memory/effector T cells were diminished in both skin and lung, suggesting this homing receptor is not a major participant in determining localization specificity in either of these sites. These results support the concept that in humans, memory T cell recruitment to the skin versus the lung involves distinct homing receptors.

Current models of airway inflammation in asthma and allergic rhinitis propose that allergen exposure leads to the activation of resident mast cells, macrophages, and TH2 lymphocytes. This results in the local release of various cytokines including IL-1, TNF, IL-4, and IL-13, which induce endothelial cells to increase their expression of ICAM-1 and VCAM-1 (25,58,59). Up-regulation of these cell adhesion molecules and secretion of appropriate chemoattractants would subsequently mediate the migration of various cells, including lymphocytes and eosinophils, into the airway.

A number of studies support a role for cell adhesion molecule expression in asthmatic airways and the nasal mucosa of patients with allergic rhinitis. In a primate model of allergen-induced airway hyperresponsiveness, Wegner et al. (60) demonstrated enhanced ICAM-1 expression on bronchial vascular endothelium following allergen challenge. More recent studies on human bronchial tissues from allergic asthmatics have demonstrated increased expression of ICAM-1 and VCAM-1 on bronchial microvessels (61,62). These investigators have also found increased numbers of LFA-1– and VLA-4–expressing leukocytes in the submucosa. Unfortunately, no attempt was made to correlate this with endothelial cell adhesion molecule expression. Thus, the relative contributions of ICAM-1 and VCAM-1 to the infiltration of lymphocytes into the airways of asthmatics are unknown. However, studies employing anti–ICAM-1 antibodies have demonstrated reduction in eosinophil influx into the airways of experimental animals (60). Similarly, immunohistochemical investigations of nasal biopsies from patients with perennial and seasonal allergic rhinitis have identified enhanced expression of VCAM-1 and ICAM-1 (63). Studies specifically targeting T lymphocytes in asthma and allergic rhinitis, however, await the identification of their primary homing receptors.

## VII. CONCLUSIONS

Taken together, allergic reactions leading to the recruitment of lymphocytes into sites of tissue inflammation results from the sequential interaction of cytokines, chemokines, and cell adhesion molecules. The generation of cytokines occurs at the tissue site of allergen deposition, resulting in the induction of endothelial adhesion molecules such as E-selectin or VCAM-1. This results in the recruitment of TH2 lymphocytes expressing specific homing receptors. In the case of skin-seeking T cells, as compared to lung-seeking T cells, there is selective expression of the CLA homing receptor. The mechanisms that regulate the expression of homing receptors are an active area of investigation. Of note, cytokines appear to have an important modulatory effect on the expression of CLA on T cells. Thus, TGF-$\beta$1 and IL-12 markedly up-regulate CLA expression on T cells (17,31), whereas IL-10 (D. Leung, unpublished observations) inhibits CLA expression.

These observations are of great interest because the capacity to develop clinical allergy is unlikely to depend exclusively on IgE responses: a positive immediate skin test is not always correlated with clinical allergy (64). Indeed, the capacity of T cells to infiltrate into a local tissue site may play a key role in controlling the magnitude of an allergic response and therefore determine whether a symptomatic reaction occurs. An understanding of the mechanisms that control recruitment of T lymphocytes to a particular organ (e.g., lung vs. skin) may therefore provide new insights into the pathogenesis of clinical allergy and lead to new directions in the treatment of this common group of illnesses.

## REFERENCES

1. D. S. Robinson, Q. Hamid, S. Ying et al., *N. Engl. J. Med.*, *326*:298 (1992).
2. Q. Hamid, M. Boguniewicz, and D. Y. M. Leung, *J. Clin. Invest.*, *94*:870 (1994).
3. L. J. Picker and E. C. Butcher, *Annu. Rev. Immunol.*, *10*:561 (1992).
4. E. C. Butcher, *Curr. Topics Microbiol. Immunol.*, *128*:85 (1986).
5. L. J. Picker, *Curr. Opin. Immunol.*, *6*:394 (1994).
6. E. C. Butcher, and L. J. Picker, *Science*, *272*:60 (1986).
7. C. R. MacKay, *Curr. Opin. Immunol.*, *5*:423 (1993).
8. Y. Shimizu, W. Newman, Y. Tanaka, and S. Shaw, *Immunol. Today*, *13*:106 (1992).
9. J. S. Berman, D. J. Beer, A. C. Theodore, H. Kornfeld, J. Bernardo, and D. M. Center, *Am. Rev. Respir. Dis.*, *142*:238 (1990).
10. M. Salmi, K. Granfors, M. Leirisalo-Repo, et al., *Proc. Natl. Acad. Sci. USA*, *89*:11436 (1992).
11. T. A. Springer, *Cell*, *76*:301 (1994).
12. R. F. Bargatze and E. C. Butcher, *J. Exp. Med.*, *178*:367 (1993).
13. D. A. Jones, L. V. McIntire, C. W. Smith, and L. J. Picker, *J. Clin. Invest.*, *94*:2443 (1994).
14. T. J. Schall, *Cytokine*, *3*:1 (1991).
15. L. J. Picker, S. A. Michie, L. S. Rott, and E. C. Butcher, *Am. J. Pathol.*, *136*:1053 (1990).
16. L. J. Picker, L. W. Terstappen, L. S. Rott, P. R. Streeter, H. Stein, and E. C. Butcher, *J. Immunol.*, *145*:3247 (1990).
17. L. J. Picker, J. R. Treer, B. Ferguson-Darnell, P. A. Collins, P. Bergstresser, and L. W. Terstappen, *J. Immunol.*, *150*:1122 (1993).
18. L. J. Picker, R. J. Martin, A. Trumble, et al., *Eur. J. Immunol.*, *24*:1269 (1994).
19. C. Pitzalis, A. Cauli, N. Pipitone, et al., *Arthritis Rheum.*, *39*:137 (1996).
20. H. Rossiter, F. van Reijsen, G. C. Muddle, et al., *Eur. J. Immunol.*, *24*:205 (1994).
21. M. J. Borowitz, A. Weidner, E. A. Olsen, and L. J. Picker, *Leukemia*, *7*:859 (1993).
22. P. W. Heald, S. L. Yan, R. L. Edelson, R. Tigelaar, and L. J. Picker, *J. Invest. Dermatol.*, *101*:222 (1993).
23. L. J. Picker, T. K. Kishimoto, C. W. Smith, R. A. Warnock, and E. C. Butcher, *Nature*, *349*:796 (1991).
24. E. L. Berg, T. Yoshino, L. S. Rott, et al., *J. Exp. Med.*, *174*:1461 (1991).
25. D. Y. M. Leung, R. S. Cotran, and J. S. Pober, *J. Clin. Invest.*, *87*:1805 (1991).
26. H. A. Waldorf, L. J. Walsh, N. M. Schechter, and G. F. Murphy, *Am. J. Pathol.*, *138*:477 (1991).
27. P. Petselbauer, J. R. Bender, J. Wilson, and J. S. Pober, *J. Immunol.*, *151*:5062 (1993).
28. W. Chu, D. H. Presky, R. A. Swerlick, and D. K. Burns, *J. Immunol.*, *153*:4179 (1994).
29. S. A. Michie, P. R. Streeter, P. A. Bolt, E. C. Butcher, and L. J. Picker, *Am. J. Pathol.*, *143*:1688 (1993).
30. L. J. Picker, J. R. Treer, B. Ferguson-Darnell, P. A. Collins, D. Buck, and L. W. Terstappen, *J. Immunol.*, *150*:1105 (1993).

31. D. Y. M. Leung, M. Gately, A. Trumble, B. Ferguson-Darnell, P. M. Schlievert, and L. J. Picker, *J. Exp. Med.*, *181*:747 (1993).
32. W. E. Paul and R. A. Seder, *Cell*, *76*:241 (1994).
33. N. E. Street and T. R. Mosmann, *FASEB J.*, *5*:171 (1991).
34. A. Sher, and R. L. Coffman, *Annu. Rev. Immunol.*, *10*:385 (1992).
35. L. J. Picker, M. K. Singh, Z. Zdraveski, et al., *Blood*, *86*:1408 (1995).
36. K. Kristensson, C. A. K. Borrebaeck, R. Carlsson, *Immunology*, *76*:103 (1992).
37. M. Salmon, G. D. Kitas, and P. A. Bacon, *J. Immunol.*, *143*:907 (1989).
38. S. Romagnani, *Int. Arch. Allergy Immunol.*, *98*:279 (1992).
39. E. A. Wierenga, M. Snoek, H. M. Jansen, J. D. Bos, R. A. W. van Lier, and M. L. Kapsenberg, *J. Immunol.*, *147*:2942 (1991).
40. P. Scott, *Curr. Opin. Immunol.*, *5*:391 (1993).
41. R. L. Coffman, K. Varkila, P. Scott, and R. Chatelain, *Immunol. Rev.*, *123*:189 (1991).
42. S. L. Swain, L. M. Bradley M. Croft, et al., *Immunol. Rev.*, *123*:114 (1991).
43. E. Schmitt, P. Hoehn, C. Huels, et al., *Eur. J. Immunol.*, *24*:793 (1994).
44. V. Brinkmann, T. Geiger, S. Alkan, and C. H. Heusser, *J. Exp. Med.*, *178*:1655 (1993).
45. A. D'Andrea, X. Ma, M. Aste-Amezaga, C. Paganin, and G. Trinchiere, *J. Exp. Med.*, *181*:537 (1995).
46. F. G. M. Snijdewint, P. Kalinski, E. A. Wierenga, J. D. Bos, and M. L. Kapsenberg, *J. Immunol.*, 150: 332 (1995).
47. R. T. Semnani, T. B. Nutman, P. Hochman, S. Shaw, and G. A. van Seventer, *J. Exp. Med.*, *180*:2125 (1994).
48. C. L. King, R. J. Stupi, M. Craighead, C. H. June, and G. Thyphronitis, *Eur. Immunol.*, *25*:587 (1995).
49. D. Y. M. Leung, *J. Allergy Clin. Immunol.*, *96*:302 (1995).
50. K. M. Müller, F. Jaunin, I. Masouye, J-H. Saurat, and C. Hauser, *J. Immunol.*, *150*:5576 (1993).
51. J. E. de Vries, *J. Invest. Dermatol.*, *102*:141 (1994).
52. L. M. Klein, R. M. Lavker, W. L. Matis, and G. F. Murphy, *Proc. Natl. Acad. Sci. USA*, *86*:8972 (1989).
53. H. Wakita, T. Sakamoto, Y. Tokura, and M. Takigawa, *J. Cutan. Pathol.*, *21*:33 (1994).
54. L. F. Santamaria Babi, R. Moser, M. T. Perez Soler, L. J. Picker, K. Blaser, and C. Hauser, *J. Immunol.*, *154*:1543 (1995).
55. K. J. Abernathy-Carver, H. A. Sampson, L. J. Picker, and D. Y. Leung, *J. Clin. Invest.*, *95*:913 (1995).
56. L. F. Santamaria Babi, L. J. Picker, M. T. Perez Soler, et al., *Exp. Med.*, *181*:1935 (1995).
57. P. A. Piletta, S. Wirth, L. Hommel, J. H. Saurat, and C. Hauser, *Arch. Dermatol.*, In press.
58. R. P. Schleimer and B. S. Bochner, *J. Immunol.*, *154*:1086 (1994).
59. B. S. Bochner, D. A. Klunk, S. A. Sterbinsky, R. L. Coffman, and R. P. Schleimer, *J. Immunol.*, *154*:799 (1995).
60. M. C. D. Wegner, R. H. Gundel, P. Reilly, M. Haynes, L. G. Letts, and R. Rothlein, *Science*, *247*:456 (1990).

61. S. Montefort, C. Gratziou, D. Goulding, et al., *J. Clin. Invest.*, *93*:1411 (1994).
62. Y. Ohkawara, K. Yamauchi, N. Maruyama, et al., *Am. J. Respir. Cell. Mol. Biol.*, *12*:4 (1995).
63. S. Montefort, I. H. Feather, S. J. Wilson, D. O. Haskard, S. T. Holgate, and P. H. Howard, *Am. J. Respir. Cell. Mol. Biol.*, *7*:393 (1992).
64. C. E. May, *J. Allergy Clin. Immunol.*, *58*:500 (1976).

# 16
# Expression of Cell Adhesion Molecules in Asthma

**Stephen Montefort**   *St. Luke's Hospital, G'Mangia, Malta*

**Stephen T. Holgate**   *University of Southampton, Southampton General Hospital, Southampton, England*

## I. EXPRESSION OF CELL ADHESION MOLECULES IN ASTHMA

It is now well recognized that inflammation plays a prominent role in the pathophysiology of asthma (1), which is reflected in the changes one encounters when viewing a microscopical section of bronchial biopsies from patients with active asthma. Characteristically, the submucosa of the bronchial wall in asthma contains many inflammatory cells that have migrated from the circulation into the tissues. It is these cells that are known to mediate the structural changes one sees in the inflamed bronchial wall of the asthmatic, the most prominent of which is shedding of the upper layer of the bronchial epithelium; this may in part explain the bronchial hyperresponsiveness one observes in this condition (2,3).

There is accumulating evidence that cell adhesion molecules are involved in these inflammatory changes (4). Leukocyte–endothelial adhesion molecules are pivotal in the early stages of recruitment and migration of circulating inflammatory leukocytes to the bronchial submucosa and epithelium in asthma (5); intraepithelial cell adhesion molecules play a key role in maintaining the integrity of the bronchial epithelial layer (6). A number of investigators, including our group, considered it important to study the expression of cell adhesion molecules in asthma in order to try and determine whether in vitro findings involving these molecules could be extrapolated to the in vivo situation in bronchial biopsies. However, when one investigates the expression of adhesion molecules in bronchial biopsies in vivo using markers that are expressed on blood vessels and other continous structures, there are problems in quantification. Fortunately, many of these have now been overcome, although despite improved methods of fixation, embedding, and tissue sectioning, quantitative comparisons between biopsies from different subjects is still problematic. Nor is it straightforward extrapolating patterns of staining to function.

## II. THE EXPRESSION OF CELL ADHESION MOLECULES IN THE BRONCHIAL MUCOSA

Up-regulation of the endothelial expression of intercellular adhesion molecule-1 (ICAM-1) after multiple allergen inhalation is important in the development of the intense eosinophil infiltration of the tracheal mucosa and a parallel increase in airway hyperresponsiveness to methacholine in *Ascaris*—sensitized monkeys (7). E-selectin has also been implicated in the neutrophilia and late-phase bronchoconstriction after a single allergen inhalation in nonhuman primates (8); in endothelial cell layers in vitro and in human skin biopsies, it has been suggested that both these adhesion molecules are instrumental in eosinophilic adherence (9). Bronchoalveolar lavage (BAL) and biopsy studies have provided overwhelming evidence for the involvement of eosinophils in the characteristic inflammation that occurs in the airways in asthma. One might speculate that E-selectin and ICAM-1 might also be important in the development and maintenance of the inflammatory cellular changes found in the mucosa of asthmatics. Corticosteroids have been proved beneficial to symptomatic asthmatics (10), and because this class of drug attenuates interleukin-1 (IL-1)-induced ICAM-1 upregulation in vitro (11), a decrease in leukocyte–endothelial adhesion might be one of the ways these drugs manage to decrease the inflammatory cell population in bronchial mucosa.

We were interested in studying the expression of these adhesion molecules on the bronchial mucosa of asthmatic volunteers without challenge and compare it with that found in normal control subjects. We also looked for any changes in expression of these adhesion molecules after a 6-week course of inhaled corticosteroids. In our first study (12), we obtained bronchial biopsies during fiberoptic bronchoscopy on 10 nonasthmatic and 10 mild asthmatic volunteers (forced expiratory volume in 1 sec [$FEV_1$] % predicted ± SEM 103.6 ± 3.3). Six of the asthmatics were rebronchoscoped, and biopsies were taken after a 6-week course of high-dose inhaled beclomethasone dipropionate, which had improved the subjects' symptomatic and spirometric indices. At the time, the biopsies were snap-frozen and 8 μm thick sections were immunostained with monoclonal antibodies (mAb) for ICAM-1 and E-selectin. We opted to use a semi-quantitative mode of analysis with two independent observers scoring the extent of staining of the epithelium or endothelium by the relevant mAb as "some" or "all," depending whether the majority or minority of the cells in question demonstrated immunoreactivity.

In biopsies from all 10 asthmatics, ICAM-1 immunostaining was constitutively expressed both in the vascular endothelium of vessels deep to the epithelium and on the ciliated epithelium itself (Fig. 1). In tissue from the normal volunteers, immunostaining for ICAM-1 was also apparent in the endothelium in seven out of the eight (two biopsies being inadequate for interpretation of endothelial staining) and in the epithelium in seven out of the 10 biopsies. In comparing ICAM-1 positive sections between the normal and asthmatic subjects, there was no apparent difference in the overall pattern of staining. The extent of staining was greater in the endothelium than in the epithelium, and although stronger on the luminal surface of the endothelial cells, staining was also present on their other aspects. In the epithelium, it was the minority of epithelial cells that were ICAM-1 positive, with most staining in the basolateral aspects of the columnar cells and surrounding the basal cells (Fig. 1), similar to that described by Wegner et al. in the *Ascaris*-sensitized monkey. In some sections, minimal staining was also noted on the apical ciliated aspect of the bronchial epithelial cells.

Immunostaining for E-selectin was observed in the endothelium of the submucosal vessels but, in contrast to ICAM-1, not on the epithelium (Fig. 2). As with ICAM-1, there was linear staining along the luminal aspect of the endothelial

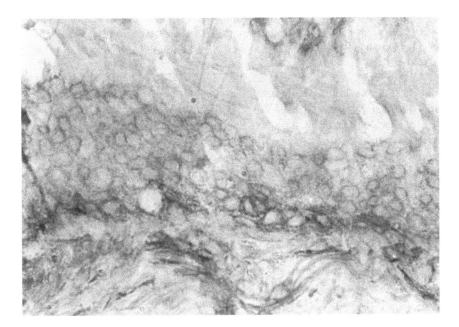

**Figure 1** High-power micrograph of intercellular adhesion molecule-1 (ICAM-1) immunostaining in the basolateral area of bronchial epithelium in a cryostat section from a bronchial biopsy of a non-asthmatic (mag. × 800).

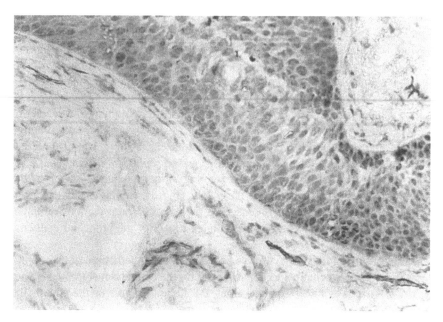

**Figure 2**   E-selectin immunostaining on an 8 μm thick frozen section of bronchial biopsy from normal subject showing staining on a few subepithelial vessels. There is no epithelial staining (mag. × 400).

cells but this was of a lesser intensity than that observed for ICAM-1 at the same site. Although only seven of the biopsies from the normal volunteers were suitable for E-selectin staining, basal expression of this immunoreactive epitope was observed in the endothelium in all but two biopsies. When present the pattern of staining did not appear different between the normal and asthmatic biopsies. The biopsies from the corticosteroid-treated asthmatics revealed a significant decrease in the number of mucosal eosinophils staining for the activated eosinophil cationic protein (ECP+) [$P<.04$]. However, despite the presence of a clear anti-inflammatory response, there was no change in the extent of immunostaining either for ICAM-1 or E-selectin.

The finding of high levels of ICAM-1 expression within the vasculature of normal airways could result from the exposure of the non-asthmatics bronchial mucosa to a multitude of inhaled reactive pollutants, pathogens, and antigens. Indeed, constitutive expression of this molecule may be an essential feature of maintaining airway homeostasis in a hostile world in which the airways are continually subjected to a variety of environmental insults. The observation that most of the T lymphocytes present in the normal bronchial mucosa are antigen-committed (CD45RO+) (13) and activated (very late activation antigen (VLA-1) (14), reinforces this view. As in the case of the endothelium, we failed to show any increase in ICAM-1 expression in the bronchial epithelium of the asthmatic subjects. It is worth noting that epithelial ICAM-1 immunostaining was absent in three out of 10 normal subjects but was present in all asthmatic subjects. The presence of ICAM-1 on the bronchial epithelium is of interest, and one can speculate that its function is to recruit and retain lymphocytes, macrophages, and eosinophils in the epithelium in a protective role, and only in asthma might this become exaggerated. ICAM-1 is also the receptor for a major subgroup of rhinoviruses (15,16). Because ICAM-1 is basally expressed on bronchial epithelial cells, a mechanism is provided for rhinovirus infection of the lower respiratory tract. As viral infection itself has been shown to increase the expression of ICAM-1 on HeLa and other cells (17), we took great care to ensure that none of the asthmatic or control subjects were studied within 6 weeks of an upper respiratory tract infection. Subsequently other investigators (18,19) have reported greater ICAM-1 expression on bronchial epithelial cells from asthmatic patients than from normal subjects. The most likely explanation for this difference from our studies relates to the increased disease severity studied.

E-selectin is reported to be expressed on endothelial cells only in the presence of an inflammatory stimulus (20). Therefore, we were surprised to find basal expression of E-selectin in the mucosal vasculature of five of seven biopsies from the normal volunteers. Basal expression may be responsible for a continous low-grade neutrophil influx into the airway submucosa as part of the normal airway defense against environmental insults. As in all previous reports in all tissues studied, E-selectin was observed only in the bronchial vasculature and not in the epithelium. E-selectin endothelial immunostaining was also observed in the asthmatic biopsies, but the intensity of staining and its distribution was not different from that observed in the normal subjects. However, in both groups E-selectin was expressed to lesser degree in the mucosal venules than was ICAM-1.

In isolated chondrosarcoma cells, cytokine-stimulated up-regulation of ICAM-1 is effectively inhibited by corticosteroids (11). In addition, corticosteroids are highly effective at inhibiting T-cell secretion of a wide array of cytokines including interferon gamma (IFN-$\gamma$) tumor necrosis factor-$\alpha$ (TNF-$\alpha$) and IL-1, cytokines all known to up-regulate ICAM-1 expression in vitro (21). Djukanovic et al. (22) reported that 6 weeks of topical beclomethasone dipropionate dramatically reduced the airway content of eosinophils in parallel with indices of clinical improvement. Because this drug also reduces the activation state of bronchial T cells, one possible mechanism is a coordinate down-regulation in production of cytokines such as granulocyte-macrophage colony-stimulating factor (GM-CSF), IL-4, and IL-5 encoded in the IL-4 gene cluster of chromosome 5, all of which are involved in eosinophil recruitment and activation. In keeping with the findings of Kaiser et al. (23), who observed that corticosteroids did not influence the induction of ICAM-1, E-selectin or VCAM-1 by TNF-$\alpha$ or of VCAM-1 by IL-4, in this study we failed to show any change in the mucosal expression of immunoreactive ICAM-1 or E-selectin following corticosteroid therapy.

Our method of biopsy processing and use of frozen sections to quantitate small differences in adhesion molecules may well have affected our results, in that the semi-quantitative analysis would have only been sensitive enough to detect gross differences in expression of any molecule between asthmatics and non-asthmatics. Another potential reason for our results was that the asthmatic subjects enrolled into our study had asthma that was not of sufficient severity to reveal any gross difference in adhesion molecule expression from non-asthmatics. A recent study has revealed a clear difference in ICAM-1, E-selectin, and vascular cell adhesion molecule-1 (VCAM-1) expression in bronchial biopsies from asthmatics recovering from a recent asthma exacerbation (24). Bentley et al. (25) compared bronchial biopsies from a group of patients with severe intrinsic asthma to a group with extrinsic asthma and a group of non-asthmatics. They could demonstrate a significant increase only in endothelial expression in ICAM-1 and E-selectin staining between the very symptomatic intrinsic asthmatic group and the non-asthmatics. As in our study no difference was noted in ICAM-1 and E-selectin expression between patients with extrinsic asthma and non-asthmatics. In a more recent study, we have noted an increase in constitutive expression of ICAM-1 and VCAM-1 in nasal mucosal biopsies from patients with severe perennial rhinitis when compared with normal control subjects (26). This further suggests that severity of the airway inflammation might be associated with an increase in adhesion molecule expression; however, none of these studies showed any correlation between symptom scores and the level of adhesion molecule expression when assessed semi-quantitatively.

## III. THE EXPRESSION OF CELL ADHESION MOLECULES ON GLYCOLMETHACRYLATE-EMBEDDED BRONCHIAL BIOPSIES

One of our main objectives was to overcome the difficulties encountered using the semi-quantitative methods employed in our previous study. An important development toward being able to achieve this was the development of glycolmethacrylate (GMA) resin embedding for tissue processing (27). This resin enabled us to cut thinner sections (2 μm), conserve morphology without disturbing the tissue, and also retain the integrity of immunoreactive epitopes. The fact that blood vessels were not distorted by crushing made it possible, with the use of an endothelial cell marker, to quantitate the proportion of vessels expressing a specific adhesion molecule. Four 2 μm thick sequential sections from the bronchial biopsies were cut, with one of the intermediate sections being immunostained with the vessel marker EN-4. The number of vessels positively staining with this anti–endothelial cell antibody was taken as being the full complement of vessels present in the biopsy; the vessels staining positively for an endothelial adhesion molecule were expressed as a percentage of this full complement (Fig. 3). This index denoted the proportion of vessels

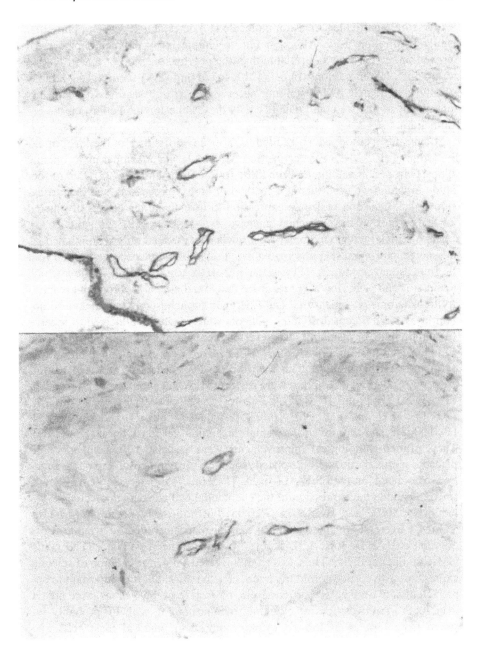

**Figure 3** Upper panel shows whole complement of vessels staining with E N4 monoclonal antibody (mAb); lower panel shows subsequent section with some vessels staining with E-selectin mAb. These were expressed as a percentage of the whole complement of vessels (mag. × 400).

expressing a particular adhesion molecule at a detectable level for a particular antibody titration used. Although this technique is by no means perfect, it proved more sensitive and reliable than the previous method of quantification using thick frozen sections. In being able to cut thin GMA sections, the already scanty epithelial ICAM-1 staining observed on frozen sections became less prominent so that we could quantify only these adhesion molecules, on the endothelium.

Using this method, we still failed to show any difference in the level of expression of ICAM-1, E-selectin, and VCAM-1 when comparing bronchial biopsies from non-asthmatics with those from mild asthmatics (28). In a further study in which we used this new method of biopsy processing and quantification of cell adhesion molecule expression to investigate the effects of inhaled corticosteroids in the bronchial mucosa of asthmatics (29), 28 patients (22 male, 6 female) with symptomatic allergic asthma treated only with inhaled β2 adrenoreceptor-agonists participated in a double-blind, placebo-controlled trial of fluticasone propionate (FP). After a 2-week run-in period, during which symptoms and peak flow were recorded daily on diary cards, bronchial reactivity to histamine was measured and fiberoptic bronchoscopy undertaken to obtain lavage and mucosal samples, patients were randomly allocated to receive FP 500 μg b.i.d. or matching placebo. Treatment was for 6 weeks, during which symptom and peak expiratory flow (PEF) recording was continued and in the sixth week further measurement was made of $PC_{20}$ histamine and repeat bronchoscopy and airway sampling were performed. When compared with placebo, FP reduced symptoms [$P<0.05$], β2-agonist use [$P<.005$], improved both A.M. PEF [$P<0.01$] and P.M. PEF [$P<0.05$], reduced diurnal variation in PEF [$P<0.0008$], and improved bronchial histamine responsiveness [$P<0.008$]. These clinico-pathological improvements were associated, in comparison to placebo, with a corticosteroid-related reduction of submucosal leukocyte function–associated antigen LFA-1+ cells [$P<0.02$], submucosal VLA4+ cells [$P<0.01$] and in the expression of the endothelial cell adhesion molecules P-selectin [$P<0.005$] and E-selectin [$P<0.05$] but noticeably not in ICAM-1 or VCAM-1 expression. These changes were also accompanied by a reduced expression of IL-2, IL-3, IL-4, IL-5, IL-6, IL-8, tumor necrosis factor (TNF), and GM-CSF but not IFN-γ. Thus, this study revealed a down-regulation of selectin expression with corticosteroid treatment of asthmatics; but it is important to appreciate that as an inhaled corticosteroid, FP is at least three times more potent than BDP. Moreover, loss of cytokine immunostaining with FP probably resulted from loss of preformed cytokine stored in mucosal mast cells and eosinophils because these cells are largely responsible for most of the cytokines observed by immunohistochemistry (30).

## IV. THE EXPRESSION OF CELL ADHESION MOLECULES AFTER ENDOBRONCHIAL ALLERGEN CHALLENGE

The next step in our studies was to investigate whether the level of expression of leukocyte–endothelial adhesion molecules are altered by an acute inflammatory stimulus such as allergen exposure in sensitized individuals (31). We opted to study this 5 to 6 hours after endobronchial local allergen instillation, a time point at which up-regulation of these adhesion molecules has been shown to begin in vitro. Expression of ICAM-1 on human umbilical vein endothelial cell culture [HUVECs] by IL-1, TNF-$\alpha$ and IFN-$\gamma$ starts within 2 to 4 hours, peaks at 24 hours, and in the presence of the inducing cytokine continues for 72 hours (32,33). In contrast, E-selectin up-regulation is more rapid: it occurs within 30 min, peaks at 4 hours, and wanes over the following 20 hours (34). In vivo, Wegner et al. (7) reported up-regulation of ICAM-1 expression on tracheal epithelial and endothelial cells in their *Ascaris*-sensitized nonhuman primate model that occurs optimally after the third of three alternate-day allergen inhalations. Kyan Aung et al. (9), on the other hand, has shown up-regulation of both ICAM-1 and E-selectin on endothelium in human skin 6 hours after intradermal allergen inoculation.

In addition to mapping adhesion molecules in bronchial biopsies, we have also quantified cellular infiltration by the various inflammatory cells in the mucosa using cell-specific markers and also the number of leukocytes bearing the LFA-1 integrin surface molecules, this being one of the key ligands for ICAM-1. In these studies we opted for local endobronchial segmental allergen instillation as a method for challenging the mucosa to be sure of the level and site of the allergen exposure while also enabling simultaneous sham-challenge of a separate lung segment for comparison. The 5 to 6-hour time point chosen for biopsy after allergen or saline instillation is also compatible with the time interval for late-phase allergic tissue response, which starts approximately 2 hours after challenge and peaks at 6 to 8 hours.

Six mildly symptomatic asthmatics (three male, three female; mean age ± SEM 30.5 ± 2.5 yrs) were recruited into the study. At the time of enrollment the asthmatic subjects had stable pulmonary function (mean % predicted $FEV_1$ ± SEM 95.9 ± 3.5%). All were using the inhaled $\beta_2$ agonist albuterol by metered dose inhaler on an as-required basis for relief of symptoms as their sole asthma medication. All subjects had hyperreactive airways to inhaled methacholine with a geometric mean $PC_{20}$ of 2.66 (range 0.25–3.52)mg/mL. On the main study day, two fiberoptic bronchoscopies were undertaken, provided that the baseline $FEV_1$ was >70% predicted and the subjects' platelet count and clotting studies were within normal limits. In the first bronchoscopy 20 mL of sterile saline was instilled into the posterior segment of the right upper lobe (as a control site) and 20 mL of specific allergen into the right middle lobe at a concentration previously determined by skin testing. Five minutes after introduction of the two solutions, the appearance of the airways was observed and photographed to record any airway narrowing. After ensuring that there was no generalized bronchoconstriction, the bronchoscope was retracted and the subject allowed to rest, with regular monitoring of his clinical status. Five to 6 hours later, a second bronchoscopy was undertaken in which endobronchial mucosal biopsies were taken from the same two sites.

When the immunostaining pattern for the three endothelial adhesion molecules between the allergen- and saline-challenged bronchi were compared, marked differences were apparent (Fig. 4). Expressed as a percentage of the total vessel complement, allergen exposure resulted in a median twofold increase in the number of vessels expressing E-selectin from 18.4% (range 3.8–35.7) after saline to 36.3% (range 15.3–68.4) [$P<.005$]. Similarly, allergen also up-regulated (to a lesser but still significant extent) the expression of ICAM-1 from a basal (saline-challenged) level of 48.8% (range 30.7–62.8) to 70.2% (range 38–87.3) ($P<.05$). In contrast, allergen provocation had no significant effect on VCAM-1

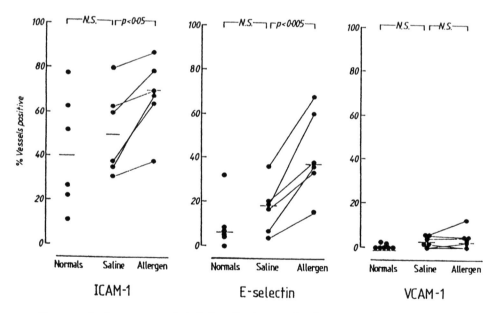

**Figure 4** Leucocyte–endothelial adhesion molecule expression on vascular endothelium of bronchial mucosa from non-asthmatics and from saline and allergen-challenged bronchial segments in asthmatic subjects. Results are expressed as percentage of total stainable vessels (lines = medians).

expression at the 5 to 6 hr time point: 2.3% (range 0–5) versus 2.4% (range 0–12). In the saline-challenged mucosa, occasional leukocytes immunostaining for the cell surface integrin LFA-1 were observed. Both in the epithelium and in the submucosa, exposure to allergen produced a four to sixfold increase in the number of infiltrating leukocytes expressing the cell surface integrin LFA-1, with cell medians increasing from 19.9 (range 2.4–48.4) to 160.5 (range 36.8–216.4)/mm$^2$ in the submucosa [$P<.01$] and 0.8 (range 0–2.4) to 3.8 (range 2.4–4.2)/mm length of basement membrane in the epithelium ($P<.01$). The increase in submucosal LFA-1+ leukocyte infiltrate following allergen correlated with the extent of up-regulation of its ligand ICAM-1 [$r_s = 0.70$], but not E-selectin or VCAM-1 [$r_s = 0.30$, $r_s = 0.41$]

The cell-specific mAbs identified a mixed inflammatory cellular infiltrate in association with the late phase response. The median neutrophil count at the allergen-exposed site was 265.5 (range 50–2043)/mm$^2$ compared to 46 (range 17–88)/mm$^2$ at the saline-challenged site ($P = .03$). This increase correlated positively with the increase in endothelial expression of ICAM-1 [$r_s = 0.59$] but not E-selectin [$r_s = 0.27$]. Eosinophils identified by their granule content of ECP, also infiltrated the allergen-challenged mucosa with numbers increasing from saline-challenged values of 8.5 (range 0–27)/mm$^2$ to a post-allergen value 45.5 (range 16–245)/mm$^2$ [$P = .025$]. No relationship was found between the increase in eosinophils and extent of expression of any the three adhesion molecules examined. Allergen exposure also stimulated an increase in submucosal mast cells, with numbers increasing in the allergen-compared to the saline-challenged mucosa from a median of 14.5 (range 8–35) to 49 (range 17–60)/mm$^2$ [$P = .03$]. Lymphocytes staining positively for CD3 also increased in the allergen-challenged site from 64.5 (range 2–333) to 107.5 (range 30–717)–/mm$^2$ [$P = .03$], but although both the CD4+ and CD8+ lymphocyte subsets showed a trend toward an increase, these failed to achieve statistical significance.

To conclude, this study has revealed an up-regulation of E-selectin and ICAM-1 but not VCAM-1 5 to 6 hours after allergen challenge. On HUVECs in vitro, E-selectin is up-regulated by IL-1 and TNF-α (34), peaking in expression at 4 hours after stimulation. Although ICAM-1 and VCAM-1 have similar stimulus-related time courses of expression in vitro, only the former showed any increase in expression with allergen in vivo, suggesting restriction of the phenomenon. This pattern of expression in the asthmatic bronchial mucosa is in agreement with that observed in the skin of atopic subjects 6 hours after intradermal allergen injection (9). We have reported an increase in E-selectin expression in the nasal mucosa of subjects with seasonal allergic rhinitis 6 hours after allergen challenge, but ICAM-1 expression did not increase (35). Interestingly, Bentley et al. (25) have reported an increased bronchial mucosal expression of VCAM-1 but not ICAM-1 or E-selectin in asthmatic bronchial biopsies obtained 24 hours after allergen inhalation and compared to biopsies taken at the same time point following sham saline inhalation. This was associated with a selective increase in eosinophil infiltration observed in the biopsies. Taken together, these observations could be interpreted as an early up-regulation of E-selectin and ICAM-1 being responsible for the mixed eosinophil, neutrophil, and T cell response at 5 to 6 hours; VCAM-1 increasing up to 24 hours may account for the more selective persistent eosinophil and T-cell responses consequent on an interaction with the leukocyte integrin VLA-4 [$\alpha 4\beta 1$] (36). These findings are similar to those of Wegner et al. (7) and Gundel et al. (8) who have shown that mAbs to ICAM-1 and E-selectin respectively attenuated leukocyte recruitment and the LAR post-allergen challenge to their *Ascaris*-sensitized monkey. Our study reveals a positive correlation between the level of adhesion molecule expression on the microvascular ed and the number of LFA-1+ cells. This probably represents most of the infiltrating leukocytes that presumably utilize this $\beta_2$ integrin as one of adhesion molecules to adhere to and transmigrate from the circulation into the bronchial mucosa.

## V.  CIRCULATING ADHESION MOLECULES IN ASTHMA

Recently, circulating forms of leukocyte–endothelial adhesion molecules have been identified in the peripheral blood and other body fluids both in normal subjects and in various disease states (37). Serum levels of cICAM-1 have been found to be increased in patients with gastrointestinal malignancies, especially in those with metastases (38), and in the serum of subjects with the leukocyte adhesion deficiency (LAD) syndrome (39). Circulating ICAM-1, cVCAM-1, and cE-selectin have also been detected in BAL fluid of patients with interstitial lung disease (40).

Utilizing an antigen capture enzyme-linked immunoassays (ELISA) for cICAM-1, cE-selectin, and cVCAM-1, we have investigated whether these adhesion molecules exist free in the serum in stable asthma and during an acute exacerbation, and whether the levels of these molecules become elevated in asthma when compared to atopic and non-atopic normal controls (41). The acute asthmatics (n = 38) presenting to the accident and emergency department of Southampton General Hospital and the Chinese University of Hong Kong Hospital were recruited for the study. A sample of venous blood was taken when first seen by the admitting physician at a time when the group's mean PEF ± SEM was 138 ± 11.8 L/min (33.4 ± 2.4% predicted). After conventional treatment of acute asthma with bronchodilators and systemic corticosteroids, a further blood sample was taken from 29 of the asthmatics who were still hospitalized 3 days after admission when their mean PEF ± SEM had risen to 247 ± 18.7 L/min (49.9 ± 3.9% predicted). Finally, following hospital discharge, 13 of these subjects agreed to return to have a further blood sample taken 28 days after discharge. The second group of asthmatics consisted of 29 stable asthmatics whose asthma was well controlled on inhaled β2 agonists and inhaled corticosteroids. Their mean % predicted PEF ± SEM was 86.4

± 1.9 and their geometric mean $PC_{20}$ histamine 1.22 mg/mL] (range 0.03–7.45 mg/mL). The control groups consisted of 13 atopic normal volunteers who had positive skin prick tests to one of five common allergens, but no history of respiratory symptoms, and 16 healthy non-atopic subjects.

Circulating ICAM-1 was detected in serum from all normal volunteers. In the non-atopic normal subjects, the median level for cICAM-1 was 14.6 U/mL (range 2.5–33.8), which was not significantly different from either the atopic normals (16.8, range 2.5–40.8 U/mL) or the patients with stable asthma (17, range 2.5–82.7 U/mL). The median cICAM-1 levels in the serum obtained from acute asthmatics on days 1, 3, and 28 were 28.9 U/mL (range 10–189), 26.8 U/mL (range 4.3–201.5), and 35.7 U/mL (range 7–162.4), respectively, which were not significantly different from each other [Fig. 5]. However these levels were significantly raised when compared to the stable asthmatics $[P = .002]$, the atopic normal $[P = .001]$, and the non-atopic normal $[P = .0002]$ controls. Overall no correlation within the acute asthma group could be found with PEF or PEF % predicted, although as expected a negative correlation was exhibited between PEF % predicted and cICAM-1 when levels in acute day 1 and stable asthmatics were combined and subjected to regression analysis ($r_s = -0.359$, $P = .002$.] In the stable asthmatics, no relationship could be found between cICAM-1 and the $PC_{20}$ histamine. In the case of cE-selectin, very similar results to cICAM-1 were

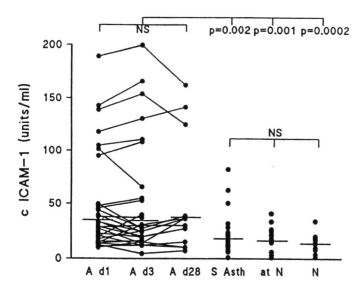

**Figure 5** Circulating intercellular adhesion molecule-1 (cICAM-1) levels in the subject groups studied (lines = medians).

observed, in that there was no significant difference between stable asthmatics (51.6, range 10.6–148.6 U/mL), atopic normals (84, range 17.9–124 U/mL) or the non-atopic normals (73.8, range 16–178 U/mL). The serum levels of this circulating selectin were also similar throughout the three acute asthma study days; day 1 (110, range 18–262 U/mL), day 3 (101.5, range 20–234 U/mL), and day 28 (117.5, range 10.7–223). These increased soluble adhesion molecule levels in acute asthma were all significantly raised when compared with the stable asthmatics ($P = .002$), the atopic normals ($P = .017$), and the non-atopic normals ($P = .02$). As in the case of cICAM-1, the leels of cE-selectin in the combined acute asthma day 1 and the stable asthmatics were negatively correlated to the % predicted PEF ($r_s = -0.34$, $P = .007$). The serum levels of cVCAM-1 in the non-atopic normals (median 13.8, range 8.7–24.8 U/mL), atopic normals (10.8, range 0.5–21.8 U/mL) and stable asthmatics (11.1, range 1.2–27.2 U/mL) did not differ significantly from the levels measured in acute asthma on days 1, 3, and 28 (medians 10.4, range 2.5–36.5 U/mL; 13.8, range 0.5–35.4 U/mL and 13.6, range 3.6–35.1 U/mL, respectively). No correlation could be found between serum samples of dCAM-1, cE-selectin and cVCAM-1 nor between cVCAM-1 and either PEF or $PC_{20}$ histamine for the asthmatic groups.

The finding of a differential rise of cICAM-1 and cE-selectin but not cVCAM-1 may be due to a mediator and cytokine profile in acute asthma that is more suited to expression of the former two adhesion molecules. The cells of origin and the mechanisms for release of the soluble components of these leukocyte-endothelial adhesion molecules are not known but could involve either shedding or enzymatic cleavage. Although the presence of circulating adhesion molecules probably reflects cellular activation in inflammation, the function of circulating adhesion molecules is not clear; they most likely reflect the inflammatory process itself. It is also possible that elevated levels of circulating adhesion molecules in inflammation could serve a protective role in blocking the complementary receptors on leukocytes and thereby decreasing inflammatory cell recruitment and migration to sites of potential inflammation within the airway wall and elsewhere. As the levels of circulating adhesion molecules failed to correlate with measures of disease severity in stable asthma, it seems unlikely that their measurement in peripheral blood will be of diagnostic use. However, it is possible that serum cICAM-1 and cE-selectin levels might be useful as markers of the duration of underlying airways inflammation responsible for an acute exacerbation of asthma.

## VI. FACTORS INFLUENCING CIRCULATING ADHESION MOLECULE LEVELS IN ASTHMA

To dissect out which factors might be involved in the increase in circulating adhesion molecules in acute asthma, we investigated the concentrations of circulating adhesion molecules in the peripheral blood of asthmatic volunteers at various time points following inhalation allergen challenge severe enough to induce a dual bronchoconstriction response. In addition we have looked for these soluble molecules in non-asthmatic and asthmatic subjects who had experimentally induced rhinoviral infections (42).

Nine atopic asthmatic volunteers who were all known to experience a dual response to allergen challenge from previous studies were recruited for the first part of the study. An allergen challenge was carried out with 5-min $FEV_1$ readings for the first 40 min after reaching at least a 20% decrease from baseline $FEV_1$, and then at hourly intervals until 8 hr after allergen inhalation. Blood samples were collected before the start of the challenge. Gradually increasing concentrations of allergen were administered by nebulizer until the baseline $FEV_1$ had fallen by at least 20%. Blood was taken 30 minutes after and then at hourly intervals to 8 hours after allergen inhalation. The subjects then returned to the laboratory 24 and 48 hours later to have further blood samples taken. Serum samples were taken from an iatrogenically rhinovirus-infected group of asthmatic and non-asthmatic subjects exposed to RV-16 (43). Briefly, these subjects had been iatrogenically infected intranasally with a cultured major group RV 16 provoking a symptomatic cold. After confirming infection, we used the blood samples taken before, during, and 8 weeks after the rhinovirus infection from four asthmatic and four non-asthmatic subjects.

In the allergen-challenge component of the study, none of the three circulating adhesion molecules studied varied significantly from baseline at any of the time points measured. Circulating E-selectin was the only of the three adhesion molecules that increased during rhinovirus infection, increasing from a median level of 203.9 to 272.7 U/mL [$P<.05$] and then returning toward baseline levels at 8 weeks after infection (204.3 U/mL) (Fig. 6). Circulating VCAM-1 also showed a trend toward increasing during RV infection (15.9 to 19.05 U/mL) but this was not significant. There was very little change in cICAM-1 levels across the three time points (medians of 30.2, 32.2, and 29.2 U/mL, respectively). As in the case of cE-selectin, four of the five highest baseline cICAM-1 levels belonged to the four asthmatics. The lack of persistent inflammatory stimuli such as those seen in acute asthma might explain the failure to demonstrate high levels of these adhesion molecules after a single dose of inhaled allergen. The decrease in spirometry seen in these controlled allergen challenges was also not as severe or prolonged as seen in the naturally occurring situation of deteriorating asthma. We were rather surprised that the cICAM-1 levels were not increased after rhinovirus infection, especially when one remembers that ICAM-1 is the receptor for the major group of rhinoviruses and is up-regulated on the epithelium during active RV infection in vitro (44). On the other hand, one could speculate that any increase in expression of this adhesion molecule that might have occurred was not detected because it was being utilized in docking to free ICAM-1-binding sites in the RV cavern of the virus capsule. Virus-mediated cE-selectin release might be initiated

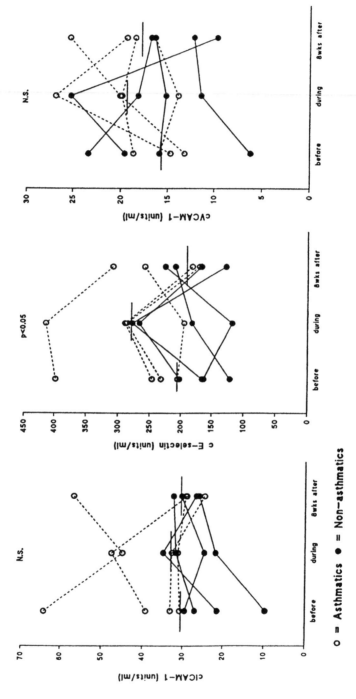

**Figure 6**  Circulating adhesion molecules before, during, and after rhinoviral infection.

by involvement of E-selectin in the adherence of neutrophils, the granulocytes recruited during a respiratory virus infection (45). In the small group of subjects enrolled in this virus study, the asthmatics did exhibit the highest baseline levels of cICAM-1 and cE-electin.

## VII. BRONCHIAL INTRAEPITHELIAL ADHESION MOLECULES

Shedding of the bronchial epithelium has been long associated with bronchial asthma, through necropsy (46) and sputum cytology (47) and, more recently, in bronchial biopsies (2,3). These studies in the living patient have allowed the investigation of the associations of this phenomenon, and recent studies have indicated that correlations exist between bronchial hyperresponsiveness and the extent of epithelial cell loss assessed by bronchalveolar lavage (48). Epithelial cell adhesion is generally held to be essential for the maintenance of normal bronchial homeostasis, and destruction of these cohesive mechanisms is believed to play an important role in bronchial asthma (49).

Various cell adhesion mechanisms are used in the maintenance of the integrity of different epithelia. Studies of tracheal epithelium have shown that ciliated and secretory columnar epithelial cells do not form junctional attachments with the underlying epithelial basement membrane. Rather, these cells appear to be attached via desmosomes on their basolateral surfaces to basal cells, which themselves then form hemidesmosomal contacts with the basement membrane (50). Thus, this respiratory epithelium is a truly stratified complex epithelium.

A study of the bronchial epithelium in asthma, based on biopsies obtained via the rigid bronchoscope, has shown partial epithelial cell denudation with residual basal cells remaining attached to the underlying basement membrane (3). Similarly, smoke damage of ovine bronchial epithelium results in columnar epithelial cell shedding, again leaving the basal cells still attached to the basement membrane (51). Similar findings have been reported in fatal bronchial asthma (46). Taken together, these findings indicate a potential plane of cleavage in the bronchial epithelium between the basal cells and the overlying columnar epithelial cells. This may be due to a difference in the cell adhesion mechanisms at this site or may reflect directed cell injury inflicted by eosinophils, smoke, and so on. Either mechanism might result in the shedding of columnar epithelial cells into the bronchial lumen. This led us to suggest that shedding of the bronchial epithelium occurs along a suprabasal plane, and thus we set up a study of epithelial cell clusters recovered from BAL of asthmatic and control subjects (52) to test this hypothesis. On studying these clusters by electron microscopy, we found very few basal cells still adherent to the clumps, which were solely made up of suprabasal cells. The cells composing the clusters were whole, confirming that they were shed from the underlying basal cells along the junction between these two layers of cells. This lack of basal cells within the epithelial cell clusters suggests that there is a potential weak plane of cleavage between the columnar and basal cell layers, which in asthmatics might be the target of an increased amount of inflammatory insults. In cytocentrifuge preparations from the same subjects, apart from an increased number of total shed epithelial cells, there were more epithelial cell clusters and a predilection for shedding in whole sheets (up to 30 cells) of columnar cells in the BAL from the asthmatics. Extrapolation from these findings would suggest that the site of adhesion between the columnar cells and the underlying basal cells may be an important target for cell damage in the bronchial epithelium. Exposure of guinea pig respiratory epithelium to eosinophil major basic protein has been shown to induce detachment of the columnar cells from the basal cell layer while the basal cells maintain their contact with the basement membrane. Increased concentrations of major basic protein are required in order to induce detachment of the basal cells (53). This suggests that the basal cell attachment to the basement membrane is more resistant to attack by eosinophils than the adhesive mechanisms that attach the columnar cells to these basal cells. This is in keeping with our findings that the epithelial cell clusters in BAL in asthma do not contain many basal cells. It would also appear that the adherence between the neighboring columnar cells is more robust than that between the columnar cells and basal cells, as the columnar cell clusters of up to 30 cells remain intact after shedding into the bronchial lumen.

In order to further investigate the mechanisms of cell adhesion in the bronchial epithelium, we have mapped the distribution of the relevant junctional and non-junctional mechanisms within this mucosal structure (54). Using various immunohistochemical techniques, we have observed various integrins present within the normal bronchial epithelium. The $\alpha_2$ integrin subunit that constitutes a component of one of the $\beta_1$ collagen/laminin receptors immunolocalized to the basal cell–basement membrane interface, where it would be expected to interact with basement membrane components. This integrin's immunostaining also extended between most of the epithelial cells, confirming the findings of Damjanovich et al. (55). We have failed to demonstrate any immunostaining for either $\alpha_1$ or $\alpha_5$, which forms a component of the classical fibronectin receptor in the intact normal epithelium, despite the reports of $\alpha_5$ being expressed in cultured human bronchial epithelial cell monolayers (56). It is possible that these integrin components are only expressed under specific conditions or in the presence of specific growth factors. The $\alpha_6$ and $\beta_4$ integrin subunits showed similar patterns of distribution, being expressed only in the sub-basal region of the epithelium as a clear linear pattern (Fig. 7). In human skin (57) and cornea (58), the basal cells are fixed firmly to the basement membrane proteins via hemidesmosomes with their constituent $\alpha6\beta_4$ integrins. However, although hemidesmosomal structures were visualized in the bronchial epithelium by transmission electron microscopy,

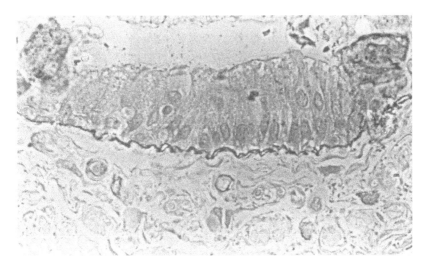

**Figure 7** Bronchial epithelial immunostaining for $\alpha_6$ at the the basal cell–basement membrane interface, on a glycolmethacrylate (GMA) embedded biopsy (mag. × 400).

they seemed less distinct than described by others in the skin. The presence of strong immunostaining for the $\alpha_6$ and $\beta_4$ components of hemidesmosomes along the inferior border of he basal cells adds further evidence to the important role played by these hemidesmosomal integrins in anchoring the epithelium to the basement membrane. Neither of these integrin subunits was evident immunohistochemically above this level, although some staining was observed around submucosal blood vessels. Thus, in addition to forming a laminin receptor with $\beta_1$, $\alpha_6$ is also probably able to interact with $\beta_4$ to form a further integrin receptor for laminin present in vascular basement membrane. A monoclonal antibody raised against the E-cadherin adhesion molecule uvomorulin stained the apical parts of the junction between neighboring columnar cells (Fig. 8) and this as well is in keeping with the observation by Boller et al. (59) of finding this adhesion molecule in the intermediate junctions of adult mouse intestinal epithelium.

The most important adhesion mechanism that operates in epithelial structures is the desmosome. This composite structure was easily identified in the bronchial epithelium between adjacent epithelial cells and the characteristic plaque and tonofilament structures as previously described in skin. Using a mAb, directed against shared epitopes expressed on desmosomal proteins 1 and 2, which are situated mainly in the desmosomal plaque, the distribution of desmosomes within the epithelium could clearly be seen as punctate immunofluorescent structures when viewed by ultraviolet microscopy. Punctate staining was present along the lateral and inferolateral borders of the suprabasal cells. This was also evident be-

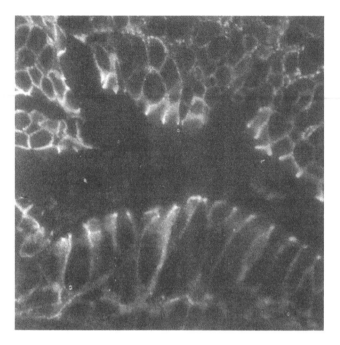

**Figure 8** Bronchial epithelium immunostained for E-cadherin on frozen section by immunofluorescence (mag. × 400).

tween adjacent basal cells, but not along their inferior border. There was a particularly dense population of desmosomes at the junction between suprabasal and basal cell layers (Fig. 9), suggesting that a strong intercellular bond is needed in this plane to maintain epithelial integrity. It is possible that a disturbance of desmosomal adhesion could account for the epithelial fragility in asthma, especially as this seems to occur between the suprabasal and basal cells, the site of high desmosomal expression. An alternative explanation is that neutral proteases—either from inflammatory cells per se (e.g., tryptase, elastase) or induced by a cognate interaction between eosinophils and epithelial cells (60)—could result in enzymatic degradation of desmosomes with a subsequent reduction in their capacity to anchor epithelial cells.

Whatever the mechanisms of epithelial damage in asthma and other disorders, the widespread distibution of desmosomes and other adhesive structures within the bronchial epithelium may be the target of inflammatory responses in which the epithelium becomes damaged. Thus, understanding of factors controlling the expression, function, and degradation of adhesion molecules is of clear importance in airway homeostasis and how it may be disturbed.

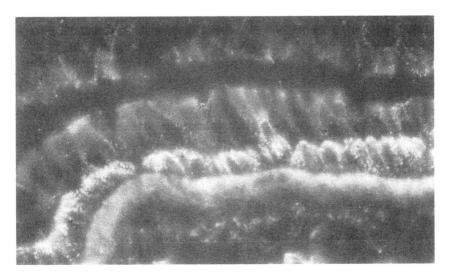

**Figure 9** Immunofluorescent punctate staining on bronchial epithelium immunostained for desmosomal proteins 1 and 2 (mag. × 200).

## VIII. CONCLUSION

Although asthma was once thought of solely as a sporadic disease of intermittent airflow obstruction, detailed studies involving biopsies and lavage samples of the airways point to destructive changes and remodeling as also part of the chronic inflammatory response. As the disease progresses through a patient's lifetime, it is important that more emphasis is placed in gaining understanding of these cellular events, for they are likely to be highly relevant to the chronicity that characterizes asthma and the sometimes incomplete response that occurs to conventional anti-asthma therapy.

## REFERENCES

1. P. H. Howarth, P. Bradding, S. Montefort, D. Peroni, R. Djukanovic, M. P. Carroll, and S. T. Holgate, *Am. J. Respir. Crit. Care Med.*, *150*:5:2:S18 (1994).
2. P. K. Jeffery, A. J. Wardlaw, F. C. Nelson, J. V. Collins, and A. B. Kay, *Am. Rev. Respir. Dis.*, *140*:1745 (1989).
3. L. A. Laitinen, M. Heino, A. Laitenen, T. Kava, and T. Hauhtela, *Am. Rev. Respir. Dis.*, *131*:599 (1985).
4. S. Montefort, S. T. Holgate, and P. H. Howarth, *Eur. Resp. J.*, *6*:1044 (1994).
5. S. Montefort and S. T. Holgate, *Resp. Med.*, *85*:91 (1991).
6. S. Montefort, C. A. Herbert, C. Robinson, and S. T. Holgate, *Clin. Exp. Allergy*, *22*:511 (1992).

7.  C. D. Wegner, R. H. Gundel, P. Reilly, N. Haynes, L. Gordon Letts, and R. Rothlein, *Science*, *247*:416 (1990).

8.  R. H. Gundel, C. D. Wegner, C. A. Torcellini, C. C. Clarke, N. Haynes, R. Rothlein, C. W. Smith, and L. G. Letts *J. Clin. Invest.*, *88* [4]:1407 (1991).

9.  U. Kyan Aung, D. O. Haskard, R. N. Poston, M. Thornhill, and T. H. Lee, *J. Immunol.*, *146*:521 (1991).

10. R. Djukanovic, S. Honeyward, D. Peroni, S. Montefort, R. Polosa, M. Judd, S. Wilson, S. T. Holgate, and P. H. Howarth, *Eur. Resp. I.*, 8 (suppl 19):302s (1995).

11. M. L. Dustin, R. Rothlein, A. K. Bhhan, C. A. Dinarello, and T. A. Springer, *J. Immunol.*, *137*:245 (1986).

12. S. Montefort, W. R. Roche, P. H. Howarth, R. Djukanovic, C. Gratziou, M. Carroll, L. Smith, K. M. Britten, D. Haskard, T. H. Lee, and S. T. Holgate, *Eur. Resp. J.*, *5*:815 (1992).

13. M. Azzawi, B. Bradley, P. K. Jeffery, A. J. Frew, A. J. Wardlaw, G. Knowles, B. Assoufi, J. V. Collins, S. Durham, and A. B. Kay, *Am. Rev. Respir. Dis.*, *142*:1403 (1990).

14. C. Saltini, M. E. Hemler, and R. G. Crystal, *Clin. Immunol. Immunopathol.*, *46*:221 (1988).

15. D. E. Staunton, V. J. Merluzzi, R. Rothlein, R. Barton, S. D. Martin, and T. A. Springer, *Cell*, *56*:849 (1989).

16. J. M. Greve, G. Davis, A. M. Meyer, C. P. Forte, S. Yost Connolly, C. W. Marlow, C. M. Kamarch, and A. McClelland, *Cell*, *56*:839 (1989).

17. J. E. Tomassini, D. Graham, C. M. Dewitt, D. W. Lineburger, J. A. Rodkey, R. J. Colonno, *Proc. Natl. Acad. Sci. U.S.A.*, *86*:4907 (1989).

18. N. Manolitsas, d'Ardenne, A. Mcaulay, K. Jones, R. J. Davies, *Br. Thor. Soc. Scientific Programme Summer Meeting 1991*, *32* [Abstr.] (1991).

19. P. Gossett, I. Tillie- Leblond, A. Janin, C. H. Marquette, M. C. Copin, B. Wallaert, and A. B. Tonnel, *Ann. N.Y. Acad. Sci.*, *725*:163 (1994).

20. T. A. Springer, *Nature*, *346*:425 (1990).

21. R. Rothlein, M. Czajkowski, M. O'Neill, M. Marlin, E. A. Mainolfi, and V. G. Merluzzi, *J. Immunol.*, *141*:1665 (1988).

22. R. Djukanovic', J. W. Wilson, K. M. Britten, S. J. Wilson, A. F. Walls, W. R. Roche, P. H. Howarth, and S. T. Holgate, *Am. Rev. Resp. Dis.*, *145*:669 (1992).

23. J. Kaiser, C. Bickel, B. S. Bochner, and R. P. Schleimer, *J. Pharmacol. Exp. Ther.*, *267*:245 (1993).

24. Y. Okhawara, K. Yamauchi, N. Maruyama, H. Hoshi, I. Ohno, H. Honma, Y. Tanno, G. Tamura, K. Shirato, H. Ohtani, *Am. J. Respir. Cell Mol. Biol.*, *12*:4 (1995).

25. A. M. Bentley, S. R. Durham, D. S. Robinson, G. Menz, O. Storz, O. Cromwell, A. B. Kay, and A. J. Wardlaw, *J. Allergy Clin. Immunol.*, *92*:857 (1993).

26. S. Montefort, I. H. Feather, S. J. Wilson, D. O. Haskard, T. H. Lee, S. T. Holgate, and P. H. Howarth, *Am. J. Resp. Cell Mol. Biol.*, *7*:393 (1992).

27. K. Britten, P. H. Howarth, and W. R. Roche, *Biotech. Histochem.*, *68*:271 (1992).

28. J. Hibbert, J. Upward, I. Feather, S. Montefort, A. E. Redington, S. T. Holgate, and P. H. Howarth, *Eur. Resp. J.*, *7*:18:239s (1994).

29. S. Montefort, I. Feather, J. Underwood, J. Madden, C. Porter, and P. H. Howarth, *Am. J. Respir. Crit. Care Med.*, *151*:4:A40 (1995).

30. P. Bradding, I. Feather, P. Howarth, R. Meuller, J. A. Roberts, K. Britten, T. Hunt, Y. Okayama, C. Heusser, G. Bullock, M. K. Church, and S. T. Holgate, *J. Exp. Med.*, *176*:1381 (1992).
31. S. Montefort, C. Gratziou, D. Goulding, R. Polosa, D. O. Haskard, P. H. Howarth, S. T. Holgate and M. P. Carroll, *J. Clin. Invest.*, *93:*1411 (1994).
32. D. Cavender, Y. Saegusa, and M. Ziff, *J. Immunol.*, *137*:1855 (1987).
33. J. S. Pober, M. A. Gimbrone, Jr., L. A. Lopierre, D. L. Mendrick, W. Fiers, R. Rothlein, and T. A. Springer, *J. Immunol.*, *137*:1893 (1986).
34. M. P. Bevilacqua, P. Stengelin, M. A. Gimbrone, and B. Seed, *Science*, *243*:1160 (1988).
35. I. Feather, S. Montefort, J. Hibbert, S. Wilson, and P. Howarth, *Eur. Resp. J.*, *7*:Suppl 18:160s (1994).
36. G. M. Walsh, J. J. Mermod, A. Hartnell, A. B. Kay, and A. J. Wardlaw, *J. Immunol.*, *146*:3419 (1991).
37. R. Seth, F. D. Raymond, and M. W. Makgoba, *Lancet*, *338*:83 (1991).
38. M. Tsujisaki K. Imai, H. Hirata, Y. Hanzawa, J. Masuya, T. Nakano, T. Sugiyama, M. Matsui, Y. Hinoda, and A. Yachi, *Clin. Exp. Immunol.*, *85*:3 (1991).
39. R. Rothlein, E. A. Mainolfi, M. Czajkowski, and S. D. Marlin, *J. Immunol.*, *147*:3788 (1991).
40. R. M. du Bois, P. G. Hellewell, I. Hemingway, and A. J. Gearing, *Am. Rev. Resp. Dis.*, *145* [4]:A190 (1992).
41. S. Montefort, C. K. Lai, P. Kapahi, J. Leung, K. Lai, H. Chan, D. O. Haskard, P. H. Howarth, and S. T. Holgate, *Am. J. Respir. Crit. Care Med.*, *149*:1149 (1994).
42. S. Montefort, I. Feather, P. Bardin, D. O. Haskard, P. H. Howarth, and S. T. Holgate, *Malta Med J.*, *3*:A09-2 (1995).
43. P. G. Bardin, S. L. Johnston, G. Sanderson, B. S. Robinson, M. A. Pickett, D. J. Fraenkel, and S. T. Holgate, *Am. J. Resp. Cell Mol. Biol.*, *10*:207 (1994).
44. A. Papi and S. L. Johnston, *J. Allergy Clin. Immunol.*, [in press].
45. M. A. Tosi, J. M. Stark, A. Hamedani, C. W. Smith, D. C. Gruenert, and Y. T. Huang, *J. Immunol.*, *149*:3345 (1992).
46. M. S. Dunnill, *J. Clin Pathol.*, *13*:27 (1960).
47. B. Naylor, *Thorax*, *17*:69 (1962).
48. C. R. Beasley, W. R. Roche, J. A. Roberts, S. T. Holgate, *Am. Rev. Respir. Dis.*, *139*:806 (1989).
49. R. Djukanovic, W. R. Roche, J. Wilson, C. R. Beasley, O. P. Twentyman, P. H. Howarth, and S. T. Holgate, *Am. Rev. Respir. Dis.*, *142*:434 (1990).
50. M. J. Evans, R. A. Cox, S. G. Shami, and C. G. Plopper, *Am. Rev. Respir. Dis.*, *3*:341 (1990).
51. S. Abdi, M. J. Evans, R. A. Cox, H. Lubbesmeyer, D. Herndon, and D. J. Traber, *Am. Rev. Respir. Dis.*, *142*:1436 (1990).
52. S. Montefort, J. A. Roberts, C. R. Beasley, S. T. Holgate, and W. R. Roche, *Thorax*, 499 (1992).
53. S. Motojima, E. Frigas, D. A. Loegering, and G. J. Gleich, *Am. Rev. Respir. Dis.*, *139*:801 (1989).
54. S. Montefort, J. Baker, W. R. Roche, and S. T. Holgate, *Clin. Exp. Allergy*, *6*:1257 (1993).

55. L. Damjanovich, S. M. Albelda, S. A. Mette, and C. A. Buck, *Am. J. Respir. Cell Mol. Biol.*, *6*:197 (1992).
56. R. J. Sapsford, J. L. Devalia, A. E. McAuley, A. J. d'Ardenne, and R. J. Davies, *J. Allergy Clin. Immunol., 87* (1):2[Abstr.653] (1991).
57. A. Sonnenberg, J. Calafat, H. Jannsen, H. Daams, L. M. H. van der Raaiji-Helmer, R. Falcioni, S. J. Kennel, J. D. Aplin, J. Baker, M. Loizidou, and D. Garrod, *J. Cell Biol.*, *113*, 4:907 (1991).
58. M. A Stepp, S. Spurr-Michaud, A. Tisdale, J. Elwell, and I. K. Gipson, *Proc. Natl. Acad. Sci. U.S.A.*, *87*:8970 (1990).
59. K. Boller, D. Vestweber, and R. Kemler, *J. Cell Biol.*, *100*:327 (1985).
60. C. A. Herbert, D. Edwards, J. R. Boot, and C. Robinson, *Br. J. Pharmacol.*, *104*:391 (1991).

# 17

# Expression of Cell Adhesion Molecules in Eosinophilic Disorders of the Skin and Nose

**Lisa A. Beck and Steve N. Georas**   *Johns Hopkins Asthma and Allergy Center, Johns Hopkins University School of Medicine, Baltimore, Maryland*

## I.  INTRODUCTION

This chapter provides an overview of the role of adhesion molecules in the trafficking of inflammatory cells to and within the skin and nose in allergic diseases, and reviews some of the inflammatory mediators found at these sites that are critical for cellular recruitment and adhesion molecule expression. The basic biology of the immunoglobulin supergene, integrin, and selectin adhesion molecule families have been covered in Chapters 1, 4, 5 and therefore will not be repeated. Several approaches have been used to determine the significance of these adhesion pathways in clinical disease. Many researchers have correlated expression of endothelial adhesion molecules with the intensity of a cellular infiltrate during inflammatory reactions in vivo. For example, studies of asthma, allergic rhinitis, nasal polyposis, and atopic dermatitis have demonstrated a significant correlation between tissue eosinophils and/or mononuclear cells and endothelial vascular cell adhesion molecule-1 (VCAM-1) expression (1–6). Others have evaluated disease severity and/or response to treatment in relationship to the degree of adhesion molecule expression (7–9). Still others have used specific anti-adhesion antibodies in a variety of animal models of experimental inflammation (10). For example, intravenous administration of an anti-CD18 (or $\beta_2$ integrin) antibody suppresses the accumulation of neutrophils into skin sites injected with C5a, LTB4, or interleukin (IL)-1 in rabbits (11). Several anti-adhesion molecule antibodies are in human clinical trials, such as anti–ICAM-1 for renal allograft recipients and anti-CD18 for multiple organ failure syndrome in cases of traumatic injury (12,13). Allergic diseases of the skin and nose have not received as much attention, possibly because good animal models are lacking and these diseases are not typically associated with the same degree of morbidity and mortality.

## II.  SKIN

The first half of this chapter focuses on cutaneous diseases with tissue eosinophilia, especially atopic dermatitis (AD). Atopic dermatitis is seen in patients with increased serum IgE, peripheral blood eosinophilia, and history of respiratory allergy (14). It is a common disease characterized histologically by infiltration of CD4+ T cells and macrophages (15,16). In AD, Unlike asthma and allergic rhinitis, intact eosinophils are absent or few in number in the skin of patients (17). Despite the absence of intact eosinophils, Leiferman et al. provided evidence for eosinophilic degranulation by demonstrating extracellular eosinophil-derived major basic protein (MBP) in the dermis of patients with AD (18). Major basic protein and other toxic eosinophil granule proteins (eosinophil-derived neurotoxin, eosinophil cationic protein, and eosinophil peroxidase) have been shown to induce wheal-and-flare reactions in human and guinea pig skin, which could be markedly diminished by pretreatment with an antihistamine (19,20). These studies, in conjunction with studies demonstrating correlations between AD disease severity and peripheral eosinophil numbers, serum eosinophil cationic protein, and serum MBP levels, suggest that the eosinophil is a critical cell in the pathogenesis of AD (21–23).

The selective recruitment of eosinophils into sites of inflammation is largely a function of eosinophil-activating cytokines, endothelial-activating cytokines, and chemotactic cytokines or chemokines. The eosinophil-activating cytokines (interleukin-3 [IL-3], IL-5, and granulocyte-macrophage colony-stimulating factor [GM-CSF]) induce degranulation and enhance cell survival, chemotactic responses, and leukotriene production (24). Eosinophils from patients with atopic dermatitis have increased migratory responses to several common chemotaxins (N-formyl-methionyl-leucyl-phenylalanine and platelet-activating factor) in comparison to eosinophils from normal individuals, suggesting that these patients were exposed to one of these cytokines in vivo (25). In fact, IL-5 treatment of normal donor eosinophils induced the migratory response seen with eosinophils from atopic dermatitis patients (25). Evidence that IL-5 and GM-CSF are produced in allergic skin disease comes from studies of cutaneous late-phase reactions (LPR) and chronic AD skin, and from studies of skin-derived aeroallergen-specific T-cell clones isolated from AD subjects (17,26–29). Interestingly, subcutaneous injection of GM-CSF in nonallergic subjects demonstrates both intact and degranulated eosinophils, suggesting that eosinophil activation by GM-CSF coupled with cutaneous trauma may be sufficient to get tissue recruitment (30). Human trials employing GM-CSF–transfected renal cell tumor vaccination for metastatic disease have also demonstrated pronounced local eosinophil recruitment and activation (31). The role of IL-3 in atopic dermatitis is more speculative because IL-3 mRNA has been found only in LPR skin biopsies (26). Leukocyte activation is just one of the steps important in the selective recruitment of eosinophils into the skin. Other critical steps include endothelial transmigration, tissue migration, and prolongation of cell survival. The role that the endothelium plays in cellular recruitment and the mechaisms responsible for cellular migration within tissues will be the focus of the remainder of this chapter.

## A. Differences Between Umbilical Vein and Dermal Microvascular Endothelial Cells

Much of what we know about the expression of adhesion molecules by human endothelial cells comes from studies of human umbilical vein endothelial cells (HUVEC), which are grown in culture (see Chap. 2). More recently, human dermal microvascular endothelial cells (HDMEC) have been isolated, and several potentially important biochemical and phenotypic differences have been found between these cells and HUVEC (32,33). For example, HDMEC are much more fastidious in that they require human serum for optimal growth (33). Furthermore, they produce primarily prostaglandin PGF2$\alpha$, in contrast to HUVEC, which produce primarily PGI$_2$, (33). Phenotypically, HDMEC have substantially higher baseline levels of ICAM-1, although with cytokine stimulation both HDMEC and HUVEC express comparable levels (34). Vascular cell adhesion molecule 1 (VCAM-1), not present constitutively on either endothelial cell, can be induced on HUVEC by either IL-1$\alpha$, tumor necrosis factor–$\alpha$ (TNF-$\alpha$), IL-4, or IL-13, whereas HDMEC respond more modestly and do so only to TNF-$\alpha$ (and not IL-1$\alpha$ or IL-4) (32,35–37).

Studies performed with skin organ cultures have yielded additional insights into the complexity of adhesion molecule expression in vivo (Table 1). In these experiments, skin explants stimulated with a variety of cytokines demonstrate VCAM-1 expression that is restricted to the arteries and veins of the deep but not superficial vascular plexus (Fig. 1). Using this same model, E-selectin can be expressed on the venular side of both superficial and deep vascular plexus (38). Explants from dermatitic skin, however, are now able to express VCAM-1 on the superficial vascular plexus in response to the combination of TNF-$\alpha$ and IL-4 (38,39). This suggests that the inflammatory infiltrate provides a signal(s) that allows the superficial dermal endothelium to express VCAM-1 in response to cytokines, and provides evidence of endothelial cell heterogeneity that can be modified by infiltrating leukocytes. In addition, the VCAM-1 expression seen on the deep vascular plexus in this model could be induced by a variety of endothelial activating cytokines including TNF-$\alpha$, IL-4, and IL-1$\alpha$, whereas only TNF-$\alpha$ was active on cultured HDMEC.

**Table 1** Cytokine-Induced Expression of Endothelial Adhesion Molecules in Skin Explant or Injection Models

| Reference | Explant or injected cytokine | VCAM-1 | ICAM-1 | E-selectin |
|---|---|---|---|---|
| Messadi (165) | Normal explant | ND | ND | + (TNF-α, IL-1α, LPS) |
| Leung (85) | Normal explant | ND | ND | + (allergen) |
| Petzelbauer (38) | Normal explant | + (TNF-α, IL-1α, IL-4)[a] | ND | + (TNF-α, IL-1α) |
| Petzelbauer (38) | Inflamed explant | + (TNF-α + IL-4)[b] | ND | ND |
| Groves (43) | IL-1α | + | – | + |
| Groves (44) | IFN-γ | ND | + | – |
| Groves (41) | IFN-γ | + | ND | ND |
| Groves (42) | TNF-α | + | + | + |
| Briscoe (40)[c] | LPS | + | + | + |
|  | TNF-α | + | + | + |
|  | IL-4 | – | – | – |
|  | TNF-α + IL-4[d] | + | – | – |

Table Key:
ND: not done
– no change in expression
+ increased expression
[a]VCAM-1 expression on deep vascular plexus only.
[b]VCAM-1 expression on the superficial vascular plexus.
[c]Injection into baboons.
[d]Dose of TNF-α ineffective in inducing adhesion molecule expression alone.

This finding has been largely confirmed by cutaneous cytokine injection studies in humans and baboons where expression of endothelial VCAM-1 was induced by injection with IL-1α, TNF-α, or interferon gamma (IFN-γ) or with the combination of IL-4 and an ineffective dose of TNF-α, but not with IL-4 alone (Table 1) (40–43).

The expression of endothelial E-selectin has been shown to correlate with neutrophil or skin-homing memory T cells that bear the phenotypic marker cutaneous lymphocyte antigen (CLA) (2,44,45) (see Chap. 15). The chronic expression of E-selectin found commonly in inflammatory skin disorders such as AD has been confusing because in vitro stimulation of HUVEC or HDMEC results in maximal E-selectin expression by 4 to 6 hours with rapid decline despite continued cytokine stimulation (33). Recent work suggests that this rapid decline may be due to the fact that E-selectin mRNA is unstable. This instability is thought to

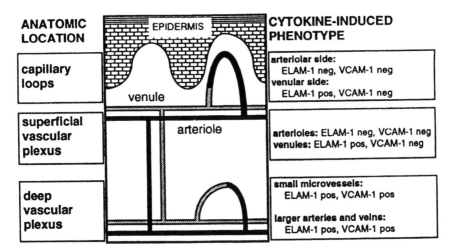

**Figure 1** Schematic diagram of the vascular system of normal human skin (From Ref. 38).

be a consequence of multiple AUUUA destabilizing elements in the 3′-untranslated region (46). Because there are three functional polyadenylation sites, it is feasible to generate three distinct E-selectin transcripts. The shortest of these transcripts lacks six of the destabilizing elements and is therefore more stable than the others. Interestingly, this transcript was not found in HUVEC or HDMEC but was found in inflammatory skin biopsies, suggesting that the presence of this short transcript may be responsible for the chronic E-selectin expression seen in cutaneous diseases (46). Collectively this work suggests that there is heterogeneity among vascular endothelial cells that is dependent not only on their location (arterial vs. venous circulation) and size, but also on their microenvironment.

## B. Role of Keratinocytes in Cell Recruitment

Keratinocytes, like respiratory epithelium, have recently been shown to play an active role in cell recruitment by virtue of their release of endothelial-activating cytokines (IL-$1\alpha$ and TNF-$\alpha$), chemotactic cytokines (IL-8, GRO-$\alpha$, $\gamma$-inducible protein-10 [IP-10] and monocyte chemoattractant peptide-1 [MCP-1]), and their ability to express ICAM-1 and $\beta_2$ integrins on their surface, which may facilitate the localization of leukocytes to the epithelium (47–49). The epidermotropism seen in many cutaneous diseases is thought to occur in part by the interaction between the adhesion ligands leukocyte function–associated antigen (LFA-1) and ICAM-1, expressed on both keratinocytes and leukocytes (50–52). Keratinocytes within normal, uninflamed skin do not express ICAM-1 or $\beta_2$ integrins. The keratinocyte expression of ICAM-1 has been well studied and can be induced by

TNF-α or IFN-γ (in vivo and in vitro), ultraviolet radiation (in vitro), or the sensitizing metal hapten nickel (in vitro) (50,53–57). Histamine alone can weakly induce keratinocyte ICAM-1 expression but when combined with TNF-α is more than additive (58). T-lymphocyte epidermotropism has been documented in association with keratinocyte ICAM-1 expression in a variety of benign cutaneous diseases including atopic dermatitis, allergic contact dermatitis, psoriasis, lichen planus, erythema multiforme, lupus erythematosus, pemphigoid, pemphigus, lichen simplex chronicus, graft-versus-host disease, alopecia areata, and fixed drug eruption (59,60). Because epithelial ICAM-1 induction is spatially and temporally correlated with the presence of intraepidermal T cells and monocytes in many of these diseases, ICAM-1 is thought to be important in the localization of these cells to and within the epithelial compartment (60). However, there are diseases such as evolving cutaneous delayed-type hypersensitivity reactions, Sjögren's syndrome, oral lichen planus, and aphthous stomatitis where the epithelial migration of LFA-1+ cells can be seen in the absence of keratinocyte ICAM-1 expression (47). Thus, keratinocyte ICAM-1 expression may be important in retaining leukocytes within the epithelium, but it is probably not the only mechanism responsible for epidermotropism.

Keratinocytes also constitutively express several members of the integrin family of adhesion molecules. For example, the integrin $\alpha_6\beta_4$, which is localized to the basal surface of basal keratinocytes, is responsible for adhesion to the basal lamina, whereas $\alpha_2\beta_1$ and $\alpha_3\beta_1$, which are enriched on the lateral margins of keratinocytes, play a crucial role in cell–cell interactions (61,62). In general, keratinocyte integrins are thought to serve a variety of functions: adhesion to extracellular matrix proteins, intercellular adhesion, stratification, lateral migration and the regulation of terminal migration. Recent work has shown that human keratinocytes from biopsies of patients with lichen planus, psoriasis, or positive tuberculin test sites demonstrated staining for the α and β subunits of CD11a (LFA-1) but not CD11b (63). No staining was observed in skin biopsies from healthy volunteers or in clinically uninvolved skin from patients with the above cutaneous processes. This work suggests that certain inflammatory conditions are capable of inducing the expression of LFA-1 on suprabasal keratinocytes, which may provide another means by which activated eosinophils and lymphocytes, which have the counter-ligand ICAM-1 on their cell surface, may become trapped in the epidermis (64–67).

## C.  Chemokines

Although $\beta_2$ integrin and ICAM-1 interactions contribute to the adherence of leukocytes within the epithelial compartment, chemotactic cytokines made by or near epithelial cells may provide the critical signal for cells to migrate toward the epithelium and may also contribute to cell-specific recruitment. These chemotactic cytokines (or chemokines) are a newly recognized class of low-molecular-weight proteins that are characterized by their potent in vitro chemotactic activity. This family is subdivided into α and β subfamilies based on the presence or absence of an intervening amino acid between the first two of four conserved cysteine residues. Members of the α subfamily contain an intervening amino acid and are often denoted by C-X-C. Probably the best-characterized C-X-C chemokine is interleukin-8, which like other members of this subfamily, such as GRO-α, is known for its effects on neutrophil chemotaxis (68). Some C-X-C members are also chemotactic for other cell types. For example, IL-8 is modestly chemotactic for T lymphocytes (68). Members of the β subfamily or C-C chemokines are known for their chemotactic activity for mononuclear cells (monocytes and lymphocytes). Some of the members such as RANTES, MCP-3, MCP-4, and eotaxin also act on eosinophils (69–75). In addition to their chemotactic properties, some C-C chemokines activate cells such as eosinophils and basophils resulting in degranulation and the release of eosinophil cationic protein and histamine, respectively (72,73). These effects have been confirmed in vivo with the injection of RANTES into human skin. A significant recruitment of eosinophils and T lymphocytes was seen with no recruitment of neutrophils (76). Both eosinophil influx and degranulation developed considerably more rapidly in allergic subjects than in nonallergic ones (e.g., within 30 min in the former group compared to 8 to 24 hours in the latter), arguing that the circulating allergic eosinophils were primed to respond to RANTES possibly because of their exposure in vivo to IL-3, IL-5, or GM-CSF. The identification of this new class of chemotactic cytokines has provided a mechanism for the selective recruitment of specific cell types into tissues. Previously, most known chemotactic agents such as C5a, leukotriene B4, or formylated Met-Leu-Phe peptide did not confer any cellular selectivity. In addition, because many of the chemokines are produced by structural cells such as fibroblasts, endothelial cells, and keratinocytes, they may promote the initial recruitment of inflammatory cells from the circulation into the tissues.

Evidence for the role of these chemokines in human diseases is just emerging. Keratinocytes have been shown to produce IL-8, GRO-α, IP-10, RANTES and MCP-1 (48,60,77,78,78b). Additionally, macrophage inflammatory protein-1β (MIP-1β) and RANTES have been shown to bind to keratinocytes in a heparin-dependent manner in which they induce the attachment of CD4+ T cells in a manner similar to what may occur on endothelium (77,79). Therefore, keratinocytes may be important in tissue trafficking not only by virtue of their production of chemokines but also by their ability to bind chemokines and display them on their surface. Patients with psoriasis have considerable amounts of IL-8 in their epithelium, which is thought to mediate the formation of neutrophil microabscesses that are characteristic of this disease (80). Interestingly, epithelial immunoreactivity for IL-8 has also been found in the skin of atopic dermatitis patients and is most pronounced adjacent to dermal mononuclear infiltrates (81). Although RANTES has been found in the respiratory epithelium of patients with asthma or nasal polyposis and is thought to play a role in the tissue eosinophilia characteristic of those two diseases, immunoreactive RANTES could not be detected in biopsies from patients with atopic dermatitis or in cutaneous allergen challenge sites (1,82). However Kay et al. have demonstrated the presence of RANTES and MCP-3 mRNA in the infiltrating leukocytes of the cutaneous late phase response (83). Interestingly, the kinetics of MCP-3 paralleled the kinetics of MBP+ cells and activated eosinophils (EG2+) whereas RANTES paralleled the kinetics of CD3+, CD4+, and CD8+ cells. Collectively, this work highlights the pro-inflammatory role that chemokines and a structural cell such as the keratinocyte play in a variety of cutaneous diseases. Clearly much more work needs to be done to dissect the keratinocyte's role in the selective recruitment, survival, and activation of eosinophils seen in aopic dermatitis.

## D. Adhesion Molecules in Clinical Diseases

Eosinophils can adhere to vascular endothelium by binding to ICAM-1, E-selectin, or VCAM-1. Because neutrophils lack very late activation antigen (VLA)-4, the counter-receptor for VCAM-1, the expression of this endothelial adhesion molecule might lead to preferential eosinophil recruitment. We will review several studies that have examined the induction of these endothelial adhesion molecules in the cutaneous LPR, atopic dermatitis, allergic contact dermatitis, eosinophilic vasculitis, and Ofuji's disease. The LPR has been used as a model of chronic allergic inflammation. Using this model, several groups have found rapid induction of E-selectin (within 3 to 6 hours after antigen), which remained elevated 24 hours later (44,84–86). Additionally, Kyan-Aung et al. noted significant increases in endothelial ICAM-1 and a trend toward increased endothelial VCAM-1 in antigen-challenged as compared to diluent-challenged sites at 6 hours (86). Interestingly, in these subjects, VCAM-1 expression correlated with the magnitude of the eosinophilic infiltrate at the antigen-challenged sites ($P <$ .02) (87). This provided the first in vivo evidence that the VCAM-1/VLA-4 adhesion pair may indeed be important in the recruitment of eosinophils in allergic skin disease. Inhibition of the eosinophilic infiltrate has been demonstrated in subjects receiving either the potent $H_1$-antagonist cetirizine, or in subjects given topical or systemic steroids before antigen challenge (88–92). How cetirizine and glucocorticoids prevent eosinophil influx in vivo is still not completely understood. Cetirizine may inhibit eosinophil chemotaxis and activation (93). Schleimer et al. have argued that glucocorticoids work by directly inhibiting eosinophil survival, as well as the synthesis of different classes of cytokines including eosinophil-activating cytokines (IL-3, IL-5, and GM-CSF), endothelial-activating cytokines (IL-1α, TNF-α, IL-4) and eosinophilic chemotactic cytokines(RANTES, MCP-3, MCP-4) (88).

In a study of skin biopsies from 16 atopic dermatitis patients, all three endothelial adhesion molecules were elevated at sites of acute AD lesions whereas only E-selectin and ICAM-1 were elevated at chronic sites (4). These investigators, unlike Hamid et al., found the greatest number of eosinophils in acute as opposed to chronic lesions (17). Moreover, they found that the staining intensity of VCAM-1 strongly correlated with the number of tissue eosinophils and the deposition of eosinophil-derived granule proteins (ECP, MBP, and EPO), whereas the staining intensity of E-selectin correlated with CD3+, CD4+, and CD45RO+ T cells (4). Although Hamid et al. did not study endothelial adhesion molecules, they did note the greatest expression of IL-4 mRNA (a cytokine capable of selective induction of endothelial VCAM-1) in biopsies of acute lesions (17). Others have confirmed the lack of VCAM-1 staining in chronic AD lesions but have demonstrated enhanced expression on endothelium from patients with other skin diseases such as psoriasis or lichen planus (41). Although no correlations were made between adhesion molecule expression and the magnitude of the cellular infiltrate, it is reasonable to assume that VCAM-1 may be playing a role in the recruitment of T lymphocytes that is characteristic of these other nonallergic skin diseases.

Only a few studies have measured circulating adhesion molecule levels in AD. Three groups have documented elevated levels of soluble ICAM-1, which decreased after treatment with topical steroids; another group of investigators failed to find a decrease after treatment with ultraviolet (UV)-A1 phototherapy (94–96). Groves et al. found elevated levels of soluble ICAM-1, E-selectin, and VCAM-1, which were comparable to those seen in patients with psoriasis (94). Unfortunately, none of these investigators correlated levels of soluble adhesion molecules with measurements of tissue or circulating eosinophils.

The kinetics of endothelial adhesion molecule expression have also been studied in allergic contact dermatitis. Although generally classified as a form of delayed hypersensitivity or type IV immunologic reaction, the histology of the elicitation phase of this condition includes variable numbers of eosinophils in addition to the more characteristic T cell– and macrophage-rich infiltrate. In experimentally induced rhus dermatitis evoked by patch testing to a poison ivy/oak extract, endothelial expression of both E-selectin and VCAM-1 was elevated at all time points tested (2 hr to 3 wk) whereas endothelial ICAM-1 expression was enhanced only during the 48-hour to 1-week period after patch testing (97). Interestingly, these investigators hypothesized that the enhanced expression of the endothelial adhesion molecules may be in part due to the immunoreactive TNF-α detected on the keratinocytes soon after placement of the patch test. Other investigators have confirmed the prominent role of endothelial VCAM-1 expression in patch test models of contact dermatitis and in de novo disease (41,98,99). None of these investigators characterized the cellular infiltrate; therefore, it is difficult to draw firm conclusions about the importance of specific adhesion molecules in the recruitment of specific cell types. However, it seems reasonable to conclude that early leukocyte recruitment in this disease involves both E-selectin and VCAM-1.

Two relatively rare cutaneous diseases characterized by often overwhelming tissue eosinophilia have also been investigated and may provide further insight into the adhesion molecules important in tissue eosinophilia. Eosinophilic vasculitis consists of a syndrome of recurrent cutaneous vasculitis and peripheral eosinophilia with no evidence of systemic involvement (100). A recent report of three such patients demonstrated prominent endothelial VCAM-1 expression in close association with VLA-4-positive eosinophils (100). Ofuji's disease or eosinophilic pustular folliculitis is a disease more commonly seen in Japan, characterized by follicular papules and pustules arranged in an annular configuration. Enhanced expression of E-selectin and VCAM-1 was noted on the perifollicular endothelial cells that were in close approximation to the lymphocytic and eosinophilic infiltrate in five Japanese patients with this disease (101). Furthermore, only the epithelial cells of the affected follicles stained for ICAM-1. Collectively, the consistent finding of endothelial VCAM-1 expression in eosinophilic cutaneous diseases suggests that it is one of the critical adhesion molecules involved in the selective recruitment of eosinophils.

## III.  NOSE

Adhesion molecule expression has been studied in several inflammatory diseases of the nose. This section will review in depth two disorders whose pathogenesis has been relatively well studied: allergic rhinitis and nasal polyposis, both characterized by a large influx of eosinophils. We will highlight specific endothelial adhesion molecules, cytokines, and chemokines that have been implicated in these disease processes.

## A.  Allergic Rhinitis

Allergic rhinitis (AR) is an inflammatory disorder of the nose characterized by symptoms of sneezing, rhinorrhea, and nasal congestion after allergen exposure (102). It is the most common form of atopic disease (102). Due to the relative ease of sampling the nasal mucosa by lavage and biopsy, the pathophysiology of AR has been well studied over the past decade. Nasal allergen challenge models have also yielded insights into this disorder. After nasal challenge with allergen, early symptoms are caused by mediators released by degranulated mast cells and basophils in response to cross-linking of surface IgE [reviewed in (102)]. Clinical disease, however, is more complicated than a single experimental allergen exposure and is more closely mimicked by the late-phase reaction that occurs in about one-half of rhinitic subjects several hours after allergen challenge. In both the late-phase response and clinical disease, there is ongoing inflammation with a marked influx of inflammatory cells into the nose. Recruited cells release mediators and cytokines that amplify the inflammatory cascade.

## 1. Eosinophil Influx Into the Nose

A hallmark of allergic rhinitis is an increased number of eosinophils in the nose of rhinitic compared with non-allergic subjects (103–106). During seasonal exposure to allergen, nasal eosinophil numbers increase further (105,107). In one study, the eosinophil influx closely correlated with nasal symptoms during the birch pollen allergy season (108). The seasonal rise is prevented by topical steroids (107) and immunotherapy (102). A striking increase in tissue and lavage eosinophils occurs following nasal allergen challenge (5,103,109,110). In tissues, eosinophils infiltrate both the nasal epithelium and submucosa (105,106). This is in contrast to the few eosinophils present in non-allergic control subjects, which are confined to the submucosa (105,106). The epithelial localization of eosinophils may reflect the ability of epithelial cells in AR to secrete factors chemotactic for eosinophils [e.g., RANTES (1)], and to retain these cells by virtue of specific adhesion molecule expression [e.g., ICAM-1 (111)].

The influx of eosinophils during allergen season and their disappearance after successful therapy both point to an important role for eosinophils in the pathogenesis of AR. Eosinophils could contribute to the pathogenesis of AR by several mechanisms. First, by synthesizing leukotrienes eosinophils could cause rhinorrhea and congestion (102). Second, eosinophil-derived cytokines (e.g., IL-4, GM-CSF) could enhance further cell recruitment, survival, and activation (112,113). Interestingly, despite the fact that eosinophil granule proteins are toxic to lower respiratory epithelium and are detectable in nasal lavages following allergen exposure (114), the nasal epithelium appears normal in AR (103). Because M2 muscarinic receptors are not highly expressed in the nasal airway (115), it is unlikely that MBP enhances nasal reactivity via this receptor, as has recently been shown in the lower airway of asthmatic subjects (116). Therefore, the role of eosinophil granule proteins in the pathophysiology of allergic rhinitis is currently unclear.

## 2. Mast Cell Migration

Mast cell (MC) numbers are also increased in the nasal mucosa of subjects with AR (105,106) and correlate with symptoms during the pollen season (117). In most studies, intra-epithelial MC are particularly prominent in AR, and stain for tryptase (but not chymase) and thus are of the mucosal type ($MC_T$) (104,105). MC-derived mediators are clearly responsible for a major component of the symptoms of both clinical AR and the early response to allergen challenge (102). Additionally, nasal MC have been implicated as a potential source of several pro-inflammatory cytokines (see below). Increased nasal mast cell numbers are due in part to local growth and differentiation of mast cell precursors (118). Interestingly, there is a migration of MC out of the subepithelial tissues toward the epithelial lining during seasonal allergen exposure (119), which is prevented by topical steroids (107). Nasal MC stain for surface IgE (104), and current evidence favors the production of IgE in regional lymph nodes; therefore, MC and their progenitors may seasonally migrate from lymph nodes to the nasal epithelium. It is not known which chemotactic factors or specific adhesion molecules might be involved in this process. Interestingly, mast cells (unlike basophils) do not express $\beta_2$ integrins (120), thus interactions of $\beta_1$ integrins with basement membrane proteins may be involved in the epithelial localization of these cells (see Chap. 8).

Other cell types implicated in the pathophysiology of AR include neutrophils, basophils, Langerhans cells, and mononuclear cells (103,109,121). Some cell types (e.g., mononuclear cells) may be compartmentalized, with numbers in lavage not adequately reflecting tissue numbers (103). Increased numbers of T cells have been noted following allergen challenge (122,123), but not consistently in seasonal disease (103,105). Interestingly, one study found that resident lymphocytes in AR were activated (as determined by staining with antibodies directed against the α subunit of the IL-2 receptor or CD25) (104), but this was not observed by another group (105). Nerve cells and neural reflexes may also play a role in nasal allergic reactions (124).

## 3. Cytokines

Because of their effects on leukocyte recruitment, activation, and survival, proinflammatory cytokines are thought to play a major role in the pathogenesis of AR. Numerous cytokines have been detected in nasal biopsies from affected subjects (Table 2). Durham et al. found a striking increase in mRNA for IL-3, IL-4, IL-5, and GM-CSF, but not IL-2 or IFN-γ, in the nasal mucosa of allergic subjects following local allergen challenge (110). This was almost identical to results obtained in the skin and lung of subjects with AD and allergic asthma, and reinforced the hypothesis that these diseases are characterized by a Th2 preponderance (26,125). Durham et al. also noted a large influx of eosinophils into the nose that correlated strongly with IL-5 mRNA production (Fig. 2). This furthered the notion that IL-5 is a key eosinophil promoting cytokine in allergic airway disorders (126). Based on studies using dual-labeling techniques, the main source of cytokines was found to be CD3+ T lymphocytes (123).

In a study of patients with perennial rhinitis, Bradding and colleagues found increased numbers of IL-4 expressing cells when compared with non-allergic control subjects (106). Using co-localization with anti-tryptase antibodies, the investigators found most of the IL-4+ cells (>90%) to be mast cells. This difference was most pronounced when they used an anti-IL-4 antibody (3H4) that produced a ring-staining pattern around the cytoplasm of MC, possibly reflecting an intracellular stored pool of this cytokine. In a subsequent study of seasonal rhinitis,

**Table 2** Expression of Cytokines in Allergic Rhinitis[a,b]

| Cytokine | Vs. non-allergics | In season | After allergen challenge | After topical steroids |
|---|---|---|---|---|
| IL-1 | ↑ (142) | | ↑ (133,134,165) | ↓ (134) |
| IL-2 | | | – (110) | |
| IL-3 | | | ↑ (110) | |
| IL-4 | ↑ (106,142) | ↑[c] (107) | ↑ (110) | ↓ (107) |
| IL-5 | – (106) | – (107) | ↑ (110) | – (107) |
| IL-6 | –, ↑ (106,142) | – (107) | ↑ (133,165) | – (107) |
| IL-8 | –, ↑ (106,142) | ↓ (135) | ↑ (134) | ↓ (134) |
| TNF-α | – (140,142) | | ↑ (140) | |
| GM-CSF | | | ↑ (110,133,134) | ↓ (134) |
| IFN-γ | | | – (110) | |
| MCP-1 | | ↑ (135) | | |
| MIP-1α | | | ↑ (134) | ↓ (134) |
| RANTES | | | ↑ (134) | ↓ (134) |

Table Key:
↑ increased
↓ decreased
– no difference/unchanged
"blank" no data available
[a]This table is a compilation of studies of nasal biopsies and nasal secretions.
[b]Numbers in parentheses refer to reference numbers.
[c]The seasonal increase in IL-4 did not reach statistical significance.

there was a trend toward increased IL-4 production during allergy season (107). In this work, mast cells co-localized with anti–IL-5 and anti–IL-6 antibodies, and eosinophils also co-localized IL-5. Thus several cell types may contribute to cytokine production in AR, including T lymphocytes, mast cells, and eosinophils. It is not clear what explains the localization of similar cytokines to different cell types by different investigators (106,107,123,127).

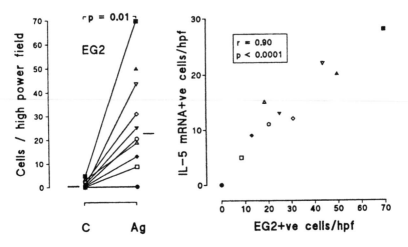

**Figure 2**   Immunohistology of nasal biopsies. EG2+ eosinophil counts 24 hours after local nasal challenge with allergen (Ag) and after control solution (C). The symbols correspond to values from individual subjects. In one subject, a biopsy for immunohistology was obtained only after allergen challenge. Median values are represented by the solid bars. The relationship between numbers of EG2+ eosinophils and cells expressing positive hybridization signals for IL-5 mRNA is shown in the right panel (From Ref. 110).

Two studies found that IL-4 mRNA expression was significantly inhibited by topical nasal corticosteroids (107,128). Given the clinical efficacy of corticosteroids in AR and their strong ability to suppress the cellular infiltrate, these studies suggest that IL-4 may be involved in cell recruitment (possibly by inducing endothelial VCAM-1 expression) and reinforce the idea that cytokine production is a major target for these drugs (129). Interestingly, the molecular mechanisms by which steroids inhibit IL-4 production in T cells is unknown (130,131).

The ability to detect cytokines in nasal lavage is beset with technical difficulties, in part due to interference by mucous proteins. Several investigators have used filter paper strips applied to the nasal mucosa to investigate cytokine and chemokine production (132,133). In one study, significant increases in IL-1β, GM-CSF, and IL-6 were detected after local allergen challenge (133). The chemokines IL-8, MIP-1α, and RANTES were also increased in a subsequent study, and were all significantly reduced by pre-treatment for one week with a topical steroid (134). Though the cellular source of the chemokines was not determined in this study, it is noteworthy that both IL-8 and RANTES are strongly expressed by nasal epithelium (1,107). In a study of seasonal rhinitis, the chemokine MCP-1 increased in nasal secretions during allergy season, whereas IL-8 actually decreased (135). A seasonal increase in nasal MCP-1 may augment basophil histamine release (136).

## 4.   Nasal Challenge Studies with Cytokines

By inducing adhesion molecule expression and providing a signal for cell chemotaxis, cytokines and chemokines are likely critical determinants of inflammation in AR. Definitive proof of the role of individual molecules in clinical disease will require studies with specific antagonists. Challenge studies have provided insights into the role of three cytokines in vivo. Terada et al. studied the effect of recombinant IL-5 applied to the nasal mucosa of allergic subjects. Both eosinophils and the response to histamine were increased in this study (137). Douglass et al. compared IL-8 challenge in a group of atopic and non-atopic subjects (138). Interestingly, pretreatment with histamine was required to elicit an inflammatory response. In this study, there was an influx of neutrophils and eosinophils in both subject groups. Rhinitic symptoms were also increased in both groups after challenge (138). Kuna and colleagues studied the effect of intranasal RANTES in a group of rhinitic subjects studied out of season. Only after priming with allergen challenge was there a significant influx of eosinophils, basophils, lymphocytes and monocytes after exposure to RANTES (139). Interestingly, RANTES challenge was accompanied by typical symptoms in two of 10 subjects (139).

## 5.   Adhesion Molecules

It is clear from the above discussion that leukocytes recruited to the nose play a critical role in the pathogenesis of AR. Therapies effective in reducing the symptoms of AR (e.g., topical corticosteroids) also diminish the inflammatory cell influx. Given the importance of leukocyte–endothelial adhesion molecules in cell recruitment (120), several investigators have studied the role of these molecules in AR. These studies have shed particular insight into the mechanisms by which eosinophils accumulate in the nose in percentages far exceeding those in peripheral blood.

Most studies to date have detected constitutive expression of endothelial adhesion molecules in nasal biopsies from normal non-allergic subjects (Table 3). Constitutive ICAM-1 expression is consistent with the detection of this molecule on endothelium from normal skin and lung. Interestingly, most studies have also found at least modestly elevated levels of E-selectin and VCAM-1 in clinically normal subjects. This likely reflects ongoing antigen exposure and immune surveillance in the nose, and is consistent with the constitutive expression of endothelial activating cytokines (e.g., TNF-$\alpha$, IL-4) in nasal biopsies from normal subjects (140,141).

Three studies to date have specifically addressed the endothelial adhesion molecules expressed in AR (Table 4). Montefort et al., using immunohistochemical techniques, studied nasal biopsies from a group of perennial allergic subjects compared with a group of non-allergic controls (2). Significantly increased staining was observed for endothelial ICAM-1 and VCAM-1, but not E-selectin (Table 4). In this study, normal controls expressed modest amounts of endothelial

**Table 3** Constitutive Expression of Endothelial Adhesion Molecules in Nasal Biopsies from Non-allergic Subjects

| Reference | VCAM-1 | ICAM-1 | E-Selectin |
|---|---|---|---|
| Beck (1) | + | | ++ |
| Montefort (2) | – | ++ | + |
| Lee (5)[a] | + | +++ | – |
| Jahnsen (6) | ++ | ++++ | ++ |

| Table Key | % of vessels expressing |
|---|---|
| "blank" | Not done |
| – | 0% |
| + | 1–25% |
| ++ | 26–50% |
| +++ | 51–75% |
| ++++ | >75% |

[a]Non-allergic subjects were challenged with allergen in Lee's study and underwent biopsy 24 hours later.

ICAM-1 and E-selectin (Table 3). Interestingly, there were no differences in tissue cell counts (eosinophils, neutrophils, or T lymphocytes) between the two groups, but overall E-selectin expression did correlate with total neutrophil influx and LFA-1 positive cells. Eosinophil numbers did not correlate with VCAM-1 staining, but the two normal subjects who expressed the greatest number of eosinophils were the only ones to express VCAM-1. Bachert et al. also found increased ICAM-1 expression when comparing rhinitic to normal subjects (142). In this study, which did not investigate VCAM-1 expression, E-selectin was also up-regulated in allergic subjects.

Lee et al. studied endothelial adhesion molecule expression in 10 seasonal rhinitic subjects studied out of season (Table 4) (5). These subjects underwent nasal allergen challenge, followed 24 hours later by biopsies of challenged sites and of the contralateral inferior turbinates as a control. A group of non-allergic subjects were also challenged and biopsied. There was a significant up-regulation of VCAM-1 expression in seven of 10 subjects following challenge, which modestly correlated with the large influx of eosinophils seen. Most of the adhesion molecules were expressed on small vessels of the lamina propria, and not on larger mucosal or submucosal vessels. Unlike Montefort, these investigators detected a small amount of VCAM-1 expression ($\cong$6%–9% positive vessels) in biopsies from both non-allergic controls and unchallenged rhinitics (Table 3) (5). Confirming the previous study, constitutive ICAM-1 expression was also detected. Reasons for the discrepancies between these two studies may relate to differences in study populations and staining techniques.

Given the ability of endothelial VCAM-1 to support eosinophil adhesion (see

**Table 4** Expression of Endothelial Adhesion Molecules in Allergic Rhinitis[a]

|  | Perennial or challenged | | |
|---|---|---|---|
| Reference | VCAM-1 | ICAM-1 | E-selectin |
| Montefort (2)[b] | +[d] | +++[d] | ++ |
| Lee (5)[c] | ++[d] | ++ | +[d] |

[a]Key as for Table 3.
[b]Montefort studied a group of patients with symptomatic perennial rhinitis.
[c]Lee studied subjects with seasonal rhinitis who were challenged outside of allergen season with relevant allergens.
[d]Significantly increased expression compared with non-allergic control groups.

Chap. 9, 10), Lee at al. hypothesized this molecule may contribute to the eosinophil recruitment in AR. This hypothesis is supported by the consistent detection of IL-4 and TNF-$\alpha$ in AR patients because these cytokines synergize to enhance endothelial VCAM-1 expression (143). Additional factors (such as cell activation by regional chemokines) must also be involved in cell recruitment because 1.) VCAM-1 expression can occur without tissue eosinophilia (2,5), 2.) eosinophil influx can occur without detectable VCAM-1 expression (5,6), and 3.) eosinophil influx also correlated with E-selectin expression after challenge (5).

In the multistep model of leukocyte–endothelial interactions (Chap. 1), circulating cells initially roll along the endothelial surface using L-selectin, then firmly adhere and transmigrate using surface $\beta_2$ integrins (e.g., CD11b/CD18). Recruited cells shed L-selectin and up-regulate CD11b. When analyzing leukocytes recruited to the nose after allergen challenge, Baroody et al. found higher levels of CD11b and absent L-selectin expression on lavage cells compared with peripheral blood cells (eosinophils and neutrophils) (144). This was similar to the phenotype of cells recruited to the lower airway in asthmatics (145), and supports the notion that these molecules are involved in leukocyte emigration in vivo.

What induces endothelial adhesion molecules seen in AR? Many of the cytokines present in AR, including IL-1$\beta$, IL-4, and TNF-$\alpha$, directly activate HUVEC adhesion molecule expression in vitro (12). For example, IL-1 induces ICAM-1, E-selectin, and VCAM-1 on HUVECs (146), and TNF-$\alpha$ and IL-4 synergistically increase VCAM-1 expression (143). Effects on HUVEC, however, do not necessarily translate into effects in tissue. Using nasal tissue explants cultured in vitro, Weinberger et al. were unable to up-regulate endothelial VCAM-1 expression with IL-1, though this cytokine strongly induced VCAM-1 expression on HUVEC (146). TNF-$\alpha$ induced VCAM-1 on endothelium from both sources. This result was reminiscent of studies using HDMEC and normal skin explants in which IL-1 could not induce VCAM-1 (above) and reflects the heterogeneity of microvascular endothelial cells. As in the skin, additional signals (possibly provided by the regional extracellular matrix) may affect endothelial responsiveness to cytokines in the nose. Terada et al. also studied nasal mucosal explants cultured ex vivo with cytokines. Interestingly, they observed IL-5–induced ICAM-1 mRNA expression in explants from allergic but not non-allergic subjects, though the source of ICAM-1 was not determined in this study (147). However, this observation supports the idea that the inflammatory state of the tissue affects the responsiveness for cytokine-induced adhesion molecule expression (39).

Despite the detection of constitutive endothelial ICAM-1 expression even in non-allergic subjects, unstimulated nasal epithelium expresses little or no ICAM-1 (148,149). Epithelial cells obtained during allergen season or after local allergen challenge, however, rapidly up-regulate this molecule (111,148,149). A soluble form of ICAM-1 (sICAM-1) is also increased in nasal washings during allergen season (150). Interestingly, the challenge-induced ICAM-1 up-regulation on epithelium is prevented by nasal immunotherapy (151). Unlike freshly isolated cells, nasal epithelial cells cultured for several days ex vivo spontaneously express high levels of ICAM-1 (148, 152). This expression can be further augmented by culture with histamine, IFN-γ, TNF-α, and the eosinophil granule proteins MBP and ECP (148, 152). Epithelial cells that express ICAM-1 may promote the retention of cells expressing counter-ligands within tissues (such as CD11a/CD18 and CD11b/CD18 positive eosinophils). Additionally, because rhinoviruses recognize this molecule, they are likely a major source of entry for this pathogen (153).

## B.  Nasal Polyps

Additional insight into mechanisms of eosinophil recruitment into the nasal mucosa has been gained by the study of nasal polyposis. This chronic inflammatory condition is characterized by an overgrowth of edematous mucosal tissue and symptoms of rhinitis and nasal obstruction (154). The relationship to atopy is somewhat controversial, although the majority of affected patients do not appear to be allergic. Characteristically, there is an enormous influx of eosinophils into polyp tissue, with numbers as high as 760 eosinophils/mm² of tissue (1). The etiology of this disorder is unknown, but it is suggested that continued mast cell degranulation and unregulated growth factor production are involved. In a series of biopsy studies, it has been demonstrated that eosinophils in nasal polyps are in a particularly activated state and express a variety of cytokines and growth factors including TNF-α, IL-4, TGF-α, TGF-β1, and platelet-derived growth factor (PDGF)-B (155–158). This eosinophil phenotype is relatively specific for nasal polyposis. For example, almost none of the IL-4 expressing cells in the lower airway are eosinophils in asthmatic subjects, but most IL-4 positive cells are eosinophils in nasal polyps (155). Secreted IgA may be responsible for eosinophil-derived IL-4 production in this disorder (112, 155). As in allergic rhinitis, topical steroids effectively reduce symptoms and cellular influx in nasal polyposis (159).

Three studies to date have analyzed the adhesion phenotype in nasal polyp tissue (Table 5). Symon et al. detected significant staining for endothelial P-selectin, E-selectin, and ICAM-1 using immunohistochemistry of nasal polyposis biopsies. In a Stamper-Woodruff style assay, eosinophil adhesion to polyp sections was strikingly dependent on P-selectin, and not other endothelial adhesion molecules (160). This result suggested that therapies directed at inhibiting P-selectin adhesion may prove clinically useful in nasal polyposis. However, because neutrophils also express the P-selectin ligand (161), it does not explain the selective influx of eosinophils in nasal polyposis.

Jahnsen and colleagues systematically analyzed endothelial adhesion molecules in 15 nasal polyps and found significantly increased VCAM-1 expression on vessels in polyp tissue when compared with lower/middle turbinates from both affected patients and control subjects (6). This difference was most notable on medium-sized vessels (10 to 30 μm in diameter). There was no significant difference in E-selectin or ICAM-1 expression between the two groups. Tissue eosinophils (but not neutrophils) correlated well with endothelial VCAM-1 expression, lending further support to the role of the VLA-4/VCAM-1 pathway in allergic cell recruitment. Since eosinophils in nasal polyps are a potential source of VCAM-1-inducing cytokines, this may represent a self-propagating inflammatory reaction, though it is unclear what the inciting events are.

Beck et al. also detected increased VCAM-1 in polyps using immunohistochemistry and ELISA of tissue homogenates (1). In this study, VCAM-1 (expressed as both percent of vessels stained and ng/gm of frozen tissue) correlated well with numbers of recruited eosinophils. These two studies have clearly implicated VCAM-1 as an important endothelial adhesion molecule in eosinophil recruitment in nasal polyposis. Specific VCAM-1 antagonists will help pinpoint the specific role of this molecule in vivo, and may prove clinically useful in this disorder.

**Table 5** Expression of Endothelial Adhesion Molecules in Nasal Polyposis[a]

| Reference | VCAM-1 | ICAM-1 | E-selectin | P-selectin |
|-----------|--------|--------|------------|------------|
| Beck (1) | + | | + | |
| Jahnsen (6) | ++[b] | ++++ | ++ | |
| Symon (160) | + | ++ | ++ | ++ |

[a]Key as for Table 3.
[b]Significantly increased compared with turbinate biopsies from the same subjects.

Because eosinophils often localize within the epithelium in both nasal polyposis and AR, Beck et al. studied the expression of the potent eosinophil-attracting chemokine RANTES in biopsies of nasal polyps (1). Airway epithelial cells produce RANTES in response to cytokines, an effect inhibitable with corticosteroids (162). RANTES was strongly expressed on nasal polyp epithelium, and less on normal turbinates or endothelium. These results suggested that RANTES may be important in localizing eosinophils within tissue after they have transmigrated using specific surface adhesion molecules.

## C. Other Diseases

It is reasonable to speculate that endothelial adhesion molecules mediate eosinophil recruitment in non-allergic nasal diseases, but these have not been well studied. Chronic sinusitis, for example, is characterized by damaged epithelium, an influx of eosinophils and deposition of eosinophil granule proteins (163). In one study, IL-1$\beta$ mRNA localized to intramucosal leukocytes of subjects with chronic sinusitis, and endothelial ICAM-1 and E-selectin expression were noted in the maxillary sinus of one subject (164).

## IV. CONCLUSIONS

Clearly, the preferential recruitment of a given cell type is the result of many separate events rather than simply the effect of a cell-specific chemoattractant or an adhesion molecule pathway. In the near future, adhesion molecule antagonists should provide insight into the function of these molecules in the pathogenesis of a variety of inflammatory disorders, including allergic diseases. Although many questions remain regarding how cells can be selectively recruited to specific sites of inflammation, it is important to realize that the endothelium is not simply a passive conduit through which inflammatory cells circulate and diapedese only at sites of endothelial cell disruption. The process of cellular trafficking requires much finer orchestration of a variety of surface proteins than one could have predicted even a decade ago.

## REFERENCES

1.  L. A. Beck, C. Stellato, L. D. Beall, T. J. Schall, D. Leopold, C. A. Bickel, F. Baroody, B. S. Bochner, and R. P. Schleimer, *J. Allergy Clin. Immunol.*, *98:766* (1996).
2.  S. Montefort, I. H. Feather, S. J. Wilson, D. O. Haskard, T. H. Lee, S. T. Holgate, and P. H. Howarth, *Am. J. Respir. Cell Mol. Biol.*, *7*:393 (1992).
3.  A. M. Bentley, S. R. Durham, D. S. Robinson, G. Menz, C. Storz, O. Cormwell, A. B. Kay, and A. J. Wardlaw, *J. Allergy Clin. Immunol.*, *92*:857 (1993).
4.  H. Wakita, T. Sakamoto, Y. Tokura, and M. Takigawa, *J. Cutan. Pathol.*, *21*:33 (1994).

5. B. J. Lee, R. M. Naclerio, B. S. Bochner, R. M. Taylor, M. C. Lim, and F. M, Baroody, *J. Allergy Clin. Immunol.*, *94*:1006 (1994).
6. F. L. Jahnsen, G. Haraldsen, J. P. Aanesen, R. Haye, and P. Brandtzaeg, Am. *J. Respir. Cell Mol. Biol.*, *12*:624 (1995).
7. J. G. Zangrilli, J. R. Shaver, R. A. Cirelli, S. K. Cho, C. G. Garlis, A. Falcone, F. M. Cuss, J. E. Fish, and S. P. Peters, *Am. J. Respir. Cell Mol. Biol.*, *151*:1346(1995).
8. T. Kobayashi, S. Hashimoto, K. Imai, E. Amemiya, M. Yamaguchi, A. Yachi, and T. Horie, *Clin. Exp. Immunol.*, *96*:110 (1994).
9. S. Montefort, W. R. Roche, P. H. Howarth, R. Djukanovic, C. Gratziou, M. Carroll, L. Smith, K. M. Britten, D. Haskard, T. H. Lee, and S. T. Holgate, *Eur. Respir. J.*, *5*:815(1992).
10. V. B. Weg, T. J. Williams, R. R. Lobb, and S. Nourshargh, *J. Exp. Med.*, *177*:561 (1993).
11. M. M. Teixeira, S. Reynia, M. Robinson, A. Xhock, W. T. J. F. M. Williams, A. G. Rossi, and P. G. Hellewell, *Br. J. Pharmacol.*, *111*:811(1994).
12. T. M. Carlos and J. M. Harlan, *Blood*, *84*:2068 (1994).
13. C. E. Haug, R. B. Colvin, F. L. Delmonico, H. Auchincloss Jr., N. Tolkoff-Rubin, F. I. Preffer, R, Rothlein, S. Norris, L. Scharschmidt, and A. B. Cosimi, *Transplantation*, *55*:766 (1993).
14. D. Y. M. Leung, *J. Allergy Clin. Immunol.*, *96*:302 (1995).
15. J. D. Bos, C. Hagenara, P. K. Das, S. R. Krieg, W. J. Voom, and M. L. Kapsenberg, *Arch. Dermatol. Res.*, *81*:24 (1989).
16. D. Y. M. Leung, A. K. Bhan, E. E. Schneeberger, and R. S. Geha, *J. Allergy Clin. Immunol. 71*:47 (1983).
17. Q. Hamid, M. Boguniewicz, and D. Y. M. Leung, *J. Clin. Invest.*, *94*:870 (1994).
18. K. M. Leiferman, S. J. Ackerman, H. A. Sampson, H. S. Haugen, P. Y. Venencie, and G. J. Gleich, *N. Engl. J. Med.*, *313*:282 (1985).
19. K. M. Leiferman, M. D. P. Davis, T. J. George, J. L. Checkel, and G. J. Gleich, *J. Invest. Dermatol.*, *106*:937 (1996).
20. G. J. Gleich, A. L. Schroeter, J. P. Marcoux, M. I. Sachs, E. J. O'Connell, and P. F. Kohler, *N. Engl. J. Med.*, *310*:1621 (1984).
21. C. Walker, M. K. Kägi, P. Ingold, P. Braun, K. Blaser, C. A. F. M. Bruijnzeel-Koomen, and B. Wütherich, *Clin. Exp. Allergy*, *23*:145 (1993).
22. N. L. Ott, G. J. Gleich, E. A. Peterson, T. Fujisawa, S. Sur, and K. M. Leiferman, *J. Allergy Clin. Immunol.*, *94*:120 (1994).
23. W. Czech, J. Krutmann, E. Schoepf, and A. Kapp, *Br. J. Dermatol.*, *126*:351 (1992).
24. P. Weller, *Clin. Immunol. Immunopathol.*, *62*:S5 (1992).
25. P. L. B. Bruijnzeel, P. H. M. Kuijper, S. Rihs, S. Betz, R. A. J. Warringa, and L. Koenderman, *J. Invest. Dermatol.*, *100*:137 (1993).
26. A. B. Kay, S. Ying, S. R. Varney, M. Gaga, S. R. Durham, R. Moqbel, A. J. Wardlaw, and Q. Hamid, *J. Exp. Med.*, *173*:775 (1991).
27. F. C. V. Reijsen, C. A. F. M. Bruijnzeel-Kooman, F. S. Kalthoff, E. Maggi, S. Romagnani, J. K. T. Westland, and G. C. Mudde, *J. Allergy. Clin. Immunol.*, *90*:184 (1992).
28. B. R. Vowels, A. H. Rook, and M. Cassin, *J. Allergy Clin. Immunol.*, *96*:92 (1995).
29. A. Tsicopoulos, Q. Hamid, A. Haczku, M. R. Jocaboson, S. R. Durham, J. North, J.

Barkans, C. J. Corrigan, Q. Meng, R. Moqbel, and A. B. Kay, *J. Allergy Clin. Immunol.*, *94*:764 (1994).

30. D. R. Mehregan, A. F. Fransway, J. H. Edmonson, and K. M. Leiferman, *Arch. Dermatol.*, *128*:1055 (1992).

31. P. T. Golumbek, A. J. Lazenby, H. I. Levitsky, L. M. Jaffee, H. Karasuyama, M. Baker, and D. M. Pardoll, *Science*, *254*:713 (1991).

32. R. A. Swerlick, K. H. Lee, L. Li, N. T. Sepp, S. W. Caughman, and T. J. Lawley, *J. Immunol.*, *149*:698 (1992).

33. R. A. Swerlick and T. J. Lawley, *J. Invest. Dermatol*, *100*:111 (1993).

34. R. A. Swerlick, E. Garcia-Gonzalez, Y. Kubota, Y. Xu, and T. J. Lawley, *J. Invest. Dermatol.*, *97*: 190 (1991).

35. B. S. Bochner, D. A. Klunk, S. A. Sterbinsky, R. L. Coffman, and R. P. Schleimer, *J. Immunol.*, *154*:799 (1995).

36. R. P. Schleimer, S. A. Sterbinsky, J. Kaiser, C. A. Bickel, D. A. Klunk, K. Tomioka, W. Newman, F. W. Luscinskas, M. A. Gimbrone Jr., B. W. McIntyre, and B. S. Bochner, *J. Immunol.*, *148*:1086 (1992).

37. A. Dobrina, R. Menegazzi, T. M. Carlos, E. Nardon, R. Cramer, T. Zacchi, J. M. Harlan, and P. Patriarca, *J. Clin. Invest.*, *88*:20 (1991).

38. P. Petzelbauer, J. R. Bender, J. Wilson, and J. S. Pober, *J. Immunol.*, *151*:5062 (1993).

39. P. Petzelbauer, J. S. Pober, A. Keh, and I. M. Braverman, *J. Invest. Dermatol.*, *103*:300 (1994).

40. D. Briscoe, R. Cotran, and J. Pober, *J. Immunol.*, *149*:2954 (1992).

41. R. W. Groves, E. L. Ross, J. N. W. N. Barker, and D. M. MacDonald, *J. Am. Acad. Dermatol.*, *29*:67 (1993).

42. R. W. Groves, M. H. Allen, E. L. Ross, J. N. W. N. Barker, and D. M. MacDonald, *Br. J. Dermatol.*, *132*:345 (1995).

43. R. W. Groves, E. Ross, J. N. W. N. Barker, J. S. Ross, R. D. R. Camp, and D. M. MacDonald, *J. Invest. Dermatol.*, *98*:384 (1992).

44. R. W. Groves, M. H. Allen, J. N. W. N. Barker, D. O. Haskard, and D. M. MacDonald, *Br. J. Dermatol.*, *124*:117 (1991).

45. A. B. Gelb, B. R. Smoller, R. A. Wamke, and L. J. Picker, *Am. J. Pathol.*, *142*:1556 (1993).

46. W. Chu, D. H. Presky, R. A. Swerlick, and D. K. Burns, *J. Immunol.*, *153*:4179 (1994).

47. L. J. Walsh and G. F. Murphy, *J. Cutan. Pathol.*, *19*:161 (1992).

48. B. J. Nickoloff and E. M. Griffiths, *J. Invest. Dermatol.*, *95*:128 (1990).

49. T. S. Kupper, *J. Clin. Invest.*, *86*:178 (1990).

50. M. L. Dustin, K. H. Singer, D. T. Tuck, and T. A. Springer, *J. Exp. Med.*, *167*:1323 (1988).

51. B. J. Nickoloff, *Arch. Dermatol.*, *124*:1835 (1988).

52. C. E. M. Griffiths and B. J. Nickoloff, *Am. J. Pathol.*, *135*:1045 (1989).

53. C. E. M. Griffiths, J. J. Voorhees, and B. J. Nickoloff, *Br. J. Dermatol.*, *120*:1 (1989).

54. C. E. M. Griffiths, J. J. Voorhees, and B. J. Nickoloff, *J. Am. Acad. Dermatol.*, *20*:617 (1989).

55. U. Behrends, R. U. Peter, R. Hintermeier-Knabe, G. EiBner, E. Holler, G. W. Bornkamm, S. W. Caughman, and K. Degitz, *J. Invest. Dermatol.*, *103*:726 (1994).

56. S. D. Bennion, M. H. Middleton, K. M. David-Bajar, S. Brice, and D. A. Norris, *J. Invest-Dermatol.*, *105*:71 (1995).

57. A. Gueniche, J. Viac, G. Lizard, M. Charveron, and D. Schmitt, *Arch. Dermatol. Res.*, *286*:466 (1994).

58. R. S. Mitra, Y. Shimizu, and B. J. Nickoloff, *J. Cell. Physiol.*, *156*:348 (1993).

59. K. H. Singer, D. T. Tuck, H. A. Sampson, and R. P. Hall, *J. Invest. Dermatol.* 92:746 (1989).

60. B. J. Nickoloff, C. E. M. Griffiths, and J. N. W. N. Barker, *J. Invest. Dermatol.*, *94*:151 (1990).

61. M. De Luca, G. Pellegrini, G. Zambruno, and P. C. Marchisio, *J. Dermatol.*, 21:821 (1994).

62. K. O. Simon and K. Burridge, *Integrins: molecular and biological responses to the extracellular matrix* (D. A. Cheresh and R. P. Mecham, eds.), Academic Press, San Diego, p. 49 (1994).

63. M. Simon, J. Hunyadi, and A. Dobozy, *Acta Derm. Venereol.*, 72:169 (1992).

64. T. T. Hansel, J. B. Braunstein, C. Walker, K. Blaser, P. L. B. Bruijnzeel, J. C. Virchow, and C. Virchow, *Clin. Exp. Immunol.*, *86*:271 (1991).

65. B. S. Bochner, K. Matsumoto, J. Appiah-Pippim, R. P. Schleimer, and L. Beck, *Clin. Res.*, (1996).

66. W. Czech, J. Krutmann, A. Budnik, E. Schopf, and A. Kapp, *J. Invest. Derm.*, *100*:417 (1993).

67. T. T. Hansel, I. J. M. Devries, J. M. Carballido, R. K. Braun, N. Carballidoperrig, S. Rihs, K. Blaser, and C. Walker, *J. Immunol.*, *149*:2130 (1992).

68. D. D. Taub and J. J. Oppenheim, *Ther. Immunol.*, *1*:229 (1994).

69. R. Alam, S. Stafford, P. Forsythe, R. Harrison, D. Faubion, M. Lett-Brown, and J. Grant, *J. Immunol.*, *150*:3442 (1993).

70. M. Ebisawa, T. Yamada, C. Bickel, D. Klunk, and R. P. Schleimer, *J. Immunol.*, *153*:2153 (1994).

71. C. A. Dahinden, T. Geiser, T. Brunner, V. von Tscharner, D. Caput, P. Ferrara, A. Minty, and M. Baggiolini, *J. Exp. Med.*, *179*:751 (1994).

72. E. A. Garcia-Zepeda, M. E. Rothenberg, R. T. Ownbey, J. Celestin, P. Leder, and A. D. Luster, *Nat. Med.*, *2*:449 (1996).

73. P. J. Jose, D. A. Griffiths-Johnson, P. D. Collins, D. T. Walsh, R. Moqbel, N. F. Totty, O. Truong, J. J. Hsuan, and T. J. Williams, *J. Exp. Med.*, *179*:881 (1994).

74. C. Stellato, L. Beck, L. Schwiebert, H. Li, J. White, and R. P. Schleimer, *J. Allergy Clin. Immunol.*, *97*:487 (1996).

75. P. D. Ponath, S. Qin, D. J. Ringler, I. Clark-Lewis, J. Wang, N. Kassam, H. Smith, X. Shi, J. Gonzalo, W. Newman, J. Gutierrez-Ramos, and C. R. Mackay, *J. Clin. Invest.*, *97*:604 (1996).

76. L. Beck, C. Bickel, S. Sterbinsky, C. Stellato, R. Hamilton, H. Rosen, B. Bochner, and R. Schleimer, *FASEB J.*, *9*:A804 (1995).

77. R. Hershkoviz, M. Marikovsky, D. Gilat, and O. Lider, *J. Invest. Dermatol.*, *106*:243 (1996).

78. M. Baggiolini, B. Deward, and A. Walz, *Inflammation: basic principles and clinical correlates* (J. I. Gallin, I. M. Goldstein and R. Snyderman, eds.), Raven Press, New York, p. 247 (1992).

78b.  J. Li, G. W. Ireland, P. M. Farthing, and M. H. Thornhill, *J. Invest. Dermatol.* *106*:661 (1996).

79.  Y. Tanaka, D. H. Adams, S. Hubscher, H. Hirano, U. Siebenlist, and S. Shaw, *Nature*, *361*:79 (1993).

80.  J. D. Bos and M. L. Kapsenberg, *Immunol. Today*, *15*:75 (1993).

81.  M. Sticherling, E. Bornscheuer, J. M. Schroder, and E. Christophers, *Arch. Dermatol. Res.*, 284:82 (1992).

82.  R. J. Davies, J. H. Wang, C. J. Trigg, and J. L. Devalia, *Int. Arch. Allergy Appl. Immunol.*, *107*:428 (1995).

83.  S. Ying, L. Taborda-Barata, Q. Meng, M. Humbert, and A. B. Kay, *J. Exp. Med.*, *181*:2153 (1995).

84.  S. Benenati, B. Bochner, T. Horn, E. Farmer, and R. Schleimer, *J. Allergy Clin. Immunol.*, *87*:304 (1991).

85.  D. Y. M. Leung, J. S. Pober, and R. S. Cotran, *J. Clin. Invest.* 87:1805 (1991).

86.  U. Kyan-Aung, D. O. Haskard, R. N. Poston, M. H. Thornhill, and T. H. Lee, *J. Immunol.*, *146*:521 (1991).

87.  R. P. Schleimer and B. S. Bochner, J. *Immunol.*, *147*:380 (1991).

88.  R. P. Schleimer, J. Kaiser, K. Tomioka, M. Ebisawa, and B. S. Bochner, *Int. Arch. Allergy Immunol.*, *99*:289 (1992).

89.  E. N. Charlesworth, A. F. Hood, N. A. Soter, A. Kagey-Sobotka, P. S. Norman, and L. M, Lichtenstein, *J. Clin. Invest.*, *83*:1519 (1989).

90.  E. N. Charlesworth, A. Kagey-Sobotka, R. P. Schleimer, P. S. Norman, and L. M. Lichtenstein, *J. Immunol.*, *146*:671 (1991).

91.  R. Fadel, B. David, N. Herpinrichard, A. Borgnon, R. Rassemont, and J. P. Rihoux, *J. Allergy Clin. Immunol.*, *86*:314 (1990).

92.  J. R. Snyman, D. K. Sommers, M. D. Gregorowski, and H. Boraine, Eur. *J. Clin. Pharmacol.*, *42*:359 (1992).

93.  C. Okada, R. Eda, H, Miyagawa, H. Sugiyama, R. J. Hopp, A. K. Bewtra, and R. G. Townley, *Int. Arch. Allergy Immunol.*, *103*:384 (1994).

94.  R. W. Groves, P. Kapahi, J. N. W. N. Barker, D. O. Haskard, and D. M. MacDonald, *J. Am. Acad. Dermatol.*, *33*:32 (1995).

95.  L. Kowalzick, A. Kleinheinz, M. Weichenthal, K. Neuber, I. Kohler, J. Grosch, and J. Ring, *J. Invest. Dermatol.*, *105*:82 (1995).

96.  B. Wuthrich, H. Joller-Jemlka, and M. K. Kagi, *Allergy*, *50*:88 (1995).

97.  C. E. M. Griffiths, J. N. W. N. Barker, S. Kunkel, and B. J. Nickoloff, *Br. J. Dermatol.*, *124*:519 (1991).

98.  P. K. Das, O. J. deBoer, A. Visser, C. E. Verhagen, J. D. Bos, and S. T. Pals, *Acta Derm. Venereol. Suppl.*, *186*:21 (1994).

99.  J. Brasch and W. Sterry, *Dermatology*, *185*:12 (1992).

100.  K. R. Chen, W. P. Su, M. R. Pittelkow, and K. M. Leiferman, *Semin. Dermatol.*, *14*:106 (1995).

101.  Y. Teraki, I. Konohana, T. Shiohara, M. Nagashima, and T. Nishikawa, *Arch. Dermatol.*, *129*:1015 (1993).

102.  R. M. Naclerio, *N. Engl. J. Med.*, *325*:860 (1991).

103.  M. C. Lim, R. M. Taylor, and R. M. Naclerio, *Am. J. Respir. Crit. Care Med.* *151*:136 (1995).

104. Y. Igarashi, M. S. Goldrich, M. A. Kaliner, A. M. Irani, L. B. Schwartz, and M. V. White, *J. Allergy Clin. Immunol.*, *95*:716 (1995).
105. A. M. Bentley, M. R. Jacobson, V. Cumberworth, J. R. Barkans, R. Moqbel, L. B. Schwartz, A. M. Irani, A. B. Kay, and S. R. Durham, *J. Allergy Clin. Immunol.*, *89*:877 (1992).
106. P. Bradding, I. H. Feather, S. Wilson, P. G. Bardin, C. H. Heusser, S. T. Holgate, and P. H. Howarth, *J. Immunol. 151*:3853 (1993).
107. P. Bradding, I. H. Feather, S. Wilson, S. T. Holgate, and P. H. Howarth, *Am. J. Respir. Crit. Care Med.*, *151*:1900 (1995).
108. U. Pipkom, K. G. and L. Enerback, *J. Allergy Clin. Immunol.*, *82*:1046 (1988).
109. R. Bascom, U. Pipkom, L. M. Lichtenstein, and R. M. Naclerio, *Am. Rev. Respir. Dis.*, *138*:406 (1988).
110. S. R. Durham, S. Ying, V. A. Varney, M. R. Jacobson, R. M. Sudderick, I. S. Mackay, A. B. Kay, and Q. A. Hamid, *J. Immunol.*, *148*:2390 (1992).
111. G. Ciprandi, C. Pronzato, V. Ricca, M. Bagnasco, and G. W. Canonica, *J. Allergy Clin. Immunol.*, *94*:738 (1994).
112. H. Nakajima, G. Gleich, and H. Kita, *J. Immunol.*, *156*:4859 (1996).
113. I. Ohno, R. Lea, S. Finotto, J. Marshall, J, Denburg, J. Dolovich, J. Gauldie, and M. Jordana, *Am. J. Respir. Cell Mol. Biol.*, *5*:505 (1991).
114. R. Bascom, U. Pipkom, D. Proud, S. Dunnette, G. J. Gleich, L. M. Lichtenstein, and R. M. Naclerio, *J. Allergy Clin. Immunol.*, *84*:338 (1989).
115. O. Okayama, J. Mullol, J. Baraniuk, J. Hausfeld, B. Feldman, M. Merida, J. Shel-hammer, and M. Kaliner, *Am. J. Respir. Cell Mol. Biol.*, *8*:176 (1993).
116. D. B. Jacoby, G. J. Gleich, and A. D. Fryer, *J. Clin. Invest.*, *91*:1314 (1993).
117. U. Pipkorn, G. Karlsson, and L. Enerback, *Int. Arch. Allergy Appl. Immunol.*, *87*:349 (1988).
118. J. A. Denburg, *Clin. Exp. Allergy. 21*:253 (1991).
119. L. Enerbäck, U. Pipkorn, and G. Granerus, *Int. Arch. Allergy Appl. Immunol.*, *80*:44 (1986).
120. B. S. Bochner and R. P. Schleimer, *J. Allergy Clin. Immunol.*, *94*:427 (1994).
121. P. H. Howarth, *Allergy*, *50*:6 (1995).
122. V. A. Varney, M. R. Jacobson, R. M. Sudderick, D. S. Robinson, A. M. Irani, L. B. Schwartz, I. S. Mackay, A. B. Kay, and S. R. Durham, *Am. Rev. Respir. Dis.*, *146*:170 (1992).
123. S. Ying, S. R. Durham, J. Barkans, K. Masuyama, M. Jacobson, S. Rak, O. Lowha-gen, R. Moqbel, A. B. Kay, and Q. A. Hamid, *Am. J. Respir. Cell Mol. Biol.*, *9*:356 (1993).
124. M. Riccio and D. Proud, *J. Allergy Clin. Immunol.*, *97*:1252 (1996).
125. D. S. Robinson, Q. Hamid, S. Ying, A. Tsicopoulos, J. Barkans, A. M. Bentley, C. Corrigan, S. R. Durham, and A. B. Kay, *N. Engl. J. Med.*, *326*:298 (1992).
126. T. Ohnishi, H. Kita, D. Weiler, S. Sur, J. B. Sedgwick, W. J. Calhoun, W. W. Busse, J. S. Abrams, and G. J. Gleich, *Am. Rev. Respir. Dis.*, *147*:901 (1993).
127. S. Ying, S. R. Durham, M. R. Jacobson, S. Rak, K. Masuyama, O. Lowhagen, A. B. Kay, and Q. A. Hamid, *Immunology*, *82*:200 (1994).
128. K. Masuyama, M. R. Jacobson, S. Rak, Q. Meng, R. M. Sudderick, A. B. Kay, O. Lowhagen, Q. Hamid, and S. R. Durham, *Immunology*, *82*:192 (1994).

129. R. P. Schleimer, *Eur. J. Clin. Pharmacol. 45*:S3 (1993).

130. N. Auphan, J. A. DiDonato, C. Rosette, A. Helmberg, and M. Karin, *Science, 270*:286 (1995).

131. V. Casolaro, S. N. Georas, Z. Song, I. D. Zubkoff, S. A. Abdulkadir, D. Thanos, and S. J. Ono, *Proc. Natl. Acad. Sci. U.S.A., 92*:11623 (1995).

132. G. Knowles, P. Townsend and M. Turner-Warwick, *Clin. Allergy, 11*:287 (1981).

133. T. C. Sim, J. A. Grant, K. A. Hilsmeier, Y. Fukuda, and R. Alam, *Am. J. Respir. Crit. Care Med., 149*:339 (1994).

134. T. C. Sim, L. M. Reece, K. A. Hilsmeier, J. A. Grant, and R. Alam, *Am. J. Respir. Crit. Care Med., 152*:927 (1995).

135. P. Kuna, M. Lazarovich, and A. Kaplan, *J. Allergy Clin. Immunol., 97*:104 (1996).

136. P. Kuna, S. R. Reddigari, T. J. Schall, D. Rucinski, M. Sadick, and A. P. Kaplan, *J. Immunol., 150*:1932 (1993).

137. N. Terada, A. Konno, H. Tada, K. Shirotori, K. Ishikawa, and K. Togawa, *J. Allergy. Clin. Immunol., 90*:160 (1992).

138. J. A. Douglass, D. Dhami, C. E. Gurr, M. Bulpitt, J. K. Shute, P. H. Howarth, I. J. Lindley, M. K. Church, and S. T. Holgate, *Am. J. Respir. Crit. Care Med., 150*:1108 (1994).

139. P. Kuna, R. Alam, P. Górski, and U. Ruta, *J. Allergy Clin. Immunol., 95*:A218 (1995).

140. P. Bradding, R. Mediwake, I. H. Feather, J. Madden, M. K. Church, S. T. Holgate, and P. H. Howarth, *Clin. Exp. Allergy, 25*:406 (1995).

141. C. Bachert, M. Wagenmann, and U. Hauser, *Int. Arch. Allergy Immunol., 107*:106 (1995).

142. C. Bachert, U. Hauser, B. Prem, C. Rudack, and U. Ganzer, *Eur. Arch. Otorhinolaryngol. Suppl., 1*:S44 (1995).

143. M. F. Iademarco, J. L. Barks, and D. C. Dean, *J. Clin. Invest., 95*:264 (1995).

144. F. M. Baroody, B.-J. Lee, M. C. Lim, and B. S. Bochner, *Eur. Arch. OtoRhinolaryngol. Suppl., 252*:S50 (1995).

145. S. N. Georas, M. C. Liu, W. Newman, W. D. Beall, B. A. Stealey, and B. S. Bochner, *Am. J. Respir. Cell Mol. Biol., 7*:261 (1992).

146. M. S. Weinberger, T. Davidson, and D. Broide, *J. Allergy Clin. Immunol., 97*:662 (1996).

147. N. Terada, A. Konno, S, Fukuda, T. Yamashita, T. Abe, H. Shimada, K. Yoshimura, K. Shirotori, K. Ishikawa, and K. Togawa, *Int. Arch. Allergy Immunol., 106*:139 (1995).

148. L. C. Altman, G. H. Ayars, C. Baker, and D. L. Luchtel, *J. Allergy Clin. Immunol., 92*:527 (1993).

149. G. Ciprandi, C. Pronzato, V. Ricca, G. Passalacqua, M. Bagnasco, and G. W. Canonica, *Am. J. Respir. Crit. Care Med., 150*:1653 (1994).

150. M. Kato, T. Hattori, M. Kitamura, R. Beppu, N. Yanagita, and I. Nakashima, *Clin. Exp. Allergy, 25*:744 (1995).

151. G. Passalacqua, M. Albano, S. Ruffoni, C. Pronzato, A. M. Riccio, L. Di Berardino, A. Scordamaglia, and G. W. Canonica, *Am. J. Respir. Crit. Care Med., 152*:461 (1995).

152. A. M. Vignola, L. Crampette, M. Mondain, G. Sauvere, W. Czarlewski, J. Bousquet, and A. M. Campbell, *Allergy, 50*:200 (1995).
153. J. M. Greve, G. Davis, A. M. Meyer, C. P. Forte, S. C. Yost, C. W. Marlor, M. E. Kamarck, and A. McClelland, *Cell, 56*:839 (1989).
154. N. Mygind, *J. Allergy Clin. Immunol., 86*:827 (1990).
155. M. Nonaka, R. Nonaka, K. Woolley, E. Adelroth, K. Miura, Y. Okhawara, M. Glibetic, K. Nakano, P. O'Byrne, J. Dolovich, et al., *J. Immunol., 155*:3234 (1995).
156. A. Elovic, D. T. Wong, P. F. Weller, K. Matossian, and S. J. Galli, *J. Allergy Clin. Immunol., 93*:864 (1994).
157. I. Ohno, Y. Nitta, K. Yamauchi, H. Hoshi, M. Honma, K. Woolley, P. O'Byrne, J. Dolovich, M. Jordana, G. Tamura, et al., *Am. J. Respir. Cell Mol. Biol., 13*:639 (1995).
158. S. Finotto, I. Ohno, J. S. Marshall, J. Gauldie, J. A. Denburg, J. Dolovich, D. A. Clark, and M. Jordana, *J. Immunol., 153*:2278 (1994).
159. N. Kanai, J. Denburg, M. Jordana, and J. Dolovich, *Am. J. Respir. Crit. Care Med., 150*:1094 (1994).
160. F. A. Symon, G. M. Walsh, S. R. Watson, and A. J. Wardlaw, *J. Exp. Med., 180*:371 (1994).
161. K. L. Moore, N. L. Stults, S. Diaz, D, F. Smith, R. D. Cummings, A. Varki, and R. P. McEver, *J. Cell Biol., 118*:445 (1992).
162. C. Stellato, L. A. Beck, G. A. Gorgone, D. Proud, T. J. Schall, S. J. Ono, L. M. Lichtenstein, and R. P. Schleimer, *J. Immunol., 155*:410 (1995).
163. S. L. Harlin, D. G. Ansel, S. R. Lane, J. Myers, G. M. Kephart, and G. J. Gleich, *J. Allergy Clin. Immunol., 81*:867 (1988).
164. E. Tokushige, K. Itoh, M. Ushikai, S. Katahira, and K. Fukuda, *Laryngoscope, 104*:1245 (1994).
165. D. V. Messadi, J. S. Pober, W. Fiers, M. A. Gimbrone, Jr., and G. F. Murphy, *J. Immunol., 139*:1557 (1987).
166. P. Gosset, F. Malaquin, Y. Delneste, B. Wallaert, A. Capron, M. Joseph, and A. B. Tonnel, *J. Allergy Clin. Immunol., 92*:878 (1993).

# 18

# Expression of Cell Adhesion Molecules in Allergic Disorders of the Eye

Giorgio Walter Canonica, Antonio Scordamaglia, Francesca Paolieri,
Nicolò Fiorino, Giovanni Passalacqua, and Giorgio Ciprandi  *University of Genoa, Genoa, Italy*

## I.  INTRODUCTION

Allergic conjunctivitis is a common disease characterized by typical clinical features caused both by mediator release (mainly histamine) from activated mast cells and by an inflammatory reaction consequent to cytokines released by inflammatory cells. The main event in allergic inflammation is the migration of inflammatory cells (such as eosinophils, neutrophils, lymphocytes and monocytes) to the site of the allergic reaction. This requires adhesion to the endothelium and, next, locomotion to the target tissue, each stage of the process depending on cell–cell and cell–matrix adhesion (1) (Fig. 1). Several mediators of the acute allergic reaction are able to induce/increase expression of selectins on endothelium and leukocytes, allowing leukocyte migration ("rolling-over"). The further stages are firm adhesion, de-adhesion, and transmigration through the blood vessel wall to the site of inflammation; these are regulated by integrins (leukocyte function–associated antigen [LFA]-1, very late activation antigen [VLA]-4) and their ligands (interecellular adhesion molecule [ICAM]-1, vascular cell adhesion molecule [VCAM]-1) in addition to other unknown factors (2). Thus, $\beta 1$ integrins and cytoadhesins allow cell–matrix adhesion (3). The growing importance of adhesion mechanisms prompted Meuer and colleagues to postulate that we can no longer evaluate inflammation by simply looking at the mononuclear cell infiltrate but instead we should consider many mediators and adhesion receptors, the "real parameters" of the inflammatory process (2).

 Clear evidence of the crucial role of adhesive interaction, in particular the ICAM-1/LFA-1 system, during allergic inflammation has been provided in a primate model of asthma: an ICAM-1 monoclonal antibody was able to inhibit

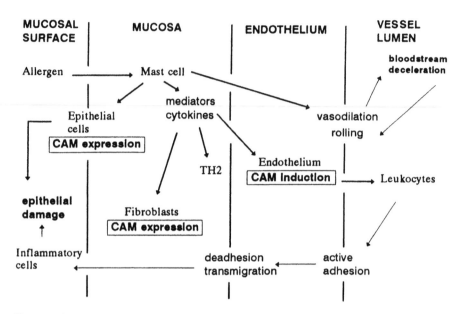

**Figure 1** Allergic reaction: the inflammatory network involving endothelium, epithelium, inflammatory cells, and cell adhesion molecules (CAM).

eosinophil migration into the bronchial mucosa and to prevent clinical symptoms (4).

Epithelial damage is a typical feature of allergic inflammation (e.g., in asthma) (5). Epithelial expression of adhesion molecules can facilitate interaction with offending inflammatory cells such as eosinophils. ICAM-1 is expressed on a wide variety of cells, including endothelium, activated lymphocytes, monocytes, eosinophils, and, under certain conditions, epithelial cells—in particular, during allergic inflammation (6–8).

A series of studies evaluated ICAM-1 epithelial expression, its modulation and possible role in allergic inflammation. We set up two different experimental models, one in vitro and one in vivo.

## II.  EXPERIMENTAL MODELS TO STUDY ICAM-1 IN ALLERGIC INFLAMMATION

### A.  In Vitro Model

Epithelial cell lines of conjunctival origin (WK), which constitutively expressed ICAM-1 to variable degrees, have been tested in culture. We evaluated whether the presence of this molecule could be up-regulated on the cell membrane by different pro-inflammatory cytokines.

The results are reported in Table 1: there are some cytokines able to up-regulate ICAM-1 expression, such as interleukin (IL)-1β, interferon gamma (IFN-γ), tumor necrosis factor–α (TNF-α), also with synergistic effect (IFN-γ+TNF-α); by contrast, other cytokines had no effect.

The presence of ICAM-1 has also been evaluated following addition of pharmacological compounds (commonly employed in the treatment of allergic diseases) to the culture medium. Some anti-allergic drugs such as lodoxamide,

**Table 1** Expression of CD54/ICAM-1 and CD29 (the common β1 integrin chain) on Cultured Conjunctival WK Cells

**Panel A**

| Stimulus (culture for 24 hr) | ICAM-1 expression | CD29 expression |
|---|---|---|
| None | + | + |
| IFN-γ 500 U/ml | ↑ | = |
| TNF-α 200 ng/ml | ↑ | = |
| IL-1β 100 pg/ml | ↑ | = |
| TGF-β1 100 mg/ml | = | = |
| IL-4 400 mg/ml | = | = |
| IL-6 100 ng/ml | = | = |
| IL-8 500 ng/ml | = | = |
| IL-10 200 ng/ml | = | = |
| TNF-α + IFN-γ | ⇑ | = |
| TNF-α + IL-1β | ↑ | = |

+ present (Flow cytometry analysis: mean fluorescence intensity MFI<20) = unchanged ↑ MFI 30–50 ⇑ MFI>90

**Panel B**

| Stimulus (culture for 24 hr) | sICAM-1 |
|---|---|
| None | + |
| IFN-γ 500 U/ml | ↑ |
| TNF-α 200 ng/ml | ↑ |
| IL-1β 100 pg/ml | = |
| TGF-β1 100 mg/ml | = |
| IL-4 400 mg/ml | = |
| IL-6 100 ng/ml | = |
| IL-8 500 ng/ml | = |
| IL-10 200 ng/ml | = |
| TNF-α + IFN-γ | ↑ |
| TNF-α + IL-1β | ↑ |

+ sICAM-1<1 ng/mL ↑ s ICAM-1>1.5 ng/mL = unchanged

levocabastine, and terfenadine as well as its active metabolite (TAM) reduced basal ICAM-1 expression on a continuously cultured conjunctival epithelial cell line, W-K; terfenadine and TAM also were able to reduce the release of its soluble form (Table 2). In contrast, nedocromil sodium did not affect ICAM-1 expression (9). The demonstration that soluble ICAM-1 can be detected in epithelial cell cultures is of potential importance.

Although the functional meaning of this shedding process is not yet clarified, one might envisage a protection mechanism exerted by epithelial cells themselves against offending viral pathogens and cells, such as the eosinophils. This is reminiscent of the protective effect exerted by melanoma cells versus cytotoxic T lymphocytes (2).

## B. In Vivo Model

This consists of direct conjunctival stimulation caused by allergen challenge or natural exposure. Upon allergen challenge, conjunctival mucosa shows an immediate hypersensitivity reaction occurring within minutes after challenge and quickly subsiding. This early-phase reaction, similar to allergen provocation in the skin and lung, is characterized by clinical symptomatology and local infiltration with inflammatory cells, mainly neutrophils (10). Bonini et al. (11) described the maintenance of the inflammatory process in the late-phase reaction peaking at 6 to 8 hours after challenge, and slowly subsiding at 24 hours or more. Ocular late-phase reactions show a predominance of eosinophils and lymphocytes. Although inflammatory cell infiltration to the site of the allergic reaction is well

**Table 2** Effect of Antiallergic Drugs on ICAM-1 Expression on Continuously Cultured Epithelial Cells (WK)

| Drugs | Basal expression[a] | IFN-γ stimulated expression (500 U/ml) | Soluble ICAM-1[b] |
|---|---|---|---|
| Levocabastine 1 μg/mL | ↓ | = | = |
| Lodoxamide 10 mg/mL | ↓ | = | = |
| Nedocromil 5μg/mL | = | = | = |
| Terfenadine 1 μg/mL | ↓ | = | ↓ |
| MDL 16455A[c] 50 μg/mL | ↓ | = | ↓ |

[a]=Unchanged.
↓ Significant decrease of ICAM-1 with respect to basal or IFN-γ up-regulated expression ($P<.05$).
[b]=Unchanged.
↓Significant decrease of the release of sICAM-1 ($P<.03$).
[c]Terfenadine—active metabolite.

documented, the activation of such cells and the interaction with epithelial cells, and the related epithelial damage, is just beginning to be understood.

By means of a sensitive immunocytochemical technique, ICAM-1 surface conjunctival epithelial expression on allergen-specific challenge has been described (7). No ICAM-1 expression occurs at baseline in normal subjects and in pollen-sensitive subjects (studies involved symptomless subjects out of the pollen season). The latter subjects expressed high levels of ICAM-1 only on allergen-specific challenge: ICAM-1 presence on conjunctival epithelial cells must therefore be related to the clinical parameters and inflammatory cell infiltration. Unexpectedly, ICAM-1 expression on epithelial cells occurs as an early event (30 min after allergen challenge) and persists for up to 48 hours when a high allergen dosage is employed (7). The mechanism of this rapid induction of ICAM-1 expression is unknown.

ICAM-1 expression has also been detected on conjunctival epithelial cells on natural exposure to allergens such as pollen or house-dust-mite allergens (12,13). Notably, "mite"-sensitive subjects, although completely symptom-free, persistently show mild expression of ICAM-1 on conjunctival epithelial cells together with a mild inflammatory cell infiltrate (Table 3) (12,13). These data are consistent with the findings in asthma, where a "minimal persistent inflammation" has been described in asthmatic subjects during clinical latency of the disease. Furthermore, we have recently demonstrated the presence of conjunctival hyperreactivity to hyperosmolar stimuli and histamine in allergic conjunctivitis due to house dust mite, when a concomitant inflammation, although minimal, is present (14,15) (Table 3). The absence of conjunctival hyperreactivity after exposure to a nonspecific stimulus in pollen-sensitive subjects is likely to derive from the study season, as they were studied outside the pollen season, when nasal and bronchial hyperreactivity is significantly reduced or absent (14).

The detection of ICAM-1 mRNA (through in situ hybridization) in epithelial

**Table 3**  ICAM-1 Expression on Epithelial Cells in Different Experimental Conditions

| | Subjects | | |
| --- | --- | --- | --- |
| Disease activity or challenge | Perennial rhinitics | Seasonal rhinitics | Normals |
| Asymptomatic or mild | + | −[a] | − |
| Symptomatic | ++ | ++ | − |
| Allergen challenge | +++ | +++ | − |
| Histamine challenge | ++ | − | − |
| Hyperosmolar solution challenge | ++ | ++ | + |

[a]=outside pollen season.

cells and the presence of soluble molecules in the supernatants of epithelial cell cultures confirm the epithelial origin of the molecule. Sometimes, healthy volunteers show a mild signal for ICAM-1 mRNA, but no modification occurs after allergen-specific challenge (Bagnasco et al., manuscript submitted). Similar results have been confirmed on nasal epithelium on allergen challenge and natural allergen exposure (8,16).

Constitutive expression of mRNA for ICAM-1 could explain the rapid kinetic of expression of ICAM-1. The possible role of mediators able to act at post-transcriptional levels is also suggested. These mediators should be rapidly released during the allergic reaction. Histamine was a possible candidate. However, it failed to induce ICAM-1 expression at the conjunctival level in vivo (15). Further studies are required to evaluate the possible role of other mediators able to induce ICAM-1 expression, such as ECP (eosinophil cationic protein) (17).

ICAM-1 expression during allergic inflammation in vivo has been evaluated following pharmacological treatment. Drugs of different families—a mast cell stabilizer (lodoxamide), antihistamines (cetirizine, oxatomide, azelastine, levocabastine), and corticosteroids (deflazacort)—can reduce ICAM-1 expression on conjunctival epithelium after allergen-specific challenge (Table 4) (18–20). On the other hand, mizolastine, a new antihistamine, is not able to reduce ICAM-1 expression twenty four.

Deflazacort, a new systemic corticosteroid, reduces ICAM-1 in the early as well in the late phase of the allergic response induced by allergen-specific challenge (20). Further studies on topical antihistamines (such as azelastine and levocabastine) have demonstrated their antiallergic activity (Table 4).

The presence of ICAM-1 on epithelial cells may reveal epithelial cell activation, with a possible direct involvement in allergic inflammation. Of note, we have provided evidence that upon specific allergen exposure, conjunctival cells express CD11a molecule, $\alpha$-chain of LFA1, the physiological ligand of ICAM1 (unpublished data). We reported a similar phenomenon in organ-specific autoim-

**Table 4** Effects of Different Drugs in Vivo on Conjunctival Expression of ICAM-1 after Allergen Challenge

| Drug (dose) | Early reaction (30 min) | Late reaction (6 hr) |
|---|---|---|
| Azelastine (single dose) | ↓ | ↓ |
| Cetirizine (20 mg × 3 days) | ↓ | ↓ |
| Levocabastine (single dose) | ↓ | ↓ |
| Mizolastine (10 mg × 7 days) | = | = |
| Oxatomide (60 mg × 7 days) | ↓ | ↓ |
| Lodoxamide (single dose) | ↓ | Not assessed |
| Deflazacort (60 mg × 3 days) | ↓ | ↓ |

munity of the thyroid (21). The coexpression of ICAM-1 and LFA-1 on epithelium during allergic inflammation could favor bidirectional interactions with activated leukocytes. To further clarify this point, the possible relationship between ICAM-1 expression and cytokine release by epithelial cells should be investigated.

Furthermore, ICAM-1 presence might result in an enhanced susceptibility of epithelial cells to damage by allergic inflammatory cells (e.g., eosinophils). This hypothesis might provide an additional pathogenic explanation for the epithelial damage that is one of the hallmarks of allergic inflammation. The presence of ICAM-1 on epithelial cells may definitely be considered a marker of inflammation caused by allergic stimulus (7,8).

We should also consider that eosinophils, upon allergic stimulation in allergic subjects, overexpress Mac-1, another ligand for ICAM-1 (1). Thus, all events contribute to eosinophil–epithelial cell adherence and perhaps epithelial damage.

Finally, recent reports have revealed additional perspectives: soluble adhesion molecules, such as soluble ICAM-1 and soluble CD44, have been described (22,23). Soluble adhesion molecules might be able to interact with adhesion receptors and with pro-inflammatory cytokines (22). All these findings underscore the role of inflammation as the major consequence of the allergic reaction and the fundamental notion that allergic diseases could be considered a "local" disease, as recently shown by in vivo observation (6).

## III.  CONCLUSIONS

The central role of adhesion molecules during conjunctival allergic inflammation has been clearly demonstrated (24). In particular, ICAM-1 has been recognized as a major hallmark of allergic inflammation (25). As reported, conjunctiva in in vivo and in vitro models provide a reliable tool for studying the inflammatory phenomena occurring during allergic reactions and for evaluating the antiallergic activity exerted by some drugs. The experimental data obtained from the reported studies confirm the role and the importance of modulating cell adhesion molecules as a new therapeutic strategy to treat allergic eye disease.

## REFERENCES

1.  E. Calderon and R. F. Lockey, *J. Allergy Clin. Immunol.*, *90* (5):852 (1992).
2.  S. C. Meuer, S. Buckhart, and Y. Samstag, *Am. Rev. Resp. Dis.*, *146*:s65 (1993).
3.  R. R. Lobb, *Adhesion: its role in inflammatory disease* (J. M. Harlan and D. Y. Liu, eds.), New York, W. H. Freeman (1992).
4.  C. D. Wegner, R. H. Gundel, P. Reilly, N. Haynes, L. G. Letts, and R. Rothlein, *Science*, *247*:456 (1990).
5.  S. Montefort, C. A. Herbert, C. Robinson, and S. T. Holgate, *Clin. Exp. Allergy*, *22*:511 (1992).

6.  C. R. Mackay and B. A. Imhof, *Immunol. Today*, *14*:99 (1993).
7.  G. Ciprandi, S. Buscaglia, G. P. Pesce, B. Villaggio, M. Bagnasco, and G. W. Canonica, *J. Allergy Clin. Immunol.*, *91*:783 (1993).
8.  G. Ciprandi, C. Pronzato, V. Ricca, G. Passalacqua, M. Bagnasco, and G. W. Canonica, *Am. J. Respir. Crit. Care Med.*, *150* (6):1653 (1994).
9.  G. Ciprandi, S. Buscaglia, C. Pronzato, V. Ricca, G. P. Pesce, B. Villaggio, N. Fiorino, M. Albano, A. Scordamaglia, M. Bagnasco, and G. W. Canonica, *New developments in the therapy of allergic disorders and asthma* (S. Z. Langer, M. K. Church, B. B. Vargatig, and S. Nicosia, eds.). Karger, Basel, p.115 (1993).
10. M. H. Friedlander, *Allergy: principles and practice* (E. Middleton, C. E. Reed, E. F. Ellis, N. F. Adkinson, J. W. Younginger, and W. W. Busse, eds.), Mosby, St Louis, p. 1649 (1993).
11. S. E. Bonini, S. T. Bonini, M. G. Bucci, A. Berruto, E. Adriani, F. Balsano, and R. Allansmith, *J. Allergy Clin. Immunol.*, *86*:869 (1990).
12. G. Ciprandi, S. Buscaglia, G. P. Pesce, M. Bagnasco, and G. W. Canonica, *Allergy* (1995).
13. G. Ciprandi, S. Buscaglia, G. P. Pesce, C. Pronzato, V. Ricca, S. Parmiani, M. Bagnasco, and G. W. Canonica, *J. Allergy Clin. Immunol.*, *96*:971 (1995).
14. G. Ciprandi, S. Buscaglia, G. P. Pesce, R. Lotti, M. Rolando, M. Bagnasco, and G. W. Canonica, *Int. Arch. Allergy Appl. Immunol.*, *104*:1. (1994).
15. G. Ciprandi, S. Buscaglia, G. P. Pesce, M. Bagnasco, and G. W. Canonica, *J. Allergy Clin. Immunol.*, *91* (6):1227 (1993).
16. G. Ciprandi, C. Pronzato, V. Ricca, M. Bagnasco, and G. W. Canonica, *J. Allergy Clin. Immunol.*, *94*:738 (1994).
17. L. C. Altman, G. H. Ayars, C. Baker, and D. L. Luchtel, *J. Allergy Clin. Immunol.*, *92*:527
18. G. Ciprandi, S. Buscaglia, G. P. Pesce, G. Passalacqua, J. P. Rihoux, M. Bagnasco, and G. W. Canonica, *J. Allergy Clin. Immunol.*, (1995).
19. G. Ciprandi, S. Buscaglia, C. Pronzato, C. Benvenuti, E. Cavalli, F. Bruzzone, and G. W. Canonica, *Ann. Allergy* (1995).
20. G. Ciprandi, S. Buscaglia, G. P. Pesce, A. Iudice, M. Bagnasco, and G. W. Canonica, *Allergy*, *48* (6):421 (1993).
21. M. Bagnasco, G. P. Pesce, A. Caretto, F. Paolieri, C. Pronzato, B. Villaggio, C. Giordano, C. Betterle, and G. W. Canonica, *J. Allergy Clin. Immunol.*, (1994).
22. A. J. H. Gearing and W. Newman, *Immunol. Today*, *14* (10):506 (1993).
23. Y. Tanaka, D. H. Adams, S. Hubscher, H. Hirano, U. Siebenlist, and S. Shaw, *Nature*, *361*:79 (1993).
24. B. S. Bochner and R. P. Schleimer, *J Allergy Clin Immunol.*, *94*(3):427 (1994).
25. G. W. Canonica, C. Pronzato, N. Fiorino, G. Ciprandi, and M. Bagnasco, *Allergy Clin. Immunol. News.*, *5*:80 (1993).

# 19

# Adhesion Molecules, Cytokines, and Chemokines in Allergic Airway Inflammation

**Nicholas W. Lukacs, Robert M. Strieter, and Steven L. Kunkel**   *University of Michigan Medical School, Ann Arbor, Michigan*

## I.  INTRODUCTION

The mechanisms involved in the activation and exacerbation of asthmatic responses have not been clearly identified. It appears that a culmination of events initiated by early mast cell activation leads to airway reactivity and bronchospasm. These physiologic responses are followed by leukocyte recruitment and peribronchial accumulation, leading to late-phase airway reactivity and damage. These late-phase events, although not completely induced by the presence of leukocytes, appear to be greatly enhanced by the localization and activation of these cells within the airway. The pathophysiology that accompanies allergic airway inflammation is likely the result of the presence of multiple leukocyte populations, including mast cells, neutrophils, lymphocytes, monocyte/macrophages, and eosinophils, as well as the expression of multiple inflammatory and chemotactic cytokines. The presence of eosinophils within the lungs during asthmatic responses is characteristic of the disease and appears to correlate well with the severity of asthma. A primary focus of research has been to inhibit the localization of leukocyte infiltration, especially the eosinophil, into and around the airways. The steps involved in the movement of cells from peripheral circulation to the peribronchial space are complex and not entirely defined. This chapter addresses the coordinated mechanisms and steps involved in leukocyte accumulation to the airway by focusing on adhesion molecules, early response cytokines, and chemokines.

## II.  ADHESION MOLECULES IN ALLERGIC AIRWAY INFLAMMATION

The activation and expression of adhesion molecules on the surface of vascular endothelial cells has been well defined. The initial up-regulation of selectin molecules, E and P, by multiple mechanisms leads to the reversible binding of peripheral leukocytes to the endothelium, facilitating slowing of leukocytes from the circulatory flow. E and P selectins are rapidly inducible and initiate "rolling" of leukocytes on activated endothelium through $Ca^{2+}$-dependent recognition of cell surface carbohydrates of both the sialyl-Lewis$^x$ family and related oligosaccharides that are expressed on glycoproteins or glycolipids (1,2). Selectin molecules function to slow the flow of leukocytes in the vascular compartment and allow transient adherence to "counter-receptors" expressed on activated vascular endothelium (3). These initial events allow the leukocytes to interact with the endothelium and facilitates firm adherence to other inflammation-induced adhesion molecules, such as intercellular adhesion molecule-1 (ICAM-1) and vascular cell adhesion molecule-1 (VCAM-1) (4–7). This firm attachment via adhesion molecules is mediated through β-integrins on the surface of leukocytes and accommodates the extravasation of leukocytes through the vascular endothelium and into the inflamed tissue, following chemotactic gradients. Normally, the expression of adhesion molecules or their counter-receptors are tightly regulated. However, inappropriate or chronic expression likely contributes to the severity and/or chronicity of inflammatory disorders. These complex events become more complicated because leukocytes interact with multiple molecules and must travel through several cell layers before reaching the foci of the inflammatory response. The elucidation of these complex events in diseases such as allergic asthma may allow targeted therapy for the alleviation of the inflammatory-induced complications.

The active participation of adhesion molecules in allergic airway inflammation and asthma has been postulated as an important mechanism to promote the response. Experiments conducted in multiple laboratories and clinics have established correlations between adhesion molecule expression and the severity of asthmatic responses. Because of the anatomical localization of the asthmatic response, the airway epithelial cells likely play an intricate role in the initiation and maintenance of the inflammatory reaction. In a prominent study, the increased expression of ICAM-1 and human leukocyte antigen, D-related (HLA-DR) was identified on bronchial epithelial cells of asthmatic, but not normal, patients. Furthermore, the expression of these molecules correlated directly with (forced expiratory volume in 1 sec) ($FEV_1$) and the clinical score of Aas (8). In further support of these observations, increased endothelial expression of ICAM-1, E-selectin, and VCAM-1 was also identified in the airway of allergic asthmatics compared to nonasthmatic patients (10). This latter study found no increase in expression of these same adhesion molecules in intrinsic asthma patients, suggesting that mechanisms involved in the two types of asthma were distinct. In an earlier study, using data from an immunohistochemical analysis, it was concluded that no significant difference in the expression of ICAM-1 or E-selectin on bronchial mucosa occurred between asthmatic and normal patients (10). More recent investigations appear to support the former conclusions of increased expression of adhesion molecules on bronchial epithelial cells in allergic asthmatics versus normal individuals (11,12). Overall, the increased expression of E-selectin, ICAM-1, and VCAM-1 in asthmatic subjects correlates well with leukocyte infiltration and clinical disease.

In addition to the above studies demonstrating an increased expression of adhesion molecules on bronchial epithelium, local and infiltrating leukocyte populations have also been shown to have increased expression of adhesion molecules. The increased expression of ICAM-1 and leukocyte function–associated antigen (LFA)-1 on alveolar macrophages has been identified in asthmatic subjects and correlates directly with the severity of disease (13). Likewise, sputum eosinophils display an increased surface expression of ICAM-1 and HLA-DR (14); the expression of $\beta2$-integrins Mac-1 and LFA-1, along with $\beta1$-integrin very late activation antigen (VLA)-4, are also increased on eosinophils and alveolar macrophages within the airway (15). Although the increased expression of adhesion molecules allows for the localization of leukocytes to the inflamed tissue, interaction of leukocytes with adhesion molecules can also induce the production of a number of activating mediators. Interaction of eosinophils with VCAM-1 and ICAM-1 has recently been shown to induce the release of granule proteins, proteases, and oxidative metabolites (17,18). In addition, the cellular interactions between structural cells and leukocytes during asthma can also drive chemokine production. Recent studies have demonstrated that endothelial cell–monocyte interactions were able to drive chemokine (macrophage inflammation protein) [MIP]-1$\alpha$, (monocyte chemoattract peptide [MCP]-1, and interleukin [IL]-8) production (19,20). These latter results indicated that both cell populations have the ability to produce one or more of the chemokines during the interaction, and each appears to have a specific requirement for activation. MIP-1$\alpha$ production was produced only by monocytes and required activation through ICAM-1 interaction (19). In contrast, MCP-1 and IL-8 production appeared to require interaction via matrix proteins (20). The expression of chemokines during cellularinteractions likely contributes to the amplification and maintenance of the inflammatory response. Altogether, these data indicate that cellular adhesion events are sufficient not only to localize cells within inflamed tissue but also to activate the leukocytes once they have arrived for the release and production of proteases, oxygen radicals, and cytokines (Fig. 1).

Much of our working knowledge of the role of adhesion molecules in disease

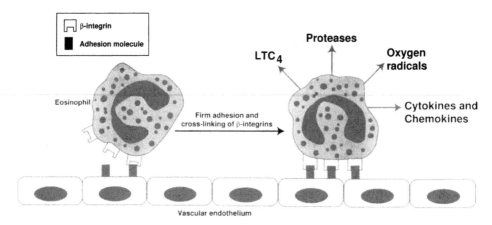

**Figure 1** Leukocyte–adhesion molecule interaction induces leukocyte activation/degranulation.

comes from multiple animal model studies in which adhesion molecule interactions have been blocked. Initial investigations were carried out in a primate model of *Ascaris*-antigen–induced allergic airway inflammation. In these seminal studies, anti–ICAM-1 treatment abrogated eosinophil infiltration and diminished airway reactivity (21). These studies demonstrated that ICAM-1 was up-regulated on both vascular endothelium and airway epithelial cells. In contrast to ICAM-1, endothelial-leukocyte adhesion molecule-1 (ELAM-1) (E-selectin) was up-regulated on endothelial, but not airway, epithelial cells and did not contribute to either accumulation of eosinophils or airway hyperresponsiveness. These studies support the conclusion reached in human asthma studies that indicated a correlation between adhesion molecule expression on airway epithelium and airway reactivity. Since these early studies, several laboratories have used other animal model systems to examine allergic airway responses. In a brown Norway rat model of ovalbumin-induced airway inflammation, anti–ICAM-1 treatment demonstrated a significant reduction in airway responsiveness without a decrease in eosinophil or lymphocyte recruitment (22). A mouse model of ovalbumin-induced airway inflammation has demonstrated that VLA-4/VCAM-1 interactions are paramount to induction of eosinophil accumulation in the airway wall. Interestingly, treatment of the mice with anti–ICAM-1 and anti–LFA-1 did not inhibit eosinophil recruitment in the lung interstitium, possibly suggesting species differences and/or the specificity of the antigen used to induce the response when compared to the primate *Ascaris*-antigen model (23). Reasons for these discrepant results are not clear. In a more recent study using a guinea pig model of ovalbumin-induced airway hyperresponsiveness, passive immunization with an anti–VLA-4 antibody decreased eosinophil accumulation and eosinophil peroxidase levels in the airway and lung, but failed to alleviate the airway hyperresponsiveness (24). This latter study suggests that eosinophil accumulation and airway hyperresponsiveness do not necessary correlate with one another. The elucidation of adhesion molecule expression pattern and location during allergic airway responses will lead to a better understanding of the mechanisms and lead to development of proper therapeutic approaches to modulate the responses.

## III.  CYTOKINES IN ALLERGIC AIRWAY INFLAMMATION

### A.  Tumor Necrosis Factor–α

The upregulation of adhesion molecules and accumulation of leukocytes is mediated by complex mechanisms that are induced by a cascade of events dependent on the production of mediators in a specified sequence. A family of soluble peptide mediators, known as cytokines, allow cell-to-cell communication, cellular activation, and leukocyte recruitment, often at great distances. One of the most potent early response cytokines has been identified as tumor necrosis factor–α (TNF-α). TNF-α has demonstrated the ability to mediate many crucial events for the initiation of both acute and chronic inflammatory responses, including the upregulation of adhesion molecules, the induction of other early cytokines (IL-1 and IL-6), and the activation of leukocyte-specific chemokines, as well as inducing anti-inflammatory molecules (24–27). Multiple studies have identified TNF-α as a central molecule in a number of responses, including allogeneic responses, granulomatous reactions, and acute endotoxemia (29–34). The correlation of TNF-α with late-phase asthma suggests a role for TNF-α as an important molecule in the allergic response (35–37). Interestingly, a number of leukocyte popu-

**Table 1**  Cytokines Implicated in Allergic Airway Inflammation

| Cytokine | Cellular source | Function |
|---|---|---|
| TNF-α | Mast cells, macrophages, lymphocytes, eosinophils | Increase adhesion molecules; augment cytokine production |
| IL-4 | Lymphocytes, mast cells, eosinophils, basophils | Induce VCAM-1 expression; IgE Ab class switching |
| IL-5 | Lymphocytes, mast cells | Eosinophil maturation; increases eosinophil chemotaxis |
| SCF | Macrophages, fibroblasts, others? | Mast cell degranulation and activation |
| Chemokines | Differentially produced by all cell types | Leukocyte recruitment; granulocyte degranulation |

lations have the ability to produce TNF-α, including macrophages, mast cells, lymphocytes, and even eosinophils. All these cell types are involved in allergic airway (asthmatic) inflammation and therefore have the potential to participate in TNF-α–mediated responses.

The correlation of increased TNF-α production and pathological events in asthma has been well supported both in vivo and in vitro using cellular populations isolated from asthmatic patients. The most convincing evidence that TNF-α plays a role in the pathophysiology of asthma was recently observed in normal subjects receiving inhaled recombinant TNF-α (33). These studies demonstrated a significant increase in airway hyperresponsiveness and decreases in $FEV_1$ within those subjects receiving TNF-α compared with the placebo group. Although the cellular infiltrate was predominantly neutrophilic, these studies do indicate a significant role for TNF-α in augmenting airway responsiveness. Production of TNF-α in asthmatic patients after IgE-dependent stimulation from both mast cells (39) and alveolar macrophages (40) demonstrate the importance of these two cellular populations in contributing to the early activation of the response. In addition, elevated release of TNF-α was observed in airway leukocytes isolated from asthmatic patients compared with leukocytes from nonasthmatic patients (41). These latter studies suggest an increased potential for TNF-α production in airway leukocytes from asthmatic patients.

In animal models of allergic airway inflammation, TNF-α has also demonstrated a significant role in airway hyperresponsiveness. The correlation of TNF-α and airway hyperreactivity in rats was established using multiple types of stimulation (42). In these studies, rats exposed to LPS demonstrated elevated airway hyperreactivity that was alleviated by antibodies to TNF-α. In addition, direct exposure by aerosolized TNF-α also increased airway hyperreactivity. In both instances, a correlation with neutrophil influx was observed, suggesting that this cell population can contribute to airway hyperresponsiveness. In a guinea pig model of ovalbumin-induced airway hyperresponsiveness, the administration of IL-1 receptor antagonist protein inhibited airway hyperreactivity, eosinophil accumulation, and TNF-α production (43). More recently, in a mouse model of allergic airway inflammation, passive immunization with a soluble TNF-α receptor construct significantly reduced both early neutrophil and later eosinophil influx in and around the airways of the challenged mice (44). In addition, the neutralization of TNF-α significantly diminished the level of eosinophil-specific chemokines, MIP-1α and regulated on activation, normal T cell expressed and secreted (RANTES), produced during an allergic response (45). Altogether, it appears that TNF-α plays a significant role during allergic airway inflammation as an overall inflammatory mediator via its ability to up-regulate adhesion molecules and chemokines specific for leukocyte recruitment.

## B. TH2 Type Cytokines

There appears to be overwhelming evidence that allergic asthmatic responses are driven by TH2-type responses. This evidence includes the presence of the appropriate cytokines, significant accumulation of eosinophils, and the prevalence of antigen-specific IgE for the activation and degranulation of mast cells. There appear to be multiple sources for the TH2-type cytokines, including lymphocytes, mast cells, and eosinophils, all of which participate in the allergic responses. The initial identification of predominantly TH2 type lymphocytes in the bronchoalveolar space in asthmatics has led investigators to the realization that these cytokines (IL-3, IL-4, IL-5, and granulocyte-macrophage colony-stimulating factor [GM-CSF]) may play the predominant role in directing the leukocyte accumulation patterns and disease severity (46). Because of their specific functions, the primary TH2 cytokines involved in driving the allergic response have been identified as IL-4 and IL-5. Of these cytokines, IL-4 may play the most critical role in directing the immune response toward an allergic event. There are several compelling reasons why IL-4 should be recognized as a key mediator molecule: 1). IL-4 has the ability to induce isotype switching in B lymphocytes to produce IgE and therefore prime mast cells for antigen-specific degranulation (47), 2.) IL-4 preferentially up-regulates VCAM-1 on the surface of endothelial cells inducing preferential eosinophil and lymphocyte recruitment (48), 3.) IL-4 appears to be the key cytokine responsible for switching of T lymphocytes to a TH2 type phenotype (49), and 4.) transfection of IL-4 into developing tumors in mice directly induced eosinophil accumulation in tumor tissue, suggesting a specific activational role of IL-4 in targeting and/or inducing eosinophil localization (50). The evidence for the role of IL-5 in allergic asthma has pointed primarily to the maturation, maintenance and recruitment of eosinophils. IL-5 appears to enhance eosinophil chemotaxis to multiple stimuli as well as prime eosinophils for degranulation (51,52). Direct immunohistochemical localization of IL-4 and IL-5 has been identified in multiple cell populations in bronchoalveolar lavage (BAL) and tissue samples from allergic asthmatics. The predominant cellular source of IL-4 and IL-5 was identified as T lymphocytes, which made up approximately $\cong70\%$ of the IL-4/IL-5–expressing cells. The remaining cellular contribution of the cytokines came from mast cells and eosinophils (53). The exact contribution of these various leukocyte populations to the total IL-4 and IL-5 protein levels within asthmatic airways has not been ascertained. Although eosinophil accumulation can be intense during allergen challenges, studies thus far have demonstrated relatively low level production of IL-4 (50 pg/$10^6$ eosinophils) (54). However, given the intensity of eosinophil influx during an allergen challenge, the contribution of eosinophils to IL-4 production within the local environment of the airway may be critical to the sustained activation and maintenance of the response. In a novel study of steroid-resistant asthmatics, a clear difference was observed in the number of IL-4 and IL-5 mRNA positive (+) cells in the BAL of steroid-sensitive and steroid-resistant asthmatics (55). After a 1-week course of oral prednisone, the steroid-sensitive asthmatics showed a significant decrease in IL-4 and IL-5 mRNA positive cells. In contrast, steroid-resistant asthmatics demonstrated no decrease in either IL-4 or IL-5 mRNA positive cells. In contrast, they did show a further diminution of interferon gamma (IFN-γ) mRNA positive cells. Overall, a clear correlation exists between the presence of IL-4 and IL-5 producing cell populations and severity of the asthmatic response.

The use of animal models of allergic airway responses has further supported the link between IL-4/IL-5 expression and airway inflammation and hyperreactivity. Early studies in allergic guinea pig models demonstrated that anti–IL-5 antibody treatment could significantly reduce eosinophil accumulation and airway hyperreactivity (56–59). In addition, instillation of recombinant IL-5 into the airways of guinea pigs significantly induced eosinophil accumulation and airway hyperresponsiveness (60), demonstrating a direct link between IL-5, eosinophils, and airway reactivity. More recently, a clear link between IL-4 and airway eosinophilia has been established within animal models. Neutralization of IL-4 by antibody treatment nearly abrogates airway eosinophil accumulation (61); the use of IL-4 deficient mice has demonstrated not only decreased eosinophilia but also significantly lower airway reactivity (62,63). Although the mechanism of IL-4–induced eosinophil accumulation is not readily apparent, IL-4 may be operative through a combination of effects, such as up-regulation of adhesion molecules (VCAM-1) as well as chemokine production (45). The regulation of the IL-4 (TH2) type responses within the airway can be cross-regulated by TH1-type cytokines, such as IL-12 and IFN-γ. The co-instillation of IL-12 along with allergen challenge has demonstrated significant reduction in airway eosinophil accumulation and airway hyperresponsiveness (64,65). This latter observation appeared to be a direct effect of IFN-γ. Likewise, the instillation of recombinant IFN-γ has demonstrated similar effects as IL-12 (66). These latter data demonstrate interesting regulatory pathways that may be advantageous for therapeutic application in human asthma.

## C.  Stem Cell Factor

A primary cytokine involved in mast cell differentiation and activation has been identified as the c-kit receptor ligand or stem cell factor (SCF) (67–70). Stem cell factor binds to its surface receptor, c-kit, which is a member of the receptor tyrosine kinase family. Endogenous SCF occurs in both transmembrane and soluble forms, and both are biologically active (71). Stem cell factor not only induces mast cell differentiation but also can directly mediate mast cell degranulation, and it appears to induce mast cell activation and cytokine production. In addition to inducing mast cell activation, the membrane form appears to have the ability to allow adhesion of mast cells to cellular surfaces. This latter observation was determined using c-kit transfected COS cells to promote mast cell adherence (72) and by inhibiting c-kit+ cell binding to fibroblasts with an anti-SCF antibody (73). Stem cell factor protein has been identified in bone marrow stromal cells (74), fetal liver stromal cells (75), and 3T3 fibroblasts (76). In addition, other cell populations may also produce SCF, such as vascular endothelial cells (77). Growth of mast cells from SI/SI (SCF-deficient/mast cell deficient) mutant mice can be induced by 3T3 fibroblasts (78), suggesting that peripheral cell populations have the ability to produce mast cell growth factors, likely SCF. Stem cell factor also appears to have a role in mast cell survival, as mast cells cultured without SCF undergo an apoptotic death, suggesting that long-term survival of mast cells may depend on constant stimulation with SCF (79). Interestingly, reconstitution of SI/SI mutant mice with recombinant SCF demonstrated increased airway inflammation and hyperreactivity to an allergen challenge (80). These latter observations may be very pertinent to the production of SCF in the periphery and, furthermore, may affect the activation and degranulation of peripheral mast cells during disease progression. In recent observations, the addition of recombinant SCF to human bronchi induced mast cell degranulation and smooth muscle cell contraction (81), suggesting a significant role for SCF in human asthmatic responses. In fact, given the ability of SCF, on its own, to cause mast cell degranulation, mast cells and their products may play an intricate role in activation and maintenance of acute and chronic inflammation.

Examination of SCF production during allergic events has only recently been investigated. Mast cells and their products are believed to be critical in the development of allergic responses, such as asthma. A recent study using mast cell deficient mice that lack proper SCF maturation pathways (W/W$^v$ mice) has demonstrated a significant decrease in eosinophil accumulation in lungs of allergic mice after allergen challenge (as compared with normal littermates) (82). These results indicate the importance of SCF in mast cell development as well as the apparent necessity for mast cells during allergic eosinophilic responses. The concept of expression of SCF in the periphery is supported by the fact that SCF can be detected in human (83) and mouse (84) serum at a constitutive level of 3–5 ng/mL. Furthermore, alveolar macrophage–derived SCF in allergic responses is driven by TNF-$\alpha$ and IL-4 and appears to be important for eosinophil accumulation, as well as a potent mast cell activator for augmentation of allergic airway responses (84). In these latter studies, neutralization of SCF during allergen challenge within the airway significantly decreased histamine release and eosinophil accumulation within and around the airway. The mechanism of this decrease may relate to the ability of SCF to directly induce mast cell activation and cytokine production. We have recently identified SCF-stimulated murine mast cells as having increased expression of IL-4, MIP-1$\alpha$, and RANTES (Fig. 2). These three cytokines have all been shown to contribute significantly to the

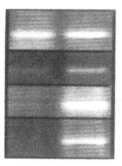

**Figure 2** Stem cell factor (10 ng/mL) induces cytokine mRNA expression in murine mast cells. Reverse transcriptase–polymerase chain reaction (RT-PCR) amplification of mRNA from 6-hr-stimulated mast cells.

overall eosinophilic airway response. Although mast cells only partially contribute to the overall inflammation, these data may help explain the mechanisms involved in mast cell–dependent airway eosinophilic inflammation.

Our concept of the activation and function of SCF during allergic inflammation is outlined in Fig. 3. The early activation and release of TNF-α and IL-4 from either IgE-degranulated mast cells, activated macrophages, and/or antigen-specific T lymphocytes serves as the stimulation to up-regulate SCF from airway macrophage populations. The subsequent production of SCF can directly and indirectly, via augmentation of IgE-mediated responses, activate mast cells, which in turn release histamine, IL-4, and eosinophil-specific chemokines. An increase in IL-4 and chemokine production can lead to an augmentation of eosinophil recruitment through the up-regulation of vascular adhesion molecules and specific recruitment of eosinophils by MIP-1α and RANTES. Disruption of SCF-induced

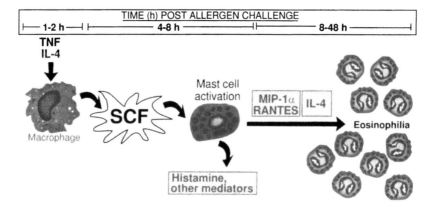

**Figure 3** Participation of stem cell factor (SCF) in allergic airway inflammation.

mast cell activation may interrupt this pathway and attenuate eosinophil accumulation and airway hyperreactivity.

## D. Chemokine Production and Leukocyte Activation

Localization of particular leukocyte populations to a site of inflammation is a crucial event that can be mediated by specialized chemotactic cytokines, the chemokines. The family of chemokines has been divided into at least two groups, C-X-C (alpha) and C-C (beta), based on their sequence homology and the location of the first two cysteine residues in their amino acid structure (Table 2). The groups appear to have distinct functional properties within immune/inflammatory responses. The C-X-C family, which includes IL-8, ENA-78, NAP-2, CTAP III, beta-thromboglobulin, GRO $\alpha,\beta,\gamma$, and GCP-2 in humans, primarily have been associated with acute inflammation because of their ability to induce neutrophil recruitment (85). Neutrophil binding and chemotaxis is conferred by a consensus sequence domain of E-L-R amino acids. Additional members of the C-X-C chemokine family, PF-4, IP-10, and MIG, lack the E-L-R amino acid motif and are not chemotactic for neutrophils. In addition, the presence of the E-L-R motif confers angiogenic activity to the C-X-C chemokines, whereas those that lack the E-L-R motif have angiostatic activity (86). The C-C chemokine family, which includes MCP-1, 2, 3, and 4, MIP-1 $\alpha$ and $\beta$, RANTES, and eotaxin in humans, all appear to be chemotactic for lymphocytes and monocytes and differentially chemotactic for eosinophils and basophils (87,88). A third family of chemotactic molecules has been suggested with a single cysteine amino acid in the N-terminal domain, known as the C family, and is represented by a T lymphocyte–specific chemoattractant, lymphotactin. The relavance of this molecule in human disease has only recently begun to be addressed. A significant attribute of chemokines is

**Table 2** Chemokines

| C-X-C (alpha) | C-C (beta) | C (gamma) |
| --- | --- | --- |
| **ELR containing** | | |
| IL-8 | MCP-1, 2, 3, 4 | Lymphotactin |
| ENA-78 | MIP-1 $\alpha$ | |
| GRO $\alpha,\beta,\gamma$ | MIP-1 $\beta$ | |
| GCP-2 | RANTES | |
| $\beta$-Thromboglobulin | Eotaxin | |
| CTAP III | | |
| NAP-2 | | |
| **Non-ELR containing** | | |
| PF-4 | | |
| IP-10 | | |
| MIG | | |

their avid heparan binding properties. Because of the ability of chemokines to bind to heparan sulfate, it is now been suggested that chemokines operate as a "solid phase" gradient, together with adhesion molecule expression, to cause leukocytes to home to a site of inflammation.

The accumulation of leukocytes into and around the airways in allergic asthmatics appears to correlate well with the expression of a number of chemotactic mediators. The release of platelet-activating factor (PAF) and leukotrienes during allergic responses can mediate nonspecific leukocyte accumulation and likely contribute to the overall inflammatory response. More importantly, these lipid mediators appear to have profound effects on the reactivity of the inflamed airways (1). However, these molecules do not explain the preferential accumulation of mononuclear and eosinophilic leukocytes within the airways of allergic asthmatics. The production of specific chemokines, which can preferentially recruit particular leukocyte populations, have been a major focus of many laboratories. Initial investigations in human populations have centered around eosinophil-specific chemokines, as the esoinophil appears to be an important molecule in the pathology of asthma. One of the most potent chemokines that promote eosinophil recruitment in vitro is RANTES (89,90). In vivo, the intradermal injection of RANTES into dogs caused the accumulation of eosinophils and mononuclear leukocytes, whereas other chemokines did not (91). The addition of RANTES to cultures of eosinophils can cause direct activation and degranulation of the cells (92). In addition, RANTES has been shown to be a potent basophil-activating protein, directly inducing release of histamine (93). The production of RANTES can be found in multiple cell types, including lymphocytes, macrophages, epithelial cells, and fibroblasts, and prepackaged in platelets (86,87). As with all C-C chemokines, RANTES is a potent chemoattractant for lymphocytes and monocytes and likely contributes to the influx of these cells during disease. These attributes may significantly contribute to the allergic asthmatic responses. In studies related to human asthma, increased production of RANTES has not been thoroughly studied. However, RANTES can be produced by TNF-$\alpha$–stimulated bronchial epithelial cells (94) and was down-regulated by inhaled steroids (95). Future studies in asthmatic patients should elucidate the increased production and cellular localization of RANTES during allergic asthmatic responses and correlate its presence with leukocyte subset recruitment and disease severity.

Other chemokines have also been investigated within allergic asthmatic populations. MIP-1$\alpha$ has chemotactic activity for eosinophils, although not as impressive as RANTES, and has been observed to be increased in asthmatics as compared to nonasthmatic patients (96). In addition, IL-8 also appears to have activity in mediating eosinophil recruitment (97) and its production can be down-regulated in bronchial epithelial cells by steroid treatment (95). Increased MCP-1 expression can also be observed in bronchial epithelial cells (98). Although MCP-1 is not chemotactic for eosinophils, it likely plays a role in other important events during allergic responses. MCP-1 is a potent basophil degranulating agent (99) and has been shown to be a strong lymphocyte chemoattractant (100). Other C-C chemokines may participate in the eosinophil-specific infiltration observed in allergic asthmatic responses. MCP-3 and eotaxin appear to have potent eosinophil chemotactic properties (101,102). Neither of these two latter chemokines have been examined within the human system. Altogether, it appears that multiple chemokines with overlapping functions are expressed during allergic asthmatic responses, and targeting a single chemokine may not prove effective in altering the clinical outcome of this disorder.

The role of chemokines in allergic airway responses have also begun to be examined in animal models. The use of these models will likely provide a more mechanistic explanation of the role of particular chemokines within allergic airway responses. During the allergic response, multiple chemokines, MIP-1α, MCP-1, and RANTES, are produced that have overlapping functions in vitro, such as monocyte and lymphocyte chemotactic properties. However, the production of these same molecules in vivo appears to have distinct functions, recruiting specific leukocytes into the lung and airway. This may partially be due to the cell type that produces the particular chemokine, as well as the location of the cytokine-producing cell within the lung parenchyma and airway. These parameters have been examined using a mouse model of allergic airway inflammation that uses a helminth parasite antigen (*Schistosoma mansoni* souble egg antigen) for the induction and elicitation of the allergic response (44,45,103). Using this parasite antigen model of airway hyperreactivity, investigators observed the expression of MIP-1α, MCP-1, and RANTES in a distinct distribution pattern. MIP-1α protein stained intensely in macrophages and along the airway epithelial cell border; MCP-1 was immunolocalized to interstitial epithelial cells and macrophages; and RANTES was localized to type II and airway epithelial cells (103; N. Lukacs, unpublished observations). These tissue/cell distributions of the various chemokines may be important to "lead" the leukocytes properly through the multiple layers of the lung and into and around the airway. When animals were passively immunized with antibodies specific for the individual chemokines, distinct patterns of leukocyte infiltration were interrupted. In vivo neutralization of MIP-1α or RANTES significantly affected the eosinophil populations (40% to 50% decrease) but did not significantly alter the influx of mononuclear cells (45,103). In contrast, when MCP-1 was neutralized in vivo, the recruitment of mononuclear cells, monocyte/macrophages and lymphocytes, was significantly reduced (N. Lukacs, unpublished observations). These latter data are consistent with in vitro data that demonstrate that MCP-1 is chemotactic for monocytes and lymphocytes (102). The ability of the chemokines to differentially affect the inflammatory response by recruitment of particular leukocytes may begin to explain why multiple chemotactic factors need to be produced during an immune response. It is also significant that even though these molecules have overlapping effects in vitro, they do not appear to function in a similar manner in vivo during an allergic response.

## IV. CONCLUSIONS

The regulation of allergic airway inflammation may be found at multiple levels, and therefore therapeutic intervention may be targeted to multiple sites along the inflammatory pathway. This pathway includes initial mast cell degranulation, release of early-response cytokine mediators, up-regulation of adhesion molecules, production of chemokines inducing preferential accumulation of specific leukocyte populations, and release of tissue-damaging products leading to the development of late-phase asthmatic responses. The continued elucidation of these responses in patient populations and animal models will allow a further understanding of the disease process and the mechanisms that promote the pathophysiology of allergic asthma.

## REFERENCES

1. B. A. Imhof and D. Dunon, *Adv. Immunol.*, *58*:345 (1995).
2. J. C. Hogg and C. M. Doerschuk, *Ann. Rev. Physiol.*, *57*:97 (1995).
3. M. S. Mulligan, S. R. Watson, C. Fennie, and P. A. Ward, *J. Immunol.*, *151*(11):6410 (1993).
4. M. S. Mulligan, J. C. Paulson, S. De Frees, Z. L. Zheng, J. B. Lowe, and P. A. Ward, *Nature*, *364*(6433):149 (1993).
5. T. T. Hansel, J. B. Braunstein, C. Walker, K. Blaser, P. L. Bruijnzeel, J. C. Virchow, Jr., and C. Virchow, Sr. *Clin. Exp. Immunol.*, *86*:271 (1991).
6. B. C. Hakkert, T. W. Kuijpers, J. F. Leeuwenberg, J. A. van Mourik, and D. Roos, *Blood*, *78*:2721 (1991).
7. T. Carlos, N. Kovach, B. Schwartz, M. Rosa, B. Newman, E. Wayner, C. Benjamin, L. Osborn, R. Lobb, and J. Harlan, *Blood*, *77*:2266 (1991).
8. A. M. Vignola, A. M. Campbell, P. Chanez, J. Bousquet, P. Paul-Lacoste, F. B. Michel, and P. Godard, *Am. Rev. Respir. Dis.*, *148*:689 (1993).
9. A. M. Bentley, S. R. Durham, D. S. Robinson, G. Menz, C. Storz, O. Cromwell, A. B. Kay, and A. J. Wardlaw, *J. Allergy Clin. Immunol.*, *92*:857 (1993).
10. S. Montefort, W. R. Roche, P. H. Howarth, R. Djudanovic, C. Gratiziou, M. Carroll, L. Smith, K. M Britten, D. Haskard, et al, *Eur. Respir. J.*, *5*:815 (1992).
11. G. W. Canonica, G. Cipandi, G. P. Pesce, S. Buscaglia, F. Paolieri, and M. Bagnasco, *Int. Arch. Allergy Immunol.*, *107*:99 (1995).
12. P. Gosset, I. Tillie-Lablond, A. Janin, C. H. Marquette, M. C. Copin, B. Wallaert, and A. B. Tonnel, *Int. Arch. Allergy Immunol.*, *106*:69 (1995).
13. P. Chanez, A. M. Vignola, P. Lacoste, F. B. Michel, P. Godard, and J. Bousquet, *Allergy*, *48*:576 (1993).
14. T. T. Hansel, J. B. Braunstein, C. Walker, K. Blaser, P. L. Brunijnzeel, J. C. Virchow, and C. Virchow, *Clin. Exp. Immunol.*, *86*:271 (1991).

15. Y. Ohkawara, K. Yamauchi, N. Maruyama, H. Hoshi, I. Ohno, M. Honma, Y. Tanno, G. Tamura, K. Shirato, and H. Ohtani, *Am. J. Respir. Cell Mol. Biol.*, *12*:4 (1995).
16. M. Nagata, J. B. Sedgwick, M. E. Bates, H. Kita, and W. W. Busse, *J. Immunol.*, *155*:2194 (1995).
17. J. Chihara, T. Kakazu, I. Higashimoto, T. Yamamoto, D. Kurachi, and S. Nakajima, *Int. Arch. Allergy Immunol.*, *108*:45 (1995).
18. J. Chihara., T. Yamamoto, D. Kurachi, T. Kakazu, I. Higashimoto, and S. Nakajima, *Int. Arch. Allergy Immunol.*, *108*:52 (1995).
19. N. W. Lukacs, R. M. Strieter, V. M. Elner, H. L. Evanoff, M. D. Burdick, and S. L. Kunkel, *Blood*, *83*:1174 (1994).
20. N. W. Lukacs, R. M. Strieter, V. Elner, H. L. Evanoff, and S. L. Kunkel, *Blood*, *86*:2767 (1995).
21. C. K. Wegner, R. Rothlein, and R. H. Gundel, *Agents Actions Suppl.*, *34*:529 (1991).
22. J. Sun, W. Elwood, A. Haczku, P. J. Barnes, P. G. Hellewell, and K. F. Chung, *Int. Arch. Allergy Immunol.*, *104*(3):291 (1994).
23. H. Nakajima, H. Sano, T. Nishimura, S. Yoshida, and I. Iwamoto, *J. Exp. Med.*, *179*:1145 (1994).
24. A. A. Miline and P. J. Piper, *Eur. J. Pharmacol.*, *282*:243 (1995).
25. T. A. Springer, M. L. Dustin, T. K. Kishimoto, and S. D. Martin, *Ann. Rev. Immunol.*, *5*:223 (1987).
26. T. A. Springer, *Nature*, *346*:425 (1990).
27. N. W. Lukacs, R. M. Strieter, and S. L. Kunkel, *Curr. Opin. Hematol.*, *1993*:26 (1993).
28. J. J. Oppenheim, C. O. C. Zachariae, N. Mukaida, and K. Matsushima, *Annu. Rev. Immunol.*, *9*:617 (1991).
29. M. R. Shalaby, T. Espevik, G. C. Rice, A. J. Ammann, I. S. Figari, G. E. Ranges, and M. A. Pallidino, Jr., *J. Immunol.*, *141*:499 (1988).
30. P. Amiri, R. M. Locksley, T. G. Parslow, M. Sadick, E. Rector, D. Ritter, and J. H. McKerrow, *Nature*, *356*:604 (1992).
31. A. L. Joseph and D. L. Boros, *J. Immunol.*, *151*:5461 (1993).
32. N. W. Lukacs, S. W. Chensue, R. M. Strieter, K. Warmington, and S. L. Kunkel, *J. Immunol.*, *152*:5883 (1994).
33. C. J. Walsh, H. J. Sugerman, P. G. Mullen, P. D. Carey, S. K. Leeper-Woodford, G. J. Jesmok, E. F. Ellis, and A. A. Fowler, *Arch. Surg.*, *127*:138 (1992).
34. D. G. Remick, R. M. Strieter, M. K. Eskandari, D. T. Nguyen, M. A. Genord, C. L. Raiford, and S. L. Kunkel, *Am. J. Pathol.*, *136*:49 (1990).
35. M. Cembryznska-Nowak, E. Szklarz, A. D. Inglot, and J. A. Teodoczyk-Injeyan, *Am. Rev. Resp. Dis.*, *147*:291 (1993).
36. K. F. Chung and P. J. Barnes, *Br. Med. Bull.*, *48*:135 (1992).
37. P. Gosset, A. Tsicopoulos, B. Wallaert, M. Joseph, A. Capron, and A. B. Tonnel, *Am. Rev. Resp. Dis.*, *146*:768 (1992).
38. P. S. Thomas, D. H. Yates, and P. J. Barnes, *Am. J. Respir. Crit. Care Med.*, *152*:76 (1995).
39. P. Bradding, J. A. Roberts, K. M. Britten, R. Montefort, R. Djukanovic, R. Mueller, C. H. Heusser, P. H. Howarth, and S. T. Holgate, *Am. J. Respir. Cell Mol. Biol.*, *10*:471 (1994).

40. P. Gosset, A. Tsicopoulos, B. Wallaert, M. Joseph, A. Capron, and A. B. Tonnel, *Am. Rev. Respir. Dis.*, *146*:768 (1992).
41. M. Cimbryzynska-Nowak, E. Szklarz, A. D. Inglot, and J. A. Teodorczyk-Injeyan, *Am. Rev. Respir. Dis.*, *147*:291 (1993).
42. J. C. Kips, J. Tavernier, and R. A. Paumels, *Am. Rev. Respir. Dis.*, *145*:332 (1992).
43. M. L. Watson, D. Smith, A. D. Bourne, R. C. Thompson, and J. Westwick, *Am. J. Respir. Cell Mol. Biol.*, *8*:365 (1993).
44. N. W. Lukacs, R. M. Strieter, S. W. Chensue, M. Widmer, and S. L. Kunkel, *J. Immunol.*, *154*:5411 (1994).
45. N. W. Lukacs. R. M. Strieter, S. W. Chensue, and S. L. Kunkel, *J. Leukoc. Biol.*, (1995).
46. D. S. Robinson, Q. Hamid, S. Ying, A. Tsicopoulos, J. Barkens, A. M. Bentley, C. Corrigan, S. R. Durham, and A. B. Kay, *N. Engl. J. Med.*, *326*:298 (1992).
47. R. L. Coffman, D. A. Lebman, and P. Rothman, *Adv. Immunol.*, *54*:229 (1993).
48. R. P. Schleimer, S. A. Sterbinsky, J. Kaiser, C. A. Bickel, D. A. Klunk, K. Tomioka, W. Newman, F. W. Luscinskas, M. A. Gmbrone, Jr., B. W. McIntyre, and B. S. Bochner, *J. Immunol.*, *148*:1086 (1993).
49. S. L. Swain, *Res. Immunol.*, *144*:616 (1993).
50. R. I. Tepper, D. A. Levinson, B. Z. Stanger, J. Campos-Torres, A. K. Abbas, and P. Leder, *Cell*, *62*:457 (1992).
51. R. Sehmi, A. J. Wardlaw, O. Cromwell, K. Kurihara, P. Waltmann, and A. B. Kay, *Blood*, *79*:2952 (1992).
52. R. A. Warringa, H. J. Mengelers, P. H. Kuijper, J. A. Raaijmakers, P. L. Bruijnzeel, and L. Koenderman, *Blood*, *79*:1836 (1992).
53. A. B. Kay, S. Ying, and S. R. Durham, *Int. Arch. Allergy Immunol.*, *107*:208 (1995).
54. M. Nonaka, R. Nonaka, K. Woolley, E. Adelroth, K. Miura, Y. Okhawara, M. Gilbetic, K. Nakano, P. O'Bryne, J. Dolovich, et al., *J. Immunol.*, *155*:3234 (1995).
55. D. Y. Leung, R. J. Martin, S. J. Szefler, E. R. Sher, S. Ying, A. B. Kay, and Q. Hamid, *J. Exp. Med.*, *181*:33 (1995).
56. P. J. Mauser, A. Pitman, A. Witt, X. Fernandez, J. Zurcher, T. Kung, H. Jones, A. S. Watnick, R. W. Egan, W. Kreutner, and A. R. Gulbenkian, *Am. Rev. Respir. Dis.*, *148*:1623 (1993).
57. A. J. Van Oosterhout, A. R. Ladenius, H. F. Savelkoul, I. Van Ark, K. C. Delsman, and F. P. Nijkamp, *Am. Rev. Respir. Dis.*, *147*:548 (1993).
58. A. R. Gulbenkian, R. W. Egan, X. Fernandez, H. Jones, W. Kreutner, T. Kung, F. Payvandi, L. Sullivan, J. A. Zurcher, and A. S. Watnick, *Am. Rev. Respir. Dis.*, *146*:263 (1992).
59. N. Chand, J. E. Harrison, S. Rooney, J. Pillar, R. Jakubicki, K. Nolan, W. Diamantis, and R. D. Sofia, *Eur. J. Pharmacol.*, *211*:121 (1992).
60. A. J. Van Oosterhaut, I. Van Ark, G. Hofman, H. F. Savelkoul, and F. P. Nijkamp, *Eur. J. Pharmacol.*, *236*:379 (1993).
61. N. W. Lukacs, R. M. Strieter, S. W. Chensue, and S. L. Kunkel, *Am. J. Respir. Cell Mol. Biol.*, *10*:526 (1994).
62. G. Brusselle, J. Kips, G. Joos, H. Bluethmann, and R. Pauwels, *Am. J. Respir. Cell. Mol. Biol.*, *12*:254 (1995).

63.  G. G. Brusselle, J. C. Kips, J. H. Tavernier, J. G. van der Heyden, C. A. Cuvelier, R. A. Pauwels, and H. Bluethmann, *Clin. Exp. Allergy*, *24*:73 (1994).
64.  S. H. Gavett, D. J. O'Hearn, X. Li, S. K. Huang, F. D. Finkleman, and M. Wills-Karp, *J. Exp. Med.*, *182*:1527 (1995).
65.  J. C. Kips, G. G. Bruselle, G. F. Joos, R. A. Peleman, R. R. Devos, J. H. Tavernier, and R. A. Pauwels. *Int. Arch. Allergy Immunol.*, *107*:115 (1995).
66.  H. Nakajima, I. Iwamoto, and S. Yoshida. *Am. Rev. Respir. Dis.*, *148*:1102 (1993).
67.  B. K. Wershil, M. Tsai, E. N. Geissler, K. M. Zsebo, and S. J. Galli, *J. Exp. Med.*, *175*:245 (1992).
68.  P. Valent, E. Spanblochl, W. R. Sperr, C. Sillaber, K. M. Zsebo, H. Agis, H. Strobl, K. Geissler, P. Bettelheim, and K. Lechner, *Blood*, *80*:2237 (1992).
69.  S. J. Galli, A. Iemura, D. S. Garlick, C. Gamba-Vitalo, K. M. Zsebo, and R. G. Andrews, *J. Clin. Invest.*, *91*:148 (1993).
70.  J. W. Coleman, M. R. Holiday, I. Kimber, K. M. Zsebo, and S. J. Galli, *J. Immunol.*, *150*:556 (1993).
71.  S. J. Galli, K. M. Zsebo, and E. N. Geissler, *Adv. Immunol.*, *55*:1 (1994).
72.  J. G. Flanagan, D. C. Chan, and P. Leder, *Cell*, *64*:1025 (1991).
73.  H. Avraham, D. T. Scadden, A. Chi, V. C. Broudy, K. M. Zsebo, and J. E. Groopman, *Blood*, *80*:1679 (1992).
74.  I. K. McNiece, K. E. Langley, and K. M. Zsebo, *J. Immunol.*, *146*:3785 (1991).
75.  K. M. Zsebo, D. A. Williams, E. N. Geissler, V. C. Broudy, F. H. Martin, H. L. Atkins, R.-Y. Hsu, N. C. Birkitt, K. H. Murdock, et al., *Cell*, *63*:213 (1990).
76.  D. M. Anderson, S. D. Lyman, A. Baird, et al., *Cell*, *63*:235 (1990).
77.  K. Nocka, J. Buck, E. Levi, and P. Besner, *EMBO J.*, *9*:3287 (1990).
78.  M. T. Aye, S. Hasemi, B. Leclair, A. Zeibdawi, E. Trudel, M. Halpenny, V. Fuller, and G. Cheng, *Exp. Hematol.*, *20*:523 (1992).
79.  M. Tsai, L.-S. Shih, G. F. Newlands, T. Takeishi, K. E. Langley, K. M. Zsebo, H. R. P. Miller, E. N. Geissler, and S. J. Galli, *J. Exp. Med.*, *174*:125 (1991).
80.  A. Ando, T. R. Martin, and S. J. Galli, *J Clin. Invest.*, *92*:1639 (1993).
81.  B. J. Undem, L. M. Lichtenstein, W. C. Hubbard, S. Meeker, and J. L. Ellis, *Am. J. Resp. Cell Mol. Biol.*, *11*:646 (1994).
82.  A. Iemura, M. Tsai, A. Ando, B. K. Wershil, and S. J. Galli, *Am. J. Pathol.*, *144*:321 (1994).
83.  K. E. Langley, L. G. Bennett, J. Wypych, S. A. Yancik, X.-D. Liu, K. R. Wescott, D. G. Chang, K. A. Smith, and K. M. Zsebo, *Blood*, *81*:656 (1993).
84.  N. W. Lukacs, R. M. Strieter, P. M. Lincoln, E. Brownell, D. M. Pullen, H. J. Schock, S. W. Chensue, D. D. Taub, and S. L. Kunkel, *J. Immunol.*, (In Press). (1996).
85.  J. J. Oppenheim, C. O. C. Zachariae, N. Mukaida, and K. Matsushima, *Annu. Rev. Immunol.*, *9*:617 (1991).
86.  R. M. Strieter, P. J. Polverini, D. A. Arenberg, et al., *J. Leukoc. Biol.*, *57*:752 (1995).
87.  S. L. Kunkel, R. M. Strieter, I. J. D. Lindley, and J. Westwick, *Immunol. Today*, *16*:559 (1995).
88.  T. J. Schall, *Cytokine*, *3*:165 (1991).
89.  A. Rot, M. Krieger, T. Brunner, S. C. Bischoff, T. J. Schall, and C. A. Dahinden, *J. Exp. Med.*, *176*:1489 (1992).

90. M. Ebisawa, T. Yamada, C. Bickel, D. Klunk, and R. P. Schleimer, *J. Immunol.*, *153*:2153 (1994).

91. R. Meurer, G. Van Riper, W. Feeney, P. Cunningham, D. Hora, M. S. Springer, D. E. MacIntyre, and H. Rosen, *J. Exp. Med.*, *178*:1913 (1993).

92. J. Chihara, N. Hayashi, T. Kakazu, T. Yamamoto, D. Kurachi, and S. Nakajima, *Int. Arch. Allergy Immunol.*, *104*:52 (1994).

93. P. Kuna, S. R. Reddigari, T. J. Schall, D. Rucinski, M. Y. Viksman, and A. P. Kaplan, *J. Immunol.*, *149*:636 (1992).

94. J. H. Wang, J. L. Devalia, C. Xia, R. J. Sapsford, and R. J. Davies, *Am. J. Respir. Cell Mol. Biol.*, *14*:27 (1996).

95. R. J. Davies, J. H. Wang, C. J. Trigg, and J. L. Devalia, *Int. Arch. Allergy Immunol.*, *107*:428 (1995).

96. W. W. Cruikshank, A. Long, R. E. Tarpy, H. Kornfeld, M. P. Carroll, L. Teran, S. T. Holgate, and D. M. Center, *Am. J. Respir. Cell Mol. Biol.*, *13*:738 (1995).

97. R. C. Schweizer, B. A. C. Welmers, J. A. M. Raaijmakers, P. Zanen, J. W. J. Lammers, and L. Koenderman, *Blood*, *83*:3697 (1994).

98. A. R. Sousa, S. J. Lane, J. A. Nakhosteen, T. Yoshimura, T. H. Lee, and R. N. Poston, *Am. J. Respir. Cell Mol. Biol.*, *10*:142 (1994).

99. S. C. Bischoff, M. Krieger, T. Brunner, A. Rot, V. von Tscharner, M. Baggiolini, and C. A. Dahinden, *Eur. J. Immunol.*, *23*:761 (1993).

100. M. W. Carr, S. J. Roth, E. Luther, S. S. Ross, and T. A. Springer, *Proc. Natl. Acad. Sci. U.S.A.*, *91*:3652 (1994).

101. C. A. Dahinden, T. Geiser, T. Brunner, V. von Tscharner, D. Caput, P. Ferrara, A. Minty, and M. Baggiolini, *J. Exp. Med.*, *179*:751 (1994).

102. P. J. Jose, D. A. Griffiths-Johnson, P. D. Collins, D. T. Walsh, R. Moqbel, N. F. Totty, O. Truong, J. J. Hsuan, and T. J. Williams, *J. Exp. Med.*, *179*:881 (1994).

103. N. W. Lukacs, R. M. Strieter, C. L. Shaklee, S. W. Chensue, and S. L. Kunkel, *Eur. J. Immunol.*, *25*:245 (1995).

# 20

# Adhesion Molecule Antagonists in Animal Models of Asthma

**Roy R. Lobb**  *Biogen, Inc., Cambridge, Massachusetts*

## I.  INTRODUCTION

Asthma is a common chronic disease causing narrowing of the bronchial tree; it is characterized by acute episodes of bronchial constriction and increased mucus production, linked to an underlying airways hypersensitivity (1–4). With the recent recognition that asthma is a disease with a significant inflammatory component (1,2), inhaled steroids have become the treatment of choice (3). However, although symptoms in many patients are well controlled with inhaled steroids and β-agonists, there remains a significant role for novel anti-inflammatories in the treatment of asthma in certain patient populations (1–5). Basic research has begun to focus on the mechanisms of infiltration and activation within the lung of eosinophils and mononuclear leukocytes, which contribute to and characterize the underlying inflammatory state. Because adhesion molecules play key roles in this process, much attention has focused on their role in leukocyte recruitment and activation in vivo. Here we review some of the key publications that document the critical role of adhesion molecules in animal models of allergic airways disease.

## II.  IN VIVO STUDIES WITH ANTAGONISTS OF α4 INTEGRINS AND THEIR LIGANDS

### A.  α4 Integrins

The two α4 (CD49d) integrins α4β1 (also called very late antigen-4: VLA-4) and α4β7 are cell surface heterodimers expressed on leukocytes (see Chap. 1 for a general description of adhesion molecules). α4β1 mediates cell–cell adhesion to the immunoglobulin superfamily member vascular cell adhesion molecule-1 (VCAM-1: CD106), and also cell–matrix adhesion to an alternately spliced form of fibronectin (6). The $\alpha4\beta1$ integrin is expressed at substantial levels on all mononuclear leukocytes. It is also found on eosinophils (7–10) and basophils (8), but is absent from neutrophils (PMN), although recent studies suggest it may be found in PMN granules (11). The integrin $\alpha4$ can associate with an alternative $\beta$ chain, first called $\beta p$ in the mouse (12) and now designated $\beta7$. This integrin, $\alpha4\beta7$, adheres to the mucosal addressin cell adhesion molecule, MAdCAM (13), as well as to VCAM-1 and fibronectin (14,15). The $\alpha4\beta7$ integrin is expressed on most lymph node T and B cells, on the gut-homing subset of CD4+ memory T cells, and on lymphocytes resident in rheumatoid synovium (16–18). Recent studies show that natural killer (NK) cells, eosinophils, and newborn-blood B and T cells show relatively homogeneous expression of $\alpha4\beta7$, whereas adult-blood B cells and CD8+ T cells, like CD4+ T cells, show more heterogeneous expression (19,20).

## B.   $\alpha4$ Integrin Monoclonal Antibodies

The majority of monoclonal antibodies (mAbs) recognizing the human $\alpha4$ subunit define three non-overlapping epitopes, designated A, B, and C (21). mAbs to epitope A weakly block $\alpha4\beta1$ adhesion to fibronectin but do not inhibit VCAM-1 adhesion, whereas epitope B mAbs completely block adhesion to both ligands. Also, mAbs to epitope A and a subset of those to epitope B trigger homotypic aggregation (21). Epitope C mAbs have no effect on either adhesion or aggregation. Antibodies that recognize the $\alpha4\beta1$ and $\alpha4\beta7$ complexes have also been described (13,18,22).

Importantly, mAbs that block adhesive function in vitro have now been characterized that work in all species, allowing in vivo studies to be performed in a variety of animal models. These include rat anti–murine $\alpha4$ mAbs R1-2 and PS/2 (12,23), murine anti–rat $\alpha4$ mAb TA-2 (24), murine anti–human $\alpha4$ mAb HP2/1 (21), which binds and blocks rat $\alpha4$ (25), and murine anti–human $\alpha4$ mAb HP1/2 (21), which binds and blocks guinea pig, rabbit, sheep, and primate $\alpha4$ (26–30).

## C.   $\alpha4$ Integrins and Eosinophil Recruitment

The demonstration that eosinophils express $\alpha4$ integrins, whereas PMN do not, suggested that these integrins might play a critical role in the selective recruitment of eosinophils into inflammatory sites and therefore play a role in allergic diseases such as asthma (7–10). To examine this question, we examined the effect of mAb HP1/2 on eosinophil recruitment into guinea pig skin in response to a variety of mediators (27), and indeed mAb HP1/2 was found to block significantly cellular recruitment into skin in response to all mediators (see Chap. 9 for detailed discussion of eosinophil recruitment mechanisms).

## D. Allergic Lung Inflammation Models

Following our demonstration of a role for α4 integrins in eosinophil recruitment into guinea pig skin, we initiated studies in a two allergic lung challenge models, in the sheep and in the guinea pig (30,31). In the sheep model of allergic airways hyperresponsiveness, where eosinophils are believed to play a significant role, sheep challenged with the parasite *Ascaris suum* undergo acute bronchoconstriction, and many animals then show a late-phase response (LPR) 6–8 hours after challenge. Importantly, since the LPR correlates with an eosinophil-rich lung infiltrate, we reasoned that treatment with HP1/2 might ameliorate the LPR via inhibition of eosinophil recruitment.

Our studies showed that indeed intravenous administration of mAb HP1/2, whether given 30 min prior to or 2 hours after antigen challenge, was highly effective at blocking both the LPR and the associated airways hyperresponsiveness (AHR) to carbachol (30). Nevertheless, inhibition of cellular recruitment could not fully explain the data, because bronchoalveolar lavage (BAL) leukocyte levels in general, and eosinophil numbers in particular, were only modestly affected. This prompted us to test mAb HP1/2 as an aerosol fomulation, and, surprisingly, aerosolized mAb HP1/2 was as effective as mAb HP1/2 given intravenously in blocking both the LPR and AHR (30). The data suggest that the therapeutic effects seen in this model are due to mechanisms operative within the lung itself.

Ovalbumin-sensitized guinea pigs show increased AHR to a variety of mediators following aerosolized antigen challenge, and in parallel studies we examined the effects of mAb HP1/2 in this model (31). MAb HP1/2 was found to block effectively AHR in response to methacholine in these animals following challenge. In this study, reduced eosinophil numbers are seen in BAL fluid. Importantly, immunohistologic studies of lung tissue show significantly reduced levels not only of eosinophils but also of both CD4+ and CD8+ T cells in the epithelial submucosa and adventitia (31).

Several other studies of the guinea pig have now been performed using mAb HP1/2 (32–35). Antigen challenge of ovalbumin-sensitized guinea pigs decreases the function of inhibitory M2 muscarinic autoreceptors on parasympathetic nerves in the lung, potentiating vagally induced bronchoconstriction (33). Loss of M2 receptor function is associated with eosinophil accumulation around airway nerves. MAb HP1/2, but not mAb LAM 1-116 to L-selectin, protected M2 receptor function and selectively inhibited eosinophil accumulation as measured by lavage and histology. In contrast, another study of the ovalbumin-sensitized gunea pig found that mAb HP1/2 did not affect AHR but strongly inhibited eosinophil recruitment (32). In this case the antigen challenge gave a significantly stronger response in the airways, so methodological differences may explain the discrepancy (32). Finally, in a distinct approach, direct administration of interleukin-5 (IL-5) to guinea pigs induced AHR to histamine, an associated eosinophilia, and an increase in BAL eosinophil peroxidase activity (34,35). MAb HP1/2 blocked the IL-5–induced AHR and blocked both eosinophil recruitment and the associated increase in lung peroxidase activity (34,35).

Ovalbumin-sensitized brown Norway rats have also been examined, and in this model system treatment with mAb TA-2 just prior to challenge significantly blocked the LPR without significant changes in BAL eosinophil composition, measured at 8 hours after challenge (36). A further study carried out at both 8 and 32 hours concluded that mAb TA-2 had little effect on eosinophil recruitment into the airways at either time, although the mAb treatment eliminated AHR to methacholine (37). These authors suggested that eosinophils do not mediate allergen-induced AHR in the rat. In contrast, a different study with mAb TA-2, also in ovalbumin-challenged brown Norway rats, demonstrated significant inhibition not only of lung tissue inflammation but also of eosinophil and T-cell infiltration into the BAL fluid (38). The reasons for these differences are unclear but are likely methodological. Moreover, the number of activated rather than total eosinophils is likely the critical parameter in AHR (see below).

Murine models of allergic airways disease are now being developed, and the roles of cytokines and adhesion molecules being evaluated (39). Blockade of α4 integrin with mAb PS/2, or VCAM-1 with mAb MK/1, significantly inhibited both eosinophil and T-cell recruitment into the tracheas of ovalbumin-sensitized and challenged mice (40), which strongly express VCAM-1 as assessed by immunohistology.

Finally, mAb PS/2 to murine α4 integrins has been found to cross-react with rabbit α4 and has been examined in rabbit models (41). Immunization of neonatal rabbits with dust-mite allergens results in the generation of allergen-specific IgE, and subsequent allergen challenge induces early- and late-phase responses and a significant AHR. MAb PS/2, given as an aerosol 90 min before antigen challenge, blocks early- and late-phase responses by >90% and blocks the associated AHR by 55%. The percentage of eosinophils was reduced by about 80% but did not reach significance (41).

## E.  Multiple Roles of α4 Integrins in Lung Pathophysiology

Taken jointly, the data argue that α4 integrins likely play multiple complex roles in lung pathobiology. In some species and models, reduced eosinophil and T-cell numbers within lung tissue are clearly observed (31–35,40), arguing that mAbs to α4 integrins can inhibit cellular recruitment into the lung. However, the ability of mAbs to inhibit airways hyperresponsiveness in the apparent absence of inhibition of eosinophil recruitment, and the ability of mAbs to block when administered as aerosols (30,36,37,41), clearly argues for important functions for these integrins within the lung parenchyma.

These observations also raise important issues about mechanism of action of these mAbs, which have all been selected for in vivo use on the basis of blockade of adhesive function in vitro. It is known that mAbs to $\alpha4$ integrins can trigger or prime leukocytes, for example, to release cytokines (42), which are known to modulate disease. In addition, the effector functions of mAbs can also play a crucial role in their in vivo efficacy profiles. To examine this question, we have generated Fab fragments of mAb HP1/2. These are monovalent in physiologic buffer as assessed by gel filtration and are almost equipotent with intact HP1/2 in blockade of adhesion in vitro. Importantly, the HP1/2 Fab shows comparable efficacy with intact mAb HP1/2 in vivo, when assessed as an aerosol formulation in the sheep model (43). Furthermore, mAb HP1/2 does not activate monocytes in vitro to release cytokines under conditions where other anti-$\alpha4$ mAbs do so [S. Haskill, L. Osborn, and R. Lobb, unpublished observations]. These results are entirely consistent with mAb HP1/2 working in vivo by blockade of $\alpha4$ integrin–dependent adhesive function within the lung parenchyma. Most importantly, small molecule antagonists of $\alpha4$ integrins are efficacious in vivo (see below), showing that the efficacy of mAbs can be mimicked by monovalent peptidomimetic antagonists.

An alternative explanation for the in vivo data is that $\alpha4$ integrin–dependent adhesion is crucial not only to leukocyte recruitment into tissues but also to leukocyte priming and activation within inflamed tissues. Adhesion-dependent enhancement of leukocyte function is well established in vitro for the $\beta2$ integrins, and numerous recent studies show the same phenomenon for $\alpha4$ integrins on eosinophils, T cells, NK cells, and mast cells (44–50). For example, eosinophils adherent to immobilized fibronectin or VCAM-1 are primed for release of leukotrienes, oxygen radicals, or granule proteins in many studies. However, it should be noted that this is not universal, and adherence of eosinophils and mast cells to fibronectin has been reported to inhibit cellular functions (51,52). It is probable that adherence to different matrix proteins has distinct functional effects. For example, it seems reasonable to assume that matrix proteins characteristic of inflammatory sites would enhance leukocyte function, whereas matrix proteins characteristic of uninflamed tissue would inhibit leukocyte function. Consistent with this, recent studies show that eosinophils are much more strongly primed when adherent to tissue fibronectin rather than plasma fibronectin (50). Tissue fibronectin is the alternately spliced form that binds $\alpha4$ integrins, and this form of fibronectin is strongly induced at sites of inflammation, such as rheumatoid synovium (53). The effects on leukocyte function of other matrix proteins characteristic of inflammatory sites, including other alternately spliced forms of fibronectin (54,55), is worth exploration.

Key recent studies on ex vivo lung explants strongly argue for a positive role for fibronectin in eosinophil priming (56). Human eosinophils activated with platelet-activating factor (PAF), but not eosinophils or PAF alone, induce a significant narrowing of human bronchial explants in an ex vivo model, due to leukotriene release from the activated eosinophils. When eosinophils are preadhered to fibronectin, then activated with PAF, bronchial explants show a further dramatic narrowing as compared to nonadherent (i.e., unprimed) eosinophils and PAF, through augmented leukotriene release (56). These data show that the priming of eosinophils through $\alpha4$ integrin/fibronectin adhesion, as observed in vitro, can be translated into pathologically significant airways bronchoconstriction ex vivo, and provide strong evidence for $\alpha4$ integrin–dependent adhesion in the modulation of eosinophil function and in induction of bronchoconstriction.

Recent in vivo studies show that eosinophil activation state rather than eosinophil number is critical to increased AHR in the ovalbumin-challenged guinea pig (57). Consistent with this hypothesis, eosinophil-derived granule proteins such as major basic protein enhance AHR (58,59). If $\alpha4$ integrin–dependent adhesion is central to eosinophil activation, then the infiltrated eosinophils of animals treated with $\alpha4$ integrin antagonists should exhibit a reduced activation state. Indeed, blockade of $\alpha4$ integrin function with mAb HP1/2 in the guinea pig model results in lung eosinophils in a reduced state of activation (M. Pretolani, D. Joseph, R. Lobb, and B. Vargaftig, manuscript submitted).

Recent evidence indicates a role for integrins in the inhibition of apoptosis of anchorage-dependent cells, such as epithelial and endothelial cells (60–64). Integrin ligation induces apoptosis-inhibitory proteins such as Bc1-2. In contrast, removal of extracellular matrix ligands, for example by proteolysis with matrix metalloproteases, results in loss of Bc1-2 and related proteins and induction of ICE-like proteases, and results in programmed cell death. These data suggest yet another possible mode of action of $\alpha4$ integrin antagonists in allergic airways challenge models, namely induction of programmed cell death. A central role for eosinophil apoptosis in allergic inflammation has been suggested (65), and consistent with a role for integrins in this process, adherence of eosinophils to fibronectin in vitro enhances their survival (44,50). Using techniques to measure cellular apoptosis in tissues, investigators are beginning to evaluate the possibility that inhibition of the apoptosis of leukocytes in inflammatory sites may exacerbate inflammation (66). Therapeutic regimens such as steroid treatment can significantly increase the numbers of apoptotic cells in animal models (67). Studies in allergic challenge models should examine this hypothesis. In particular, the role of leukocyte integrin ligation to ligands within the lung parenchyma in the inhibition of apoptosis, and thereby in prolongation of chronic lung inflammation, deserves significant attention. The importance of this concept cannot be overstated, because it implies that short-term treatment with integrin-blocking drugs should have long-term therapeutic effects, by driving inflammatory cells irreversibly down a pathway of programmed cell death. In particular, short-term treatment with integrin blockers delivered as aerosolized drugs might have profound anti-inflammatory effects in asthma, reducing the chronic underlying inflammation characteristic of this disease by inducing apoptosis of long-lived inflammatory cells within the lung parenchyma.

In summary, there is greater sophistication in our understanding of how integrins may modulate leukocyte function in inflammatory disease in general, and in lung inflammation in particular. Integrins may work at multiple levels, including leukocyte recruitment into the tissue itself (endothelial/leukocyte adhesion); migration of cells within the inflamed tissue; priming and activation via ligation to matrix, particularly inflammatory matrix such as alternately spliced fibronectin; and more speculatively through inhibition of apoptosis. These possibilities, illustrated from the perspective of the α4 integrins on eosinophils, are summarized in Fig. 1, and should clearly apply to other leukocyte subsets, as well as the β2 integrins.

## III. IN VIVO STUDIES WITH ANTAGONISTS OF β2 INTEGRINS AND THEIR LIGANDS

The β2 (CD18) integrins are a group of four leukocyte-restricted integrins known to play a major role in leukocyte biology (68–70). Three members of this subclass have been extensively studied, namely CD11a/CD18, or LFA1; CD11b/CD18, or Mac1/CR3; and CD11c/CD18, or gp150/95. Recently a fourth member, CD11d/CD18, has been characterized, although its functions are not yet

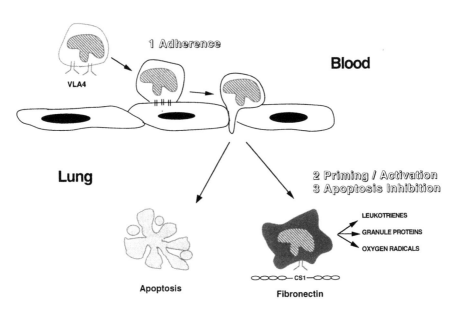

**Figure 1** Eosinophils, α4 integrins, and asthma. α4 integrins, such as VLA-4, expressed on eosinophils play multiple roles in the inflammatory response within the asthmatic lung, including 1. *adherence* to vascular endothelium at sites of inflammation; 2. *priming and activation* of eosinophils for lipid, granule, and oxygen radical release; and 3. inhibition of eosinophil *apoptosis* within the lung parenchyma.

well defined (71). The best-defined ligands for these integrins are the intercellular adhesion molecules (ICAMs), which are members of the immunoglobulin superfamily. There are now at least four distinct ICAMs, designated ICAM-1 through ICAM-4 (68,70,72). The roles of ICAM/β2 integrin adhesion in lung pathophysiology have been well reviewed recently (73) and will be summarized here only briefly.

ICAM-1 appears to be the dominant ligand for β2 integrins in lung pathophysiology (73). It is a major adhesion molecule for leukocyte recruitment on endothelium, but unlike VCAM-1, it is also present on airway and alveolar epithelium and is involved in antigen-specific immune responses by virtue of its presence on lymphocytes. Studies with mAbs to either ICAM-1 or to the leukointegrins have been used in several in vivo asthma models. In cynomolgus monkeys, repeated inhalation of *Ascaris suum* extract results in a prolonged airway eosinophilia and airways hyperresponsiveness, associated with increased levels of ICAM-1 on airway and alveolar epithelium and on lung endothelium (73–75). Intravenous administration of mAb to ICAM-1 inhibited both eosinophil accumulation and AHR (74). Delivery of mAb to ICAM-1 via aerosol also attenuated AHR, consistent with an important role for ICAM-1 on airway epithelium in the generation of bronchial hyperreactivity (73). A mAb to Mac-1 did not affect eosinophil infiltration, but also inhibited hyperresponsiveness, implicating this integrin in lung pathology. Finally, although systemic dexamethasone treatment reduced the persistent airway eosinophilia, eosinophil activation, and hyperresponsiveness in antigen-challenged animals, daily treatment with mAb to ICAM-1 for 7 days failed to do so. Nevertheless, following steroid therapy, the ICAM-1 mAb was effective at inhibiting reestablishment of these parameters (73). Taken jointly, the data suggest a critical role for ICAM-1 in the primate asthma model.

Blocking mAbs to the ICAM/β2 integrin pathway have also been used in several rodent models. Pretreatment of brown Norway rats with mAb to ICAM-1 blocked ovalbumin-induced bronchial hyperresponsiveness without affecting eosinophil or lymphocyte recruitment (76). In other studies in this model, a combination of mAbs to LFA-1 and Mac-1 blocked both early- and late-phase responses and inhibited PMN recruitment (36,37), and mAbs to ICAM-1 and leukocyte function–associated antigen (LFA)-1 significantly inhibited both early- and late-phase responses (77). In a murine model of eosinophil and T-cell recruitment into the tracheas of allergic animals, blockade of ICAM-1 and LFA-1 did not affect eosinophil recruitment, and had only a modest effect on T-cell infiltration, in contrast to blockade of VCAM-1 and α4 integrins (40). In addition, in a murine model of nonallergic asthma, utilizing skin sensitization and airway challenge with dinitrofluorobenzene, anti–LFA-1 but not anti–Mac-1 mAbs inhibited completely the tracheal hyperreactivity observed 24 hours after challenge (78).

Taken jointly, the data from multiple species argue for a significant role for the β2 integrins and their ICAM ligands in the lung pathophysiology associated with allergic challenge.

## IV. IN VIVO STUDIES WITH ANTAGONISTS OF SELECTINS AND THEIR LIGANDS

The selectins and their ligands have been reviewed in Chap. 1. There has been only one report of the use a selectin antagonist in an animal model of asthma (79). In chronically challenged cynomolgus monkeys in which ICAM-1 mAbs are efficacious (see above) a mAb to E-selectin had no effect on either AHR or lung eosinophilia. In contrast, in animals exposed to a single inhalation of antigen, mAb to E-selectin blocked the influx of neutrophils into the lung, which correlated with E-selectin expression on lung vasculature, and blocked the late-phase response (79). Further studies in this or other animal models are warranted but have not been pursued.

## V. SMALL MOLECULE ANTAGONISTS OF ADHESION PATHWAYS

Despite intensive study, there have been no reports of potent small molecule antagonists of the β2 integrins. However, several groups have begun to generate small molecule peptidomimetic antagonists of α4 integrins, based on the CS1 recognition sequence (L-leucine–L-aspartate–L-valine: LDV) within the alternately spliced form of fibronectin (41,80–82; S. Adams, R. Lobb, K. Lin, A. Gill, et al., unpublished observations). Importantly, these monovalent small molecule antagonists of the VLA-4/fibronectin interaction are efficacious in vivo in several animal models, including both the rabbit allergic challenge model (41) and the sheep allergic challenge model described above (82; W. Abraham, S. Adams, R. Lobb, A. Gill, et al., unpublished observations). These data show that potent selective small molecule antagonists of α4 integrins can be generated; that distinct molecular entities are efficacious in two lung allergic challenge models in different species; and that such small molecules can be efficacious when delivered in aerosolized form. Finally, some antagonists are highly α4β1 selective, showing no blockade of α4β7 integrin function, yet still are efficacious in the sheep allergic challenge model, arguing that at least in this in vivo setting it is α4β1 rather than α4β7 that is the critical integrin (W. Abraham, S. Adams, R. Lobb, A. Gill, K. Lin, et al., unpublished observations).

## VI. SUMMARY

There is rapidly mounting evidence for a central role for leukocyte adhesion, and in particular for leukocyte integrin–dependent adhesion, in lung pathophysiology. Based largely on in vivo studies with blocking mAbs, a clear role for both α4 and β2 integrins and their ligands, and to a much lesser extent the selectins and their ligands, in allergic lung challenge models has been demonstrated. Importantly, the recent generation of monovalent peptidomimetic antagonists of the α4 integrins, and their efficacy as aerosolized drugs in allergic challenge models, suggest that blockade of adhesion pathways with small molecules is a feasible approach to human asthma. Although acute allergic challenge models are not models of chronic human asthma, the efficacy in such models of drugs that are efficacious in the asthmatic patient, including corticosteroids, β-agonists, and leukotriene antagonists, provides hope that adhesion blockers such as α4 integrin antagonists will also be capable of down-modulating the underlying chronic inflammatory response characteristic of human asthma.

## ACKNOWLEDGMENTS

I would like to thank Bruce Bochner, Mariano Elices, Alan Leff, Ivan Richards, and Antoon Van Oosterhout for access to information prior to publication, and Diane Leone for help with the figure.

## REFERENCES

1. W. W. Busse, W. F. Calhoun, and J. D. Sedgwick, *Am. Rev. Respir. Dis.*, *147*:S20 (1993).
2. W. Busse, *Ann. Allergy*, *69*:261 (1992).
3. P. J. Barnes, *N. Engl. J. Med.*, *332*:868 (1995).
4. E. R. McFadden and I. A. Gilbert, *N. Engl. J. Med.*, *327*:1928 (1992).
5. P. J. Barnes, A. P. Greening, and G. K. Crompton, *Am. J. Respir. Crit. Care Med.*, *152*:S125 (1995).
6. R. R. Lobb and M. E. Hemler, *J. Clin. Invest.*, *94*:1722 (1994).
7. A. Dobrina, R. Menegazzi, T. M. Carlos, E. Nardon, R. Cramer, T. Zacchi, J. M. Harlan, and P. Patriarca, *J. Clin. Invest.*, *88*:20 (1991).
8. B. S. Bochner, F. W. Luscinskas, M. A. Gimbrone, W. Newman, S. A. Sterbinsky, D.-A. C. P., D. Klunk, and R. P. Schleimer, *J. Exp. Med.*, *173*:1553 (1991).
9. G. M. Walsh, J. J. Mermod, A. Hartnell, A. B. Kay, and A. J. Wardlaw, *J. Immunol.*, *146*:3419 (1991).
10. P. F. Weller, T. H. Rand, S. E. Goelz, G. Chi-Rosso, and R. R. Lobb, *Proc. Natl. Acad. Sci. U. S. A.*, *88*:7430 (1991).
11. P. Kubes, X.-F. Niu, C. W. Smith, M. E. Kehrli, Jr., P. H. Reinhardt, and R. C. Woodman, *FASEB J.*, *9*:1103 (1995).
12. B. Holzmann, B. W. McIntyre, and I. L. Weissman, *Cell*, *56*:37 (1989).
13. C. Berlin, E. L. Berg, M. J. Briskin, D. P. Andrew, P. J. Kilshaw, B. Holzmann, I. L. Weissman, A. Hamann, and E. C. Butcher, *Cell*, *74*:185 (1993).
14. B. M. Chan, M. J. Elices, E. Murphy, and M. E. Hemler, *J. Biol. Chem.*, *267*:8366 (1992).
15. C. Ruegg, A. A. Postigo, E. E. Sikorski, E. C. Butcher, R. Pytela, and D. J. Erle, *J. Cell Biol.*, *117*:179 (1992).
16. P. J. Kilshaw and S. J. Murant, *Eur. J. Immunol.*, *20*:2201 (1990).
17. A. I. Lazarovits and J. Karsh, *J. Immunol.*, *151*:6482 (1993).

18. T. Schweighoffer, Y. Tanaka, M. Tidswell, D. J. Erle, K. J. Horgan, G. E. Luce, A. I. Lazarovits, D. Buck, and S. Shaw, *J. Immunol.*, *151*:717 (1993).
19. D. J. Erle, M. J. Briskin, E. C. Butcher, A. Garcia-Pardo, A. I. Lazarovits, and M. Tidswell, *J. Immunol.*, *153*:517 (1994).
20. A. A. Postigo, P. Sanchez-Mateos, A. I. Lazarovits, F. Sanchez-Madrid, and M. O. de Lanzaduri, *J. Immunol.*, *151*:2471 (1993).
21. R. Pulido, M. J. Elices, M. R. Campanero, L. Osborn, S. Schiffer, A. Garcia-Pardo, R. Lobb, M. E. Hemler, and F. Sanchez-Madrid, *J. Biol. Chem.*, *266*:10241 (1991).
22. J. L. Bednarczyk, M. C. Szabo, J. N. Wygant, A. I. Lazarovits, and B. W. McIntyre, *J. Biol. Chem.*, *269*:8348 (1994).
23. K. Miyake, I. L. Weissman, J. S. Greenberger, and P. W. Kincade, *J. Exp. Med.*, *173*:599 (1991).
24. T. B. Issekutz, *J. Immunol.*, *147*:4178 (1991).
25. T. A. Yednock, C. Cannon, L. C. Fritz, F. Sanchez-Madrid, L. Steinman, and N. Karin, *Nature*, *356*:63 (1992).
26. R. K. Winn and J. M. Harlan, *J. Clin. Invest.*, *92*:1168 (1993).
27. V. B. Weg, T. J. Williams, R. R. Lobb, and S. Nourshargh, *J. Exp. Med.*, *177*:561 (1993).
28. D. K. Podolsky, R. Lobb, N. King, C. D. Benjamin, B. Pepinsky, P. Sehgal, and M. deBeaumont, *J. Clin. Invest.*, *92*:372 (1993).
29. T. Papayannopoulou and B. Nakamoto, *Proc. Natl. Acad. Sci. U.S.A.*, *90*:9374 (1993).
30. W. M. Abraham, M. W. Sielscak, A. Ahmed, A. Cortes, I. T. Lauredo, J. Kim, B. Pepinsky, C. D. Benjamin, D. L. Leone, R. R. Lobb, and P. F. Weller, *J. Clin. Invest.*, *93*:776 (1994).
31. M. Pretolani, C. Ruffie, J.-R. Lapa e Silva, D. Joseph, R. R. Lobb, and B. Vargaftig, *J. Exp. Med.*, *180*:795 (1994).
32. A. A. Y. Milne and P. J. Piper, *Eur. J. Pharmacol.*, *282*:243 (1995).
33. A. D. Fryer, R. W. Costello, R. R. Lobb, T. T. Tedder, and B. S. Bochner, *J. Clin. Invest.*, *In Press.* (1997).
34. A. Kraneveld, I. Van Ark, H. J. Van der Linde, D. Fattah, F. P. Nijkamp, and A. J. M. Van Oosterhout, *J. Allergy Clin. Immunol.*, *In Press.* (1996).
35. A. D. Kraneveld, I. Van Ark, H. J. Van der Linde, D. Fattah, F. P. Nijkamp, and A. J. M. Van Oosterhout, *Am. J. Respir. Mol. Cell Biol.*, *In Press.* (1996).
36. H. A. Rabb, R. Olivenstein, T. B. Issekutz, P. M. Renzi, J. G. Martin, R. Pantano, and S. Seguin, *Am. J. Respir. Crit. Care Med.*, *149*:1186 (1994).
37. S. Laberge, H. Rabb, T. B. Issekutz, and J. G. Martin, *Am. J. Respir. Crit. Care Med.*, *151*:822 (1995).
38. I. M. Richards, K. P. Kolbasa, C. A. Hatfield, G. E. Winterrowd, S. L. Vonderfecht, S. F. Fidler, R. L. Griffin, J. R. Brashler, R. F. Krzesicki, L. M. Sly, K. A. Ready, N. D. Staite, and J. E. Chin, *Am. Rev. Respir. Cell. Mol. Biol.*, *15*:172 (1996).
39. I. M. Richards, *Clin. Exp. Allergy*, *26*:618 (1996).
40. H. Nakajima, H. Sano, T. Nishimura, S. Yoshida, and I. Iwamoto, *J. Exp. Med.*, *179*:1145 (1994).
41. W. J. Metzger, *Springer Semin. Immunopathol.*, *16*:467 (1995).
42. A. D. Yurochko, D. Y. Liu, D. Eierman, and S. Haskill, *Proc. Natl. Acad. Sci. U.S.A.*, *89*:9034 (1992).

43. R. R. Lobb, B. Pepinsky, D. R. Leone, and W. M. Abraham, *Eur. Respir. J.*, *9*:1045 (1996).
44. A. R. F. Anwar, R. Moqbel, G. M. Walsh, A. B. Kay, and A. J. Wardlaw, *J. Exp. Med.*, *177*:839 (1993).
45. A. R. E. Anwar, G. M. Walsh, O. Cromwell, A. B. Kay, and A. J. Wardlaw, *Immunology*, *82*:222 (1994).
46. M. Nagata, J. B. Sedgwick, M. E. Bates, H. Kita, and W. W. Busse, *J. Immunol.*, *155*:2914 (1995).
47. S. P. Neeley, K. J. Hamann, T. L. Dowling, K. T. McAllister, S. R. White, and A. R. Leff, *Am. J. Resp. Cell Mol. Biol.*, *11*:206 (1994).
48. G. Palmieri, A. Serra, R. De Maria, A. Gismondi, M. Milella, M. Piccoli, L. Frati, and A. Santoni, *J. Immunol.*, *155*:5314 (1995).
49. A. M. Romanic and J. A. Madri, *J. Cell Biol.*, *125*:1165 (1994).
50. G. M. Walsh, F. A. Symon, and A. J. Wardlaw, *Clin. Exp. Allergy*, *25*:1128 (1995).
51. H. Kita, S. Horie, and G. J. Gleich, *J. Immunol.*, *156*:1174 (1996).
52. S. E. Lavens, K. Goldring, L. H. Thomas, and J. A. Warner, *Am. J. Respir. Cell Mol. Biol.*, *14*:95 (1996).
53. M. J. Elices, V. Tsai, D. Strahl, A. S. Goel, V. Tollefson, T. Arrhenius, E. A. Wayner, F. C. A. Gaeta, J. D. Fikes, and G. S. Firestein, *J. Clin. Invest.*, *93*:405 (1994).
54. J. E. Schwarzbauer, *Bioessays*, *13*:527 (1991).
55. W. R. Jarnagin, D. C. Rockey, V. E. Koteliansky, S.-S. Wang, and D. M. Bissell, *J. Cell Biol.*, *127*:2037 (1994).
56. N. M. Munoz, K. F. Rabe, S. P. Neeley, A. Herrnreiter, X. Zhu, K. McAllister, D. Mayer, H. Magnussen, S. Galens, and A. R. Leff, *Am. J. Physiol.*, *270*:L587 (1996).
57. M. Pretolani, C. Ruffie, D. Joseph, M. G. Campos, M. K. Church, J. Lefort, and B. Vargaftig, *Am. J. Respir. Crit. Care Med.*, *149*:1167 (1994).
58. R. H. Gundel, L. G. Letts, and G. J. Gleich, *J. Clin. Invest.*, *87*:1470 (1991).
59. J. Lefort, M.-A. Nahori, C. Ruffie, B. B. Vargaftig, and M. Pretolani, *J. Clin. Invest.*, *97*:1117 (1996).
60. Z. Zhang, K. Vuori, J. C. Reed, and E. Ruoslahti, *Proc. Natl. Acad. Sci. U.S.A.*, *92*:6161 (1995).
61. J. E. Meredith, Jr., B. Fazeli, and M. A. Schwartz, *Mol. Biol. Cell*, *4*:953 (1993).
62. N. Boudreau, C. J. Sympson, Z. Werb, and M. J. Bissell, *Science*, *267*:891 (1995).
63. P. C. Brooks, A. M. P. Montgomery, M. Rosenfeld, R. A. Reisfield, T. Hu, G. Klier, and D. A. Cheresh, *Cell*, *79*:1157 (1994).
64. A. M. P. Montgomery, R. A. Resifield, and D. A. Cheresh, *Proc. Natl. Acad. Sci. U.S.A.*, *91*:8856 (1994).
65. H.-U. Simon and K. Blaser, *Immunol. Today*, *16*:53 (1995).
66. M. Schmeid, H. Breitschopf, R. Gold, H. Zischler, G. Rothe, H. Wekerle, and H. Lassmann, *Am. J. Pathol.*, *143*:446 (1993).
67. U. K. Zettl, R. Gold, K. V. Toyka, and H.-P. Hartung, *J. Neuropathol. Exp. Neurol.*, *54*:540 (1995).
68. T. A. Springer, *Cell*, *76*:301 (1994).
69. R. O. Hynes, *Cell*, *69*:11 (1992).
70. T. M. Carlos and J. M. Harlan, *Blood*, *84*:2068 (1994).

71. M. Van der Vieren, H. Le Trong, C. L. Wood, P. F. Moore, T. St. John, D. E. Staunton, and W. M. Gallatin, *Cell, 3*:683 (1995).
72. P. Bailly, E. Tontti, P. Hermand, J.-P. Cartron, and C. G. Gahmberg, *Eur. J. Immunol., 25*:3316 (1995).
73. C. D. Wegner, *Lung Biology in Health and Disease, 89*:243 (1996).
74. C. D. Wegner, R. H. Gundel, P. Reilly, N. Haynes, L. G. Letts, and R. Rothlein, *Science, 247*:456 (1990).
75. C. D. Wegner, C. A. Torcellini, C. C. Clarke, L. G. Letts, and R. H. Gundel, *J. Allergy Clin. Immunol., 87*:776 (1991).
76. J. Sun, W. Elwood, A. Haczku, P. J. Barnes, P. G. Hellewell, and K. F. Chung, *Int. Arch. Allergy Immunol., 104*:291 (1994).
77. T. Nagase, Y. Fukuchi, T. Matsuse, E. Sudo, H. Matsui, and H. Orimo, *Am. J. Respir. Crit. Care Med., 151*:1244 (1995).
78. P. G. M. Bloemen, T. L. Buckley, M. C. van den Tweel, P. A. J. Henricks, F. A. M. Redegeld, A. S. Koster, and F. P. Nijkamp, *Am. J. Respir. Crit. Care Med., 153*:521 (1996).
79. R. H. Gundel, C. D. Wegner, C. A. Torcellini, C. C. Clarke, N. Haynes, R. Rothlein, C. W. Smith, and L. G. Letts, *J. Clin. Invest., 88*:1407 (1991).
80. P. M. Cardarelli, R. R. Cobb, D. M. Nowlin, W. Scolz, F. Gorcsan, M. Moscinski, M. Yasuhara, S.-L. Chiang, and T. J. Lobl, *J. Biol. Chem., 269*:18668 (1994).
81. S. Molossi, M. Elices, T. Arrhenius, R. Diaz, C. Coulber, and M. Rabinovitch, *J. Clin. Invest., 95*:2601 (1995).
82. M. J. Elices, T. Arrhenius, and W. M. Abraham, *American Academy of Allergy, Asthma and Immunology and American Thoracic Society*, (1995).

# Index

α-actinin, 87, 89, 90–92, 96
Acid phosphatase, 228
Actin, 3, 33, 37, 81, 89, 91
Adherence junctions, 36
Adhesion
  antagonists, 17–21, 393–406
    treatment of rheumatoid
      arthritis, 19–21
  cascade, 13
  molecules, 12, 14, 16, 315
    in allergic airway inflammation,
      376–379
    bronchial epithelial, 377
    correlations with cell influx,
      324–325
    expression after endobronchial
      allergen challenge,
      323–326
    expression in asthma, 315
    expression in nasal mucosa,
      320, 325
    leukocyte activation and,
      377–378
    in nasal biopsies, 353
    in nasal polyps, 357

[Adhesion]
  in sinusitis, 358
  intraepithelial, 315, 331–335
  soluble, 16
    after experimental rhinovirus
      infection, 329–330
    in allergic rhinitis, 356
    in atopic dermatitis and
      psoriasis, 347
    from conjunctival epithelium,
      369
    levels in asthma or after
      allergen challenge,
      326–331
    therapeutic targets, 14–21
Airway hyperresponsiveness, 202,
    209, 213, 378, 380, 395,
    400
  effect of adhesion molecule
    antibodies on, 209–210,
    395–396
Allopurinol, 290
Alveolar cells, 43
Anaphylaxis, passive cutaneous, 164
Antamanide, 37

*407*

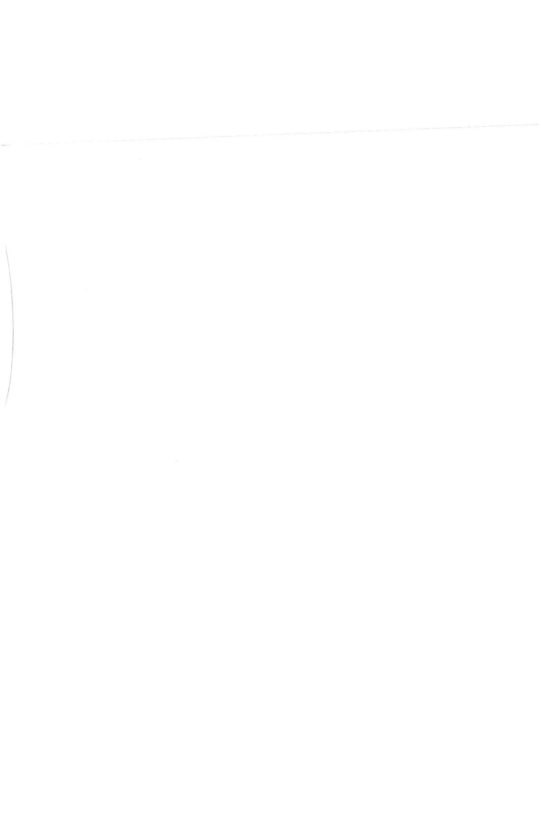